Recent Advances in

Human Retroviruses:

Principles of Replication and Pathogenesis

Advances in Retroviral Research

Recent Advances in Human Retroviruses: Principles of Replication and Pathogenesis

Advances in Retroviral Research

edited by

Andrew ML Lever
University of Cambridge, UK

Kuan-Teh Jeang
National Institutes of Health, USA

Ben Berkhout
University of Amsterdam, The Netherlands

World Scientific

NEW JERSEY • LONDON • SINGAPORE • BEIJING • SHANGHAI • HONG KONG • TAIPEI • CHENNAI

Published by

World Scientific Publishing Co. Pte. Ltd.
5 Toh Tuck Link, Singapore 596224
USA office: 27 Warren Street, Suite 401-402, Hackensack, NJ 07601
UK office: 57 Shelton Street, Covent Garden, London WC2H 9HE

British Library Cataloguing-in-Publication Data
A catalogue record for this book is available from the British Library.

RECENT ADVANCES IN HUMAN RETROVIRUSES: PRINCIPLES OF REPLICATION AND PATHOGENESIS
Advances in Retroviral Research

ISBN-13 978-981-4295-30-7
ISBN-10 981-4295-30-2

Typeset by Stallion Press
Email: enquiries@stallionpress.com

Printed by FuIsland Offset Printing (S) Pte Ltd. Singapore

Contents

CHAPTER 1

The Role of Hematopoietic Progenitor Cells in Retroviral Pathogenesis

Prabal Banerjee[†,‡], Elizabeth Samuelson[†] and Gerold Feuer[*,†,‡]

INTRODUCTION

Hematopoiesis is the process by which hematopoietic stem cells (HSCs) give rise to all the cellular components of the hematopoietic system. HSCs have the potential to proliferate indefinitely, and can differentiate into mature hematopoietic lineage-specific cells. The bone marrow is the principal site for hematopoiesis in adults. HSCs proliferate within the bone marrow and differentiate to produce the requisite number of highly specialized cells of the hematopoietic system. HSCs give rise to two different populations of cells: (1) a common lymphoid progenitor (CLP) population, that generates B cells, T cells and NK lineage cells; and (2) a common myeloid progenitor (CMP) population that generates granulocytes, neutrophils, eosinophils, macrophages and erythrocytes. Lineage commitment by HSCs is a complex process that can be induced in response to a variety of factors, including

*Corresponding author. E-mail: feuerg@upstate.edu.

†Department of Microbiology and Immunology, SUNY Upstate Medical University, Syracuse, NY.

‡Center for Humanized SCID Mice and Stem Cell Processing Laboratory, SUNY Upstate Medical University, Syracuse, NY.

relative levels of hematopoiesis associated cytokines and transcription factors. These factors not only help maintain the pluripotency of HSCs but can also actively induce commitment and differentiation of HSCs.[1–10] By virtue of their importance in cell proliferation and differentiation, any deregulations in the cytokine milieu or modulation of hematopoiesis-associated transcriptional factors may result in the disruption of cell cycle or lineage commitment of these cells leading to perturbation of hematopoiesis.

One of the major causes of disruption of this hematopoietic network is viral infection.[11] Viruses such as human immunodeficiency virus-1 (HIV-1), human cytomegalovirus (CMV) and human herpes virus-8 (HHV-8) have been previously shown to disrupt the normal pattern of hematopoiesis either by infecting bone marrow resident stromal cells or by direct infection of hematopoietic progenitor cells which can result in alteration in the cytokine milieu/transcription factor levels and subsequent perturbation of hematopoiesis.[12–14] Viral infection within the bone marrow results not only in the perturbation of hematopoiesis, but infected HSCs may also become a source for the generation of infected lineage-specific cells that may differentiate into various hematologic malignancies. Retroviruses can establish latent infection in HPCs (hematopoietic progenitor cells) and HSCs, resulting in perturbation of hematopoiesis and induction of viral pathogenesis.

VIRAL INFECTION IN HP/HSCs

A primary infection of HSCs with virus requires the expression of corresponding cellular receptors for binding and internalization of that particular virus. It has now been well established that HP/HSCs express cellular receptors and as a result they become susceptible to various viral infections. However, since destruction of infected cells by various effector cells of the immune system leads to the clearance of infection, viruses have evolved elaborate strategies to evade the immune system and thus to persist within the host. One of the most prominent mechanisms of viral evasion is through the infection of progenitor cells within the bone marrow, which is a relatively immune-privileged site. Viral infections in HSCs can have deleterious effects, including induction of cytolytic and apoptotic cell death resulting in the suppression of hematopoiesis as well as dysregulation of normal development of HSCs, leading to subsequent outgrowth of malignant cells pertaining to a particular hematopoietic lineage. The marrow microenvironment is a complex system comprising many cell types, including stromal cells that produce cytokine/growth factors as well as adhesion molecules that are vital for the maintenance, differentiation and maturation of HP/HSCs.[1,3] Infection of bone marrow stromal cells can make these cells incapable of supporting hematopoiesis, resulting in multilineage hematopoietic failure.[15] To this end a variety of viruses have clinically important effects on the hematopoietic system. $CD34^+$ HPCs have been shown to be susceptible to infection with a number of viruses: HIV-1, hepatitis C virus, JC virus, Parvovirus, HCMV, HHV-5, HHV-6, HHV-7, HHV-8 and HTLV-1[12,16–28] (Fig. 1). Suppression of hematopoiesis has been documented to occur following infection of $CD34^+$ HPCs with HCMV/HHV-5, HHV-6, HIV-1, and measles virus either by direct infection of $CD34^+$ HPCs or by indirect mechanisms, such

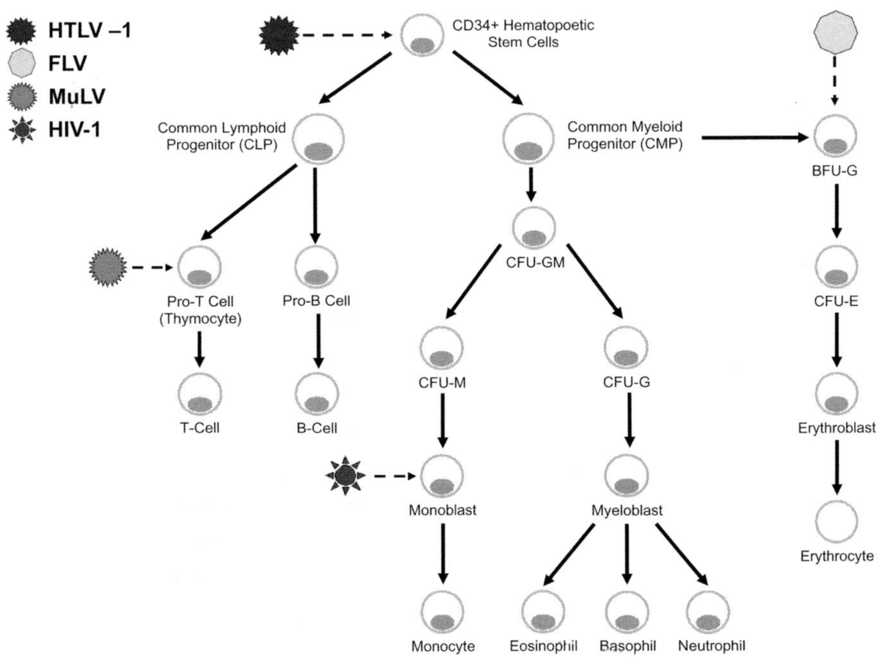

Fig. 1. Susceptibility of progenitor cells in the hematopoietic system to retroviral infection.

as the disruption of cytokine expression via infection of bone marrow stromal cells. The authors' laboratory has reported that HTLV-1 and KSHV infection of $CD34^+$ HPCs results in suppression of hematopoiesis both *in vitro* and *in vivo* in humanized SCID mice.[28–30] The range of effects resulting from retroviral infection of hematopoietic progenitor cells leading to dysregulation of hematopoiesis and manifestation of various hematological malignancies are discussed in this chapter.

RETROVIRUSES AND CANCER: SIGNIFICANCE OF INFECTION OF HEMATOPOIETIC PROGENITOR CELLS

The first association of retroviruses with malignancies was reported with the observation that spontaneous erythroleukemia in chickens was infectious, which was later shown to be associated with avian erythroblastosis virus (AEV) infection.[31] In 1911, Peyton Rous showed that some avian sarcomas could be transmitted through inoculation of cell-free extracts, proving for the first time that viruses can be linked to cancer. However, these early studies were largely ignored, and it was not until observations made by Bittner and Gross in mice with mammary carcinoma and thymic lymphoma that these malignancies resulted from viral infections, that an association between cancers and retroviral infections was firmly established.[32–34] Since the 1960s, retroviruses have been shown to cause leukemia, lymphoma and other forms of cancer in a wide variety of vertebrate animals, ranging from fish to humans. The first oncogenic human retrovirus, human T cell leukemia virus-type 1 (HTLV-1) was discovered in 1980 and has subsequently been associated with the development of adult T-cell leukemia/lymphoma (ATL).[35,36] HIV-1 was discovered and characterized a few years later and the study of both animal and human retroviruses has lead to the fundamental understanding of the multistep process of oncogenesis and has provided insights into complex human biological processes such as hematopoiesis, since retroviruses often target cells of the hematopoietic system. In this chapter, we will discuss the effects of retroviral infection on hematopoietic progenitor cells and their role in the development of malignancies. During the process of hematopoiesis, HSCs commit to decide between the self-renewal and the differentiation/maturation pathways to give rise to mature lineage-committed hematopoietic cells. Under normal conditions the ability of the HSC to undergo self-renewal is irreversibly lost, but in leukemia, progenitor cells can undergo unlimited self-renewal without entering the differentiation pathway.[37] On the contrary, dysregulation of hematopoiesis-associated transcription factors and cytokines can lead to cell-cycle arrest and induction of apoptosis in HSCs. Retroviruses have evolved a variety of mechanisms to both induce continuous self-renewal of infected hematopoietic progenitor cells as well as suppression of hematopoiesis by induction of cell cycle arrest or apoptosis in hematopoietic progenitor cells.

RETROVIRUS INDUCED LEUKEMOGENESIS

Avian Erythroblastosis Virus (AEV)

Avian erythroblastosis virus (AEV), which was isolated from a spontaneous erythroleukemia from chickens in 1908, was the first retrovirus discovered and has subsequently been associated with cancer.[31] AEV is an acute leukemogenic retrovirus that belongs to the alpha retrovirus family and causes fatal erythroleukemia when injected into the ova or in young chickens.[38] Studies done *in vitro* indicate that development of malignancy is associated with co-infection of both AEV and a helper virus and that the target cell for infection is a committed erythroid progenitor cell, the burst forming units-erythroid (BFU-E), which becomes terminally blocked for further differentiation following infection.[39,40] AEV encodes for two viral oncoproteins, *v-Erb-A*, which represents a mutated, oncogenic thyroid hormone receptor α (c-Erb-A/TRα), and *v-Erb-B*, which is a mutated constitutively active transmembrane receptor for epidermal growth factor (EGF).[41,42] The AEV oncoprotein *v-Erb-A*, unless expressed at high levels, has no transformation capacity on its own but the co-expression of both *v-Erb-A* and *v-Erb-B*, leads to a severe leukemic phenotype.[43–48] It is currently believed that *v-Erb-A* actively cooperates both with ligand activated receptor tyrosine kinases such as stem-cell factor (SCF) activated c-kit as well as constitutively activated viral oncoprotein *v-Erb-B*, to arrest differentiation and induce leukemia in erythroid progenitor cells.[48,49] The binding of the thyroid hormone (T3) to the thyroid hormone receptor

(TRα) leads to the transcriptional activation of a multitude of target genes, some of which are required for normal erythropoiesis.[50] Since *v-Erb-A* is a mutant form of TRα that retains the DNA binding capability of the receptors but cannot bind to its ligand, T3, it has been suggested that *v-Erb-A* mimics TRα which, in the absence of the ligand, induces transcriptional repression of a host of target genes including those involved in differentiation of erythropoietic progenitors.[51] The process involves c-Kit as well as downstream signal tranducers, including the other viral oncoprotein *v-Erb-B* which has been shown to induce proliferation and self-renewal of erythroid progenitors. Ultimately, this leads to long-term proliferation and a block in differentiation of primary erythroid progenitor cells.[52]

Murine Leukemia Virus

The discovery and characterization of murine leukemia viruses (MuLV) derived from mouse strains that develop spontaneous leukemia has led to the development of useful models for studying the critical events related to leukemogenesis.[53] Leukemogenesis by MuLV is a multistep process and the virus-induced changes during hematopoiesis are important in the initiation of disease. Several related viruses are grouped under MuLV, many of which were initially isolated from serial passage of murine tumors, including Gross murine leukemia virus (G-MuLV), Friend leukemia virus (FLV), Moloney murine leukemia virus (M-MuLV) and Rauscher murine leukemia virus (R-MuLV). While FLV and R-MuLV induce leukemia in erythroid and myeloid lineages, G-MuLV and M-MuLV cause T-cell leukemia in susceptible mouse strains.

Induction of erythroleukemia by Friend leukemia virus (FLV)

Since the discovery of acute erythroleukemia (acute form of myeloid leukemia caused by erythroblastic precursors) in mice infected by FLV[54] in the early 1950s, FLV-induced disease models have evolved as powerful tools to study the susceptibility of hematopoietic progenitor cells to retroviral infections.[55,56] FLV is known to have two distinct viral components: A replication-competent Friend murine leukemia virus (F-MuLV) and a replication defective component know as the Friend spleen focus forming virus (F-SFFV). The replication-competent F-MuLV is the non-pathogenic component of the viral complex and serves as a helper virus to SFFV, which is pathogenic by itself and responsible for the induction of acute erythroleukemia in mice.[57–59] Infection of susceptible mouse strains with FLV leads to the development of acute erythroleukemia which is characterized by a rapid proliferation of pre-erythroblastic cells in the spleen that can be detected as early as 30 hours following viral infection, which is followed by the appearance of distinct foci of proliferating cells on the spleen, 10–12 days post-infection.[54,60,61] These events represent the early hyperplastic (proliferative) phase of the disease and are a direct result of FLV infection of erythroid precursor cells. Two to three weeks post-infection, a clonally transformed population of pre-erythroblastic cells emerges in the spleen, causing hepatosplenomegaly and metastasis into peripheral organs such as the liver.[62,63] The infected animal ultimately dies due to splenic rupture. These events represent the late oncogenic phase of the infection, which occurs as a consequence of FLV-mediated transformation of a small subpopulation of erythroid progenitor cells.

SFFV represents the pathogenic component of FLV and can infect a variety of hematopoietic cells, though early erythroid lineage-specific precursors are the primary targets of infection.[60,64] The effect of SFFV can be divided into two distinct phases, correlating with the two phases of disease development: The early phase characterized by rapid proliferation of erythroid precursors (erythroid hyperplasia), and the late phase characterized by a clonal outgrowth of a subpopulation of the erythroid precursors. Erythroid hyperplasia is mediated by SFFV-induced alterations of the normal growth and differentiation profile of erythroid progenitor cells. The proliferation and differentiation of erythroid cells is tightly controlled by the erythropoietin (Epo) signaling pathway, which is initiated via binding of Epo with its receptor, Epo-R.[65] Erythroid hyperplasia is induced by SFFV in the absence of Epo.[60,66,67] The envelope gene of SFFV, which encodes for a 55 kDa glycoprotein (gp55),[68,69] has been shown to be crucial for the induction of erythroid hyperplasia during the early phase of SFFV infection in murine models.[70] Following SFFV infection of erythroid precursors, gp55 binds to the erythropoietin receptor, Epo-R,[71] which is ubiquitously expressed in all erythroid progenitors, including the main targets for SFFV, the late BFU-E and CFU-E cells.[60,64] The binding of Epo-R by gp55, mediated by their respective transmembrane domains, results in Epo-R activation and promotes Epo-independent proliferation, differentiation and expansion of the erythroid progenitor cells.[72–74] Several signaling pathways and molecules are activated downstream of the Epo-R and many of these, such as the JAK/STAT, Ras/Raf/mitogen-activated protein kinase (MAPK), and phosphatidylinositol 3-kinase (PI3K)/Akt pathways, are constitutively activated in Epo-R-expressing cells infected with SFFV.[65,75–77] It was initially thought that the constitutive activation of the Epo-R signaling pathway by gp55 was primarily responsible for SFFV-induced hyperplasia. However, gp55 has also been shown to interact and constitutively activate a truncated form of the stem-cell receptor kinase (sf-Stk).[78,79] Although the exact role of sf-Stk in normal erythropoiesis is unknown, studies indicate that sf-Stk activation, through its interaction with gp55, appears to contribute to the development of erythroleukemia in mice.[80] The importance of the sf-Stk downstream signaling pathway in the initiation and maintenance of SFFV-mediated transformation has been established in SFFV-infected fibroblasts, outlining the importance of this pathway in SFFV leukemogenesis.[81,82] Thus, both activation of Epo-R and sf-Stk pathway by binding of gp55 leads to a block in the normal differentiation pattern of erythroid progenitor cells, resulting in hyperproliferation and subsequent transformation in infected animals.

Subsequent to the formation of erythroid hyperplasia, the later leukemic phase of SFFV infection is manifested by clonal proliferation of a small subpopulation of the malignant erythroid precursor cells. These clonal subpopulations of erythroid precursor cells are characterized by the integration of SFFV into a specific cellular gene, *PU.1*, resulting in the alteration of its transcriptional activities.[83,84] Although the initial interactions of SFFV gp55^{env} with the Epo-R and sf-Stk signaling pathways are important for the early deregulation and proliferation of erythroid progenitors, these events are not sufficient for inducing malignant transformation of erythroid precursor cells. SFFV-mediated transformation of erythroid precursors is achieved through insertional mutagenesis, a method usually employed by non-acute transforming retroviruses. Analysis of malignant cells from SFFV-infected mice indicates that the provirus predominantly integrates into a specific region of murine DNA known as *sp1* (SFFV-proviral-integration-site-1).[84,85] Integration leads to the re-arrangement and transcriptional activation of the *sp1* gene by the SFFV-LTR, resulting in *sp1* overexpression.[86,87] The *sp1* gene encodes for the transcription factor *PU.1*, which is a member of the *ets* family of transcription factors expressed in all hematopoietic cells with particularly high expression levels in B cells and macrophages. *PU.1* expression has been shown to play an important role in the regulation of hematopoiesis.[88,89] Dysregulation of *PU.1* expression had been linked to the development of hematopoietic malignancies, including the transformation of myeloid cells.[90] During hematopoiesis, *PU.1* is required for hematopoietic development along both the lymphoid and myeloid lineages but is down-regulated during erythropoiesis. Studies done in knockout mice have shown that perturbation of *PU.1* expression results not only in the loss of B lymphoid and macrophage development but also delayed T lymphopoiesis.[91,92] Additionally, *PU.1* also supports self-renewal of hematopoietic stem cells by regulating the multilineage commitment of multipotent hematopoietic progenitor cells, thereby maintaining a pool of pluripotent HSCs within the bone marrow.[93,94] Importantly, *PU.1*, which is expressed in low levels in erythroid progenitor cells, needs to be down-regulated for terminal erythroid development.[95,96] Thus, retroviral activation of *sp1/PU.1* transcription results in the blockage of erythroid development and outgrowth of a clonal population of erythroid precursor cells.

The induction of multistage erythroleukemia by FLV is a two-stage process: A pre-leukemic stage known as "erythroid hyperplasia" and a leukemic phase referred to as "erythroid cell transformation." The pre-leukemic stage is characterized by the infection and random integration of the SFFV virus into erythroid precursor cells, followed by the expression and subsequent surface transportation of the viral envelope glycoprotein gp55 to the surface of the infected precursors. Gp55 activates both the Epo-R and the sf-Stk signaling pathways, leading to a constitutive active signal for proliferation of undifferentiated erythroid progenitor cells independent of erythropoietin (Epo). Within the proliferating erythroid progenitor cell population are infected cells where the virus randomly integrates into the *sp-1* locus, leading the activation and overexpression of *PU.1*. The overexpression of *PU.1* in erythroid precursor cells leads to a block in erythroid differentiation, and in conjunction with an inactivation of p53, leads to clonal expansion of these leukemic cells in susceptible mice.

Induction of T-Lymphoma by Moloney Murine Leukemia Virus

Moloney murine leukemia virus (M-MuLV) is a non-acute retrovirus that typically induces T-cell lymphoma after a latency period of 3–6 months.[53] Typically, the tumor cells have the phenotype of immature T cells ($CD4^-$/$CD8^-$ or $CD4^+$/$CD8^+$) although some tumors show a more mature surface phenotype ($CD4^+$/$CD8^-$ or $CD4^-$/$CD8^+$).[97,98] This has led to the hypothesis that the virus might originally infect a hematopoietic progenitor or an immature T-cell that continues to differentiate post-infection, initially in the bone marrow and then in the thymus.[53,99] Because T lymphocytes develop in the thymus from bone marrow-derived immature precursors (pro-thymocytes), it has been proposed by several investigators that a bone-marrow-thymus axis plays a very important role in the development of T-cell lymphoma induced by M-MuLV.[97,100–102] One of the characteristic features of M-MuLV leukemogenesis is the formation of mink cell focus-inducing recombinant viruses (MCF), which arise by the recombination between the envelope gene of an infecting M-MuLV and endogenous MuLV proviruses present in susceptible mice strains.[103,104] Several studies have indicated that MCF recombination is important for M-MuLV leukemogenesis and efficient disease induction was correlated with efficient early infection of the bone marrow and the appearance of MCF recombinants.[105,106] This association also suggested that MCF recombinant formation and initial propagation may take place in the bone marrow of the infected mice,[107] and that establishment of pre-leukemic changes including defects in bone marrow hematopoiesis, thymic atrophy, and hematopoietic hyperplasia in the spleen, might be induced by MCF recombinants.[108–110] During the preleukemic phase of the disease, preleukemic cells can be detected in the bone marrow and spleen but not in the thymus. This is highly suggestive of the fact that the first preleukemic events may arise in the bone marrow and that the initial target cells for M-MuLV infection might be bone marrow-resident progenitor cells such as pro-thymocytes rather than mature thymocytes. A subsequent study also found increased levels of immature splenocytes during the preleukemic phase of M-MuLV infection in mice.[111] Collectively, this suggests that murine hematopoietic progenitors and other bone marrow-resident cells are infected during the early preleukemic phase of infection and the infected cells subsequently migrate into the spleen and thymus, a hypothesis supported by data from M-MuLV murine models such as Akr mice.[101] It was also shown that injection of bone marrow cells from AKr mice leads to a more rapid development of leukemia in irradiated secondary mice, confirming that the virus first infects bone marrow resident cells rather than infection of mature thymocytes or splenocytes.[101]

Studies involving Gross murine leukemia virus (G-MuLV) which induces lymphatic leukemia after a long latency have also shown that the bone marrow-thymus axis is crucial for the development of leukemogenesis.[34] G-MuLV-induced leukemia cells are predominantly immature, lymphoblastic cells ($CD4^+/CD8^+$ or $CD4^-/CD8^-$). Removal of the thymus prior to injection of virus prevents disease development, thus suggesting a role of the thymic micro-environment in development and manifestation of leukemia.[112] In early studies involving G-MuLV, the removal of the thymus a month post-infection resulted in the development of a myeloid leukemia instead of a T-cell lymphoma, implying that bone marrow resident progenitor cells are initial targets for infection.[33,34] Subsequent studies have shown that the end targets for G-MuLV infection may be the immature thymocytes that house the outer thymic cortex.[113] In contrast to other retroviruses such as AEV or FIV, where the target cells associated with infection have been conclusively determined as pro-erythroid precursors, the early target cells associated with both G-MuLV and M-MuLV infection have not been definitively identified. Studies in mice indicate that hematopoietic progenitor and/or bone marrow-resident stromal cells may, indeed, be the first targets for infection.[106] When bone marrow cells from BALB.C mice were infected with a particular strain of M-MuLV and subsequently analyzed using hematopoietic colony forming assays, it was found that the hematopoietic progenitor cells were the most likely target for infection.[114] However, other reports have suggested that osteoclasts or osteoclast progenitors in the bone marrow are the primary targets for M-MuLV infection.[115] Nonetheless, these studies detected a subsequent spread of infection to hematopoietic progenitors, indicating that the infection of hematopoietic progenitors in the bone marrow may result from a secondary spread of the virus from osteoclasts or other directly infected cells.

Although the identity of the initial target cell in the bone marrow for M-MuLV infection is still unknown, a two-stage leukemogenesis model for the development of M-MuLV-induced leukemia has been proposed.[53] In this model, the virus infects an animal twice, in the preleukemic (early) phase as well as the leukemic (late) phase of infection. During the early phase of infection, recombination and formation of MCF occur in the bone marrow, which results in the generation of virus that progressively infects various bone marrow-resident progenitor cells, including hematopoietic progenitors, stromal cells and osteoclast progenitors resulting in defects in the bone marrow stromal microenvironment as well as dysregulation of normal hematopoiesis. This is followed by a distinctive hyperplasia of the spleen at 4 to 8 weeks after infection.[108] The early infection of the bone marrow is thought to be essential for the establishment of the preleukemic state because hyperplasia of the spleen is thought to occur as a result of perturbation of normal hematopoiesis in the bone marrow. The splenic hyperplasia is the result of a compensatory extramedullary hematopoiesis because of diminished normal hematopoiesis in the bone marrow and plays an integral role in the establishment of the malignant process.[105,110,116]

It recently has been shown that M-MuLV induces a significant reduction in B-lymphoid differentiation in the bone marrow of infected mice by reducing the viability of differentiating B lymphoid progenitor cells.[117] This suppression was found to be selective because the myeloid differentiation potential of the hematopoietic progenitors was found to be enhanced. This correlates with previously published reports on the disruption of normal bone marrow hematopoiesis and thymopoiesis in M-MuLV-related SL-3 murine leukemia virus infected mice, leading to a significant increase in myeloid but not lymphoid progenitor cells.[118] This reduction in viability of differentiated lymphoid cells may confer a selective advantage to undifferentiated lymphoid progenitors in the bone marrow of MuLV-infected animals, contributing to the establishment of a preleukemic state.

On the whole, this supports a model of M-MuLV leukemogenesis in which the virus actively replicates in dividing hematopoietic progenitors as well as other bone marrow-resident cells. Bone marrow hematopoiesis is thereby altered, particularly with the generation and maturation of increasing numbers of hematopoietic progenitors and increasing numbers of both myeloid and lymphoid progenitor cells that exit the bone marrow and migrate to the spleen. Eventually the pro-thymocytic population migrates into the thymus, where the environment is more suitable for their continued expansion. This triggers the leukemic phase of the infection, whereby these pro-thymocytes are re-exposed to the virus during thymic differentiation. Secondary infection of these progenitors is followed by M-MuLV LTR-mediated activation of proto-oncogenes, including *c-myc*, *bmi-1 gfi-1*, *tpl-1* and *tpl-2*.[119–123] Apart from proto-oncogene activation, other important events occur during the later phase of infection, such as autocrine stimulation of the IL-2 receptor, creating IL-2-independent T cells,[110] and chromosomal trisomy in chromosome 15, which perturbs the *c-myc* proto-oncogene.[124]

Studies of M-MuLV-induced leukemia have proven very useful in understanding multistep processes associated with viral leukemogenesis and also identification of key physiological and virological events associated with tumor progression. These include, most notably, the elucidation of the activation of proto-oncogenes in tumorigenesis and the critical insight into the role of infection of bone marrow-resident hematopoietic progenitor and stromal cells in the establishment of the preleukemic state. Infection of these cells leads to dysregulation of normal hematopoiesis, and subsequent proliferation of malignant hematopoietic progenitors. Interestingly, defects in hematopoiesis in humans frequently result in the establishment of a leukemic state.[125] Murine retroviral models that induce leukemic states, particularly FLV and M-MuLV, have emerged as powerful models to study leukemogenesis as evident in recent studies done in AML.[126–132] These murine models can be successfully used to study the complex processes of signaling and transcriptional activation of key hematopoietic regulatory genes that ultimately lead to malignancies.

Human Immunodeficiency Virus Type-1

Human immunodeficiency virus type-1 (HIV-1) infection is usually associated with depletion of $CD4^+$ T cells. However, patients with HIV-1 often manifest a wide range of hematopoietic abnormalities, such as cytopenia, leucopenia, anemia, neutropenia, and dysplasia.[133–138] One intriguing question is whether the presence and replication of HIV-1 in the bone marrow of infected patients directly or indirectly impacts the normal proliferation and differentiation of hematopoietic progenitor cells. For more than two decades, researchers have attempted to establish if HIV-1 is capable of directly infecting hematopoietic stem or progenitor cells. It has now been convincingly demonstrated that hematopoietic progenitor cells largely remain uninfected by HIV-1 *in vivo*.[139–142] Moreover, even if a subset of hematopoietic progenitor cells is susceptible to HIV-1 infection, the effective suppression of multilineage hematopoiesis does not require latent or active infection; instead it is primarily mediated by cytokine dysregulation induced by the interaction of HIV-1 gene products with hematopoietic progenitor cells or bone marrow stromal cells.[143–153] Nonetheless, the presence and replication of HIV-1 within the bone marrow of infected individuals not only may result in bone marrow failure and subsequent progression to AIDS, but can also lead to the generation of progeny cells that are more susceptible to infection as it facilitates the rapid spread of the infection.[154–156]

HIV infection of hematopoietic progenitor cells: In vivo and in vitro

Regardless of their source and differentiation state, a significant population of bone marrow-resident $CD34^+$ hematopoietic progenitor cells express both the receptor and coreceptors for HIV-1, CD4 and CXCR4, although the expression of the other HIV coreceptor, CCR-5 is variable.[157–161] Since the first reported study of $CD34^+$ HPC infection with HIV-1, there have been conflicting reports in the literature about the susceptibility of these cells to HIV-1 infection, both *in vitro* and *in vivo*.[162] Analysis of hematopoietic progenitors in the bone marrow of HIV-1 seropostive patients at various stages of infection has shown that these cells remain substantially uninfected, suggesting that suppression of hematopoiesis in HIV-1-infected individuals is probably related to other factors.[139–141] Furthermore, in rare cases where there was detectable infection of hematopoietic progenitors, HIV-1 infection was at levels insufficient to explain its direct role in the development of AIDS-associated hematological malignancies.[142,163,164] In the initial reports of detection of HIV-1-infected hematopoietic progenitor cells within the bone marrow cells of HIV-1-infected patients, there were concerns regarding the contamination of the CD34$^+$ cell preparations with either HIV-1 infected T-lymphocytes or macrophages leading to false positive results.[165,166] Use of animal models that resemble human HIV infection, such as the SCID-hu mouse (Thy/Liv model) or simian immunodeficiency virus infection in macaques, have shown that HIV-1 infection disrupts both lineage-restricted and multilineage hematopoiesis *in vivo*, as indicated by the rapid and severe decrease of human progenitor cells capable of differentiation into both erythroid and myeloid lineages.[167–172] These studies also suggested that depletion of early hematopoietic progenitor cells occurs in the absence of direct viral infection, indicating that the suppression of hematopoiesis during HIV-1 infection may result from indirect effects of viral replication in hematopoietic progenitor cells.[140,164,167,173–182] It is possible that infection of CD34$^+$ HPCs is a rare event *in vivo*, and even when it does occur, it might only infect a more mature subset of HPCs in patients who are in a more advanced stage of the disease. It can be inferred that direct infection of hematopoietic progenitor cells is not sufficient to explain the suppression of hematopoiesis observed in HIV-1-infected patients. Other factors, such as abnormal stromal microenvironment, including infection of stromal cells and alteration in the cytokine milieu resulting from these infection mediated events, may have an important role to play in the suppression of hematopoiesis.

It has been established that HIV-1 can infect bone marrow stromal cells and that subsequently these infected cells fail to provide the optimal milieu of cytokines and adhesion molecules required for hematopoiesis.[15] HIV-1-infected bone marrow cells secrete large amounts of proinflammatory cytokines and chemokines, including TNF-α, IL-1α, IL-6, MIP-1 and RANTES, all of which can have deleterious effects on hematopoiesis.[143,145,146] TNF-α, in particular, is a potent negative regulator of hematopoiesis and can induce expression of functional Fas on CD34$^+$ cells, resulting in apoptosis and a compromised ability to reconstitute both short- and long-term multilineage hematopoiesis.[183–186] Thus, the reduced population of hematopoietic progenitors observed in the bone marrow of HIV-1-infected patients might be associated with hematopoietic suppression resulting from viral infection of accessory cells, leading to impaired stromal function and alteration of the hematopoietic cytokine network.[180] Since some of these effects, particularly the restoring to normal levels of TNF-α and Fas expression, are reversible with antiretroviral therapy, it can be inferred that HIV-1 infection of bone marrow stromal cells plays an important role in hematopoietic suppression as these cells serve to maintain the hematopoietic cytokine network. A recent study has suggested that very low levels of infection in a cell that is highly pluripotent and proliferative such as hematopoietic progenitors, can serve as an important reservoir of the virus, and may become a major obstacle in the complete eradication of the virus, which has a tendency to re-emerge.[187,188] In a recent review by Alexaki *et al.*, the importance of HIV infection of hematopoietic progenitor cells is highlighted and implicated in viral trafficking and dissemination, resulting in the development of HIV-1-mediated dementia, among other pathologies.[155] The authors also speculate that HIV-1-mediated dysregulation of hematopoiesis might result in the outgrowth of a monocytic subpopulation that is more highly susceptible to infection and can potentially traffic the virus into various organs and the CNS.

Similar to *in vivo* infection of hematopoietic progenitor cells with HIV-1, the infection of hematopoietic progenitor cells with HIV-1 *in vitro* has also been controversial. Hematopoietic progenitor cells, including stem cells, express both the receptor and coreceptor for HIV-1 on their surface[159–161,189,190] and some studies have reported successful *in vitro* infection of $CD34^+$ HPCs with HIV-1.[162,191] Other groups, however, have failed to infect hematopoietic progenitors with either primary or laboratory-derived strains of HIV-1.[163,178,192] A number of possibilities could account for these differences, including the purity of the $CD34^+$ HPCs cell population as well as the time of exposure and the particular virus isolate employed. The purity of the primary hematopoietic progenitor cells is an important consideration. In early studies, $CD34^+$ HPCs were a heterogeneous population of cells also containing more mature differentiated lineage-specific progenitor cells which are more susceptible to HIV-1 infection.[175,190,191] Studies done using clonogenic colony forming assays of HIV-1-infected $CD34^+$ hematopoietic progenitor cells have shown that the more mature hematopoietic progenitors (including $CD34^+/CD38^+$ cells) are more susceptible to HIV-1 infection, indicating that primitive hematopoietic stem cells might be resistant to HIV-1 infection.[193] Later studies demonstrated that a small subpopulation of $CD34^+/CD38^-$ hematopoietic stem cells may retain susceptibility to HIV-1 infection, but true hematopoietic stem cells that are mitotically inactive are resistant to infection even though they express both the receptor and coreceptor for viral entry.[194] Collectively, these data indicate that $CD34^+$ human hematopoietic progenitors are not susceptible to HIV-1 infection either *in vitro* or *in vivo*, and suggest that defects in hematopoiesis observed as a result of HIV-1 infection are related to an alteration of bone marrow and peripheral blood microenvironments by soluble HIV-1-specific gene products.

HIV-1-mediated hematosuppression: Role of viral protein gp120-^{env}, Gag and Vpr

Most of the *in vitro* and *in vivo* studies involving HIV-1 and hematopoietic progenitor cells point to the fact that cytopenia and other hematological malignancies exhibited in HIV-1-infected patients is multifactorial and is primarily due to indirect effects of HIV-1 infection on progenitor cells. Studies done *in vitro* with HIV-1-infected $CD34^+$ HPCs and in SCID-hu mouse Thy/Liv models have shown that HIV-1 can affect both myelopoiesis and erythropoiesis from HPCs without productive or latent infection.[171,172] In addition, human bone marrow cells exposed to HIV-1 *in vitro* show suppression of erythroid burst-forming units (BFU-E) and granulocyte-macrophage (CFU-GM) colony formation, although the defects in colony formation were not attributed to a productive HIV-1 infection of $CD34^+$ cells. These data indicate that HIV-1 can impair bone - marrow hematopoiesis, acting in part at the level of hematopoietic progenitor cells but without the need of a productive or latent infection.[140,142,170,171,175,178] Heat-inactivated virus can induce similar effects on hematopoietic progenitor cell activity,[150,151] confirming the observation that suppression of hematopoiesis does not require productive or latent infection. The ability of antibodies against HIV-1 surface envelope glycoprotein gp120 to negate some of harmful effects of HIV-1 infection on hematopoietic progenitor cells has highlighted the potential role of this viral glycoprotein in hematopoietic dysregulation.[148–151] Exposure of hematopoietic progenitor cells to gp120 leads to the upregulation of TNF-α in these cells, which induces cells to undergo apoptosis through activation of the Fas pathway. Upregulation of TNF-α expression can potently inhibit the growth and proliferation of hematopoietic progenitor cells,[195,196] and may result in a loss of hematopoietic progenitor cells, as reported in HIV-1 seropostive patients.[180] Other HIV-1 proteins, such as Gag, have been shown to suppress colony formation from hematopoietic progenitor cells, while intracellular Vpr expression in $CD34^+$ HPCs induces G_2/M cell-cycle arrest.[29,30,152] Another HIV-1 protein Tat, which is actively secreted by infected cells and can be taken up by a broad range of uninfected cells, including hematopoietic progenitor cells, has been shown to induce secretion of various cytokines important for regulation of hematopoiesis, including TNF-α and TGF-β.[197–201] Tat has been speculated to cause hematosuppression, most likely in conjunction with gp120, as they induce the secretion of a similar pattern of cytokines from affected cells. Nevertheless, since studies have failed to establish whether inductions of these cytokines occur as a result of infection or in response to conventional antiretroviral therapy, the role of these pro-inflammatory cytokines on the development of AIDS-related cytopenias has yet to be conclusively established.

A clear paradigm has only now begun to emerge on the impact of HIV-1 infection on hematopoietic progenitor cells arising from the research conducted over the last two decades. The ambiguous evidence for the existence of both HIV-1-infected progenitor cells in *in vitro* cultures and from HIV-1-infected patients could suggest that *in vivo* infection of progenitor cells occurs rarely, if ever. However, it is clear that viral infection of auxiliary cells such as stromal cells and differentiated lineage-specific hematopoietic cells as well as exposure to viral gene products can indirectly influence survival and growth of hematopoietic progenitors by adversely affecting the microenvironment within the bone marrow.[135,181] The greatest impact of HIV infection on growth and differentiation of hematopoietic progenitor cells results from the capacity of the virus to infect and perturb the hematopoietic regulatory function of supportive cells such as bone marrow stromal cells, and not from its capacity to infect progenitors and stem cells themselves. Further studies in this area are warranted to understand the role of these factors in the manifestation of HIV-1-mediated hematopoietic malignancies.

Human T cell Leukemia Virus Type-1 and the Development of Adult T Cell Leukemia/Lymphoma (ATLL): Potential Role of Infectious Leukemic Stem Cells

Human T-cell leukemia/lymphoma virus type-1 (HTLV-1) is the causative agent of adult T-cell leukemia (ATL), an aggressive $CD4^+$ leukemia.[36] ATL is a rare T-cell malignancy characterized by hypercalcemia, hepatomegaly, splenomegaly, lymphadenopathy, the presence of proliferating $CD4^+$-$CD25^+$ the leukemia cells and infiltration of lymphocytes into the skin and liver. HTLV-1 causes ATL in a small percentage of infected individuals after a prolonged latency period of up to 20–40 years.[202] ATL is characterized by a monoclonal expansion of malignant $CD4^+CD25^+$ T cells which evolve from a polyclonal subpopulation of HTLV-1 infected $CD4^+$ T cells. Although HTLV-1 can replicate through reverse transcription during the initial phase of infection, the viral genome is effectively replicated during proliferation of the infected cells.[203] Typically HTLV-1 infected cells can persist for decades in patients and the infected cell population transitions from a polyclonal phase into a monoclonal expansion. There are four ATL subtypes: acute, lymphomatous, chronic, and smoldering. The first two subtypes are associated with a rapidly progressing clinical course with a mean survival time of 5–6 months. Smoldering and chronic ATL have a more indolent course and may represent transitional states towards acute ATL. The clinical features of ATL include leukemic cells with multi-lobulated nuclei called "flower cells" which infiltrate various tissues, including the skin and the liver with abnormally high blood calcium levels and concurrent opportunistic infections.

Although considerable progress has been made in understanding ATL biology, the exact sequence of events occurring during initial stages of malignancy, including cell types infected with HTLV-1, remains unclear. The primary target cells for HTLV-1 infection may not only influence HTLV pathogenesis, but sequestration of these cell types may allow the virus to effectively evade the primary immune response against infection. It has been previously reported by authors' laboratory and other investigators that HTLV-1 can infect hematopoietic progenitor cells.[204,205] It was hypothesized that HTLV-1 can specifically induce a latent infection in $CD34^+$ HPCs and can initiate preleukemic events in these progenitor cells.[206] These cells could potentially provide a durable reservoir for latent virus in infected individuals. HTLV-1 infection of $CD34^+$ HPCs may also induce perturbation of normal hematopoiesis, ultimately resulting in the outgrowth of malignant clones and development of ATL. It has been hypothesized that HTLV-1 infection of $CD34^+$ HPCs may result in the generation of an "infectious leukemic stem cell." Data from the authors laboratory indicate that HTLV-1 infection of $CD34^+$ HPCs induces G_0/G_1 cell-cycle arrest and causes perturbation of hematopoiesis *in vitro*, an effect also mediated by the transduction of HTLV-1 oncoprotein Tax1.[29,30,207] This is in stark contrast to the growth stimulation and transformation following HTLV-1 infection of mature $CD4^+$ T cells. Infection of hematopoietic progenitor cells by HTLV-1 may establish important preleukemic events in HPCs, which ultimately manifests in HTLV-1 pathogenesis. *Ex vivo* infection of $CD34^+$ HPCs and reconstitution of hematopoiesis in humanized SCID mice results in recapitulation of ATL, implicating the role of infected HPCs in the manifestation of HTLV-1 mediated leukemia.[208]

Tax1: The role of the HTLV-1 oncogene in ATL development

The molecular events that cause HTLV-1 infection to progress from clinical latency to T-cell malignancy are not fully understood, but it is generally believed that they involve the critical viral transactivator protein Tax1. HTLV-1 carries no cellular proto-oncogenes and the oncogenic potential of the virus is linked to Tax1, which is a 40 kDa protein that functions as a trans-activator of viral gene expression and as a key component of HTLV-1-mediated transformation.[209] Apart from regulating viral gene expression through the 5′-long terminal repeat (LTR), Tax1 can induce or repress the expression of a large variety of cellular genes, including those encoding for cytokines, apoptosis inhibitors, cell cycle regulators, transcription factors and intracellular signaling molecules.[210–212] Tax1 usually induces cellular gene expression through activation of transcription factors such as NF-κB and cyclic AMP response element-binding protein/activating transcription factor (CREB/ATF).[213] Tax1 has also been shown to *trans*-repress transcription of certain cellular genes, including *bax*,[214] human β-polymerase,[211] cyclin A,[215] *lck*,[216]

MyoD,[217] INK4,[218] and p53.[219] Tax1 can induce immortalization and transformation of T cells *in vitro*[220,221] as well as induce formation of a variety of malignancies in transgenic mice.[222,223] It is noteworthy, however, that none of the Tax-transgenic mouse models showed mature CD4 T cell leukemia. Although Tax1 has been implicated in having a pivotal role in ATL development, precisely how Tax1 influences ATL development is not completely understood. One of the emerging views of HTLV-1 leukemogenesis and ATL development correlates with neonatal transmission and that HTLV-1 targets HPCs and immature human thymocytes which ultimately results in perturbation of normal hematopoiesis.[29,204,205] Infection of HPCs results in skewing of hematopoiesis towards distinct cellular lineages and outgrowth of malignant clones with an eventual manifestation of ATL. HPCs may serve as a crucial target for HTLV-1 latency and an *in vivo* reservoir of infected cells, thus playing a pivotal role in the development of ATL.[204,206] This hypothesis is supported by the development of lymphoma in Tax1 transgenic mice in which Tax1 expression was restricted to immature thymocytes through the use of the lymphocyte-specific protein tyrosine kinase (LCK) promoter.[223] More importantly, this data indicates that Tax1 expression in early HPCs has a unique distinguishing role, which contributes to the induction of lymphoproliferative disease. The role of Tax1 in progenitor cells may include, but is not restricted to, cell cycle dysregulation and perturbation of hematopoiesis as have previously been reported (Fig. 2).[29,207]

Cell cycle arrest in HPCs: Dysregulation of $p21^{cip1/waf1}$ (p21), $p27^{kip1}$ (p27) and survivin

Eukaryotic cell cycle progression in cells is regulated by sequential activation and inactivation of a series of cyclin-dependent kinases (CDKs) at different stages of the cell cycle. D-type cyclins (D1, D2 and D3) and cyclin E are involved in regulating G_1 progression and entry into the S phase. Tax1 plays a significant role in the dysregulation of cell cycle in HTLV-1-infected $CD4^+$ T cells, especially in accelerating G_1 to the S phase transition by modulation of several cellular activities.[224] Tax1 increases the levels of G_1 D cyclins and also activates transcription of CDK2 and CDK4,[225] leading to the stabilization of cyclin D/CDK complexes.[226] Tax1 has also been shown to bind to the hyperphosporylated form of retinoblastoma protein (pRB), facilitating its

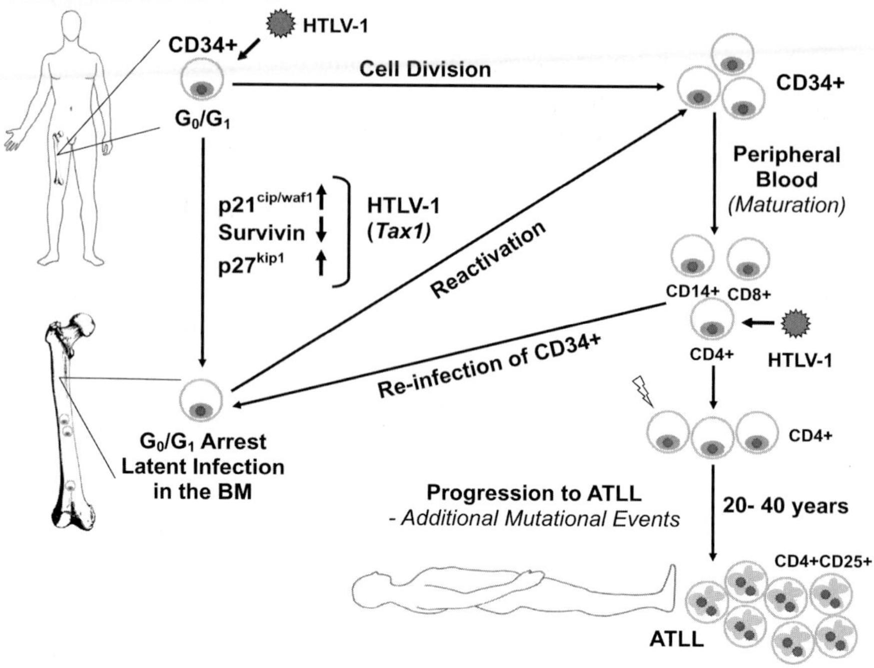

Fig. 2. Model of HTLV-1 latency within the bone marrow of infected patients.

proteosomal degradation.[227] The phosphorylation and degradation of RB frees the E2F1 transcription factor, accelerating cell-cycle transition from G1 to S. Thus, the ability of Tax1 to generate an abundance of activated cyclin D/CDK complexes and to promote RB phosphorylation and subsequent E2F release, results in committed G1 egress in HTLV-1-infected $CD4^+$ T cells.

Paradoxically, Tax1 expression also induces G_1/S cell cycle arrest, leading to a senescence-like state in both cultured mammalian cell lines and primary human $CD34^+$ HPCs.[207,228] Likewise, expression of Tax1 in *Saccharomyces cerevisiae* leads to growth arrest and loss of cell viability.[229,230] Intriguingly, in addition to increasing the levels of cyclins and CDKs, Tax1 also increases the levels of CDK inhibitors $p16^{Ink4}$, $p21^{cip1/waf1}$ and $p27^{kip}$ in infected cells.[29,30,228,231] Overexpression of p21 inhibits two critical checkpoints in mammalian cell cycle, namely G_1/S and S/G_2, through p53-independent and dependent pathways.[232] Moreover, p21 and p27 are the key contributors in cell-cycle regulation of $CD34^+$ HPCs.[233–235]

Cell cycle and differentiation is highly regulated in $CD34^+$ HPCs, with a majority of $CD34^+$ HPCs residing in quiescence and demonstrating unique expression patterns of CDKs, cyclins, and CDK inhibitors. The CDK inhibitors $p21^{cip1/waf1}$ (p21) and $p27^{kip1}$ (p27), in particular, have been shown to be key contributors in restricting cell cycle entry from G_0 and maintaining quiescence in $CD34^+$ HPCs.[233–235] We have previously shown that during HTLV-1 infection, induction of G_0/G_1 cell cycle arrest and suppression of multilineage hematopoiesis in HPCs is attributed to the concomitant activation of $p21^{cip1/waf1}$ and $p27^{kip}$ in these cells by Tax1.[29,30,207] Although Tax1 usually induces cellular gene expression by activation of transcription factors such as NF-κB and CREB/ATF, it has recently been suggested that Tax1 dysregulation of p21 and p27 might be mediated by mechanisms independent of NF-κB activation.[236] The reported absence of NF-κB activity in $CD34^+/CD38^-$ hematopoietic stem cells[237] allows speculation that the hematopoietic stem cells may provide a unique microenvironment for HTLV-1 infection, which stands in stark contrast to the cellular environment provided by mature $CD4^+$ T lymphocytes. It may be inferred that Tax1-mediated cell cycle dysregulation is cell-type specific, inducing cell cycle arrest in HPCs while concurrently maintaining the ability to activate cell proliferation in $CD4^+$ T cells.

Survivin, originally identified as a member of the inhibitor of apoptosis protein family, has recently been implicated in regulating hematopoiesis, cell cycle control and transformation.[238–241] Survivin is expressed in normal adult bone marrow and in $CD34^+$ HPCs and its expression is upregulated by hematopoietic growth factors.[242] Notably, survivin has been shown to be a key mediator of early cell cycle entry in $CD34^+$ HPCs and regulates progenitor cell proliferation through p21-dependent and -independent pathways.[243] These data implicates survivin as an integral cellular factor which regulates multiple aspects of hematopoiesis. The authors have recently demonstrated that HTLV-1 infection of $CD34^+$ HPCs suppresses hematopoiesis as a result of induction of G_0/G_1 cell cycle arrest by modulation of not only p21 but survivin gene expression.[30] Notably, $CD34^+/CD38^-$ hematopoietic stem cells demonstrate sensitivity to cell-cycle arrest following HTLV-1 infection in comparison to more mature $CD34^+/CD38^+$ hematopoietic progenitor cells, suggesting that HTLV-1 may facilitate a latent infection in these cell types *in vivo* by arresting cell cycle and inducing quiescence.

CD34+ HPCs have previously been shown to be cellular targets for viral infection, including CMV, HHV-6, HHV-8, measles virus, and MuLV.[12,18,20,25,26,117,187,244] Cells residing in the bone marrow, HPCs and HSCs, have previously been proposed as sites of retroviral latency.[206] It has been demonstrated that CMV selectively establishes latency in CD34+/CD38− stem cells, resulting in altered cellular gene expression patterns and inhibition of clonogenic colony formation activity (CFA) *in vitro*.[20] These results are precedents for modeling the role of HTLV-1-infected HPCs in the manifestation of T cell leukemia. It has been demonstrated that high levels of proviral HTLV-1 DNA exist in the bone marrow (BM) of HTLV-1 infected patients.[245] Exposure of peripheral blood to HTLV-1 results in CD4+ T cell infection and this cell population routinely traffics to the BM as part of normal immune surveillance. Trafficking of infected CD4+ T cells into the BM during the acute phase of infection may result in HTLV-1 infection of the CD34+ hematopoietic progenitor cells, including the bone marrow resident HSCs. It can be speculated that human HPCs provide a unique microenvironment in contrast to the cellular environment provided by transformed cell lines for cell cycle arrest and suppression of hematopoiesis.

ATL is etiologically linked with neonatal or perinatal transmission of HTLV-1 infection and the disease develops decades after the initial infection. HTLV-1 mediated suppression of multilineage hematopoiesis and cell-specific cell cycle arrest in CD34+ HPCs has been reported and identifies a unique mechanism by which this human retrovirus establishes viral latency and avoids immune surveillance in humans, thus accounting for the relatively long persistence of HTLV-1 infection demonstrated in ATL patients. The sequestration and concomitant differentiation of HTLV-1-infected HPCs may be a mechanistic pathway which allows the virus to evade immune surveillance while concurrently providing a continuing supply of infected cells *in vivo*. This is also in accordance with previous reports that the bone marrow, a site enriched with CD34+ HPCs, is a target for HTLV-1 infection.[206,245] Ongoing studies in our laboratory have outlined the relevance of HTLV-1 infection of CD34+ HPCs *in vivo*. HTLV-1 infection of CD34+ HPCs recapitulates ATL disease in humanized SCID mice, showing concurrent hyperproliferation infected HSCs in the bone marrow. This suggests that

HPCs and HSCs represent target cells for maintaining HTLV-1 infection for extended periods of time *in vivo*, leading to the development of lymphoproliferative disease and may result in the establishment of infectious leukemic stem cells.

CONCLUSIONS

Retroviruses can establish latent infection in HPCs and HSCs, resulting in perturbation of hematopoiesis and induction of viral pathogenesis as have been discussed here. Retroviral infection of HP/HSCs results in wide-ranging consequences, including suppression of hematopoiesis and establishment of viral latency in the relatively immune-privileged site of the bone marrow. Moreover, some retroviral infections specifically target more mature lineage-specific hematopoietic progenitors, including those of the erythroid or lymphoid lineages, leading to the development of hematological malignancies. Study of retroviral pathogenesis has helped scientists decipher some of the basic molecular events associated with hematopoiesis and differentiation. More research is warranted in the future to further our understanding of these mechanisms and to better understand how retroviruses infect, survive and cause disease in humans and other vertebrates.

ACKNOWLEDGMENTS

This work was supported with grants from the National Institutes of Health (CA124595) and the Empire State Stem Cell Fund through New York State Department of Health Contract (NYSTEM #C023059 and #N08G-127) to G.F. Opinions expressed here are solely those of the author and do not necessarily reflect those of the Empire State Stem Cell Board, the New York State Department of Health, or the State of New York.

REFERENCES

1. Allen TD, Dexter TM. (1984) The essential cells of the hemopoietic microenvironment. *Exp Hematol* **12**: 517–521.
2. Laiosa CV, Stadtfeld M, Graf T. (2006) Determinants of lymphoid-myeloid lineage diversification. *Annu Rev Immunol* **24**: 705–738.

3. Ogawa M. (1993) Differentiation and proliferation of hematopoietic stem cells. *Blood* **81**: 2844–2853.
4. Galy A, Travis M, Cen D, Chen B. (1995) Human T, B, natural killer, and dendritic cells arise from a common bone marrow progenitor cell subset. *Immunity* **3**: 459–473.
5. Ikawa T, Kawamoto H, Fujimoto S, Katsura Y. (1999) Commitment of common T/Natural killer (NK) progenitors to unipotent T and NK progenitors in the murine fetal thymus revealed by a single progenitor assay. *J Exp Med* **190**: 1617–1626.
6. Akashi K, Traver D, Miyamoto T, Weissman IL. (2000) A clonogenic common myeloid progenitor that gives rise to all myeloid lineages. *Nature* **404**: 193–197.
7. Weissman IL, Shizuru JA. (2008) The origins of the identification and isolation of hematopoietic stem cells, and their capability to induce donor-specific transplantation tolerance and treat autoimmune diseases. *Blood* **112**: 3543–3553.
8. Metcalf D. (1989) The molecular control of cell division, differentiation commitment and maturation in haemopoietic cells. *Nature* **339**: 27–30.
9. Chao MP, Seita J, Weissman IL. (2008) Establishment of a normal hematopoietic and leukemia stem cell hierarchy. *Cold Spring Harb Symp Quant Biol.*
10. Wright DE, Wagers AJ, Gulati AP, *et al.* (2001) Physiological migration of hematopoietic stem and progenitor cells. *Science* **294**: 1933–1936.
11. Kolb-Maurer A, Goebel W. (2003) Susceptibility of hematopoietic stem cells to pathogens: Role in virus/bacteria tropism and pathogenesis. *FEMS Microbiol Lett* **226**: 203–207.
12. Sindre H, Tjoonnfjord GE, Rollag H, *et al.* (1996) Human cytomegalovirus suppression of and latency in early hematopoietic progenitor cells. *Blood* **88**: 4526–4533.
13. Luppi M, Barozzi P, Morris C, *et al.* (1999) Human herpesvirus 6 latently infects early bone marrow progenitors *in vivo. J Virol* **73**: 754–759.
14. Bahner I, Kearns K, Coutinho S, *et al.* (1997) Infection of human marrow stroma by human immunodeficiency virus-1 (HIV-1) is both required and sufficient for HIV-1-induced hematopoietic suppression *in vitro*: Demonstration by gene modification of primary human stroma. *Blood* **90**: 1787–1798.

15. Moses AV, Williams S, Heneveld ML, *et al.* (1996) Human immunodeficiency virus infection of bone marrow endothelium reduces induction of stromal hematopoietic growth factors. *Blood* **87**: 919–925.
16. Segovia JC, Guenechea G, Gallego JM, *et al.* (2003) Parvovirus infection suppresses long-term repopulating hematopoietic stem cells. *J Virol* **77**: 8495–8503.
17. Sansonno D, Lotesoriere C, Cornacchiulo V, *et al.* (1998) Hepatitis C virus infection involves CD34(+) hematopoietic progenitor cells in hepatitis C virus chronic carriers. *Blood* **92**: 3328–3337.
18. Manchester M, Smith KA, Eto DS, *et al.* (2002) Targeting and hematopoietic suppression of human CD34+ cells by measles virus. *J Virol* **76**: 6636–6642.
19. Monaco MC, Atwood WJ, Gravell M, *et al.* (1996) JC virus infection of hematopoietic progenitor cells, primary B lymphocytes, and tonsillar stromal cells: Implications for viral latency. *J Virol* **70**: 7004–7012.
20. Goodrum F, Jordan CT, Terhune SS, *et al.* (2004) Differential outcomes of human cytomegalovirus infection in primitive hematopoietic cell subpopulations. *Blood* **104**: 687–695.
21. Maciejewski JP, Bruening EE, Donahue RE, *et al.* (1992) Infection of hematopoietic progenitor cells by human cytomegalovirus. *Blood* **80**: 170–178.
22. Movassagh M, Gozlan J, Senechal B, *et al.* (1996) Direct infection of CD34+ progenitor cells by human cytomegalovirus: Evidence for inhibition of hematopoiesis and viral replication. *Blood* **88**: 1277–1283.
23. Zhuravskaya T, Maciejewski JP, Netski DM, *et al.* (1997) Spread of human cytomegalovirus (HCMV) after infection of human hematopoietic progenitor cells: Model of HCMV latency. *Blood* **90**: 2482–2491.
24. Khaiboullina SF, Maciejewski JP, Crapnell K, *et al.* (2004) Human cytomegalovirus persists in myeloid progenitors and is passed to the myeloid progeny in a latent form. *Br J Haematol* **126**: 410–417.
25. Isomura H, Yamada M, Yoshida M, *et al.* (1997) Suppressive effects of human herpesvirus 6 on *in vitro* colony formation of hematopoietic progenitor cells. *J Med Virol* **52**: 406–412.
26. Isomura H, Yoshida M, Namba H, *et al.* (2000) Suppressive effects of human herpesvirus-6 on thrombopoietin-inducible megakaryocytic colony formation *in vitro*. *J Gen Virol* **81**: 663–673.

27. Mirandola P, Secchiero P, Pierpaoli S, *et al.* (2000) Infection of CD34(+) hematopoietic progenitor cells by human herpesvirus 7 (HHV-7). *Blood* **96**: 126–131.
28. Wu W, Vieira J, Fiore N, Banerjee P, *et al.* (2006) KSHV/HHV-8 infection of human hematopoietic progenitor (CD34+) cells: Persistence of infection during hematopoiesis *in vitro* and *in vivo. Blood* **108**: 141–151.
29. Tripp A, Banerjee P, Sieburg M, *et al.* (2005) Induction of cell cycle arrest by human T-cell lymphotropic virus type 1 Tax in hematopoietic progenitor (CD34+) cells: Modulation of p21cip1/waf1 and p27kip1 expression. *J Virol* **79**: 14069–14078.
30. Banerjee P, Sieburg M, Samuelson E, Feuer G. (2008) Human T-cell lymphotropic virus type 1 infection of CD34+ hematopoietic progenitor cells induces cell cycle arrest by modulation of p21(cip1/waf1) and survivin. *Stem Cells* **26**: 3047–3058.
31. Payne LN. (1992) Developments in avian leukosis research. *Leukemia* **6**(*Suppl* 3): 150S–152S.
32. Bittner JJ. (1936) Some possible effects of nursing on the mammary gland tumor incidence in mice. *Science* **84**: 162.
33. Gross L. (1960) Development of myeloid (Chloro-) leukemia in thymectomized C3H mice following inoculation of lymphatic leukemia virus. *Proc Soc Exp Biol Med* **103**: 509–514.
34. Gross L. (1960) Biological and pathogenic properites of a mouse leukemia virus. *Acta Haematol* **23**: 259–275.
35. Uchiyama T, Yodoi J, Sagawa K, *et al.* (1977) Adult T-cell leukemia: Clinical and hematologic features of 16 cases. *Blood* **50**: 481–492.
36. Poiesz BJ, Ruscetti FW, Gazdar AF, *et al.* (1980) Detection and isolation of type C retrovirus particles from fresh and cultured lymphocytes of a patient with cutaneous T-cell lymphoma. *Proc Natl Acad Sci USA* **77**: 7415–7419.
37. Keller G. (1992) Hematopoietic stem cells. *Curr Opin Immunol* **4**: 133–139.
38. Beard JW. (1963) Avian virus growths and their etiologic agents. *Adv Cancer Res* **7**: 1–127.
39. Moscovici C, Gazzolo L, Moscovici MG. (1975) Focus assay and defectiveness of avian myeloblastosis virus. *Virology* **68**: 173–181.
40. Samarut J, Gazzolo L. (1982) Target cells infected by avian erythroblastosis virus differentiate and become transformed. *Cell* **28**: 921–929.

41. Downward J, Yarden Y, Mayes E, *et al.* (1984) Close similarity of epidermal growth factor receptor and v-erb-B oncogene protein sequences. *Nature* **307**: 521–527.
42. Sap J, Munoz A, Damm K, *et al.* (1986) The c-erb-A protein is a high-affinity receptor for thyroid hormone. *Nature* **324**: 635–640.
43. Casini T, Graf T. (1995) Bicistronic retroviral vector reveals capacity of v-erbA to induce erythroleukemia and to co-operate with v-myb. *Oncogene* **11**: 1019–1026.
44. Miles BD, Robinson HL. (1985) High-frequency transduction of c-erbB in avian leukosis virus-induced erythroblastosis. *J Virol* **54**: 295–303.
45. Robinson HL, Miles BD, Catalano DE, *et al.* (1985) Susceptibility to erbB-induced erythroblastosis is a dominant trait of 151 chickens. *J Virol* **55**: 617–622.
46. Kahn P, Frykberg L, Brady C, *et al.* (1986) v-erbA cooperates with sarcoma oncogenes in leukemic cell transformation. *Cell* **45**: 349–356.
47. Forrest D, Munoz A, Raynoschek C, *et al.* (1990) Requirement for the C-terminal domain of the v-erbA oncogene protein for biological function and transcriptional repression. *Oncogene* **5**: 309–316.
48. Beug H, Metz T, Mullner EW, Hayman MJ. (1996) Self renewal and differentiation in primary avian hematopoietic cells: An alternative to mammalian *in vitro* models? *Curr Top Microbiol Immunol* **211**: 29–39.
49. Beug H, Bauer A, Dolznig H, *et al.* (1996) Avian erythropoiesis and erythroleukemia: Towards understanding the role of the biomolecules involved. *Biochim Biophys Acta* **1288**: M35–M47.
50. Angelin-Duclos C, Domenget C, Kolbus A, *et al.* (2005) Thyroid hormone T3 acting through the thyroid hormone alpha receptor is necessary for implementation of erythropoiesis in the neonatal spleen environment in the mouse. *Development* **132**: 925–934.
51. Bauer A, Mikulits W, Lagger G, *et al.* (1998) The thyroid hormone receptor functions as a ligand-operated developmental switch between proliferation and differentiation of erythroid progenitors. *Embo J* **17**: 4291–4303.
52. von Lindern M, Deiner EM, Dolznig H, *et al.* (2001) Leukemic transformation of normal murine erythroid progenitors: v- and c-ErbB act through signaling pathways activated by the EpoR and c-Kit in stress erythropoiesis. *Oncogene* **20**: 3651–3664.

53. Fan H. (1997) Leukemogenesis by Moloney murine leukemia virus: A multistep process. *Trends Microbiol* **5**: 74–82.
54. Friend C. (1957) Cell-free transmission in adult Swiss mice of a disease having the character of a leukemia. *J Exp Med* **105**: 307–318.
55. Ruscetti SK. (1995) Erythroleukaemia induction by the Friend spleen focus-forming virus. *Baillieres Clin Haematol* **8**: 225–247.
56. Moreau-Gachelin F. (2008) Multi-stage Friend murine erythroleukemia: Molecular insights into oncogenic cooperation. *Retrovirology* **5**: 99.
57. Wolff L, Tambourin P, Ruscetti S. (1986) Induction of the autonomous stage of transformation in erythroid cells infected with SFFV: Helper virus is not required. *Virology* **152**: 272–276.
58. Berger SA, Sanderson N, Bernstein A, Hankins WD. (1985) Induction of the early stages of Friend erythroleukemia with helper-free Friend spleen focus-forming virus. *Proc Natl Acad Sci USA* **82**: 6913–6917.
59. Bestwick RK, Hankins WD, Kabat D. (1985) Roles of helper and defective retroviral genomes in murine erythroleukemia: Studies of spleen focus-forming virus in the absence of helper. *J Virol* **56**: 660–664.
60. Tambourin P, Wendling F. (1971) Malignant transformation and erythroid differentiation by polycythaemia-inducing Friend virus. *Nat New Biol* **234**: 230–233.
61. Axelrad AA, Steeves RA. (1964) Assay for Friend leukemia virus: Rapid quantitative method based on enumeration of macroscopic spleen foci in mice. *Virology* **24**: 513–518.
62. Tambourin P, Wendling F, Moreau-Gachelin F. (1981) Friend leukemia as a multiple-step disease. *Blood Cells* **7**: 133–144.
63. Ben-David Y, Bernstein A. (1991) Friend virus-induced erythroleukemia and the multistage nature of cancer. *Cell* **66**: 831–834.
64. Fredrickson T, Tambourin P, Wendling F, *et al.* (1975) Target cell of the polycythemia-inducing Friend virus: Studies with myleran. *J Natl Cancer Inst* **55**: 443–446.
65. Richmond TD, Chohan M, Barber DL. (2005) Turning cells red: Signal transduction mediated by erythropoietin. *Trends Cell Biol* **15**: 146–155.
66. Mirand EA. (1967) Virus-induced erythropoiesis in hypertransfused-polycythemic mice. *Science* **156**: 832–833.
67. Liao SK, Axelrad AA. (1975) Erythropoietin-independent erythroid colony formation *in vitro* by hemopoietic cells of mice infected with friend virus. *Int J Cancer* **15**: 467–482.

68. Linemeyer DL, Menke JG, Ruscetti SK, *et al.* (1982) Envelope gene sequences which encode the gp52 protein of spleen focus-forming virus are required for the induction of erythroid cell proliferation. *J Virol* **43**: 223–233.
69. Ruta M, Bestwick R, Machida C, Kabat D. (1983) Loss of leukemogenicity caused by mutations in the membrane glycoprotein structural gene of Friend spleen focus-forming virus. *Proc Natl Acad Sci USA* **80**: 4704–4708.
70. Wolff L, Ruscetti S. (1988) The spleen focus-forming virus (SFFV) envelope gene, when introduced into mice in the absence of other SFFV genes, induces acute erythroleukemia. *J Virol* **62**: 2158–2163.
71. Li JP, D'Andrea AD, Lodish HF, Baltimore D. (1990) Activation of cell growth by binding of Friend spleen focus-forming virus gp55 glycoprotein to the erythropoietin receptor. *Nature* **343**: 762–764.
72. Zon LI, Moreau JF, Koo JW, *et al.* (1992) The erythropoietin receptor transmembrane region is necessary for activation by the Friend spleen focus-forming virus gp55 glycoprotein. *Mol Cell Biol* **12**: 2949–2957.
73. Li JP, Hu HO, Niu QT, Fang C. (1995) Cell surface activation of the erythropoietin receptor by Friend spleen focus-forming virus gp55. *J Virol* **69**: 1714–1719.
74. Chung SW, Wolff L, Ruscetti SK. (1989) Transmembrane domain of the envelope gene of a polycythemia-inducing retrovirus determines erythropoietin-independent growth. *Proc Natl Acad Sci USA* **86**: 7957–7960.
75. Muszynski KW, Ohashi T, Hanson C, Ruscetti SK. (1998) Both the polycythemia- and anemia-inducing strains of Friend spleen focus-forming virus induce constitutive activation of the Raf-1/mitogen-activated protein kinase signal transduction pathway. *J Virol* **72**: 919–925.
76. Muszynski KW, Thompson D, Hanson C, *et al.* (2000) Growth factor-independent proliferation of erythroid cells infected with Friend spleen focus-forming virus is protein kinase C dependent but does not require Ras-GTP. *J Virol* **74**: 8444–8451.
77. Nishigaki K, Hanson C, Ohashi T, *et al.* (2000) Erythroid cells rendered erythropoietin independent by infection with Friend spleen focus-forming virus show constitutive activation of phosphatidylinositol 3-kinase and Akt kinase: Involvement of insulin receptor substrate-related adapter proteins. *J Virol* **74**: 3037–3045.
78. Nishigaki K, Thompson D, Hanson C, *et al.* (2001) The envelope glycoprotein of Friend spleen focus-forming virus covalently interacts with and

constitutively activates a truncated form of the receptor tyrosine kinase Stk. *J Virol* **75**: 7893–7903.

79. Zhang J, Randall MS, Loyd MR, *et al.* (2006) Role of erythropoietin receptor signaling in Friend virus-induced erythroblastosis and polycythemia. *Blood* **107**: 73–78.
80. Rulli K, Yugawa T, Hanson C, *et al.* (2004) *Ex vivo* and *in vivo* biological effects of a truncated form of the receptor tyrosine kinase Stk when activated by interaction with the friend spleen focus-forming virus envelope glycoprotein or by point mutation. *J Virol* **78**: 4573–4581.
81. Jelacic TM, Thompson D, Hanson C, *et al.* (2008) The tyrosine kinase sf-Stk and its downstream signals are required for maintenance of Friend spleen focus-forming virus-induced fibroblast transformation. *J Virol* **82**: 419–427.
82. Nishigaki K, Hanson C, Jelacic T, *et al.* (2005) Friend spleen focus-forming virus transforms rodent fibroblasts in cooperation with a short form of the receptor tyrosine kinase Stk. *Proc Natl Acad Sci USA* **102**: 15488–15493.
83. Moreau-Gachelin F, Robert-Lezenes J, Wendling F, *et al.* (1985) Integration of spleen focus-forming virus proviruses in Friend tumor cells. *J Virol* **53**: 292–295.
84. Moreau-Gachelin F, Tavitian A, Tambourin P. (1988) Spi-1 is a putative oncogene in virally induced murine erythroleukaemias. *Nature* **331**: 277–280.
85. Paul R, Schuetze S, Kozak SL, Kabat D. (1989) A common site for immortalizing proviral integrations in Friend erythroleukemia: Molecular cloning and characterization. *J Virol* **63**: 4958–4961.
86. Moreau-Gachelin F, Ray D, de Both NJ, *et al.* (1990) Spi-1 oncogene activation in Rauscher and Friend murine virus-induced acute erythroleukemias. *Leukemia* **4**: 20–23.
87. Okuno Y, Huang G, Rosenbauer F, *et al.* (2005) Potential autoregulation of transcription factor PU.1 by an upstream regulatory element. *Mol Cell Biol* **25**: 2832–2845.
88. Fisher RC, Scott EW. (1998) Role of PU.1 in hematopoiesis. *Stem Cells* **16**: 25–37.
89. Klemsz MJ, McKercher SR, Celada A, *et al.* (1990) The macrophage and B cell-specific transcription factor PU.1 is related to the ets oncogene. *Cell* **61**: 113–124.
90. Kastner P, Chan S. (2008) PU.1: A crucial and versatile player in hematopoiesis and leukemia. *Int J Biochem Cell Biol* **40**: 22–27.

91. McKercher SR, Torbett BE, Anderson KL, *et al.* (1996) Targeted disruption of the PU.1 gene results in multiple hematopoietic abnormalities. *Embo J* **15**: 5647–5658.
92. Celada A, Borras FE, Soler C, *et al.* (1996) The transcription factor PU.1 is involved in macrophage proliferation. *J Exp Med* **184**: 61–69.
93. Iwasaki H, Somoza C, Shigematsu H, *et al.* (2005) Distinctive and indispensable roles of PU.1 in maintenance of hematopoietic stem cells and their differentiation. *Blood* **106**: 1590–1600.
94. Dakic A, Metcalf D, Di Rago L, *et al.* (2005) PU.1 regulates the commitment of adult hematopoietic progenitors and restricts granulopoiesis. *J Exp Med* **201**: 1487–1502.
95. Back J, Dierich A, Bronn C, *et al.* (2004) PU.1 determines the self-renewal capacity of erythroid progenitor cells. *Blood* **103**: 3615–3623.
96. Nutt SL, Metcalf D, D'Amico A, *et al.* (2005) Dynamic regulation of PU.1 expression in multipotent hematopoietic progenitors. *J Exp Med* **201**: 221–231.
97. Brightman BK, Chandy KG, Spencer RH, *et al.* (1988) Characterization of lymphoid tumors induced by a recombinant murine retrovirus carrying the avian v-myc oncogene. Identification of novel (B-lymphoid) tumors in the thymus. *J Immunol* **141**: 2844–2854.
98. Brightman BK, Davis BR, Fan H. (1990) Preleukemic hematopoietic hyperplasia induced by Moloney murine leukemia virus is an indirect consequence of viral infection. *J Virol* **64**: 4582–4584.
99. Lazo PA, Klein-Szanto AJ, Tsichlis PN. (1990) T-cell lymphoma lines derived from rat thymomas induced by Moloney murine leukemia virus: Phenotypic diversity and its implications. *J Virol* **64**: 3948–3959.
100. Buckheit RW Jr, Kurtzberg J, Bolognesi DP, Weinhold KJ. (1988) The coenrichment of stem cells, prothymocytes, and stromal elements with ecotropic retrovirus-producing cells from the bone marrow of leukemia-prone AKR mice. *Virology* **162**: 354–361.
101. Haran-Ghera N. (1980) Potential leukemic cells among bone marrow cells of young AKR/J mice. *Proc Natl Acad Sci USA* **77**: 2923–2926.
102. Haran-Ghera N, Rubio N, Leef F, Goldstein G. (1978) Characteristics of preleukemia cells induced in mice. *Cell Immunol* **37**: 308–314.
103. Hartley JW, Wolford NK, Old LJ, Rowe WP. (1977) A new class of murine leukemia virus associated with development of spontaneous lymphomas. *Proc Natl Acad Sci USA* **74**: 789–792.

104. Stoye JP, Moroni C, Coffin JM. (1991) Virological events leading to spontaneous AKR thymomas. *J Virol* **65**: 1273–1285.

105. Belli B, Fan H. (1994) The leukemogenic potential of an enhancer variant of Moloney murine leukemia virus varies with the route of inoculation. *J Virol* **68**: 6883–6889.

106. Belli B, Patel A, Fan H. (1995) Recombinant mink cell focus-inducing virus and long terminal repeat alterations accompany the increased leukemogenicity of the Mo+PyF101 variant of Moloney murine leukemia virus after intraperitoneal inoculation. *J Virol* **69**: 1037–1043.

107. Lander JK, Chesebro B, Fan H. (1999) Appearance of mink cell focus-inducing recombinants during *in vivo* infection by Moloney murine leukemia virus (M-MuLV) or the Mo+PyF101 M-MuLV enhancer variant: Implications for sites of generation and roles in leukemogenesis. *J Virol* **73**: 5671–5680.

108. Davis BR, Brightman BK, Chandy KG, Fan H. (1987) Characterization of a preleukemic state induced by Moloney murine leukemia virus: Evidence for two infection events during leukemogenesis. *Proc Natl Acad Sci USA* **84**: 4875–4879.

109. Davis BR, Chandy KG, Brightman BK, *et al.* (1986) Effects of nonleukemogenic and wild-type Moloney murine leukemia virus on lymphoid cells *in vivo*: Identification of a preleukemic shift in thymocyte subpopulations. *J Virol* **60**: 423–430.

110. Li QX, Fan H. (1990) Combined infection by Moloney murine leukemia virus and a mink cell focus-forming virus recombinant induces cytopathic effects in fibroblasts or in long-term bone marrow cultures from preleukemic mice. *J Virol* **64**: 3701–3711.

111. Storch TG, Arnstein P, Manohar V, *et al.* (1985) Proliferation of infected lymphoid precursors before Moloney murine leukemia virus-induced T-cell lymphoma. *J Natl Cancer Inst* **74**: 137–143.

112. Gaulton GN. (1998) Viral pathogenesis and immunity within the thymus. *Immunol Res* **17**: 75–82.

113. Marshall DJ, Park BH, Korostoff JM, Gaulton GN. (1995) Manipulation of the immune response by foreign gene expression in the thymus. *Leukemia* **9**(Suppl 1): S128–S132.

114. Tumas-Brundage KM, Garret W, Blank K, Prystowsky MB. (1996) Murine leukemia virus infects early bone marrow progenitors in immunocompetent mice. *Virology* **224**: 573–575.

115. Okimoto MA, Fan H. (1999) Identification of directly infected cells in the bone marrow of neonatal moloney murine leukemia virus-infected mice by use of a moloney murine leukemia virus-based vector. *J Virol* **73**: 1617–1623.
116. Fan H, Brightman BK, Belli B, *et al.* (1997) Early (preleukemic) events in Moloney murine leukemia virus leukemogenesis. *Leukemia* **11**(Suppl 3): 149–151.
117. Finstad SL, Rosenberg N, Levy LS. (2007) Diminished potential for B-lymphoid differentiation after MuLV infection *in vivo* and in EML hematopoietic progenitor cells. *J Virol.*
118. Rulli K, Lenz J, Levy LS. (2002) Disruption of hematopoiesis and thymopoiesis in the early premalignant stages of infection with SL3-3 murine leukemia virus. *J Virol* **76**: 2363–2374.
119. van Lohuizen M, Verbeek S, Scheijen B, *et al.* (1991) Identification of cooperating oncogenes in E mu-myc transgenic mice by provirus tagging. *Cell* **65**: 737–752.
120. Bear SE, Bellacosa A, Lazo PA, *et al.* (1989) Provirus insertion in Tpl-1, an Ets-1-related oncogene, is associated with tumor progression in Moloney murine leukemia virus-induced rat thymic lymphomas. *Proc Natl Acad Sci USA* **86**: 7495–7499.
121. Patriotis C, Makris A, Bear SE, Tsichlis PN. (1993) Tumor progression locus 2 (Tpl-2) encodes a protein kinase involved in the progression of rodent T-cell lymphomas and in T-cell activation. *Proc Natl Acad Sci USA* **90**: 2251–2255.
122. Makris A, Patriotis C, Bear SE, Tsichlis PN. (1993) Genomic organization and expression of Tpl-2 in normal cells and Moloney murine leukemia virus-induced rat T-cell lymphomas: Activation by provirus insertion. *J Virol* **67**: 4283–4289.
123. Gilks CB, Bear SE, Grimes HL, Tsichlis PN. (1993) Progression of interleukin-2 (IL-2)-dependent rat T cell lymphoma lines to IL-2-independent growth following activation of a gene (Gfi-1) encoding a novel zinc finger protein. *Mol Cell Biol* **13**: 1759–1768.
124. Wiener F, Ohno S, Spira J, *et al.* (1978) Cytogenetic mapping of the trisomic segment of chromosome 15 in murine T-cell leukaemia. *Nature* **275**: 658–660.
125. Lichtman MA. (1983) Hemopoietic stem cell disorders. *Am J Med Technol* **49**: 97–102.

126. Speck NA, Gilliland DG. (2002) Core-binding factors in haematopoiesis and leukaemia. *Nat Rev Cancer* **2**: 502–513.
127. Gilliland DG, Griffin JD. (2002) The roles of FLT3 in hematopoiesis and leukemia. *Blood* **100**: 1532–1542.
128. Mueller BU, Pabst T, Osato M, *et al.* (2003) Heterozygous PU.1 mutations are associated with acute myeloid leukemia. *Blood* **101**: 2074.
129. Tenen DG. (2003) Disruption of differentiation in human cancer: AML shows the way. *Nat Rev Cancer* **3**: 89–101.
130. Reilly JT. (2005) Pathogenesis of acute myeloid leukaemia and inv(16)(p13;q22): a paradigm for understanding leukaemogenesis? *Br J Haematol* **128**: 18–34.
131. Koschmieder S, Rosenbauer F, Steidl U, *et al.* (2005) Role of transcription factors C/EBPalpha and PU.1 in normal hematopoiesis and leukemia. *Int J Hematol* **81**: 368–377.
132. Rosenbauer F, Koschmieder S, Steidl U, Tenen DG. (2005) Effect of transcription-factor concentrations on leukemic stem cells. *Blood* **106**: 1519–1524.
133. Zauli G, Davis BR. (1993) Role of HIV infection in the hematologic manifestations of HIV seropositive subjects. *Crit Rev Oncol Hematol* **15**: 271–283.
134. Roder W, Muller H, Muller WE, Merz H. (1992) HIV infection in human bone. *J Bone Joint Surg Br* **74**: 179–180.
135. Moses A, Nelson J, Bagby GC Jr. (1998) The influence of human immunodeficiency virus-1 on hematopoiesis. *Blood* **91**: 1479–1495.
136. Harbol AW, Liesveld JL, Simpson-Haidaris PJ, Abboud CN. (1994) Mechanisms of cytopenia in human immunodeficiency virus infection. *Blood Rev* **8**: 241–251.
137. Scadden DT, Zon LI, Groopman JE. (1989) Pathophysiology and management of HIV-associated hematologic disorders. *Blood* **74**: 1455–1463.
138. Zon LI, Arkin C, Groopman JE. (1987) Haematologic manifestations of the human immune deficiency virus (HIV). *Br J Haematol* **66**: 251–256.
139. Davis BR, Schwartz DH, Marx JC, *et al.* (1991) Absent or rare human immunodeficiency virus infection of bone marrow stem/progenitor cells *in vivo*. *J Virol* **65**: 1985–1990.
140. Zauli G, Re MC, Davis B, *et al.* (1992) Impaired *in vitro* growth of purified (CD34+) hematopoietic progenitors in human immunodeficiency virus-1 seropositive thrombocytopenic individuals. *Blood* **79**: 2680–2687.

141. von Laer D, Hufert FT, Fenner TE, *et al.* (1990) CD34+ hematopoietic progenitor cells are not a major reservoir of the human immunodeficiency virus. *Blood* **76**: 1281–1286.
142. Stanley SK, Kessler SW, Justement JS, *et al.* (1992) CD34+ bone marrow cells are infected with HIV in a subset of seropositive individuals. *J Immunol* **149**: 689–697.
143. Isgro A, Aiuti A, Mezzaroma I, *et al.* (2002) Improvement of interleukin 2 production, clonogenic capability and restoration of stromal cell function in human immunodeficiency virus-type-1 patients after highly active antiretroviral therapy. *Br J Haematol* **118**: 864–874.
144. Isgro A, Aiuti F, Mezzaroma I, *et al.* (2002) Interleukin 7 production by bone marrow-derived stromal cells in HIV-1-infected patients during highly active antiretroviral therapy. *Aids* **16**: 2231–2232.
145. Wang L, Mondal D, La Russa VF, Agrawal KC. (2002) Suppression of clonogenic potential of human bone marrow mesenchymal stem cells by HIV type 1: Putative role of HIV type 1 tat protein and inflammatory cytokines. *AIDS Res Hum Retroviruses* **18**: 917–931.
146. Isgro A, Aiuti A, Leti W, *et al.* (2005) Immunodysregulation of HIV disease at bone marrow level. *Autoimmun Rev* **4**: 486–490.
147. Isgro A, Leti W, De Santis W, *et al.* (2008) Altered clonogenic capability and stromal cell function characterize bone marrow of HIV-infected subjects with low CD4+ T cell counts despite viral suppression during HAART. *Clin Infect Dis* **46**: 1902–1910.
148. Zauli G, Vitale M, Gibellini D, Capitani S. (1996) Inhibition of purified CD34+ hematopoietic progenitor cells by human immunodeficiency virus 1 or gp120 mediated by endogenous transforming growth factor beta 1. *J Exp Med* **183**: 99–108.
149. Banda NK, Tomczak JA, Shpall EJ, *et al.* (1997) HIV-gp120 induced cell death in hematopoietic progenitor CD34+ cells. *Apoptosis* **2**: 61–68.
150. Maciejewski JP, Weichold FF, Young NS. (1994) HIV-1 suppression of hematopoiesis *in vitro* mediated by envelope glycoprotein and TNF-alpha. *J Immunol* **153**: 4303–4310.
151. Gibellini D, Vitone F, Buzzi M, *et al.* (2007) HIV-1 negatively affects the survival/maturation of cord blood CD34(+) hematopoietic progenitor cells differentiated towards megakaryocytic lineage by HIV-1 gp120/CD4 membrane interaction. *J Cell Physiol* **210**: 315–324.

152. Rameshwar P, Denny TN, Gascon P. (1996) Enhanced HIV-1 activity in bone marrow can lead to myelopoietic suppression partially contributed by gag p24. *J Immunol* **157**: 4244–4250.
153. Kulkosky J, Laptev A, Shetty S, *et al.* (1999) Human immunodeficiency virus type 1 Vpr alters bone marrow cell function. *Blood* **93**: 1906–1915.
154. Alexaki A, Liu Y, Wigdahl B. (2008) Cellular reservoirs of HIV-1 and their role in viral persistence. *Curr HIV Res* **6**: 388–400.
155. Alexaki A, Wigdahl B. (2008) HIV-1 infection of bone marrow hematopoietic progenitor cells and their role in trafficking and viral dissemination. *PLoS Pathog* **4**: e1000215.
156. Davis BR, Zauli G. (1995) Effect of human immunodeficiency virus infection on haematopoiesis. *Baillieres Clin Haematol* **8**: 113–130.
157. Zauli G, Furlini G, Vitale M, *et al.* (1994) A subset of human CD34+ hematopoietic progenitors express low levels of CD4, the high-affinity receptor for human immunodeficiency virus-type 1. *Blood* **84**: 1896–1905.
158. Louache F, Debili N, Marandin A, *et al.* (1994) Expression of CD4 by human hematopoietic progenitors. *Blood* **84**: 3344–3355.
159. Carr JM, Ramshaw HS, Li P, Burrell CJ. (1998) CD34+ cells and their derivatives contain mRNA for CD4 and human immunodeficiency virus (HIV) co-receptors and are susceptible to infection with M- and T-tropic HIV. *J Gen Virol* **79**(Pt 1): 71–75.
160. Deichmann M, Kronenwett R, Haas R. (1997) Expression of the human immunodeficiency virus type-1 coreceptors CXCR-4 (fusin, LESTR) and CKR-5 in CD34+ hematopoietic progenitor cells. *Blood* **89**: 3522–3528.
161. Aiuti A, Turchetto L, Cota M, *et al.* (1999) Human CD34(+) cells express CXCR4 and its ligand stromal cell-derived factor-1. Implications for infection by T-cell tropic human immunodeficiency virus. *Blood* **94**: 62–73.
162. Folks TM, Kessler SW, Orenstein JM, *et al.* (1988) Infection and replication of HIV-1 in purified progenitor cells of normal human bone marrow. *Science* **242**: 919–922.
163. Molina JM, Scadden DT, Sakaguchi M, *et al.* (1990) Lack of evidence for infection of or effect on growth of hematopoietic progenitor cells after *in vivo* or *in vitro* exposure to human immunodeficiency virus. *Blood* **76**: 2476–2482.
164. Louache F, Henri A, Bettaieb A, *et al.* (1992) Role of human immunodeficiency virus replication in defective *in vitro* growth of hematopoietic progenitors. *Blood* **80**: 2991–2999.

165. Kojouharoff G, Ottmann OG, von Briesen H, *et al.* (1991) Infection of granulocyte/monocyte progenitor cells with HIV1. *Res Virol* **142**: 151–157.
166. Kaczmarski RS, Davison F, Blair E, *et al.* (1992) Detection of HIV in haemopoietic progenitors. *Br J Haematol* **82**: 764–769.
167. Yamakami K, Honda M, Takei M, *et al.* (2004) Early bone marrow hematopoietic defect in simian/human immunodeficiency virus C2/1-infected macaques and relevance to advance of disease. *J Virol* **78**: 10906–10910.
168. Caufour P, Le Grand R, Cheret A, *et al.* (2001) Longitudinal analysis of CD8(+) T-cell phenotype and IL-7, IL-15 and IL-16 mRNA expression in different tissues during primary simian immunodeficiency virus infection. *Microbes Infect* **3**: 181–191.
169. Thiebot H, Louache F, Vaslin B, *et al.* (2001) Early and persistent bone marrow hematopoiesis defect in simian/human immunodeficiency virus-infected macaques despite efficient reduction of viremia by highly active antiretroviral therapy during primary infection. *J Virol* **75**: 11594–11602.
170. Koka PS, Fraser JK, Bryson Y, *et al.* (1998) Human immunodeficiency virus inhibits multilineage hematopoiesis *in vivo. J Virol* **72**: 5121–5127.
171. Koka PS, Jamieson BD, Brooks DG, Zack JA. (1999) Human immunodeficiency virus type 1-induced hematopoietic inhibition is independent of productive infection of progenitor cells *in vivo. J Virol* **73**: 9089–9097.
172. Jenkins M, Hanley MB, Moreno MB, *et al.* (1998) Human immunodeficiency virus-1 infection interrupts thymopoiesis and multilineage hematopoiesis *in vivo. Blood* **91**: 2672–2678.
173. Lunardi-Iskandar Y, Georgoulias V, Bertoli AM, *et al.* (1989) Impaired *in vitro* proliferation of hemopoietic precursors in HIV-1-infected subjects. *Leuk Res* **13**: 573–581.
174. Banda NK, Simon GR, Sipple JD, *et al.* (1999) Depletion of CD34+ CD4+ cells in bone marrow from HIV-1-infected individuals. *Biol Blood Marrow Transplant* **5**: 162–172.
175. Steinberg HN, Crumpacker CS, Chatis PA. (1991) *In vitro* suppression of normal human bone marrow progenitor cells by human immunodeficiency virus. *J Virol* **65**: 1765–1769.
176. Zauli G, Re MC, Furlini G, *et al.* (1992) Human immunodeficiency virus type 1 envelope glycoprotein gp120-mediated killing of human haematopoietic progenitors (CD34+ cells). *J Gen Virol* **73**(Pt 2): 417–421.
177. Zauli G, Capitani S. (1996) HIV-1-related mechanisms of suppression of CD34+ hematopoietic progenitors. *Pathobiology* **64**: 53–58.

178. Cen D, Zauli G, Szarnicki R, Davis BR. (1993) Effect of different human immunodeficiency virus type-1 (HIV-1) isolates on long-term bone marrow haemopoiesis. *Br J Haematol* **85**: 596–602.
179. Costantini A, Giuliodoro S, Mancini S, *et al.* (2006) Impaired *in vitro* growth of megakaryocytic colonies derived from CD34 cells of HIV-1-infected patients with active viral replication. *Aids* **20**: 1713–1720.
180. Marandin A, Katz A, Oksenhendler E, *et al.* (1996) Loss of primitive hematopoietic progenitors in patients with human immunodeficiency virus infection. *Blood* **88**: 4568–4578.
181. Re MC, Furlini G, Zauli G, La Placa M. (1994) Human immunodeficiency virus type 1 (HIV-1) and human hematopoietic progenitor cells. *Arch Virol* **137**: 1–23.
182. Zauli G, Vitale M, Re MC, *et al.* (1994) *In vitro* exposure to human immunodeficiency virus type 1 induces apoptotic cell death of the factor-dependent TF-1 hematopoietic cell line. *Blood* **83**: 167–175.
183. Nagafuji K, Shibuya T, Harada M, *et al.* (1995) Functional expression of Fas antigen (CD95) on hematopoietic progenitor cells. *Blood* **86**: 883–889.
184. Sato T, Selleri C, Anderson S, *et al.* (1997) Expression and modulation of cellular receptors for interferon-gamma, tumour necrosis factor, and Fas on human bone marrow CD34+ cells. *Br J Haematol* **97**: 356–365.
185. Dybedal I, Bryder D, Fossum A, *et al.* (2001) Tumor necrosis factor (TNF)-mediated activation of the p55 TNF receptor negatively regulates maintenance of cycling reconstituting human hematopoietic stem cells. *Blood* **98**: 1782–1791.
186. Bryder D, Ramsfjell V, Dybedal I, *et al.* (2001) Self-renewal of multipotent long-term repopulating hematopoietic stem cells is negatively regulated by Fas and tumor necrosis factor receptor activation. *J Exp Med* **194**: 941–952.
187. Bailey JR, Sedaghat AR, Kieffer T, *et al.* (2006) Residual human immunodeficiency virus type 1 viremia in some patients on antiretroviral therapy is dominated by a small number of invariant clones rarely found in circulating CD4+ T cells. *J Virol* **80**: 6441–6457.
188. Kaneda T, Murakami T, Hagiwara T, *et al.* (2001) Defective HIV-1 provirus found in peripheral T lymphocytes and granulocytes in an AIDS patient imply viral infection of progenitor cells. *Aids* **15**: 939–940.

189. Nicholls SE, Lucas G, Graham GJ, *et al.* (1999) Macrophage-inflammatory protein-1alpha receptor expression on normal and chronic myeloid leukemia CD34+ cells. *J Immunol* **162**: 6191–6199.
190. Ruiz ME, Cicala C, Arthos J, *et al.* (1998) Peripheral blood-derived CD34+ progenitor cells: CXC chemokine receptor 4 and CC chemokine receptor 5 expression and infection by HIV. *J Immunol* **161**: 4169–4176.
191. Kitano K, Abboud CN, Ryan DH, *et al.* (1991) Macrophage-active colony-stimulating factors enhance human immunodeficiency virus type 1 infection in bone marrow stem cells. *Blood* **77**: 1699–1705.
192. Zauli G, Re MC, Visani G, *et al.* (1992) Evidence for a human immunodeficiency virus type 1-mediated suppression of uninfected hematopoietic (CD34+) cells in AIDS patients. *J Infect Dis* **166**: 710–716.
193. Weichold FF, Zella D, Barabitskaja O, *et al.* (1998) Neither human immunodeficiency virus-1 (HIV-1) nor HIV-2 infects most-primitive human hematopoietic stem cells as assessed in long-term bone marrow cultures. *Blood* **91**: 907–915.
194. Shen H, Cheng T, Preffer FI, *et al.* (1999) Intrinsic human immunodeficiency virus type 1 resistance of hematopoietic stem cells despite coreceptor expression. *J Virol* **73**: 728–737.
195. Rusten LS, Jacobsen FW, Lesslauer W, *et al.* (1994) Bifunctional effects of tumor necrosis factor alpha (TNF alpha) on the growth of mature and primitive human hematopoietic progenitor cells: Involvement of p55 and p75 TNF receptors. *Blood* **83**: 3152–3159.
196. Jacobsen SE, Jacobsen FW, Fahlman C, Rusten LS. (1994) TNF-alpha, the great imitator: role of p55 and p75 TNF receptors in hematopoiesis. *Stem Cells* **12**(Suppl 1): 111–126; discussion 126–118.
197. Ensoli B, Barillari G, Salahuddin SZ, *et al.* (1990) Tat protein of HIV-1 stimulates growth of cells derived from Kaposi's sarcoma lesions of AIDS patients. *Nature* **345**: 84–86.
198. Frankel AD, Pabo CO. (1988) Cellular uptake of the tat protein from human immunodeficiency virus. *Cell* **55**: 1189–1193.
199. Viscidi RP, Mayur K, Lederman HM, Frankel AD. (1989) Inhibition of antigen-induced lymphocyte proliferation by Tat protein from HIV-1. *Science* **246**: 1606–1608.
200. Buonaguro L, Barillari G, Chang HK, *et al.* (1992) Effects of the human immunodeficiency virus type 1 Tat protein on the expression of inflammatory cytokines. *J Virol* **66**: 7159–7167.

201. Zauli G, Davis BR, Re MC, *et al.* (1992) Tat protein stimulates production of transforming growth factor-beta 1 by marrow macrophages: A potential mechanism for human immunodeficiency virus-1-induced hematopoietic suppression. *Blood* **80**: 3036–3043.
202. Matsuoka M, Jeang KT. (2007) Human T-cell leukaemia virus type 1 (HTLV-1) infectivity and cellular transformation. *Nat Rev Cancer* 7: 270–280.
203. Mortreux F, Gabet AS, Wattel E. (2003) Molecular and cellular aspects of HTLV-1 associated leukemogenesis *in vivo. Leukemia* **17**: 26–38.
204. Feuer G, Fraser JK, Zack JA, *et al.* (1996) Human T-cell leukemia virus infection of human hematopoietic progenitor cells: Maintenance of virus infection during differentiation *in vitro* and *in vivo. J Virol* **70**: 4038–4044.
205. Wencker M, Sausse C, Derse D, *et al.* (2007) Human T-cell leukemia virus type 1 Tax protein down-regulates pre-T-cell receptor alpha gene transcription in human immature thymocytes. *J Virol* **81**: 301–308.
206. Grant C, Barmak K, Alefantis T, *et al.* (2002) Human T cell leukemia virus type I and neurologic disease: Events in bone marrow, peripheral blood, and central nervous system during normal immune surveillance and neuroinflammation. *J Cell Physiol* **190**: 133–159.
207. Tripp A, Liu Y, Sieburg M, *et al.* (2003) Human T-cell leukemia virus type 1 tax oncoprotein suppression of multilineage hematopoiesis of CD34+ cells *in vitro. J Virol* **77**: 12152–12164.
208. Banerjee P, Tripp A, Crawford L, *et al.* (2010) Adult T cell leukemia/lymphoma development in HTLV-1-infected humanized SCID mice. *Blood* Feb 1, 2010 (Epub ahead of print).
209. Jeang KT, Giam CZ, Majone F, Aboud M. (2004) Life, death, and tax: Role of HTLV-I oncoprotein in genetic instability and cellular transformation. *J Biol Chem* **279**: 31991–31994.
210. Sun SC, Ballard DW. (1999) Persistent activation of NF-kappaB by the tax transforming protein of HTLV-1: Hijacking cellular IkappaB kinases. *Oncogene* **18**: 6948–6958.
211. Jeang KT, Widen SG, Semmes OJt, Wilson SH. (1990) HTLV-I trans-activator protein, tax, is a trans-repressor of the human beta-polymerase gene. *Science* **247**: 1082–1084.
212. Kasai T, Jeang KT. (2004) Two discrete events, human T-cell leukemia virus type I Tax oncoprotein expression and a separate stress stimulus, are required for induction of apoptosis in T-cells. *Retrovirology* **1**: 7.

213. Yoshida M. (2001) Multiple viral strategies of HTLV-1 for dysregulation of cell growth control. *Annu Rev Immunol* **19**: 475–496.
214. Brauweiler A, Garrus JE, Reed JC, Nyborg JK. (1997) Repression of bax gene expression by the HTLV-1 Tax protein: implications for suppression of apoptosis in virally infected cells. *Virology* **231**: 135–140.
215. Kibler KV, Jeang KT. (2001) CREB/ATF-dependent repression of cyclin a by human T-cell leukemia virus type 1 Tax protein. *J Virol* **75**: 2161–2173.
216. Lemasson I, Robert-Hebmann V, Hamaia S, *et al.* (1997) Transrepression of lck gene expression by human T-cell leukemia virus type 1-encoded p40tax. *J Virol* **71**: 1975–1983.
217. Riou P, Bex F, Gazzolo L. (2000) The human T cell leukemia/lymphotropic virus type 1 Tax protein represses MyoD-dependent transcription by inhibiting MyoD-binding to the KIX domain of p300. A potential mechanism for Tax-mediated repression of the transcriptional activity of basic helix-loop-helix factors. *J Biol Chem* **275**: 10551–10560.
218. Suzuki T, Narita T, Uchida-Toita M, Yoshida M. (1999) Down-regulation of the INK4 family of cyclin-dependent kinase inhibitors by tax protein of HTLV-1 through two distinct mechanisms. *Virology* **259**: 384–391.
219. Uittenbogaard MN, Giebler HA, Reisman D, Nyborg JK. (1995) Transcriptional repression of p53 by human T-cell leukemia virus type I Tax protein. *J Biol Chem* **270**: 28503–28506.
220. Grassmann R, Berchtold S, Radant I, *et al.* (1992) Role of human T-cell leukemia virus type 1 X region proteins in immortalization of primary human lymphocytes in culture. *J Virol* **66**: 4570–4575.
221. Akagi T, Ono H, Shimotohno K. (1995) Characterization of T cells immortalized by Tax1 of human T-cell leukemia virus type 1. *Blood* **86**: 4243–4249.
222. Nerenberg M, Hinrichs SH, Reynolds RK, *et al.* (1987) The tat gene of human T-lymphotropic virus type 1 induces mesenchymal tumors in transgenic mice. *Science* **237**: 1324–1329.
223. Hasegawa H, Sawa H, Lewis MJ, *et al.* (2006) Thymus-derived leukemia-lymphoma in mice transgenic for the Tax gene of human T-lymphotropic virus type I. *Nat Med* **12**: 466–472.
224. Lemoine FJ, Marriott SJ. (2001) Accelerated G(1) phase progression induced by the human T cell leukemia virus type I (HTLV-I) Tax oncoprotein. *J Biol Chem* **276**: 31851–31857.

225. Iwanaga R, Ohtani K, Hayashi T, Nakamura M. (2001) Molecular mechanism of cell cycle progression induced by the oncogene product Tax of human T-cell leukemia virus type I. *Oncogene* **20**: 2055–2067.
226. Haller K, Wu Y, Derow E, *et al.* (2002) Physical interaction of human T-cell leukemia virus type 1 Tax with cyclin-dependent kinase 4 stimulates the phosphorylation of retinoblastoma protein. *Mol Cell Biol* **22**: 3327–3338.
227. Schmitt I, Rosin O, Rohwer P, *et al.* (1998) Stimulation of cyclin-dependent kinase activity and G1- to S-phase transition in human lymphocytes by the human T-cell leukemia/lymphotropic virus type 1 Tax protein. *J Virol* **72**: 633–640.
228. Liu M, Yang L, Zhang L, *et al.* (2008) HTLV-1 infection leads to arrest in the G1 phase of the cell cycle. *J Virol.*
229. Liu B, Liang MH, Kuo YL, *et al.* (2003) Human T-lymphotropic virus type 1 oncoprotein tax promotes unscheduled degradation of Pds1p/securin and Clb2p/cyclin B1 and causes chromosomal instability. *Mol Cell Biol* **23**: 5269–5281.
230. Liu B, Hong S, Tang Z, *et al.* (2005) HTLV-I Tax directly binds the Cdc20-associated anaphase-promoting complex and activates it ahead of schedule. *Proc Natl Acad Sci USA* **102**: 63–68.
231. Kuo YL, Giam CZ. (2006) Activation of the anaphase promoting complex by HTLV-1 tax leads to senescence. *Embo J* **25**: 1741–1752.
232. Macleod KF, Sherry N, Hannon G, *et al.* (1995) p53-dependent and independent expression of p21 during cell growth, differentiation, and DNA damage. *Genes Dev* **9**: 935–944.
233. Furukawa Y. (2002) Cell cycle control genes and hematopoietic cell differentiation. *Leuk Lymphoma* **43**: 225–231.
234. Cheng T, Rodrigues N, Dombkowski D, *et al.* (2000) Stem cell repopulation efficiency but not pool size is governed by p27(kip1). *Nat Med* **6**: 1235–1240.
235. Cheng T, Rodrigues N, Shen H, *et al.* (2000) Hematopoietic stem cell quiescence maintained by p21cip1/waf1. *Science* **287**: 1804–1808.
236. Merling R, Chen C, Hong S, *et al.* (2007) HTLV-1 Tax mutants that do not induce G1 arrest are disabled in activating the anaphase promoting complex. *Retrovirology* **4**: 35.
237. Guzman ML, Neering SJ, Upchurch D, *et al.* (2001) Nuclear factor-kappaB is constitutively activated in primitive human acute myelogenous leukemia cells. *Blood* **98**: 2301–2307.

238. Ambrosini G, Adida C, Altieri DC. (1997) A novel anti-apoptosis gene, survivin, expressed in cancer and lymphoma. *Nat Med* **3**: 917–921.
239. Li F, Ackermann EJ, Bennett CF, *et al.* (1999) Pleiotropic cell-division defects and apoptosis induced by interference with survivin function. *Nat Cell Biol* **1**: 461–466.
240. Li F, Ambrosini G, Chu EY, *et al.* (1998) Control of apoptosis and mitotic spindle checkpoint by survivin. *Nature* **396**: 580–584.
241. Fukuda S, Pelus LM. (2002) Elevation of survivin levels by hematopoietic growth factors occurs in quiescent CD34+ hematopoietic stem and progenitor cells before cell cycle entry. *Cell Cycle* **1**: 322–326.
242. Fukuda S, Foster RG, Porter SB, Pelus LM. (2002) The antiapoptosis protein survivin is associated with cell cycle entry of normal cord blood CD34(+) cells and modulates cell cycle and proliferation of mouse hematopoietic progenitor cells. *Blood* **100**: 2463–2471.
243. Fukuda S, Mantel CR, Pelus LM. (2004) Survivin regulates hematopoietic progenitor cell proliferation through p21WAF1/Cip1-dependent and -independent pathways. *Blood* **103**: 120–127.
244. MacKintosh FR, Adlish J, Hall SW, *et al.* (1993) Suppression of normal human hematopoiesis by cytomegalovirus *in vitro*. *Exp Hematol* **21**: 243–250.
245. Jacobson S, Krichavsky M, Flerlage N, Levin M. (1997) Immunopathogenesis of HTLV-I associated neurologic disease: Massive latent HTLV-I infection in bone marrow of HAM/TSP patients. *Leukemia* **11**(Suppl 3): 73–75.

CHAPTER 2

Human T-Cell Leukemia Virus Type 1, Cellular Transformation, and Adult T-Cell Leukemia

Junichiro Yasunaga[†] and Kuan-Teh Jeang[*,†]

INTRODUCTION

Human T-cell leukemia virus type 1 (HTLV-1) belongs to the delta type retroviruses, which also include bovine leukemia virus (BLV), human T-cell leukemia virus type 2 (HTLV-2) and simian T-cell leukemia virus (STLV).[1] HTLV-1 is the first retrovirus that was identified as a causative agent of several human diseases, including adult T-cell leukemia (ATL), HTLV-1 associated myelopathy (HAM)/tropical spastic paraparesis (TSP), and HTLV-1 associated uveitis.[2,3] ATL is a neoplastic disease of CD4-positive T lymphocytes, which is characterized by pleomorphic tumor cells with hypersegmented nuclei, called "flower cells." HTLV-1 encodes in its pX region a potent oncoprotein, Tax. Tax is a transcriptional activator protein which has been reported to activate and inactivate

*Corresponding author.
†Molecular Virology Section, Laboratory of Molecular Microbiology, the National Institute of Allergy and Infectious Diseases, National Institutes of Health, Bethesda, Maryland, USA. Building 4, Room 306, 9000 Rockville Pike, Bethesda, MD 20892-0460, USA. E-mail: kj7e@nih.gov.

the transcription of many cellular genes. Ectopic over-expression of the HTLV-1 Tax has been found to transform rodent cells and immortalize human cells *ex vivo*.[4–8] Moreover, transgenic mice engineered to express Tax also form tumors readily.[9–12] Taken together, these findings strongly support a physiological role for Tax in ATL leukemogenesis. Nevertheless, there are reports that *tax* transcripts are detected in only ~40 percent of transformed ATL cells, and that only a small proportion of HTLV-1 carriers (6.6 percent for males and 2.1 percent for females in Japan) develop ATL after a long latency period (about 60 years in Japan, and 40 years in Jamaica) from the initial infection.[1] These observations argue that besides Tax, other factors such as additional viral genes, host cell and immune factors, may also contribute to the development of ATL. Compatible with the above thinking, it was recently discovered that the HTLV-1 bZIP factor (HBZ), which is encoded by the minus strand of the provirus, is ubiquitously expressed in all ATL cells and possesses cell proliferative function in T cells.[13] HBZ may also contribute to leukemogenesis.

HTLV-1 PROVIRUS AND INFECTION OF CELLS

Structure of HTLV-1 Provirus

The HTLV-1 proviral genome has *gag*, *pol* and *env* genes, flanked by long terminal repeat (LTR) sequences at both ends (Fig. 1). The 5′ LTR serves as the virus' transcriptional promoter; Pol provides reverse transcription and integration functions to copy viral RNA into DNA and to integrate the DNA into the host cell's chromosomes; Gag and Env are structural proteins for the HTLV-1 virion. In the HTLV-1 genome, a unique region is located between *env* and the 3′ LTR; this region is named pX. The plus strand of the pX region encodes viral regulatory proteins Tax, Rex, p12, p13, p30 and p21, which are implicated in viral infectivity and the proliferation of infected cells. The HBZ protein is encoded by the minus strand of pX.[14] There are spliced and unspliced forms of HBZ.[13,15]

Clonal Proliferation of HTLV-1-Infected Cells

HTLV-1 integrates randomly into the host's chromosomes.[16] Sequential analyses of integration sites in HTLV-1-infected subjects verify that the

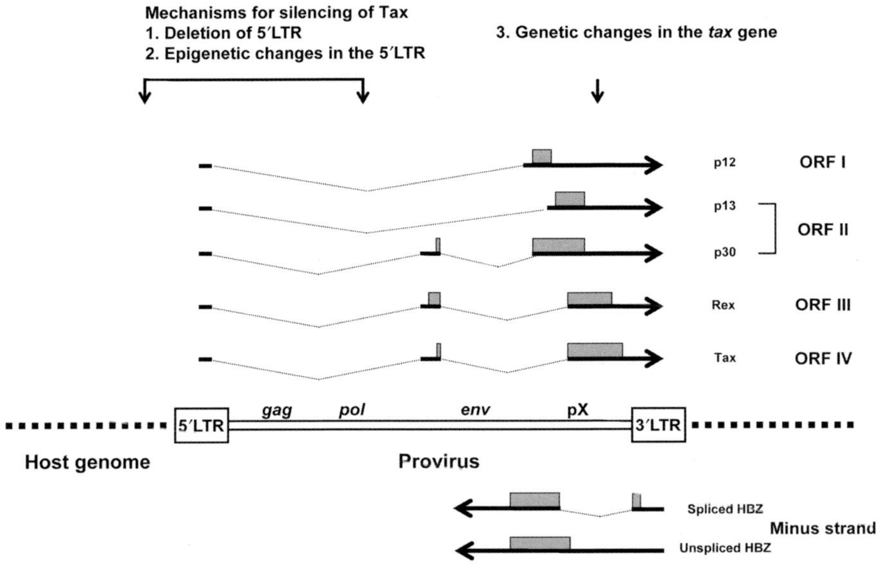

Fig. 1. Structure of the HTLV-1 provirus. The *gag*, *pol*, and *env* structural genes are flanked by 5′ and 3′ long terminal repeats (LTRs). Regulatory genes [p12(I), p13(II), p30(II), Rex(III), Tax(IV)] are encoded by the plus strand of the pX region in four different open reading frames (ORFs). The HTLV-1 basic leucine zipper factor (HBZ) gene is located on the minus strand of the provirus. Tax is frequently inactivated in ATL cells by (i) deletion and (ii) epigenetic changes in the 5′ LTR, and (iii) genetic alteration of the *tax* gene. Roman numerals indicate the respective ORF used for the translation of the indicated protein.

proliferation of HTLV-1-infected cells is clonal and persistent.[17–19] Experiments using a mouse model of initial HTLV-1 infection indicate that clonal expansion of primary infected cells predominates over secondary *de novo* infection of previously uninfected cells.[20] HTLV-1 induces the clonal proliferation of infected cells. These clonal cells survive durably; over time, it is envisioned that one of these clones become fully transformed and proliferate as ATL.

THE ONCOGENIC FUNCTIONS OF TAX

In order to transform cells, HTLV-1 must overcome the tendency of virus-infected cells to undergo apoptosis and/or senescence. The virus must also defeat the cellular checkpoints that censor genetic damage, and the virus must stimulate proliferative factors that induce cell-cycle progression and cellular division. Tax is a central player in cellular transformation with pleiotropic functions used to conquer those cellular machineries (Fig. 2).

Cell Survival

Tax activates two major cellular pathways for quelling apoptosis; the first is the Akt pathway, and the second is the NF-κB pathway. Recent findings suggest that HTLV infection may also upregulate cellular microRNAs, which could function to prevent cellular apoptosis.

Activation of the Akt pathway

Akt is a serine/threonine kinase that influences cell survival and proliferation, and is regulated by phosphatidylinositol 3-kinase (PI3K) through site-specific phosphorylation, primarily on Ser473.[21] Activated Akt signals through downstream transcription factors such as activator protein 1 (AP1), which is highly expressed in many invasive human cancers, including ATL.[22] Tax promotes Akt phosphorylation by directly binding PI3K, and consequently the survival and proliferation of virus-infected cells are increased.[23]

Activation of the NF-κB pathway

NF-κB is a second major survival pathway engaged by HTLV-1. Many human cancers have activated NF-κB, and although NF-κB is tightly regulated in normal T cells, this pathway is constitutively active in HTLV-1-infected cells.[24] Activation of NF-κB by Tax occurs predominantly in the cytoplasm. Cytoplasmic Tax was shown to bind IKKγ. This binding triggers the phosphorylation of IKKα and IKKβ, which form a complex with IKKγ. Subsequently, this activated IKK complex phosphorylates IκBα, leading to its proteasome-mediated degradation, which frees IκBα-sequestered cytoplasmic NF-κB to migrate into the nucleus, where it activates the transcription of NF-κB-responsive genes.[25] Tax can also stimulate a non-canonical NF-κB pathway through the IKKα-dependent processing of the NF-κB p100 precursor protein to its active p52 form.

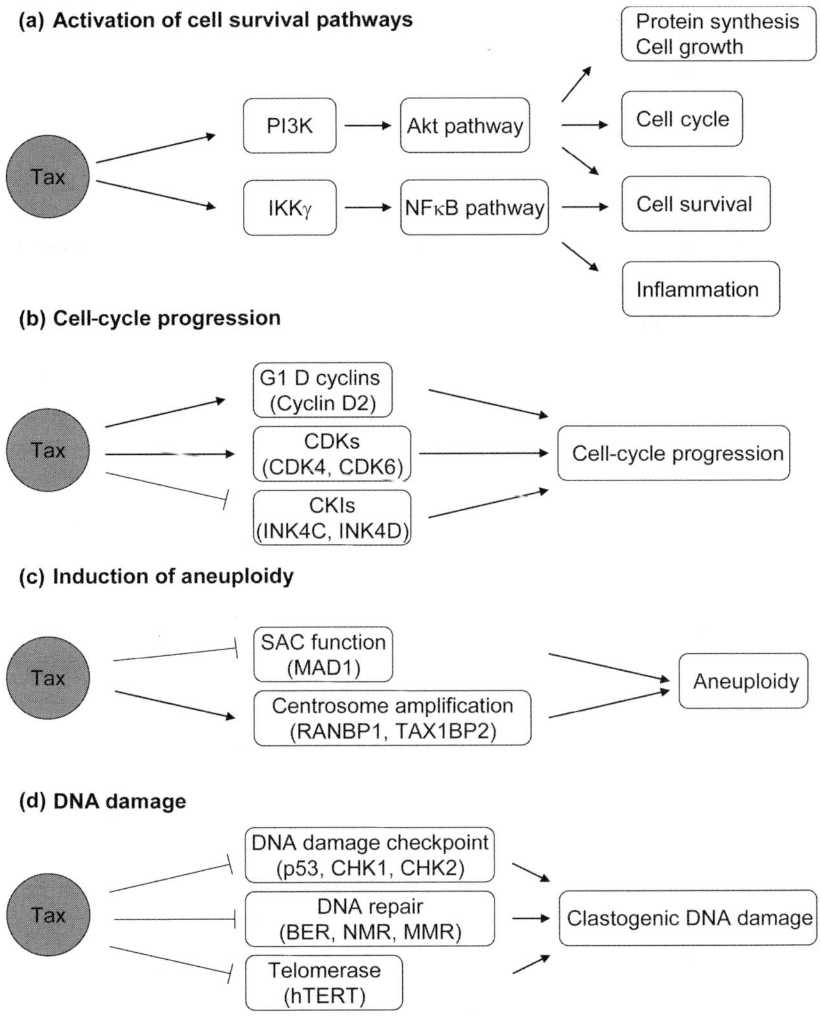

Fig. 2. Tax functions that affect cellular metabolism. Tax has multiple functions inside cells. Tax inhibits apoptosis of HTLV-1-infected cells by activating Akt and NF-κB pathways. Tax accelerates cell-cycle progression by activating cyclins, cyclin dependent kinases (CDKs), and suppressing CDK inhibitors (CKIs). Aneuploidy is induced by Tax through a loss of spindle assembly checkpoint (SAC) function. Tax induces supernumeral centrosomes, leading to multipolar mitosis. Finally, Tax causes clastgenic DNA damage by suppressing DNA damage checkpoints, DNA repair, and inhibiting hTERT transcription. BER: base excision repair; NER: nucleotide excision repair; MMR: DNA mismatch repair.

This alternative pathway is activated by Tax binding to IKKγ and p100, in an IKKα–IKKγ–p100 complex that lacks IKKβ.[26]

Upregulation of miRNAs that inhibit apoptosis

Recently, miRNA expression in HTLV-1-transformed cell lines and primary ATL cells has been profiled using microarray techniques. It was found that six miRNAs are consistently upregulated in these ATL cells.[27] Among them, miR-93 and miR-103b target a tumor suppressor protein, tumor protein 53-induced nuclear protein 1 (TP53INP1). TP53INP1 activates an apoptosis pathway that guards against cellular transformation. It was postulated that the HTLV-1 Tax oncoprotein can transcriptionally activate both miR-93 and miR-103b to reduce the expression of TP53INP1 and thus suppress cellular apoptosis.

Cell-Cycle Progression

Tax also provides significant mitogenic activity, especially at the G1–S-phase transition,[28–30] by provoking several cellular activities.

Upregulation of G1 D cyclins

Increased cyclin D2 expression occurs through direct activation of its promoter by Tax. Cyclin D2 expression is also enhanced by interleukin 2 (IL2) receptor signaling,[31] which is observed in Tax-expressing cells. Indeed, it has been shown that high IL2 secretion occurs in HTLV-1-infected T lymphocytes.

Activation of cyclin-dependent kinases (CDKs)

Tax can activate CDKs, such as CDK4, through direct protein binding.[32,33] Activated CDK4 hyperphosphorylates the retinoblastoma (RB) protein. Tax can also interact with RB directly, leading to its proteosomal degradation.[34] The phosphorylation and degradation of RB free the otherwise sequestered E2F1 transcription factor. Unfettered E2F accelerates the cell's cell-cycle transition from G1 to S.[30] There is also evidence that Tax can activate the transcription of the *E2F1* gene.[35]

Downregulation of CDK inhibitors (CKIs)

Tax has been reported to transcriptionally repress CKIs such as INK4C,[28] INK4D and KIP1.[36] The mechanism for this activity is through the direct binding of Tax to E-box proteins, which would otherwise activate the promoters for INK4C and INK4D.[37,38] A secondary mechanism may be through Tax binding to, and inactivating, INK4A and INK4B.[39,40]

Others

Although a cellular factor p21/WAF1 was identified as a CDK inhibitor that can act to suppress cell-cycle progression, subsequent findings show that it can also be an assembly factor which promotes G1-S transition via formation of an active p21/WAF1-cyclin D2-CDK4 complex. This complex can phosphorylate RB. Because Tax can increase the levels of WAF1 in cells, this is another way that Tax can contribute to increased cell-cycle progression.[41] DLG1 is the human homologue of the *Drosophila melanogaster* discs large tumor-suppressor protein, containing PDZ domains.[42] DLG1 signals downstream of Wnt and frizzled,[42] and binds the C-terminus of the adenomatous polyposis complex (APC); binding of DLG1 and APC regulates cellular proliferation and cell-cycle transition.[43] Tax has a PDZ-binding motif (PBMs) in its C-terminus. Tax has been reported to interact with DLG1 through this motif, and inactivates DLG1 by inducing its hyperphosphorylation and subcellular mislocalization.[44,45]

Cellular Aneuploidy

Structurally damaged DNA and chromosomal numerical abnormalities (aneupoidy) are common in cancers. Because most human cancer cells, including ATL cells, are aneuploid, aneuploidy has been proposed to be causal of transformation. It has been reported that Tax can induce aneuploidy by several mechanisms.

Induction of multipolar mitoses

Aneuploidy can arise from multipolar mitoses, which happen when more than two spindle poles (supernumerary centrosomes) emerge in a single cell. Tax creates over-amplification of centrosomes by targeting the cellular TAX1BP2 protein, which normally blocks centriole replication.[46] On the other hand, Tax engages RANBP1 during mitosis and fragments spindle poles, provoking multipolar segregation.[47] These complementary mechanisms can explain the long-standing observations of aneuploidy and frequent multipolar spindles in ATL cells.

Inhibition of mitotic spindle assembly checkpoint (SAC)

Cells have a SAC machinery that preserves the correct number of chromosomes by monitoring the fidelity of chromosomal segregation.[48] SAC proteins include MAD1, MAD2, MAD3/BUBR1, BUB1, BUB3, and MSP1. These proteins function at kinetochores to check for proper microtubule attachment and orderly mitotic chromosome partitioning.[48] The emergence of aneuploid ATL cells implies that these cells lack intact SAC function. Indeed, the loss of SAC function has been experimentally verified in several ATL cell lines.[49] Tax can bind MAD1 and inactivate its function, thus creating a loss of SAC activity.[50] Tax has also been shown to bind and activate the anaphase-promoting complex/cyclosome (APC/C), which functions downstream of the SAC.[51] By interaction with the APC/C, Tax is thought to promote premature mitotic exit and contribute to aneuploidy.

DNA Structural Damage

The chromosomes in ATL cells have clastogenic DNA damage.[52] Although other oncoproteins can induce direct DNA damage through increased reactive oxygen species, such activity has yet to be demonstrated for Tax. However, HTLV-1 infection appears to abrogate the cellular checkpoints and DNA repair functions that monitor and censor ambient structural DNA damage.

Inhibition of DNA damage checkpoints

It has been suggested that structurally damaged DNA is prevalent in cancer cells because ~50 percent of human cancers have checkpoint-disabling mutations in *TP53*. Interestingly, mutation of *TP53* is infrequent in ATL cells.[53] Nevertheless, the p53 checkpoint is functionally inactivated in ATL cells by the Tax protein.[54] How Tax inactivates p53 is still unclear since several competing mechanisms have been reported. These mechanisms include Tax abrogation of p53 function by competition for the binding of the p300/CBP transcriptional coactivator,[55] Tax acting through an NF-κB/RelA(p65) pathway to perturb p53 function[56]; and/or Tax inactivating p53 through another as yet uncharacterized pathway. Finally, Tax can also physiologically interact with CHK1[57] and CHK2[58] kinases to attenuate their activities. These two kinases contribute to the proper execution of G1 and G2/M checkpoints.

Perturbation of DNA repair

Tax represses the expression of the DNA polymerase β enzyme used for base excision repair (BER).[59,60] In addition, Tax suppresses an alternate repair process, nucleotide excision repair (NER), following UV irradiation through its induction of PCNA.[61,62] Further, it has been reported that the expression of human DNA mismatch repair (MMR) genes, such as human MutS homologue 2 (hMSH2), are reduced in primary ATL cells.[63] These findings support Tax inhibition of DNA repair by several means.

Suppression of telomerase

Chromosome end-to-end fusions and shortened telomeres are common in cancer cells; these findings are also prevalent in ATL cells.[64] It has been demonstrated that Tax suppresses the expression of human telomerase reverse transcriptase (hTERT)[65] and targets the DNA-end-binding and protective activity of the KU80 protein, which normally has the capacity in cells to protect newly formed double-strand breaks as well as normal chromosome ends.[66]

The Requirements for Tax in ATL Leukemogenesis

Tax is required for the virus to transform cells;[67] however, since Tax is a major target of host immunity, ATL cells frequently (in about 60 percent of cases) lose Tax expression by several mechanisms (Fig. 1). The first is through genetic changes in the *tax* gene (~10 percent of ATLs).[68,69] In some cases, ATL cells mutate the class I MHC recognition site of the Tax protein, resulting in an escape from immune recognition.[68] The second mechanism is an epigenetic change in the viral promoter/enhancer in the 5′ LTR; in this setting, DNA hypermethylation and histone modifications silence the transcription of viral genes (~15 percent of ATLs).[69–71] The third mechanism is a deletion of 5′ LTR (~27 percent of ATL cases).[69,72] Although the significance of Tax in leukemogenesis has not been fully clarified, it is recognized that Tax is needed early after infection to initiate cellular transformation, but it may not be required after cells are fully transformed and have acquired Tax-independent proliferative ability (Fig. 3).

HBZ RNA AND PROTEIN IN LEUKEMOGENESIS

As noted above, the proviral 5′ LTR is often deleted or epigenetically inactivated in ATL cells. Therefore, transcription of viral genes, including *tax*, encoded on the plus HTLV-1 strand, is frequently suppressed. Conversely, the 3′ LTR is conserved and hypomethylated in all stages of ATL developments,[71] suggesting its significance in the initiation and/or maintenance of leukemia. HBZ was identified to be encoded by the minus strand of the provirus as a protein that interacts with CREB-2 through its bZIP domain; HBZ was originally reported to suppress Tax-mediated viral transcription.[14] It also interacts with and modulates the transcriptional activities of other bZIP proteins, such as CREB, CREM-Ia, ATF-1,[73] c-Jun,[74,75] JunB,[76] and JunD.[77] The *HBZ* gene is transcribed from the 3′ LTR and expressed in all ATL cells.[13] Because there are no mutations, such as nonsense stop codons or deletions in its sequence, *HBZ* has been suggested to be indispensable for ATL leukemogenesis. The findings that knocking down *HBZ* transcription by short hairpin RNA inhibits ATL cell proliferation and that over-expression of *HBZ* promotes proliferation of a human T-cell line

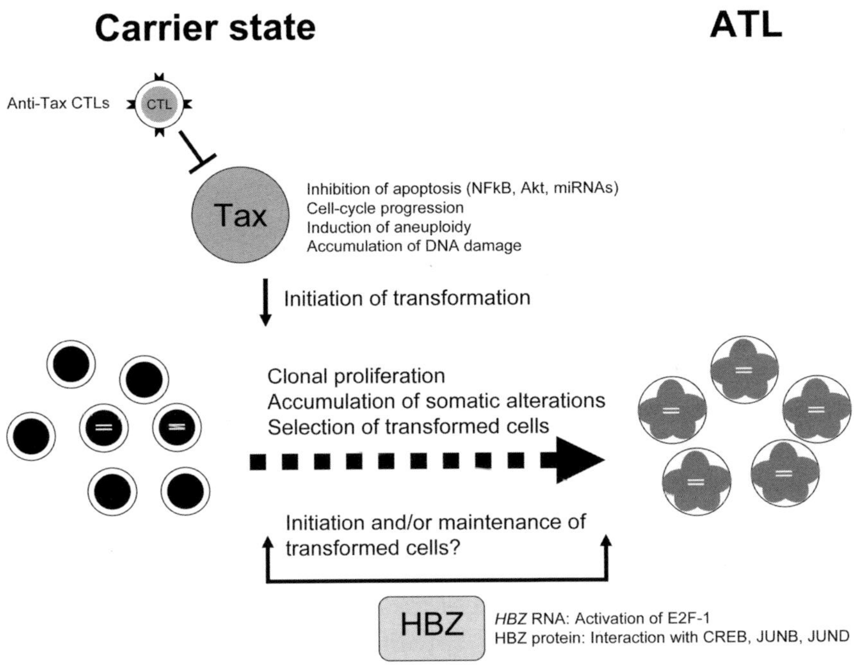

Fig. 3. The natural history of ATL development. HTLV-I-infected cells, marked by the "=" symbol, proliferate clonally and require Tax and HBZ. In the carrier state, proliferation of HTLV-I-infected cells is controlled by the host immune system. During the long latency period, Tax initiates cellular transformation in several ways, and in later stages, fully transformed ATL cells do not appear to need Tax expression. The suppression of Tax expression in these cells allows cells to evade immune surveillance and to proliferate *in vivo*, resulting in ATL. HBZ has a cell proliferative function and is expressed constitutively, supporting its role in leukemogenesis.

support the conclusion that HBZ enhances T-cell growth.[13] Mutational analysis of *HBZ* genes shows that *HBZ* as an RNA form can support T-cell proliferation by regulating the E2F-1 pathway. The *HBZ* RNA is predicted to form a stem-loop structure that may be essential for its growth-promoting activity, perhaps through interactions with uncharacterized cellular factors. Thus, the *HBZ* gene may have bimodal functions both as an RNA and a protein. In these regards, the RNA form of *HBZ* is likely to be important in the proliferation of HTLV-1-infected cells while the protein suppresses viral transcription by Tax.

Several animal models have been established to evaluate the functions of HBZ *in vivo*. Transgenic mice expressing HBZ under control of the mouse CD4 promoter/enhancer, compared with wild-type littermates, showed that their percentage of CD4+ T lymphocytes increased in splenocytes, and their proliferation induced by cross-linking with an immobilized anti-CD3 antibody was augmented in thymocytes.[13] It has been reported that when an HTLV-1 with a mutation in the leucine zipper domain of HBZ was inoculated into rabbits, the proviral load was reduced.[78] These data indicate that *HBZ* promotes cellular proliferation and support the idea that *HBZ* plays an important role in leukemogenesis of ATL (Fig. 3).

OTHER VIRAL GENES

p12^{I}

p12^{I}, which is encoded by ORF I of the pX region, localizes to the endoplasmic reticulum and Golgi apparatus.[79] p12^{I} interacts with the interleukin 2 receptor (IL-2R) β and γc chains, modulating their surface expression. p12^{I} activates signal transducers and activators of transcription 5 (STAT5),[80,81] and it increases intracellular calcium levels through interaction with calnexin and calreticulin and leads to activation of nuclear factor of activated T cells (NFAT)-mediated transcription.[79,82] p12^{I} also binds the major histocompatibility complex I heavy chain (MHC I Hc), and this complex is degraded by the proteasome.[83] These findings suggest that p12^{I} may support proliferation of HTLV-1-infected cells and contribute to escape from surveillance by the host immune system.

p27Rex

The pX ORF III encodes Rex in a doubly spliced *tax/rex* mRNA. Rex increases nuclear export of mRNA that encode viral structural genes, such as *gag*, *pol*, and *env*, and Rex functions to enhance viral production.[84] Rex also suppresses the expression of fully spliced mRNA encoding regulatory proteins, and it negatively regulates viral transcription.[85]

p30^{II} and p13^{II}

ORF II of the pX region encodes both p30^{II}, a nuclear protein, and the p13^{II} mitochondrial protein.[79] p30^{II} interacts with CREB binding protein (CBP)/p300 and affects Tax-mediated viral transcription.[86] On the other hand, p30^{II} inhibits the transport of doubly spliced mRNA encoding Tax and Rex proteins to the cytoplasm, resulting in inhibition of their translation.[87] p30^{II} can repress Tax/Rex expression in HTLV-1-infected cells both transcriptionally and post-transcriptionally. It has been suggested that strict control of viral expression by p30^{II} contributes to viral persistence and escape from immune elimination. Also, it has been shown that p30^{II} enhances the transforming potential of c-Myc and can transcriptionally activate the human cyclin D2 promoter through interaction with the c-Myc-60-kDa Tat-interacting protein (TIP60) complex.[88] These findings suggest a contribution of p30^{II} to cellular proliferation. The p13^{II} protein interacts with farnesyl pyrophosphate synthetase (FPPS), a protein that functions in the post-translational modification of Ras. p13^{II} alters Ras-mediated apoptosis in T lymphocytes.[89,90]

OTHER HOST FACTORS

Familial clustering of ATL cases suggests that the host's genetic background influences ATL onset.[91] Polymorphism of the promoter region of tumor necrosis factor alpha (TNF-α) was reported to be associated with ATL susceptibility.[92] HLA haplotypes are also candidate genetic factors controlling the immune response against viral antigens. Specific HLA alleles are reportedly associated with an increased risk of ATL onset.[93] Such HLA alleles are genetically defective in their recognition of HTLV-1 Tax peptides and allow HTLV-1-infected cells to escape from immunity.

CONCLUDING REMARKS

Advances in our understanding of the mechanisms for ATL leukemogenesis have been made since the first discovery of HTLV-1. Intensive studies on Tax biology have revealed molecular strategies used by HTLV-1 for cellular transformation. Tax appears to be required for the initiation of transformation but may be dispensable in later stages of leukemogenesis. Therefore, other factor(s) could contribute to the additional steps needed for transformation. Emerging evidence concerning HBZ indicates its proliferative capacity. Further investigation of HBZ function may shed light on its role in leukemogenesis. Despite the progress in our molecular understanding of ATL, the prognosis for this disease is still poor. Future research is needed to uncover new strategies for treatment and prophylaxis; hopefully, these applications will be based on the accumulated knowledge of the virus' leukemogenic mechanisms.

REFERENCES

1. Matsuoka M, Jeang KT. (2007) Human T-cell leukaemia virus type 1 (HTLV-1) infectivity and cellular transformation. *Nat Rev Cancer* 7: 270–280.
2. Gallo RC. (2005) The discovery of the first human retrovirus: HTLV-1 and HTLV-2. *Retrovirology* 2: 17.
3. Takatsuki K. (2005) Discovery of adult T-cell leukemia. *Retrovirology* 2: 16.
4. Akagi T, Shimotohno K. (1993) Proliferative response of Tax1-transduced primary human T cells to anti-CD3 antibody stimulation by an interleukin-2-independent pathway. *J Virol* **67**: 1211–1217.
5. Grassmann R, Dengler C, Muller-Fleckenstein I, *et al.* (1989) Transformation to continuous growth of primary human T lymphocytes by human T-cell leukemia virus type I X-region genes transduced by a *Herpesvirus saimiri* vector. *Proc Natl Acad Sci USA* **86**: 3351–3355.
6. Grassmann R, Berchtold S, Radant I, *et al.* (1992) Role of human T-cell leukemia virus type 1 X region proteins in immortalization of primary human lymphocytes in culture. *J Virol* **66**: 4570–4575.
7. Rosin O, Koch C, Schmitt I, *et al.* (1998) A human T-cell leukemia virus Tax variant incapable of activating NF-kappaB retains its immortalizing potential for primary T-lymphocytes. *J Biol Chem* **273**: 6698–6703.
8. Tanaka A, Takahashi C, Yamaoka S, *et al.* (1990) Oncogenic transformation by the *tax* gene of human T-cell leukemia virus type I *in vitro. Proc Natl Acad Sci USA* **87**: 1071–1075.
9. Green JE, Baird AM, Hinrichs SH, *et al.* (1992) Adrenal medullary tumors and iris proliferation in a transgenic mouse model of neurofibromatosis. *Am J Pathol* **140**: 1401–1410.

10. Grossman WJ, Kimata JT, Wong FH, *et al.* (1995) Development of leukemia in mice transgenic for the *tax* gene of human T-cell leukemia virus type I. *Proc Natl Acad Sci USA* **92**: 1057–1061.
11. Hasegawa H, Sawa H, Lewis MJ, *et al.* (2006) Thymus-derived leukemia-lymphoma in mice transgenic for the Tax gene of human T-lymphotropic virus type I. *Nat Med* **12**: 466–472.
12. Hinrichs SH, Nerenberg M, Reynolds RK, *et al.* (1987) A transgenic mouse model for human neurofibromatosis. *Science* **237**: 1340–1343.
13. Satou Y, Yasunaga J, Yoshida M, Matsuoka M. (2006) HTLV-I basic leucine zipper factor gene mRNA supports proliferation of adult T cell leukemia cells. *Proc Natl Acad Sci USA* **103**: 720–725.
14. Gaudray G, Gachon F, Basbous J, *et al.* (2002) The complementary strand of the human T-cell leukemia virus type 1 RNA genome encodes a bZIP transcription factor that down-regulates viral transcription. *J Virol* **76**: 12813–12822.
15. Cavanagh MH, Landry S, Audet B, *et al.* (2006) HTLV-I antisense transcripts initiating in the 3′ LTR are alternatively spliced and polyadenylated. *Retrovirology* **3**: 15.
16. Doi K, Wu X, Taniguchi Y, *et al.* (2005) Preferential selection of human T-cell leukemia virus type I provirus integration sites in leukemic versus carrier states. *Blood* **106**: 1048–1053.
17. Cavrois M, Leclercq I, Gout O, *et al.* (1998) Persistent oligoclonal expansion of human T-cell leukemia virus type 1-infected circulating cells in patients with tropical spastic paraparesis/HTLV-1 associated myelopathy. *Oncogene* **17**: 77–82.
18. Etoh K, Tamiya S, Yamaguchi K, *et al.* (1997) Persistent clonal proliferation of human T-lymphotropic virus type I-infected cells *in vivo. Cancer Res* **57**: 4862–4867.
19. Wattel E, Vartanian JP, Pannetier C, Wain-Hobson S. (1995) Clonal expansion of human T-cell leukemia virus type I-infected cells in asymptomatic and symptomatic carriers without malignancy. *J Virol* **69**: 2863–2868.
20. Miyazato P, Yasunaga J, Taniguchi Y, *et al.* (2006) *De novo* human T-cell leukemia virus type 1 infection of human lymphocytes in NOD-SCID, common gamma-chain knockout mice. *J Virol* **80**: 10683–10691.
21. Song G, Ouyang G, Bao S. (2005) The activation of Akt/PKB signaling pathway and cell survival. *J Cell Mol Med* **9**: 59–71.

22. Jeong SJ, Pise-Masison CA, Radonovich MF, *et al.* (2005) Activated AKT regulates NF-kappaB activation, p53 inhibition and cell survival in HTLV-1-transformed cells. *Oncogene* **24**: 6719–6728.
23. Peloponese JM, Jr., Jeang KT. (2006) Role for Akt/protein kinase B and activator protein-1 in cellular proliferation induced by the human T-cell leukemia virus type 1 *Tax* oncoprotein. *J Biol Chem* **281**: 8927–8938.
24. Sun SC, Yamaoka S. (2005) Activation of NF-kappaB by HTLV-I and implications for cell transformation. *Oncogene* **24**: 5952–5964.
25. Iha H, Kibler KV, Yedavalli VR, *et al.* (2003) Segregation of NF-kappaB activation through NEMO/IKKgamma by Tax and TNFalpha: Implications for stimulus-specific interruption of oncogenic signaling. *Oncogene* **22**: 8912–8923.
26. Xiao G, Cvijic ME, Fong A, *et al.* (2001) Retroviral oncoprotein Tax induces processing of NF-kappaB2/p100 in T cells: Evidence for the involvement of IKKalpha. *EMBO J* **20**: 6805–6815.
27. Yeung ML, Yasunaga J, Bennasser Y, *et al.* (2008) Roles for microRNAs, miR-93 and miR-130b, and tumor protein 53-induced nuclear protein 1 tumor suppressor in cell growth dysregulation by human T-cell lymphotrophic virus 1. *Cancer Res* **68**: 8976–8985.
28. Akagi T, Ono H, Shimotohno K. (1996) Expression of cell-cycle regulatory genes in HTLV-I infected T-cell lines: Possible involvement of Tax1 in the altered expression of cyclin D2, p18Ink4 and p21Waf1/Cip1/Sdi1. *Oncogene* **12**: 1645–1652.
29. Neuveut C, Low KG, Maldarelli F, *et al.* (1998) Human T-cell leukemia virus type 1 Tax and cell cycle progression: Role of cyclin D-cdk and p110Rb. *Mol Cell Biol* **18**: 3620–3632.
30. Schmitt I, Rosin O, Rohwer P, *et al.* (1998) Stimulation of cyclin-dependent kinase activity and G1- to S-phase transition in human lymphocytes by the human T-cell leukemia/lymphotropic virus type 1 Tax protein. *J Virol* **72**: 633–640.
31. Santiago F, Clark E, Chong S, *et al.* (1999) Transcriptional up-regulation of the cyclin D2 gene and acquisition of new cyclin-dependent kinase partners in human T-cell leukemia virus type 1-infected cells. *J Virol* **73**: 9917–9927.
32. Fraedrich K, Muller B, Grassmann R. (2005) The HTLV-1 Tax protein binding domain of cyclin-dependent kinase 4 (CDK4) includes the regulatory PSTAIRE helix. *Retrovirology* **2**: 54.

33. Haller K, Wu Y, Derow E, *et al.* (2002) Physical interaction of human T-cell leukemia virus type 1 Tax with cyclin-dependent kinase 4 stimulates the phosphorylation of retinoblastoma protein. *Mol Cell Biol* **22**: 3327–3338.
34. Kehn K, Fuente CL, Strouss K, *et al.* (2005) The HTLV-I Tax oncoprotein targets the retinoblastoma protein for proteasomal degradation. *Oncogene* **24**: 525–540.
35. Lemasson I, Thebault S, Sardet C, *et al.* (1998) Activation of E2F-mediated transcription by human T-cell leukemia virus type I Tax protein in a p16(INK4A)-negative T-cell line. *J Biol Chem* **273**: 23598–23604.
36. Iwanaga R, Ohtani K, Hayashi T, Nakamura M. (2001) Molecular mechanism of cell cycle progression induced by the oncogene product Tax of human T-cell leukemia virus type I. *Oncogene* **20**: 2055–2067.
37. Riou P, Bex F, Gazzolo L. (2000) The human T cell leukemia/lymphotropic virus type 1 Tax protein represses MyoD-dependent transcription by inhibiting MyoD-binding to the KIX domain of p300. A potential mechanism for Tax-mediated repression of the transcriptional activity of basic helix-loop-helix factors. *J Biol Chem* **275**: 10551–10560.
38. Suzuki T, Narita T, Uchida-Toita M, Yoshida M. (1999) Down-regulation of the INK4 family of cyclin-dependent kinase inhibitors by Tax protein of HTLV-1 through two distinct mechanisms. *Virology* **259**: 384–391.
39. Low KG, Dorner LF, Fernando DB, *et al.* (1997) Human T-cell leukemia virus type 1 Tax releases cell cycle arrest induced by p16INK4a. *J Virol* **71**: 1956–1962.
40. Suzuki T, Kitao S, Matsushime H, Yoshida M. (1996) HTLV-1 Tax protein interacts with cyclin-dependent kinase inhibitor p16INK4A and counteracts its inhibitory activity towards CDK4. *EMBO J* **15**: 1607–1614.
41. Kehn K, Deng L, de la FC, *et al.* (2004) The role of cyclin D2 and p21/waf1 in human T-cell leukemia virus type 1 infected cells. *Retrovirology* **1**: 6.
42. Woods DF, Bryant PJ. (1991) The discs-large tumor suppressor gene of Drosophila encodes a guanylate kinase homolog localized at septate junctions. *Cell* **66**: 451–464.
43. Matsumine A, Ogai A, Senda T, *et al.* (1996) Binding of APC to the human homolog of the Drosophila discs large tumor suppressor protein. *Science* **272**: 1020–1023.
44. Hirata A, Higuchi M, Niinuma A, *et al.* (2004) PDZ domain-binding motif of human T-cell leukemia virus type 1 Tax oncoprotein augments the transforming activity in a rat fibroblast cell line. *Virology* **318**: 327–336.

45. Suzuki T, Ohsugi Y, Uchida-Toita M, *et al.* (1999) Tax oncoprotein of HTLV-1 binds to the human homologue of Drosophila discs large tumor suppressor protein, hDLG, and perturbs its function in cell growth control. *Oncogene* **18**: 5967–5972.
46. Ching YP, Chan SF, Jeang KT, Jin DY. (2006) The retroviral oncoprotein Tax targets the coiled-coil centrosomal protein TAX1BP2 to induce centrosome overduplication. *Nat Cell Biol* **8**: 717–724.
47. Peloponese JM, Jr., Haller K, Miyazato A, Jeang KT. (2005) Abnormal centrosome amplification in cells through the targeting of Ran-binding protein-1 by the human T cell leukemia virus type-1 Tax oncoprotein. *Proc Natl Acad Sci USA* **102**: 18974–18979.
48. Musacchio A, Hardwick KG. (2002) The spindle checkpoint: Structural insights into dynamic signalling. *Nat Rev Mol Cell Biol* **3**: 731–741.
49. Kasai T, Iwanaga Y, Iha H, Jeang KT. (2002) Prevalent loss of mitotic spindle checkpoint in adult T-cell leukemia confers resistance to microtubule inhibitors. *J Biol Chem* **277**: 5187–5193.
50. Jin DY, Spencer F, Jeang KT. (1998) Human T cell leukemia virus type 1 oncoprotein Tax targets the human mitotic checkpoint protein MAD1. *Cell* **93**: 81–91.
51. Liu B, Hong S, Tang Z, *et al.* (2005) HTLV-I Tax directly binds the Cdc20-associated anaphase-promoting complex and activates it ahead of schedule. *Proc Natl Acad Sci USA* **102**: 63–68.
52. Marriott SJ, Lemoine FJ, Jeang KT. (2002) Damaged DNA and miscounted chromosomes: Human T cell leukemia virus type I Tax oncoprotein and genetic lesions in transformed cells. *J Biomed Sci* **9**: 292–298.
53. Yasunaga J, Matsuoka M. (2007) Leukaemogenic mechanism of human T-cell leukaemia virus type I. *Rev Med Virol* **17**: 301–311.
54. Tabakin-Fix Y, Azran I, Schavinky-Khrapunsky Y, *et al.* (2006) Functional inactivation of p53 by human T-cell leukemia virus type 1 Tax protein: Mechanisms and clinical implications. *Carcinogenesis* **27**: 673–681.
55. Ariumi Y, Kaida A, Lin JY, *et al.* (2000) HTLV-1 Tax oncoprotein represses the p53-mediated trans-activation function through coactivator CBP sequestration. *Oncogene* **19**: 1491–1499.
56. Pise-Masison CA, Mahieux R, Jiang H, *et al.* (2000) Inactivation of p53 by human T-cell lymphotropic virus type 1 Tax requires activation of the

NF-kappaB pathway and is dependent on p53 phosphorylation. *Mol Cell Biol* **20**: 3377–3386.

57. Park HU, Jeong JH, Chung JH, Brady JN. (2004) Human T-cell leukemia virus type 1 Tax interacts with Chk1 and attenuates DNA-damage induced G2 arrest mediated by Chk1. *Oncogene* **23**: 4966–4974.
58. Park HU, Jeong SJ, Jeong JH, *et al.* (2006) Human T-cell leukemia virus type 1 Tax attenuates gamma-irradiation-induced apoptosis through physical interaction with Chk2. *Oncogene* **25**: 438–447.
59. Jeang KT, Widen SG, Semmes OJ, Wilson SH. (1990) HTLV-I trans-activator protein, Tax, is a trans-repressor of the human beta-polymerase gene. *Science* **247**: 1082–1084.
60. Philpott SM, Buehring GC. (1999) Defective DNA repair in cells with human T-cell leukemia/bovine leukemia viruses: Role of *Tax* gene. *J Natl Cancer Inst* **91**: 933–942.
61. Kao SY, Marriott SJ. (1999) Disruption of nucleotide excision repair by the human T-cell leukemia virus type 1 Tax protein. *J Virol* **73**: 4299–4304.
62. Lemoine FJ, Kao SY, Marriott SJ. (2000) Suppression of DNA repair by HTLV type 1 Tax correlates with Tax trans-activation of proliferating cell nuclear antigen gene expression. *AIDS Res Hum Retrovir* **16**: 1623–1627.
63. Morimoto H, Tsukada J, Kominato Y, Tanaka Y. (2005) Reduced expression of human mismatch repair genes in adult T-cell leukemia. *Am J Hematol* **78**: 100–107.
64. Bellon M, Datta A, Brown M, *et al.* (2006) Increased expression of telomere length regulating factors TRF1, TRF2 and TIN2 in patients with adult T-cell leukemia. *Int J Cancer* **119**: 2090–2097.
65. Gabet AS, Mortreux F, Charneau P, *et al.* (2003) Inactivation of hTERT transcription by Tax. *Oncogene* **22**: 3734–3741.
66. Majone F, Luisetto R, Zamboni D, *et al.* (2005) Ku protein as a potential human T-cell leukemia virus type 1 (HTLV-1) Tax target in clastogenic chromosomal instability of mammalian cells. *Retrovirology* **2**: 45.
67. Akagi T, Ono H, Shimotohno K. (1995) Characterization of T cells immortalized by Tax1 of human T-cell leukemia virus type 1. *Blood* **86**: 4243–4249.
68. Furukawa Y, Kubota R, Tara M, *et al.* (2001) Existence of escape mutant in HTLV-I Tax during the development of adult T-cell leukemia. *Blood* **97**: 987–993.

69. Takeda S, Maeda M, Morikawa S, *et al.* (2004) Genetic and epigenetic inactivation of *Tax* gene in adult T-cell leukemia cells. *Int J Cancer* **109**: 559–567.
70. Koiwa T, Hamano-Usami A, Ishida T, *et al.* (2002) 5′-long terminal repeat-selective CpG methylation of latent human T-cell leukemia virus type 1 provirus *in vitro* and *in vivo*. *J Virol* **76**: 9389–9397.
71. Taniguchi Y, Nosaka K, Yasunaga J, *et al.* (2005) Silencing of human T-cell leukemia virus type I gene transcription by epigenetic mechanisms. *Retrovirology* **2**: 64.
72. Tamiya S, Matsuoka M, Etoh K, *et al.* (1996) Two types of defective human T-lymphotropic virus type I provirus in adult T-cell leukemia. *Blood* **88**: 3065–3073.
73. Lemasson I, Lewis MR, Polakowski N, *et al.* (2007) Human T-cell leukemia virus type 1 (HTLV-1) bZIP protein interacts with the cellular transcription factor CREB to inhibit HTLV-1 transcription. *J Virol* **81**: 1543–1553.
74. Basbous J, Arpin C, Gaudray G, *et al.* (2003) The HBZ factor of human T-cell leukemia virus type I dimerizes with transcription factors JunB and c-Jun and modulates their transcriptional activity. *J Biol Chem* **278**: 43620–43627.
75. Matsumoto J, Ohshima T, Isono O, Shimotohno K. (2005) HTLV-1 HBZ suppresses AP-1 activity by impairing both the DNA-binding ability and the stability of c-Jun protein. *Oncogene* **24**: 1001–1010.
76. Hivin P, Basbous J, Raymond F, *et al.* (2007) The HBZ-SP1 isoform of human T-cell leukemia virus type I represses JunB activity by sequestration into nuclear bodies. *Retrovirology* **4**: 14.
77. Thebault S, Basbous J, Hivin P, *et al.* (2004) HBZ interacts with JunD and stimulates its transcriptional activity. *FEBS Lett* **562**: 165–170.
78. Arnold J, Yamamoto B, Li M, *et al.* (2006) Enhancement of infectivity and persistence *in vivo* by HBZ, a natural antisense coded protein of HTLV-1. *Blood* **107**: 3976–3982.
79. Franchini G, Fukumoto R, Fullen JR. (2003) T-cell control by human T-cell leukemia/lymphoma virus type 1. *Int J Hematol* **78**: 280–296.
80. Mulloy JC, Crownley RW, Fullen J, *et al.* (1996) The human T-cell leukemia/lymphotropic virus type 1 p12I proteins bind the interleukin-2 receptor beta and gammac chains and affects their expression on the cell surface. *J Virol* **70**: 3599–3605.

81. Nicot C, Mulloy JC, Ferrari MG, *et al.* (2001) HTLV-1 p12(I) protein enhances STAT5 activation and decreases the interleukin-2 requirement for proliferation of primary human peripheral blood mononuclear cells. *Blood* **98**: 823–829.
82. Ding W, Albrecht B, Luo R, *et al.* (2001) Endoplasmic reticulum and *cis*-Golgi localization of human T-lymphotropic virus type 1 p12(I): Association with calreticulin and calnexin. *J Virol* 75: 7672–7682.
83. Johnson JM, Nicot C, Fullen J, *et al.* (2001) Free major histocompatibility complex class I heavy chain is preferentially targeted for degradation by human T-cell leukemia/lymphotropic virus type 1 p12(I) protein. *J Virol* 75: 6086–6094.
84. Inoue J, Seiki M, Yoshida M. (1986) The second pX product p27 chi-III of HTLV-1 is required for *gag* gene expression. *FEBS Lett* **209**: 187–190.
85. Hidaka M, Inoue J, Yoshida M, Seiki M. (1988) Post-transcriptional regulator (rex) of HTLV-1 initiates expression of viral structural proteins but suppresses expression of regulatory proteins. *EMBO J* 7: 519–523.
86. Zhang W, Nisbet JW, Albrecht B, *et al.* (2001) Human T-lymphotropic virus type 1 p30(II) regulates gene transcription by binding CREB binding protein/p300. *J Virol* 75: 9885–9895.
87. Nicot C, Dundr M, Johnson JM, *et al.* (2004) HTLV-1-encoded p30II is a post-transcriptional negative regulator of viral replication. *Nat Med* **10**: 197–201.
88. Awasthi S, Sharma A, Wong K, *et al.* (2005) A human T-cell lymphotropic virus type 1 enhancer of Myc transforming potential stabilizes Myc-TIP60 transcriptional interactions. *Mol Cell Biol* 25: 6178–6198.
89. Hiraragi H, Michael B, Nair A, *et al.* (2005) Human T-lymphotropic virus type 1 mitochondrion-localizing protein p13II sensitizes Jurkat T cells to Ras-mediated apoptosis. *J Virol* 79: 9449–9457.
90. Lefebvre L, Vanderplasschen A, Ciminale V, *et al.* (2002) Oncoviral bovine leukemia virus G4 and human T-cell leukemia virus type 1 p13(II) accessory proteins interact with farnesyl pyrophosphate synthetase. *J Virol* **76**: 1400–1414.
91. Yamaguchi K, Yul LS, Shimizu T, *et al.* (1985) Concurrence of lymphoma type adult T-cell leukemia in three sisters. *Cancer* **56**: 1688–1690.
92. Tsukasaki K, Miller CW, Kubota T, *et al.* (2001) Tumor necrosis factor alpha polymorphism associated with increased susceptibility to development of

adult T-cell leukemia/lymphoma in human T-lymphotropic virus type 1 carriers. *Cancer Res* **61**: 3770–3774.

93. Yashiki S, Fujiyoshi T, Arima N, *et al.* (2001) HLA-A*26, HLA-B*4002, HLA-B*4006, and HLA-B*4801 alleles predispose to adult T cell leukemia: The limited recognition of HTLV type 1 Tax peptide anchor motifs and epitopes to generate anti-HTLV type 1 Tax CD8(+) cytotoxic T lymphocytes. *AIDS Res Hum Retrovir* **17**: 1047–1061.

CHAPTER 3

Application of Proteomics to HTLV-1: Understanding Pathogenesis and Enhancing Diagnostics

O. John Semmes*,†, Mehdi S. Belgnaoui†, Robert Mitchell† and Julius O. Nyalwidhe†

SUMMARY

Although the structure of viral genes is a critically important aspect of virus biology, the study of virus proteins reigns supreme in the understanding of viral pathogenesis and diagnostics. Because of the relatively low number of encoded viral gene products, retroviruses have been ideal targets for proteomics research. The human T-cell leukemia virus, for example, encodes just 14 proteins. These include structural proteins (envelope, capsid, nucleocapsid, matrix), enzymes (protease, integrase, reverse transcriptase), and regulatory proteins (Tax, Rex, p21, p12, p13, p30, HBZ). The reader is referred to several excellent recent reviews of HTLV.[1–4] The goal of this chapter is to overview the field of proteomics with respect to the impact it has had on understanding the biology of HTLV. No attempt has been made

*Corresponding author. Email: semmesoj@EVMS.EDU.

†Department of Microbiology and Molecular Cell Biology, Eastern Virginia Medical School, Norfolk, Virginia, USA.

to exhaustively list all instances of when proteomics was used to study this virus, but rather it is our hope that the reader is provided with an appreciation for the role proteomics has played in virology.

WHAT IS PROTEOMICS?

The field of proteomics can be broken down into four major areas with obvious overlap: protein structure, protein function and regulation, protein–protein interaction, and global protein expression analysis. As one might guess, the field has been around for quite some time, although the term "proteomics" was coined to encompass the research activity deriving from functional genomics. In addition, as genomes have become fully sequenced, this information has been used to transform the way researchers approach proteomic inquiry.[5] The most dramatic example of this transformation is in the application of mass spectrometry, which with knowledge of protein primary structure is able to render detailed analysis of complex protein mixtures. The advances in mass spectrometry have subsequently allowed for the pursuit and development of high-throughput studies of samples with known and unknown protein content. Parallel to the emergence of high-throughput mass spectrometry based approaches was the equally impressive expansion in antibody and protein array platforms. Excellent reviews of the recent developments in proteomics are available and thus are not the focus of this chapter.[6,7] The objective of this chapter is to present the reader with an appreciation of the impact proteomics has had on understanding the biology of HTLV with specific reference to pathogenesis and diagnostics.

Protein Structure Studies

The following discussion of protein structure benefits from an understanding of the conceptual hierarchy employed in structural elucidation. Specifically, primary structure refers to the molecular constituents and their sequential order within the parent protein. Primary structure then is the protein "sequence" as well as any chemical modifications that occur to the protein following translation. The actual three-dimensional shape of a protein is driven by the inherent secondary and tertiary structure involving the amino acid backbone and side chain interactions, respectively. Quaternary structure refers to protein–protein interaction but is historically limited to those interactions that reside within a single protein structure, for instance, between separable polypeptide chains. Elucidation of these structure characteristics of viral proteins can aid in the determination of protein function. Thus, structural and functional studies are generally conducted hand in hand. Furthermore, since viral proteins are the primary interface between the virus and the host, understanding function will also reveal mechanism of pathogenesis and has often provided attractive therapeutic targets. As will be seen in the following examples, uncovering the structure of a protein has lead to increased understanding of the proteins involved in viral-cell interaction, viral packaging and viral replication of HTLV. The most elegant tools at the disposal of protein structural biologists are NMR and crystallography. However, significant improvements in mass spectrometry have prompted an evolution of protein chemistry techniques that have moved structural analysis into the hands of the protein biochemist.

Primary Structure Analysis

One of the earliest structural studies conducted on an HTLV protein was the determination of the primary structure of p24 (matrix). As was routine for the structural biochemist, this study involved a total reduction of the protein to individual amino acids via incubation in "constant boiling" HCl. This mixture was then processed for amino acid analysis and a composition library was developed. When sufficient quantities of the protein could be developed, C-terminal amino acid analysis was done followed by N-terminal sequencing via the Edman degradation reaction. This lengthy and time-consuming task was none-the-less able to reveal the sequence of the N-terminal 134 amino acids of p24, and it typifies the effort and material needed to accomplish primary structure analysis in the earlier ages of protein chemistry.[8] It was because of the difficulty of primary structure determination that a gene first strategy was developed which was especially successful for the study of completely sequenced virus such as HTLV. In these strategies, open reading frames are deduced from the viral genome and the primary sequence determined without arduous protein chemistry approaches. The advent of sophisticated techniques for the manipulation of genes and the growth of modern molecular biology dictated much of the subsequent studies aimed at primary structure analysis of viral proteins.

One aspect of primary structure that cannot be predicted by genetic analysis is that of post-translation modification (PTM). The modification of protein structure following translation is a common strategy for regulating the function of a protein. Phosphorylation of proteins can activate or repress enzymatic activity by alteration of the secondary structure of a protein and the coordinated structural changes generated by phosphorylation is clearly pivotal to the regulation of signal transduction among cellular kinases.[9] In addition to phosphorylation, proteins are acylated, sulfonated, glycosylated and amidated. Also, larger structures can be added through ubiquitylation and sumoylation processes that can direct proteins to undergo proteolytic degradation, target specific subcellular sites, or nucleate protein aggregation.[10,11] Specific to the following discussion, the determination of PTMs in viruses has uncovered critical regulator steps in structural and regulatory viral proteins. The methodology used for determining PTMs ranges from mutational analysis of theoretical PTM sites, SDS-PAGE of modified and unmodified forms of the protein and direct assessment via mass spectrometry.

One common strategy has been the utilization of *in silico* determination of potential PTM sites followed by functional assessment of amino acid substitution. This strategy has been applied to the characterization of the proteolytic processing of p12. In this study, the authors examined the role of post-translational processing of p12 into an alternative 8 kDa form.[12] The study describes the "maturation" of p12 as a two-step process. The first step is proteolytic cleavage to remove an ER retention signal. The cleavage results in the 8 kDa protein that can leave the ER and localize to the cell surface. In a second step at the surface, the 8 kD species binds to T-cell receptor (TCR) at the immunologic synapse and down-regulates TCR signaling. The 12 kDa species, which resides in the ER, interacts with IL-2R, MHC class I and calreticulin. The authors used *in silico* methods to locate a potential proteolytic cleavage site in the p12 sequence. When specific amino acid substitutions were introduced to "destroy" the cleavage site, the appearance of the 8 kD protein following SDS-PAGE was ablated.

In a similar strategy, the role of glycosylation in Env processing was examined. HTLV-1 Env is expressed as a gp61 precursor that is cleaved into a gp45 exterior protein and the gp21 subunit that anchors Env to the membrane. Initial suggestion that glycosylation played a role in gp61 processing was gleaned from observations that when cells were grown in the presence of an N-glycosylation inhibitor, the maturation of gp61 was also inhibited.[13] *In silico* analysis revealed five potential N-glycosylation sites within the predicted gp61 sequence. The authors showed that mutation of the first or fifth sites resulted in the loss of maturation as revealed by western blot analysis for the presence of gp45 and gp21. Interestingly, mutation of any of the five sites resulted in the inhibition of syncitia formation. Thus, glycosylation is critical to maturation of Env as well as to the regulation of post-maturation processes that contribute to a "normal" cell-surface presentation and/or viral assembly.

More recently, a role for glycosylation in achieving molecular mimicry between HTLV-1 and cellular proteins was established using a combination of affinity isolation and mass spectrometry.[14] The authors hypothesized that molecular mimicry between HTLV-1 proteins and host autoantigens was based on glycosylation. This study isolated membrane bound protein from MT-2 cells and separated them by SDS-PAGE. An immunoblot analysis using IgG enriched from HAM/TSP patients revealed a reactive species at 22–24 kDa in both cell types. This same band was not reactive to IgG enriched from uninfected control patients. The band was excised and identified by mass spectrometry to be peroxiredoxin-1 and p24 (Gag). Confirmation of differential glycosylation was accomplished by the use of glycan-specific lectin affinity isolation of proteins from MT-2 and Jurkat cells. The lectin-enriched proteins were subjected to 2D-PAGE and immunoblotted with HAM/TSP IgG. Only the peroxiredoxin-1 from MT-2 cells reacted with HAM/TSP IgG, suggesting that the glycosylation is cell type specific. The authors suggest that changes in both viral and host proteins contribute to cross-reactive immune response in HAM/TSP patients.

Perhaps due to the variety of functions attributed to the protein, Tax has been the subject of extensive primary sequence analysis. The majority of these studies have also been conducted using the mutation-function approach without the use of direct structural assessment. Most of the effort to determine primary structure has focused on the functional significance of PTM of Tax. Although phosphorylation has been examined extensively, there is a recent fascination with ubiquitylation and sumoylation due to the profound effect these modifications have on protein localization. The basic approach to analysis of ubiquitylation/sumoylation has been to first demonstrate that the protein possesses a specific modification and then mutate in combination all and/or individual lysine residues until the modification is reduced. The result generated from this approach is notoriously difficult to interpret because of the promiscuity of ubiquitin ligases for alternative modification sites. When this uncertainty of interpretation is coupled with the inherent flaws of amino acid substitution mutational analysis, assumptions of structure become very problematic. Specifically, mutations that result in loss of activity do not directly reflect functional specificity of the introduced mutation since global structural changes cannot be ruled out. Thus, the failure of a protein with a select mutation to become post-translationally modified does not insure that the mutated amino acid is the site of modification. In addition, the introduction of a mutation at a site that is in reality the preferred enzymatic modification site might be rescued by enzymatic modification of an alternative site, thus masking the detection of the actual preferred modification site. In spite of these shortcomings, considerable insight has been gained with respect to the functional role of ubiquitylation/sumoylation of the Tax protein.

The first report of ubiquitylation of Tax utilized a combination of protein isolation, co-expression of tagged ubiquitin and metabolic regulators to establish a functional role for this modification.[15,16] Subsequently, by incorporating lysine to arginine substitutions at all 10 potential sites, a functional analysis revealed that Tax is also sumoylated.[17] These authors established that ubiquitylated Tax is cytoplasmic and that sumoylation is required for its nuclear retention. A follow-up study further clarified that the sites of ubiquitylation and sumoylation are potentially the same as those residing at amino acids K280/K284 and that ubiquitylation of Tax is critical for interaction with IKK complexes in the cytoplasm.[18] Interestingly, ubiquitylation can be forced upon Tax at K280 or K284 in response to DNA damage, resulting in the dissociation of Tax from nuclear Tax speckle structures (TSS) and export to the cytoplasm.[19] Analysis of one of the potential ubiquitin modification sites in Tax by a combination of K > R substitution and detection by an antibody specific for acetylated lysine revealed that K346 is actually acetylated.[20] Inferences were made with respect to the role of acetylation in NF-κB activation by Tax because the K > R mutant showed reduced activity. Similarly, the authors showed that when S300/301 was mutated to alanines and K280/284 was mutated to arginines, acetylation of K346 was reduced. Thus, although no direct assessment of the actual site of modification has been conducted through techniques such as mass spectrometry, significant structure-function insight regarding the role of these modifications was obtained.

Improvements in data analysis specific to mass spectrometry have made direct assessment of PTMs practically routine. This analysis relies upon the assignment of ion fragmentation spectra to expected masses that are generated *in silico* for target proteins. The availability of so-called tandem mass spectrometry allows for the unambiguous assignment of amino acid modifications. This is largely due to the ability to obtain masses from the products of ionic fragmentation of the tryptic peptide ion. Since the fragmentation process is not complete, information on the masses of all possible peptide fragments is theoretically possible. When appropriate liquid chromatography separation of the peptides is performed prior to analysis in a modern ion trap instrument, slightly enriched preparations of the whole protein can yield extensive structural coverage. There are several outstanding reviews on the subject of mass spectrometry of proteins.[21,22] In the case of phosphorylation, the mass associated with the addition of a phosphate moiety is included in the potential *in silico* solutions for serine, threonine and tyrosine containing peptides from the generated spectra. We recently conducted a tandem MS analysis of mammalian expressed Tax protein in which we were able to unambiguously assign primary structure corresponding to 80 percent coverage of the protein. We identified a much greater number of phosphorylation sites than previously known and following selective substitution analysis identified phosphorylation events that were both activated and repressed Tax function[23] (Fig. 1). Although the direct assessment of primary structure with mass spectrometry has no parallel, there are shortcomings to the approach. Specifically, null data are not the same as confirmation of specific sequence structure. For instance, the lack of data indicating that a specific serine is phosphorylated is not

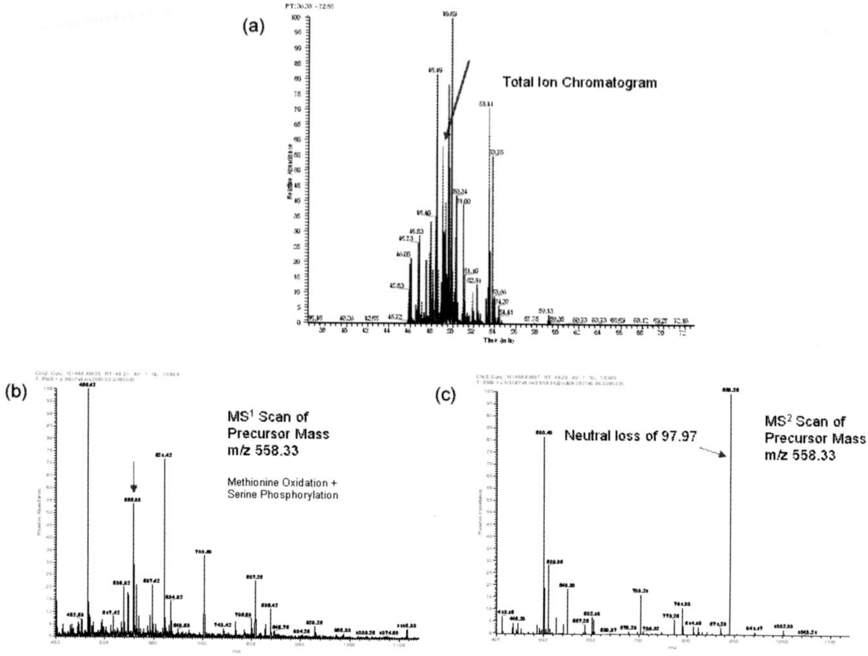

Fig. 1. An example of a tandem mass spectrometry experiment. (a) The total ion chromatogram is shown. The indicated eluting ion is then targeted for analysis. (b) An MS scan of the designated elution is shown with a complete mass scan of all ions present. The ion targeted for additional cleavage and analysis is indicated. (c) The MS scan for the products of the ionic cleavage is shown. Indicated is a neutral loss event, which is indicative of the presence of a phosphate group. This data was collected from an analysis of the Tax protein.

conclusive evidence that the specific serine is unphosphorylated. However, properly controlled experiments using a technique called multiple reaction monitoring (MRM) can unambiguously determine the stoichiometry of a selected PTM so as to enable an assessment of the absolute abundance of the modification.

Crystal Structure

When biomedical scientists think of crystallography, they are specifically referring to techniques in X-ray scattering used to deduce the structure of a crystal. More specifically, these approaches use X-ray diffraction to reveal elastic scattering of X-rays from the crystal surface.[24] The clear advantage for X-ray crystallography is the unmatched resolution of structure obtained. However, these techniques are not commonplace laboratory protocols. There is specific expertise needed in the ability to form protein crystals, and in fact, not all native proteins can form stable crystal structures. Secondly, the required data acquisition instrumentation and data analysis infrastructure represent significant investment to the extent that these capabilities lie in the hands of select numbers of laboratories. As representative of what can be accomplished by crystallography, we discuss studies in which the structure for protease and envelope protein were resolved.

Protease (PR) functions to cleave Gag into p19, p24 and p17, the capsid and matrix proteins.[25] In fact, gag-pro is first cleaved to release PR and to initiate further processing of Gag. Clearly, the function of PR is critical to the production of virus particles and in the targeting of proteins to specific subcellular compartments. By extension then the characterization of the structure of this protein would have direct implications in both pathogenesis and as potential therapeutic targets. In order to overcome the hurdle of protein mass needed to form stable crystals, the HTLV-1 PR ORF was cloned into an appropriate expression vector and expressed in *E. coli.* Following expression and purification of the PR protein, the investigators then demonstrated that the preparation retained biological activity. Following successful crystal formation, a resolution of 2.6 Å was achieved. Since the HIV PR crystal structure had been previously solved, these authors were able to compare the structural characteristics of these two similarly functioning proteins. They found that, although there was clear structural homology between the two proteins, the HTLV-PR possessed specific structural characteristics that could present steric hindrance in the accommodation of existing HIV PR inhibitors.[26] These observations helped to explain why the application of HIV PR inhibitor drugs failed to inhibit HTLV PR activity. The resolution of this region was the foundation for new drug development efforts. In addition, such findings were used to justify the continued efforts at structural elucidation of viral proteins to enhance drug development.

Envelope (Env), as the outermost coating in a virus, regulates the first step in cellular tropism. Since the recognition of the host cell via binding of viral Env to cellular receptor represents a critical selective step in viral entry, this interface has historically been a hot spot for antiviral research efforts. In addition, aspects of Env structure contribute to viral packaging and syncitia formation, each of which is fundamental to successful transmission of HTLV.[27] Not surprisingly then Env was an early target for structural elucidation via crystallography. The Env protein consists of two non-covalently bound mature subunits, the surface exposed gp46 that dictates cell receptor binding, and a transmembrane (TM) gp21 subunit that initiates viral-cell fusion. Because of the critical role in viral entry, the gp21 subunit was targeted for structure analysis. The TM domain of Env was crystallized and structure resolved to approximately 3 Å. As was the case for PR, the TM protein was expressed in bacteria. However, in this approach the authors fused gp21 to maltose binding protein since the crystal structure for this molecule has been determined and could be used as a crystal catalyst.[28] The resolved structure of TM consisted of an N-terminal trimeric coiled-coil and neighboring loop region typical of most retroviruses. The variable C-terminal region was structurally unique to HTLV. The information gained from the structure was used to design two peptides that could self-assemble into a gp21 mimic.[29] The mimic was injected into mice that could subsequently mount an effective immune response to whole virus. The antibodies raised were able to block syncitia formation. Thus, by using the structural data obtained from crystallography, peptide mimics can be designed to develop neutralizing antivirals.

Solution Structure

Although crystallography has been very successful at the elucidation of protein structure, not all proteins can form crystals. In addition, there is an experimental assumption in crystallography that the crystal state represents a biologically relevant structure. Since proteins exist in solutions among other proteins, understanding structure in a more dynamic state is beneficial. The use of nuclear magnetic resonance (NMR) determines structure by resolving information gained via atomic spin.[30] Under very high magnetic fields, the nuclei in an atom line up with the magnetic moment and by pulsing the magnetic field the nuclei will spin. Since the physical parameters regulating the threshold for spin are determined by the nuclei and its relationship with surrounding atoms, deconvolution of the data from enough such nuclei allows for structure determination. The fact that these analyses take place in the liquid state may provide information not available to crystal diffraction. The main limitations of NMR studies are the large amounts of protein required for the analyses and the fact that larger proteins are typically more difficult to characterize, and the use of the approach has been limited to proteins that are 30 kDa or less.[31] The NMR analysis equipment is also not readily accessible to most laboratories.

Rex functions to regulate the cytoplasmic appearance of single and unspliced viral mRNA.[32] As such Rex indirectly regulates the switch between virus production and viral protein production. Although there was success in the crystal structure for HIV rev, the structure for HTLV Rex was not forthcoming. Following a concerted effort, researchers resolved the solution structure of an HTLV-1 Rex peptide that spanned the arginine-rich domain and was bound to a Rex responsive element aptamer mimic.[33] These authors observed specific conformational extension (ability to relax) in the regions of Rex that intercalated the RNA structure. They also observed an unexpected alteration in the aptamer structure to accommodate the peptide. These observations have helped establish a conceptual paradigm for adaptive recognition in which both protein and RNA utilize tertiary structure to accommodate binding.

Matrix performs critical functions in viral particle assembly and packaging. Initially, matrix serves to facilitate the localization of the forming viral particle to the cellular membrane. Somewhere during the assembly, matrix also becomes the foundation for a stable structural layer at the inner surface of the viral membrane. The mechanisms by which these duties are performed are inherently tied to matrix structure and defining such structure has significant biological merit. In one of the earlier applications of NMR to solve solution structure of an HTLV protein, matrix was resolved to <1 Å. These studies determined structure in solution to be globular with four α-helices and a 3(10) helix.[34] The difference in these structures is that 3(10) helices contain three residues per helical revolution whereas α-helices contain four residues. Clear surface exposed structure and helical arrangement resembled HIV matrix, suggesting similar function. The conservation of structure among the retroviral matrix proteins underscores the similarity in function.

Nucleocapsid (NC) is a critical component of the mature virus particle and like matrix, is involved in directing the packaging process. In the mature virion, nucleocapsid is stably bound to the retroviral RNA copies and the quaternary structure of nucleocapsid–RNA determines the viral core shape.[35] The ability to form a zinc finger, a trait shared by all retroviral nucleocapsid proteins, presents a specific structure required for recognition of the viral RNA packaging signal. Two-dimensional NMR was used to determine the structure of the zinc finger region.[36] An 18-amino-acid peptide corresponding to the minimum sequence of the N-terminal Z-finger domain of HTLV-1 NC was structurally characterized. The ability to bind dicationic metals was established by combined UV-visible spectroscopy and mass spectrometry. These authors were able to conclude that even though the finger can bind several cations, only Zn^{++} binding resulted in a stable conformation.

As might be expected, HTLV capsid has been an exciting target for solution NMR based analysis. Extensive efforts at structural elucidation of capsid also serve to highlight improvements in NMR analysis that impact specific structural efforts. Capsid structure has been separated into an N-terminal domain (NTD) and a C-terminal domain (CTD), which are generally viewed as the "core" and "dimerization" domains, respectively. The 145-amino-acid NTD is resistant to proteolysis and the 85-amino-acid CTD facilitates multimerization of capsid. It was perhaps fitting that a solution structure for capsid was the first reported study of structural elucidation of any HTLV protein.[37] In this study, the authors produced capsid in an *in vitro* transcription system with a His-tag to facilitate purification. They observed that HTLV capsid in solution was a monomer, unlike what was observed for HIV capsid. Similar to HIV, however, was the presence of two independent N and C terminal domains connected by a flexible linker sequence. Interestingly, the dimerization domain of the HTLV capsid differs from that of HIV in that several hydrophobic residues believed to be critical to dimerization are buried. In addition, key cysteine residues that form a disulfide bond in the HIV capsid were reduced in the HTLV capsid. These observations clearly established a route for HTLV virus maturation that involves an oxidizing environment and distinguishes it from HIV. In a second study, NMR was used for the structural elucidation of the 134-amino-acid NTD.[38] In this analysis, the authors showed that, as expected, there was a high degree of structural conservation in that retroviral NTDs have a β-hairpin and a coiled-coil consisting of six α-helices. However, they established that the overall orientation of the HTLV NTD was dramatically different from that seen in HIV NTD. This result suggested that these higher-ordered structural features may explain why the HIV viral core is conical whereas the HTLV core is spherical. The same group followed up this study with a quantitative ^{15}N NMR relaxation analysis in conjunction with nuclear overhauser effect (NoE) to assess the backbone dynamic properties of the NTD.[39] Relaxation studies involve stopping the pulsed field applications and determining the time required for the atomic spin to become random. The authors compared the data from the analysis of HTLV NTD to a similar analysis conducted on the HIV NTD and revealed specific structural characteristics that stem from a difference in overall hydrodynamic character. These results confirmed their earlier suspicions and provided a structural basis for the virus-specific core shapes.

PROTEIN FUNCTION STUDIES

Clearly, the characterization of protein structure is ultimately concerned with the impact of structure on function. This section attempts to expose the reader to the types of functional assays that have been successful in proteomic studies of HTLV. By necessity all studies in HTLV biology that utilize functional assays cannot practically be reviewed and so the purpose behind this section is to present an awareness of the general types of approaches available. These approaches include measurement of transcriptional activation and subcellular localization of viral proteins as well as cell adhesion assays and the analysis of the products of protein function.

Transcription Assays

Transcription assays basically measure transcription via either direct assessment of message or indirect analysis of the message product. In fact, the incorporation of fluorescent proteins that are soluble and secretable has led to the development of rapid high-throughput screening platforms for protein function via transcriptional activation.[40,41] These same assays, minus the platform, have been conveniently used to assess the activity of HTLV proteins such as Tax that function as transcriptional activators.[42,43] However, the same systems can be modified to measure other activities such as protein–protein interactions, as described in a later section.

Protein Localization

Often, the first clue about the activities associated with novel proteins is the cellular location. Just as cellular proteins are subject to functional compartmentalization, viral proteins must operate under these same constraints. By extension, the loss of appropriate localization can be a measure of loss of function of the viral protein. Microscopy has reigned supreme for these types of studies to date. However, as in the case of transcription assays, protein localization has been successfully integrated into high-throughput screening approaches primarily for drug discovery. One can envision the likelihood of selection of functional partners or coenzymes using the same approach.

As a group, the so-called "accessory" proteins of HTLV are a more recent discovery, and as such the structural and functional elucidation has lagged behind the other proteins. These proteins are encoded in the ORF-I and ORF-II of the virus and consist of $p12^{I}$, $p13^{II}$, $p27^{I}$, and $p30^{II}$. The role of these proteins in HTLV biology has been discussed in some detail elsewhere.[44] In an attempt to decipher the functional role of one of the HTLV-1 accessory proteins, p13, studies were conducted that established a mitochondrial expression pattern.[45] These authors used dual labeling confocal microscopy to establish the subcellular localization of p13. They utilized the same localization assay to define a minimal mitochondrial targeting sequence within p13 that when fused to GFP could direct the fusion protein to mitochondria. Over-expression of p13 resulted in disruption of mitochondrial structure and prompted the investigators to hypothesize that p13 induces apoptosis. This same group later combined structural analysis using circular dichroism to establish the role of a cluster of surface arginine residues in the interaction with the inner face of mitochondria.[46] Thus, the combination of structural elucidation and assessment of subcellular localization could define mechanism of action for p13.

One of the first observations made regarding the cellular expression of HTLV-1 Tax was that the protein localizes to discrete nuclear speckles, which we termed Tax speckle structures (TSS).[47] The co-localization of TSS with proteins involved in transcription supported the notion of compartmentalization of a viral transcriptional activator. Very recently, our laboratory used a combined mutation and localization approach to define a 25-amino-acid domain in the N-terminus of Tax that when fused to GFP can target the fusion protein to TSS.[48] In a separate but related effort, we combined mutation/localization with heterokaryon formation to distinguish between nuclear and cytoplasmic Tax localization and established that Tax shuttles between the nucleus and cytoplasm.[49] In this study, cells were fused via heterokaryon formation and Tax protein was assayed for shuttling between nuclear TSS. We attributed this activity to a consensus NES site in the C-terminal region of Tax. A subsequent study using similar approaches was conducted to finely map the NES sequence and to demonstrate a mechanism behind the delayed kinetics observed in the first study.[50]

Cell Adhesion Assay

Somewhat unique among the retroviruses, HTLV requires cell–cell contact for the spread of infectious virus. Because of this aspect, the regulation of cell adhesion may play an important role in the infectivity of HTLV. Early efforts to define the function of p12 had established that the protein localizes to ER and subsequently activates release of calcium.[51] The authors correlated calcium release with T-cell activation. Inspired by the observation of ER-mediated calcium release, in this study they investigated the possibility that p12 might stimulate T-cell adhesion in a calcium-dependent manner. These authors found that HTLV-1 infected cells that do not express p12 have reduced LFA-mediated adhesion compared to wild type.[52] They employed an adhesion assay in which ICAM was adhered to 96 well plates and exposed to cells. In this manner, cell adhesion mediated by the ICAM receptor, LFA-1, can be directly measured. HTLV-1 infected cells were added and the percentage bound was measured. The assay also allowed for the determination of the role of calcium in the adhesion process. These studies established a compelling cause and effect relationship between p12 expression, cell adhesion and viral infectivity.

Analysis of Functional Products

A common approach in defining protein function is the analysis of potential products that are logically expected to be altered given a defined function. This approach has been most successful when analyzing proteins with a functional role in cell cycle progression, checkpoint activation or other similar molecular cascade. One more recent example of the application of this strategy is in the establishment of possible functional consequences of the expression of p30.[53] This study established that following expression of p30, the infected cells accumulated at the G2/M border. Verification of this activity was achieved by measuring the differential expression of cellular proteins for expression patterns indicative of G2/M arrest. Specifically, they observed that p30 expression resulted in increased phosphorylation of Cdc25C and Chk1 along with decreased expression of PLK1 and total Cdc25C.

Another approach for determining function via product analysis is to measure a single product, but to do so in a high-throughput platform so that multiple assay conditions can be rapidly and accurately assessed for production of the single product. In the establishment of a Rex functional defect in a viral clone, the authors established a stable cell line to build a single-product analytical model. The cell line produced significantly reduced levels of p19 Gag as determined by specific ELISA. Upon addition of Rex, p19 production was restored. These data were correlated with viral spread and persistence in a rabbit model.[54]

Multiplexed ELISA platforms pose a further technological improvement to the assessment of functional products strategy. Specifically, entire signaling pathways can be measured accurately and reproducibly. A modification of this strategy was used to determine if the ORF I and ORF II viral reading frame proteins p12, p13, p27 and p30 were capable of generating a CTL response.[55] The assay was constructed by synthesizing peptides for Tax and the accessory proteins. Standard 96 well plates were coated with antibody for IFNγ. Purified CD8$^+$ T cells and the specific peptide were added to each well. The mixture was incubated and time allowed for activation via release of IFNγ, which was detected by addition of a second antibody to IFNγ. The CD8$^+$ cells were derived from HTLV-1 infected individuals representing a full spectrum of disease. The peptide-specific responses established that (i) the proteins have been expressed during infection, and (ii) HTLV-1 carriers have $CD8^+$ effector cells for accessory proteins. They also developed an *in vitro* assay to show that each of the proteins alone can generate a CTL response in naïve lymphocytes.

PROTEIN–PROTEIN INTERACTION STUDIES

The vast majority of proteins display protein–protein interaction and a significant portion of these rely on such properties to accomplish specific functions. Enzymes must contact substrate proteins in order to catalyze an inherent reaction. In fact, many enzymes function as multiprotein complexes with indispensible protein partners that act as co-enzymes, conformational chaperones, and subcellular targeting factors. Indeed, it is difficult to enumerate examples in which protein function is not in some way impacted by interaction with other proteins. It is for this reason that the identification of protein-binding partners has allowed for associating function with novel proteins. Many of the proteomic tools used for these analyses have been commonplace in the laboratory for some time. However, current improvement in methods for isolation of biologically active proteins and the analysis of protein complexes has made protein-centric interactome analysis fairly routine although still challenging. As is given through the example below, the process of identifying functionally significant protein–protein interactions has progressed from yeast two-hybrid systems that allowed for the interrogation of theoretically all possible proteins encoded in a genomic library to mass spectrometry based direct analysis of isolated biologically active complexes.

Two-Hybrid Analysis

The two-hybrid assay is actually a transcription assay in which two essential domains of a transcriptional activator, the DNA binding domain and the promoter activating domain, are brought together by the protein–protein interaction of fused heterologous proteins. Using this system one can identify unknown protein partners, evaluate the binding of known proteins, and map protein–protein interaction structure. Several manifestations of the two-hybrid approach have been used to solve HTLV protein–protein interaction.

The yeast two-hybrid has often been constructed to analyze the binding of specific proteins. One common application of the two-hybrid in this manner is the analysis of dimerization. For example, this approach was used to establish the dimer interface for Tax.[56] This study defined a large central domain of Tax as critical for dimerization. Similar targeted strategies were used to map the interaction domain in serum response factor [57] and topoisomerase 1[58] for Tax binding and to determine which clathrin adapter subunits of AP1/AP2 bind to the Env TM-CD.[59]

Concerns that the expression of mammalian proteins in yeast might not be faithfully duplicated lead to the transfer of two-hybrid assays to mammalian cell-based systems. The core aspects of the assay remain unchanged. Such a mammalian-based two-hybrid system was used to characterize the multimerization of Rex and to help establish dominant negative mutations that have been very useful in subsequent functional studies.[60] When used in combination with *in vitro* binding assays and immunoprecipitation, the mammalian two-hybrid approach was successfully used to uncover a 40-amino-acid stretch in the N-terminus of Tax that is required for binding to CDK4.[61] This combined analytical strategy also established the requirement for both N- and C-terminal regions of CDK4; the N-terminus contains the PSTAIRE helix that regulates kinase activity.

However, the more common application of the two-hybrid assay is for exploratory analysis for binding of a library of proteins to a selected bait protein. This strategy has been used extensively to identify potential Tax-binding proteins. Several investigators, including ourselves, used yeast two-hybrid as a first step in establishing that Tax binds to a variety of cellular proteins with defined functional relevance, such as MAD1,[62] ATF4,[63] hDLG,[64] NF-Y[65] and ATF5.[66] The strategy was also responsible for the identification of a novel HTLV protein, HBZ, which is encoded from the minus strand RNA.[67] In this study, the investigators were attempting to identify proteins that would bind to the CREB-2 bZIP domain using a library from the MT-2 HTLV-1-infected cell line. Other notable instances of successful application of the two-hybrid were in the identification of Env binding to hDLG[68] and the isolation of cellular proteins that bound to p13.[69]

Electrophoretic Mobility Shift Assay (EMSA)

The assembly of proteins on and around an active promoter is a prerequisite to function. Thus, analysis of protein binding to DNA sequence can correlate with protein function. The use of EMSA is a classic approach to measure the binding of proteins to nucleic acids, an event that results in reduced mobility of the complex during native PAGE separation. The application of EMSA as a functional measurement of protein function was used to demonstrate the existence of an inhibitory complex of TBP at the Tax responsive element (TxRE) within the PCNA promoter.[70] The assay was further used to show a competitive interaction between Tax and TRB for the TxRE.

Immunoprecipitation

Just as EMSA is the mainstay for protein–DNA interaction, the use of immunoprecipitation is the most common approach to defining protein–protein interaction. The basic principles of antibody directed isolation of a target protein has been used to define the binding of known proteins, identify new protein binding partners, determine protein domains involved in protein–protein interactions and to assay for factors that regulate protein–protein interaction. Again a comprehensive listing of all instances that immunoprecipitation has been used even within HTLV-related studies is out of the scope of this chapter. However, the reader is referred to several examples in which Tax and CREB interaction,[71] HBZ and CREB interaction,[72] Tax and IKKγ interaction,[73] and Tax and calreticulin interaction[74] have been described.

When DNA binding and immunoprecipitation approaches are combined, it is possible to characterize entire protein complexes that are bound to DNA. The technique involves immunoprecipitation of a targeted protein from partially digested chromatin fraction followed by PCR amplification of specific protein complex-protected DNA sequence. These assays have been referred to as chromatin immunoprecipitation (ChIP) assays and have been used in a variety of HTLV studies. An excellent example of the application of this technology is the use of the ChIP strategy to show that HBZ and JunD interact on the hTERT promoter.[75] This study established that HBZ can up-regulate hTERT transcription and does so by targeting JunD to a promoter proximal region that is devoid of known AP-1 binding sites. A similar approach was employed to demonstrate that SRF binds to the consensus SRE within the HTLV LTR.[76] This observation was a first step in establishing that the HTLV-1 SRE is functional. The ChIP approach was also used to determine a general expression profile of the proteins that occupy space on the viral LTR.[72] In this study, the authors found that the presence of HDACs competes with Tax on the viral LTRs. They also showed that the occupation of sites along the 5′ and 3′ LTR were similar, suggesting that each promoter is functionally equivalent. Thus, for viral transcription factors, ChIP can establish protein–DNA complexes with functional significance.

Comprehensive Protein Complex Characterization Using Mass Spectrometry

A notable improvement on the immunoprecipitation followed by western analysis is the combination of affinity purification with tandem mass spectrometry. The immediate advantage is that no knowledge of the binding partners is required and the state-of-the-art instrumentation is capable of identifying hundreds of bound proteins in a single complex. An example of workflow for this approach is shown in Fig. 2.

There have been two reported studies in which this technique was employed in analysis of HTLV proteins. The first study was conducted by Kashanchi and colleagues.[77] In this study, total lysates from an HTLV-1-infected cell line known to express Tax at high levels was used for the chromatographic isolation of native Tax complexes. Tax complexes were then subjected to immunoprecipitation with anti-Tax antibody and subjected to 2D PAGE analysis. The 2D gel pattern resulting from the separation of the Tax complexes was compared to the 2D pattern of complexes isolated with control antibody. The differentially expressed "spots" were excised and the sequence was identified by mass spectrometry. Tax associated complexes were found to the tune of 32 proteins. As is common in such strategies, the differentially expressed identified proteins were confirmed by western blotting. Several proteins known to bind Tax were identified, such as TXBP151, and several proteins were logical candidates

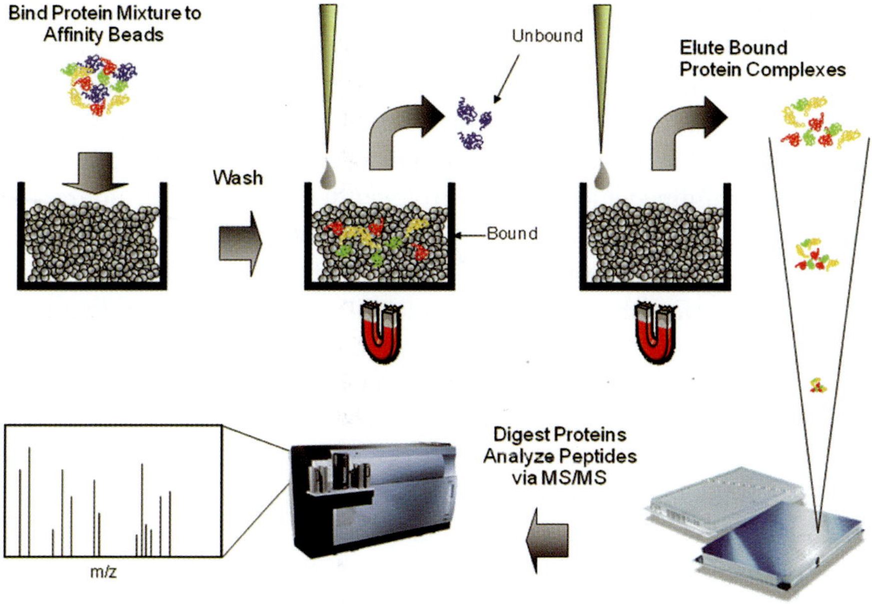

Fig. 2. Example workflow for analysis of protein complexes. A single target protein fused to an affinity tag can be introduced into living cells and bound protein complexes can be isolated. Following standard affinity isolation methods, purified protein complexes are completely digested to constituent peptides and analyzed via tandem mass spectrometry.

for explaining known Tax function such as Rac1, RhoA and Cdc42. However, the authors focused on members of the SWI/SNF complex to propose a role for this interaction in Tax transcription and chromatin remodeling.

In a subsequent study, our group utilized a mammalian expressed tandem affinity tagged Tax protein to isolate protein complexes.[78] The complexes were subjected to LC-MS/MS to achieve the identification of approximately 250 proteins. The increased numbers of proteins identified in this analysis is due to the use of so-called bottom-up proteomics in which the entire protein complex is digested to peptides that are then chromatographically separated in the LC component. This allows for an effective increase in the duty cycle of the MS/MS analysis. In addition, the use of a fast ESI-TRAP instrument allows for greater speed in analysis that when combined with the LC separation results in dramatic improvement in the ability to conduct MS/MS on each peptide. That being said, only time will determine if identification of 250 proteins yields greater understanding of Tax function than the identification of 32 proteins. One disadvantage of increased numbers of potentially important binding partners is determining which protein–protein interactions merit resources to verify functional significance. In the above study, we conducted a combined *in silico* analysis of published studies involving Tax-binding proteins and secondarily, pathway extension analysis of known Tax-binding proteins to generate a list of candidate proteins that are expected to factor heavily in Tax function. These were then narrowed down to those with expected impact for DNA damage response related functions and applied to the 250 candidates to select DNA-PKcs, Ku70 and Ku86 (DNA-PK) as the top candidate protein–protein interactions for analysis. The combination of proteomics and genomics studies represents a growing trend in "Integromics" science that promises to help guide decision making in high data content studies.

GLOBAL PROTEOMICS

Global assessment of the proteome has nearly come within reach as a direct result of technological improvements in the ability to assess protein structure, expression and function in increasingly high-throughput fashion. Each year technological breakthroughs allow for larger and larger numbers of proteins or greater and greater amounts of structure to be analyzed in single experiments. The field is dominated by two major technologies, mass spectrometry and planar array platforms.

Antibody Arrays

An extremely powerful approach to the analysis of large segments of the proteome is the use of antibody arrays.[79,80] The arrays basically consist of defined and optimized presentation of antibodies that have been covalently attached to a solid surface. The basic format is a direct array of antibodies with which unknown protein mixtures are reacted. The sensitivity and specificity of the array can be dramatically improved by employing an ELISA-like sandwich approach in which the signal can be amplified by derivatizing the second antibody. Common amplification approaches are the use of rolling circle amplification, resonance light scattering, chemiluminescence, tyramide signal amplification and streptavidin-conjugation. The use of suspensions of beads as opposed to planar surfaces can also greatly improve sensitivity and throughput when the beads are spectrally coded to an individual antibody. In addition, there are reverse antibody arrays in which the unknown protein mixture is fixed to the surface and defined primary antibodies are applied. There are a number of recent reviews of this technology.[81]

One example of a simple antibody array targeted to specific cytokines expression was developed to investigate the ability of Tax to elicit cytokine release.[82] The array was a direct antibody array in which the 16 antibodies were attached to a slide upon which immature dendritic cells (DC) along with either Tax or lipopolysaccharide (LPS) were then incubated. In response to LPS or Tax protein, the DC cells released cytokines that were then bound by antibody and detected. Under these conditions, Tax induced the secretion of IFNγ, IL-12, TNF-α, Eotaxin, MCP-1, and MCP-3. The authors further determined a role for NF-κB signaling, in that a Tax mutant defective for NF-κB activation was unable to elicit the same response.

The application of an indirect antibody array was used to establish the interaction of Tax with proteins in the cellular secretory pathway.[83] In this analysis, over 725 antibodies were arrayed and reacted with cellular extracts and Tax protein. The Tax protein was fused to GFP and used as the signal for binding of Tax to the antibody captured cellular protein. Thus, this antibody array platform was used to determine potential interactions between Tax and cellular proteins. The interaction data from this large array was combined with a smaller array targeted to proteins known to be involved with protein secretion. The combined analysis supported a conclusion that Tax interacted with a full spectrum of secretory proteins. Then, having identified cellular proteins that could potentially regulate Tax secretion, the authors selectively knocked down the targets via siRNA. The results defined the protein–protein interactions required for Tax secretion.

Protein Arrays

Protein arrays are distinguished from antibody arrays in that the proteins are known and fully characterized and expression on the platform is optimized. The standard format is to array many hundreds and thousands of proteins or overlapping peptides onto a planar surface.[84] As with antibody arrays, there are systems that utilize beads and so-called nanodots instead of a planar surface and in fact, the support technology for both antibody and protein arrays are identical. One very attractive use for protein arrays is in the high-throughput mapping of protein–protein interaction interfaces. Although this has been done successfully for many proteins, there are no published accounts of the application of this approach to HTLV proteins.

The single example of the use of protein arrays for HTLV analysis has employed what can be termed a reverse protein array.[85] In this study, known expressed and quantified viral proteins were fused to Ruc (luciferase enzyme). The selected fusion protein was then mixed with patient serum (containing serum antibodies) and allowed to bind. The resulting antibody antigen mixture was transferred to a second plate that had been pre-treated with bound A/G anti-human antibodies. The tagged proteins that bound to patients produced antibodies were trapped and quantitated by measurement of luciferase activity. The authors determined that measurement of the patient antibody response to Gag antibodies could discriminate between HTLV infected and uninfected individuals. However, even more interesting was that patient antibody responses to Env and Tax were higher in HAM/TSP patients when compared to asymptomatic carriers and that antibody responses to Env could discriminate HAM/TSP from ATLL. Thus, the analysis of patient sera using a reverse protein array has potential toward the development of a diagnostic platform.

HLA Class I Peptide Profiling

Hunt and colleagues developed a technique for the identification of peptides directly isolated from isolated HLA class I molecules.[86–88] The rationale behind this approach is that one could determine critical antigens as a means of defining immunotherapeutic strategies. In addition, the technique has also been used to both characterize specific disease states in patients and as a means of defining potential tumor-specific antigens with diagnostic value.

In the application of this strategy to HTLV, HLA class I was purified from ATL derived cells.[89] An extended chromatographic separation combined with LC-MS/MS was performed to sequence identify the HLA-bound peptides. In all, 188 peptides were successfully identified. When expression of the corresponding genes was examined, five of the proteins giving rise to the HLA-bound peptides were shown to be overexpressed in ATL cells. One of these proteins was PRAME, a known tumor antigen. The authors were also able to show that PRAME-specific cytotoxic T cells could be derived *in vitro*. The results strongly support the contention that surveying HLA-peptides could be an approach to tailor immunotherapy for hematologic malignancies.

SERUM PROTEIN PROFILING

One of the early promises of clinical proteomics was the ability to interrogate complex clinical tissues such as serum in a rapid and thorough manner. The expectation was that after accomplishing this, one would have a list of several proteins that together would present an expression profile that could act as a fingerprint for underlying disease. The earliest approaches to this technique were conducted in our laboratory and others' using a modification of a MALDI-TOF mass spectrometer.[90,91] The basic principle was to obtain spectral images comprised of peptide ion peaks for which the intensities could be used as a measure of relative intensity. Subsequent to our early findings, the identification of such patterns has occurred in abundance although none has passed clinical validation and acceptance. With respect to HTLV-1, we first showed that a serum profile was able to distinguish a clinical outcome of a human retroviral infection.[92] Several of the proteins that comprised the signature were identified. These were alpha antitrypsin inhibitor and haptoglobin-2. In a separate follow-up effort, using similar strategies, expression of proteins comprising the complement system were implicated in disease.[93] These two studies demonstrated that hematologic diseases result in dramatic systemic changes in relatively abundant proteins. The progression of these earlier studies to clinical validation is partly hampered by the rarity of this disease and uncertainty of treatment regimes.

FUTURE DIRECTIONS

Targeted quantitative proteomics based on MRM-MS is a new area of proteomics that is being applied to verify and validate global proteomics data, to discover low abundant proteins, to identify and validate PTMs, and to discriminate protein isoforms. This approach addresses several limitations of global expression proteomics analyses including sensitivity and selectivity. Although the application of MRM-MS in proteomics is a relatively recent development, it is a version of selected reaction monitoring or single reaction monitoring (SRM), which has been used extensively in metabolite profiling and quantitation in pharmacology. SRM experiments are conducted in triple quadrupole mass spectrometers in which transitions are designed to allow the first quadrupole (Q1) to transmit a specific parent ion and the third quadrupole (Q3) is set to transmit a specific fragment ion that is produced in the second quadrupole (Q2), which acts as a collision cell. The generation and detection of two unique ions greatly enhances the sensitivity and specificity of the approach. In MRM-MS several SRM transitions for the same protein or a complex protein mixture can be monitored in one LC-MS experiment. The MRM-MS approach is currently been used for protein quantitation using known concentrations of isotopically labeled peptides.[94] In this approach, depending on the experimental strategy, the labeled internal standard peptide is mixed with the protein sample prior to or during the protease treatment. The labeled internal standard and the unlabeled endogenous peptides are analyzed by LC-MRM-MS, and the peak areas of the labeled and unlabeled versions of the peptides are used for the quantitation. The two versions of the peptide are chemically identical, co-elute chromatographically and have similar ionization, but their mass difference allows their differentiation by the mass spectrometer. The concentration of the internal standard is known and the ratio between the internal standard and the endogenous peptide is determined and subsequently used to determine the absolute amount of the endogenous peptide.

MRM-MS is gaining utility in the analysis of different categories of PTMs, including phosphorylation, ubiquitylation, acetylation and glycosylation sites in proteins. The combination of targeted MS-based approaches with traditional affinity-based isolation will greatly improve the sensitivity and specificity of this approach in proteomics studies. The ability to follow multiple post-translational modifications in a single protein over time is now becoming practical. An example of MRM-MS analysis for the semi-quantitative assessment of huChk2 phosphorylation is shown Fig. 3. Equally exciting is the ability to access changes in one type

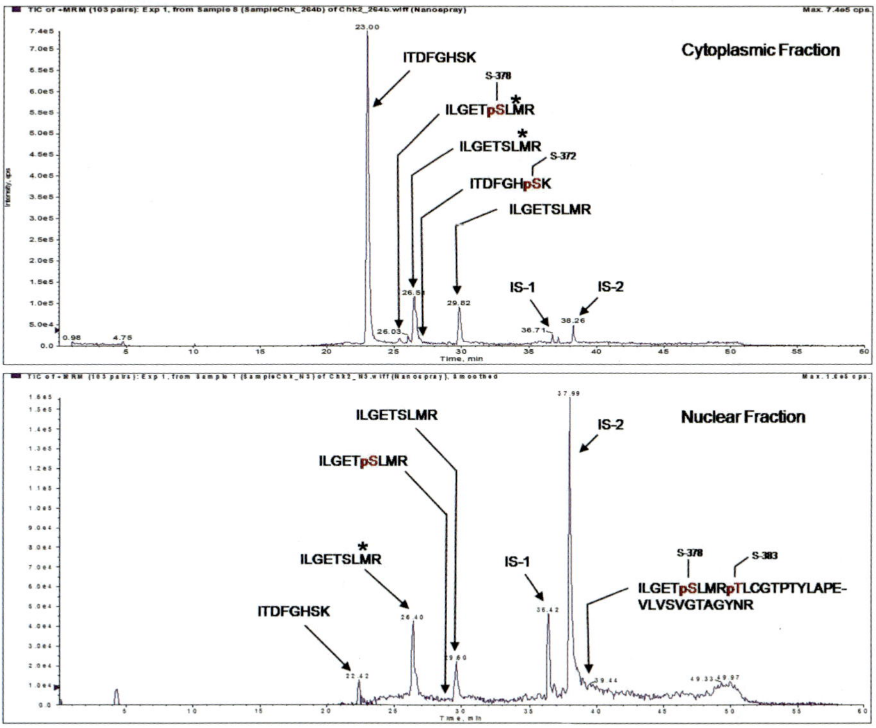

Fig. 3. Multiple *reaction monitoring of phosphorylation in the activation loop of the serine/threonine-protein kinase (Chk2)*. The Total ion chromatograms (TIC) of the MRM show the detection of phosphorylated and non-phosphorylated versions of some of the peptides in the activation loop of the Chk2 using specific Q1/Q3 transition masses. The upper panel shows the detection of the transitions in the cytoplasmic fraction and the bottom panel the same transitions from the nuclear fraction. Two non-phosphorylated Chk2 peptides are used as internal standards (IS1, IS2) for relative quantitation of the phosphorylation of serine and threonine residues in the two compartments. The sequences of the peptides are derived in MS/MS experiments that are triggered upon the detection of the specific transition masses. The methionine residues marked with an asterisk (*) are oxidized.

of PTM, for instance ubiquitylation, with another type of PTM in absolute stoichiometric quantities. These types of methodologies promise the more rapid completion of study objectives that would otherwise involve years of mutational analysis and antibody development.

One step beyond structure and closer then to actual phenotype is the understanding of protein–protein interaction on a comprehensive scale. The application of large-scale platforms such as protein/antibody arrays in which entire critical cellular pathways are represented will help to create interaction maps for all HTLV proteins. The eventual incorporation of direct analyte resonance measurements as utilized in surface plasmon resonance instrumentation will allow for real-time dynamic assessment of many hundreds of target protein interactions. Integrating this data with that derived from orthogonal technologies such as gene arrays and pathway analysis will ultimately provide a better understanding of how viruses operate.

REFERENCES

1. Legros S, Boxus M, Dewulf JF, *et al.* (2009) Protein–protein interactions and gene expression regulation in HTLV-1 infected cells. *Front Biosci* **14**: 4138–4148.
2. Matsuoka M. (2005) Human T-cell leukemia virus type I (HTLV-I) infection and the onset of adult T-cell leukemia (ATL). *Retrovirology* **2**: 27.
3. Matsuoka M, Jeang KT. (2007) Human T-cell leukaemia virus type 1 (HTLV-1) infectivity and cellular transformation. *Nat Rev Cancer* **7**: 270–280.
4. Taylor G. (2007) Molecular aspects of HTLV-I infection and adult T-cell leukaemia/lymphoma. *J Clin Pathol* **60**: 1392–1396.
5. Shadforth I, Bessant C. (2006) Genome annotating proteomics pipelines: Available tools. *Expert Rev Proteomic* **3**: 621–629.
6. Wingren C, James P, Borrebaeck CA. (2009) Strategy for surveying the proteome using affinity proteomics and mass spectrometry. *Proteomics* **9**: 1511–1517.
7. Wodak SJ, Pu S, Vlasblom J, Seraphin B. (2009) Challenges and rewards of interaction proteomics. *Mol Cell Proteomic* **8**: 3–18.
8. Oroszlan S, Sarngadharan MG, Copeland TD, *et al.* (1982) Primary structure analysis of the major internal protein p24 of human type C T-cell leukemia virus. *Proc Natl Acad Sci USA* **79**: 1291–1294.

9. Bogoyevitch MA, Fairlie DP. (2007) A new paradigm for protein kinase inhibition: Blocking phosphorylation without directly targeting ATP binding. *Drug Discov Today* **12**: 622–633.
10. Gao M, Karin M. (2005) Regulating the regulators: Control of protein ubiquitination and ubiquitin-like modifications by extracellular stimuli. *Mol Cell* **19**: 581–593.
11. Hunter T. (2007) The age of crosstalk: Phosphorylation, ubiquitination, and beyond. *Mol Cell* **28**: 730–738.
12. Fukumoto R, Andresen V, Bialuk I, *et al.* (2009) *In vivo* genetic mutations define predominant functions of the human T-cell leukemia/lymphoma virus p12I protein. *Blood* **113**: 3726–3734.
13. Pique C, Pham D, Tursz T, Dokhelar MC. (1992) Human T-cell leukemia virus type I envelope protein maturation process: Requirements for syncytium formation. *J Virol* **66**: 906–913.
14. Lee S, Shin Y, Marler J, Levin MC. (2008) Post-translational glycosylation of target proteins implicate molecular mimicry in the pathogenesis of HTLV-1 associated neurological disease. *J Neuroimmunol* **204**: 140–148.
15. Chiari E, Lamsoul I, Lodewick J, *et al.* (2004) Stable ubiquitination of human T-cell leukemia virus type 1 tax is required for proteasome binding. *J Virol* **78**: 11823–11832.
16. Peloponese JM, Jr., Iha H, Yedavalli VR, *et al.* (2004) Ubiquitination of human T-cell leukemia virus type 1 tax modulates its activity. *J Virol* **78**: 11686–11695.
17. Nasr R, Chiari E, El-Sabban M, *et al.* (2006) Tax ubiquitylation and sumoylation control critical cytoplasmic and nuclear steps of NF-kappaB activation. *Blood* **107**: 4021–4029.
18. Lamsoul I, Lodewick J, Lebrun S, *et al.* (2005) Exclusive ubiquitination and sumoylation on overlapping lysine residues mediate NF-kappaB activation by the human T-cell leukemia virus Tax oncoprotein. *Mol Cell Biol* **25**: 10391–10406.
19. Gatza ML, Dayaram T, Marriott SJ. (2007) Ubiquitination of HTLV-I Tax in response to DNA damage regulates nuclear complex formation and nuclear export. *Retrovirology* **4**: 95.
20. Lodewick J, Lamsoul I, Polania A, *et al.* (2009) Acetylation of the human T-cell leukemia virus type 1 Tax oncoprotein by p300 promotes activation of the NF-kappaB pathway. *Virology* **386**: 68–78.

21. Ahmed FE. (2008) Utility of mass spectrometry for proteome analysis: Part I. Conceptual and experimental approaches. *Expert Rev Proteomic* **5**: 841–864.
22. Han X, Aslanian A, Yates JR, 3rd. (2008) Mass spectrometry for proteomics. *Curr Opin Chem Biol* **12**: 483–490.
23. Durkin SS, Ward MD, Fryrear KA, and Semmes OJ. (2006) Site-specific phosphorylation differentiates active from inactive forms of the human T-cell leukemia virus type 1 Tax oncoprotein. *J Biol Chem* **281**: 31705–31712.
24. Azaroff LV, Kaplow R, Kato N, *et al.* (1974) *X-Ray Diffraction.* McGraw-Hill.
25. Shuker SB, Mariani VL, Herger BE, Dennison KJ. (2003) Understanding HTLV-I protease. *Chem Biol* **10**: 373–380.
26. Li M, Laco GS, Jaskolski M, *et al.* (2005) Crystal structure of human T cell leukemia virus protease, a novel target for anticancer drug design. *Proc Natl Acad Sci USA* **102**: 18332–18337.
27. Delamarre L, Rosenberg AR, Pique C, *et al.* (1996) The HTLV-I envelope glycoproteins: Structure and functions. *J Acquir Immune Defic Syndr Hum Retrovirol* **13**(Suppl 1): S85–S91.
28. Kobe B, Center RJ, Kemp BE, Poumbourios P. (1999) Crystal structure of human T cell leukemia virus type 1 gp21 ectodomain crystallized as a maltose-binding protein chimera reveals structural evolution of retroviral transmembrane proteins. *Proc Natl Acad Sci USA* **96**: 4319–4324.
29. Sundaram R, Lynch MP, Rawale SV, *et al.* (2004) *De novo* design of peptide immunogens that mimic the coiled coil region of human T-cell leukemia virus type-1 glycoprotein 21 transmembrane subunit for induction of native protein reactive neutralizing antibodies. *J Biol Chem* **279**: 24141–24151.
30. Mittermaier A, Kay LE. (2006) New tools provide new insights in NMR studies of protein dynamics. *Science* **312**: 224–228.
31. Clore GM, Gronenborn AM. (1998) NMR structure determination of proteins and protein complexes larger than 20 kDa. *Curr Opin Chem Biol* **2**: 564–570.
32. Baydoun HH, Bellon M, Nicot C. (2008) HTLV-1 Yin and Yang: Rex and p30 master regulators of viral mRNA trafficking. *AIDS Rev* **10**: 195–204.
33. Jiang F, Gorin A, Hu W, *et al.* (1999) Anchoring an extended HTLV-1 Rex peptide within an RNA major groove containing junctional base triples. *Structure* **7**: 1461–1472.
34. Christensen AM, Massiah MA, Turner BG, *et al.* (1996) Three-dimensional structure of the HTLV-II matrix protein and comparative analysis of matrix

proteins from the different classes of pathogenic human retroviruses. *J Mol Biol* **264**: 1117–1131.
35. Darlix JL, Cristofari G, Rau M, *et al.* (2000) Nucleocapsid protein of human immunodeficiency virus as a model protein with chaperoning functions and as a target for antiviral drugs. *Adv Pharmacol* **48**: 345–372.
36. Bertola F, Manigand C, Picard P, *et al.* (2000) Human T-lymphotrophic virus type I nucleocapsid protein NCp15: Structural study and stability of the N-terminal zinc-finger. *Biochem J* **352** (Part 2): 293–300.
37. Khorasanizadeh S, Campos-Olivas R, Summers MF. (1999) Solution structure of the capsid protein from the human T-cell leukemia virus type-I. *J Mol Biol* **291**: 491–505.
38. Cornilescu CC, Bouamr F, Yao X, *et al.* (2001) Structural analysis of the N-terminal domain of the human T-cell leukemia virus capsid protein. *J Mol Biol* **306**: 783–797.
39. Cornilescu CC, Bouamr F, Carter C, Tjandra N. (2003) Backbone (15)N relaxation analysis of the N-terminal domain of the HTLV-I capsid protein and comparison with the capsid protein of HIV-1. *Protein Sci* **12**: 973–981.
40. Williams C. (2004) cAMP detection methods in HTS: Selecting the best from the rest. *Nat Rev Drug Discov* **3**: 125–135.
41. Willingham AT, Deveraux QL, Hampton GM, Aza-Blanc P. (2004) RNAi and HTS: Exploring cancer by systematic loss-of-function. *Oncogene* **23**: 8392–8400.
42. Semmes OJ, Jeang KT. (1992) Mutational analysis of human T-cell leukemia virus type I Tax: Regions necessary for function determined with 47 mutant proteins. *J Virol* **66**: 7183–7192.
43. Smith MR, Greene WC. (1990) Identification of HTLV-I Tax transactivator mutants exhibiting novel transcriptional phenotypes. *Genes Dev* **4**: 1875–1885.
44. Bindhu M, Nair A, Lairmore MD. (2004) Role of accessory proteins of HTLV-1 in viral replication, T cell activation, and cellular gene expression. *Front Biosci* **9**: 2556–2576.
45. Ciminale V, Zotti L, D'Agostino DM, *et al.* (1999) Mitochondrial targeting of the p13II protein coded by the x-II ORF of human T-cell leukemia/lymphotropic virus type I (HTLV-I). *Oncogene* **18**: 4505–4514.

46. D'Agostino DM, Ranzato L, Arrigoni G, *et al.* (2002) Mitochondrial alterations induced by the p13II protein of human T-cell leukemia virus type 1. Critical role of arginine residues. *J Biol Chem* **277**: 34424–34433.
47. Semmes OJ, Jeang KT. (1996) Localization of human T-cell leukemia virus type 1 Tax to subnuclear compartments that overlap with interchromatin speckles. *J Virol* **70**: 6347–6357.
48. Fryrear KA, Durkin SS, Gupta SK, *et al.* (2009) Dimerization and a novel Tax speckled structure localization signal are required for Tax nuclear localization. *J Virol* **83**: 5339–5359.
49. Burton M, Upadhyaya CD, Maier B, *et al.* (2000) Human T-cell leukemia virus type 1 Tax shuttles between functionally discrete subcellular targets. *J Virol* **74**: 2351–2364.
50. Alefantis T, Barmak K, Harhaj EW, *et al.* (2003) Characterization of a nuclear export signal within the human T cell leukemia virus type I transactivator protein Tax. *J Biol Chem* **278**: 21814–21822.
51. Ding W, Albrecht B, Kelley RE, *et al.* (2002) Human T-cell lymphotropic virus type 1 p12(I) expression increases cytoplasmic calcium to enhance the activation of nuclear factor of activated T cells. *J Virol* **76**: 10374–10382.
52. Kim SJ, Nair AM, Fernandez S, *et al.* (2006) Enhancement of LFA-1-mediated T cell adhesion by human T lymphotropic virus type 1 p12I1. *J Immunol* **176**: 5463–5470.
53. Datta A, Silverman L, Phipps AJ, *et al.* (2007) Human T-lymphotropic virus type-1 p30 alters cell cycle G2 regulation of T lymphocytes to enhance cell survival. *Retrovirology* **4**: 49.
54. Ye J, Silverman L, Lairmore MD, Green PL. (2003) HTLV-1 Rex is required for viral spread and persistence *in vivo* but is dispensable for cellular immortalization *in vitro*. *Blood* **102**: 3963–3969.
55. Pique C, Ureta-Vidal A. Gessain A, *et al.* (2000) Evidence for the chronic *in vivo* production of human T cell leukemia virus type I Rof and Tof proteins from cytotoxic T lymphocytes directed against viral peptides. *J Exp Med* **191**: 567–572.
56. Jin DY, Jeang KT. (1997) HTLV-I Tax self-association in optimal trans-activation function. *Nucl Acids Res* **25**: 379–387.

57. Fujii M, Chuhjo T, Minamino T, *et al.* (1995) Identification of the Tax interaction region of serum response factor that mediates the aberrant induction of immediate early genes through CArG boxes by HTLV-I Tax. *Oncogene* **11**: 7–14.
58. Suzuki T, Uchida-Toita M, Andoh T, Yoshida M. (2000) HTLV-1 Tax oncoprotein binds to DNA topoisomerase I and inhibits its catalytic activity. *Virology* **270**: 291–298.
59. Berlioz-Torrent C, Shacklett BL, Erdtmann L, *et al.* (1999) Interactions of the cytoplasmic domains of human and simian retroviral transmembrane proteins with components of the clathrin adaptor complexes modulate intracellular and cell surface expression of envelope glycoproteins. *J Virol* **73**: 1350–1361.
60. Bogerd H, Greene WC. (1993) Dominant negative mutants of human T-cell leukemia virus type I Rex and human immunodeficiency virus type 1 Rev fail to multimerize *in vivo. J Virol* **67**: 2496–2502.
61. Fraedrich K, Muller B, Grassmann R. (2005) The HTLV-1 Tax protein binding domain of cyclin-dependent kinase 4 (CDK4) includes the regulatory PSTAIRE helix. *Retrovirology* **2**: 54.
62. Jin DY, Spencer F, Jeang KT. (1998) Human T cell leukemia virus type 1 oncoprotein Tax targets the human mitotic checkpoint protein MAD1. *Cell* **93**: 81–91.
63. Reddy TR, Tang H, Li X, and Wong-Staal F. (1997) Functional interaction of the HTLV-1 transactivator Tax with activating transcription factor-4 (ATF4). *Oncogene* **14**: 2785–2792.
64. Suzuki T, Ohsugi Y, Uchida-Toita M, *et al.* (1999) Tax oncoprotein of HTLV-1 binds to the human homologue of Drosophila discs large tumor suppressor protein, hDLG, and perturbs its function in cell growth control. *Oncogene* **18**: 5967–5972.
65. Pise-Masison CA, Dittmer J, Clemens KE, Brady JN. (1997) Physical and functional interaction between the human T-cell lymphotropic virus type 1 Tax1 protein and the CCAAT binding protein NF-Y. *Mol Cell Biol* **17**: 1236–1243.
66. Forgacs E, Gupta SK, Kerry JA, Semmes OJ. (2005) The bZIP transcription factor ATFx binds human T-cell leukemia virus type 1 (HTLV-1) Tax and

repesses HTLV-1 long terminal repeat-mediated transcription. *J Virol* **79**: 6932–6939.

67. Gaudray G, Gachon F, Basbous J, *et al.* (2002) The complementary strand of the human T-cell leukemia virus type 1 RNA genome encodes a bZIP transcription factor that down-regulates viral transcription. *J Virol* **76**: 12813–12822.
68. Blot V, Delamarre L, Perugi F, *et al.* (2004) Human Dlg protein binds to the envelope glycoproteins of human T-cell leukemia virus type 1 and regulates envelope mediated cell-cell fusion in T lymphocytes. *J Cell Sci* **117**: 3983–3993.
69. Hou X, Foley S, Cueto M, Robinson MA. (2000) The human T-cell leukemia virus type I (HTLV-I) X region encoded protein p13(II) interacts with cellular proteins. *Virology* **277**: 127–135.
70. Edwards DC, Marriott SJ. (2008) Human T-cell leukemia virus type 1 Tax relieves repression of proliferating cell nuclear antigen gene expression. *J Virol* **82**: 11714–11722.
71. Ramirez JA, Nyborg JK. (2007) Molecular characterization of HTLV-1 Tax interaction with the KIX domain of CBP/p300. *J Mol Biol* **372**: 958–969.
72. Lemasson I, Polakowski NJ, Laybourn PJ, Nyborg JK. (2004) Transcription regulatory complexes bind the human T-cell leukemia virus 5′ and 3′ long terminal repeats to control gene expression. *Mol Cell Biol* **24**: 6117–6126.
73. Harhaj EW, Good L, Xiao G, Sun SC. (1999) Gene expression profiles in HTLV-I-immortalized T cells: Deregulated expression of genes involved in apoptosis regulation. *Oncogene* **18**: 1341–1349.
74. Alefantis T, Flaig KE, Wigdahl B, Jain P. (2007) Interaction of HTLV-1 Tax protein with calreticulin: Implications for Tax nuclear export and secretion. *Biomed Pharmacother* **61**: 194–200.
75. Kuhlmann AS, Villaudy J, Gazzolo L, *et al.* (2007) HTLV-1 HBZ cooperates with JunD to enhance transcription of the human telomerase reverse transcriptase gene (hTERT). *Retrovirology* **4**: 92.
76. Wycuff DR, Yanites HL, Marriott SJ. (2004) Identification of a functional serum response element in the HTLV-I LTR. *Virology* **324**: 540–553.
77. Wu K, Bottazzi ME, de la Fuente C, *et al.* (2004) Protein profile of Tax-associated complexes. *J Biol Chem* **279**: 495–508.
78. Ramadan E, Ward M, Guo X, *et al.* (2008) Physical and *in silico* approaches identify DNA-PK in a Tax DNA-damage response interactome. *Retrovirology* **5**: 92.

79. Haab BB. (2006) Applications of antibody array platforms. *Curr Opin Biotechnol* **17**: 415–421.
80. Huang RP. (2007) An array of possibilities in cancer research using cytokine antibody arrays. *Expert Rev Proteomic* **4**: 299–308.
81. Sanchez-Carbayo M. (2006) Antibody arrays: Technical considerations and clinical applications in cancer. *Clin Chem* **52**: 1651–1659.
82. Ahuja J, Lepoutre V, Wigdahl B, *et al.* (2007) Induction of pro-inflammatory cytokines by human T-cell leukemia virus type-1 Tax protein as determined by multiplexed cytokine protein array analyses of human dendritic cells. *Biomed Pharmacother* **61**: 201–208.
83. Jain P, Mostoller K, Flaig KE, *et al.* (2007) Identification of human T cell leukemia virus type 1 tax amino acid signals and cellular factors involved in secretion of the viral oncoprotein. *J Biol Chem* **282**: 34581–34593.
84. Spurrier B, Honkanen P, Holway A, *et al.* (2008) Protein and lysate array technologies in cancer research. *Biotechnol Adv* **26**: 361–369.
85. Burbelo PD, Meoli E, Leahy HP, *et al.* (2008) Anti-HTLV antibody profiling reveals an antibody signature for HTLV-I-associated myelopathy/tropical spastic paraparesis (HAM/TSP). *Retrovirology* **5**: 96.
86. Henderson RA, Cox AL, Sakaguchi K, *et al.* (1993) Direct identification of an endogenous peptide recognized by multiple HLA-A2.1-specific cytotoxic T cells. *Proc Natl Acad Sci USA* **90**: 10275–10279.
87. Skipper JC, Hendrickson RC, Gulden PH, *et al.* (1996) An HLA-A2-restricted tyrosinase antigen on melanoma cells results from posttranslational modification and suggests a novel pathway for processing of membrane proteins. *J Exp Med* **183**: 527–534.
88. Slingluff CL, Jr., Cox AL, Stover JM, Jr., *et al.* (1994) Cytotoxic T-lymphocyte response to autologous human squamous cell cancer of the lung: Epitope reconstitution with peptides extracted from HLA-Aw68. *Cancer Res* **54**: 2731–2737.
89. Kawahara M, Hori T, Matsubara Y, *et al.* (2006) Identification of HLA class I-restricted tumor-associated antigens in adult T cell leukemia cells by mass spectrometric analysis. *Exp Hematol* **34**: 1496–1504.
90. Adam BL, Qu Y, Davis JW, *et al.* (2002) Serum protein fingerprinting coupled with a pattern-matching algorithm distinguishes prostate cancer from benign prostate hyperplasia and healthy men. *Cancer Res* **62**: 3609–3614.

91. Petricoin EF, Ardekani AM, Hitt BA, *et al.* (2002) Use of proteomic patterns in serum to identify ovarian cancer. *Lancet* **359**: 572–577.
92. Semmes OJ, Cazares LH, Ward MD, *et al.* (2005) Discrete serum protein signatures discriminate between human retrovirus-associated hematologic and neurologic disease. *Leukemia* **19**: 1229–1238.
93. Ishida Y, Yamashita K, Sasaki H, *et al.* (2008) Activation of complement system in adult T-cell leukemia (ATL) occurs mainly through lectin pathway: A serum proteomic approach using mass spectrometry. *Cancer Lett* **271**: 167–177.
94. Gerber SA, Rush J, Stemman O, *et al.* (2003) Absolute quantification of proteins and phosphoproteins from cell lysates by tandem MS. *Proc Natl Acad Sci USA* **100**: 6940–6945.

CHAPTER 4

Antisense Transcription in Human T-Cell Leukemia Virus Type 1: Discovery of a New Viral Gene

Benoit Barbeau[†], Christian Devaux[‡,§,¶]
and Jean-Michel Mesnard[*,‡,§,¶]

INTRODUCTION

Early studies conducted on murine and avian retroviruses have been essential in enhancing understanding of retroviral replication. For this reason, most of our understanding of the replicative cycle of human retroviruses such as human immunodeficiency virus type 1 (HIV-1), human T-cell leukemia virus type 1 (HTLV-1), and type 2 (HTLV-2) and other complex non-human retroviruses stem from murine and avian

*Corresponding author. CS 69033, Institut de Biologie, Centre d'études d'agents Pathogènes et Biotechnologies pour la Santé (CPBS), 4 Bd Henri IV, 34965 Montpellier Cedex 2, France, E-mail: jean-michel.mesnard@univ-montp1.fr.

†Université du Québec à Montréal, Département des sciences biologiques et Centre de recherche BioMed, Montréal, Canada.

‡Université Montpellier 1, Centre d'études d'agents Pathogènes et Biotechnologies pour la Santé (CPBS).

§CNRS, UM5236, CPBS, F-34965 Montpellier, France.

¶Université Montpellier 2, CPBS, F-34095 Montpellier, France.

retroviruses. Based on these early studies on these retroviruses, it has been presumed that retroviral gene expression relies on a single transcript which is in turn either left unspliced or is singly or multiply spliced. This transcript initiates from the 5′ long terminal repeat (LTR) region, which harbors most of the necessary binding sites for important transcription factors regulating the retroviral genes in its U3 segment and terminates in the other bordering 3′ LTR region.

Few studies have addressed the possible existence of transcripts initiated at other positions in the retroviral genome. A number of reports have, however, suggested an interesting and unexpected complexity attributed to retroviral transcription. In a few complex retroviruses, including HIV-1 and HTLV-1, a transcript located in the antisense strand was detected and subsequently suggested to have the potential to encode for a protein.[1–4] Although these findings have been debated and rigorously contested, results obtained with the HTLV-1 virus led by an initial important discovery, i.e. the identification of an unexpected protein encoded on the antisense strand[5] have revived the issue of retroviral antisense transcription over the last few years. The goal of this chapter is to summarize previous results obtained on retroviral antisense transcripts and encoded proteins and to expand on results that we and others have recently gathered on HTLV-1, which is the most well-studied virus in terms of this type of retroviral expression pattern.

INITIAL STUDIES ON ANTISENSE TRANSCRIPTION IN RETROVIRUSES

The existence of antisense transcription in retroviruses was initially suggested based on the identification of the conserved open reading frame (ORF) in the antisense strand of their genomes. Work has mostly focused on HIV-1, including an early study by Miller.[2] A 180 amino acid-long ORF was maintained opened on the antisense strand and was generally well conserved in all the analyzed proviral DNA sequences. Similar ORFs were subsequently identified in the antisense strand of the feline immunodeficiency virus (FIV) and HTLV-1 retroviruses.[1,6] Initial studies, although limited in number, have searched for the existence of antisense transcripts in HIV-1. One early report suggested that antisense RNA in HIV-1 infected cells could be detected by northern blot analysis. Using a strand-specific RNA probe, three signals (1.0, 1.1 and 1.6 kb) were detected and appeared concomitantly with early sense transcripts.[7] However, these transcripts were thought to be infrequent, and no other studies detected them by northern blot analysis. More sensitive approaches were undertaken for the identification of the HIV-1 antisense transcript. Reverse transcription polymerase-chain reaction (RT-PCR) approaches were designed to selectively amplify antisense transcripts. By means of a selective primer complementary to the antisense strand, antisense strand-derived cDNAs were synthesized and used for PCR amplification. Two studies used this approach, demonstrating signals that were specific to the antisense transcript.[3,4] These analyses were carried out in activated latently infected cells, freshly infected cell lines, and cells derived from HIV-1-infected patients. Most of these experiments included controls in which RNA samples were directly tested for DNA contamination, but lacked an important control (PCR testing of RNA samples subjected to RT in absence of RT primer), which would probably have indicated that these signals were RT-PCR artefacts. This false positive RT-PCR signal is representative of what is termed endogenous RT priming and is caused by undesirable RT priming. Several possibilities could account for such a signal and could stem from either priming during the RT step by a stem loop formed by the sense RNA or by small DNA or RNA fragments originating from degraded molecules and acting as primers in the RT step. Although the exact mechanism behind this process has not yet been fully explained, in this type of artefact, PCR signals can be obtained upon the addition of RT reactions conducted in the absence of primers and this PCR reaction is thus the ultimate control. A more robust RT-PCR approach using higher temperature conditions and a modified RT enzyme have been used by Vaquero's team and has revealed that for FIV, specifically *in vivo* and *in vitro*, signals could be detected.[6] We have recently been very active in demonstrating the existence of antisense transcription in HIV-1.[8] We have optimized a RT-PCR approach whereby endogenous RT priming-derived artefacts are greatly reduced and do not mask the expected signal for the antisense transcript. Indeed, through this protocol, we have been able to demonstrate for the first time a specific signal attributed to antisense transcription in HIV-1-infected cell lines and 293T cells transfected with proviral DNA.

The characterization of the antisense transcript has been an important issue to resolve. In one study, cDNA isolation from the antisense transcript derived from HIV-1 suggested that the antisense transcripts in HIV-1 were polyadenylated and that the 3′ end processing of this transcript occurred after the stop codon of the predicted antisense protein (ASP) ORF.[3] In our recent work, we have not been able to confirm the functionality of this proposed polyadenylation site. However, through 3′ RACE, we have rather positioned the polyA site on the antisense strand segment of the *pol* gene next to a conserved polyA signal.[8] Studies presenting the position of the CAP sites have also caused confusion: one study located these transcription initiations at the 5′ border of the 3′ LTR (in the U3 region) while two other studies, including a recent study by Ludwig *et al.*,[9] suggested that antisense transcription was initiated in the U5 region. Using two different 5′ RACE approaches, our results have highlighted the presence of numerous CAP sites for antisense transcriptions, most of which seem to be located next to the U3 LTR region and thus concur with the results from Peeters *et al.*[10]

A certain number of reports have focused on the promoter region of the HIV-1 antisense transcript. These studies have led to the suggestion that antisense transcription is inducible by NF-κB-inducing agents such as phorbol 12-myristate 13-acetate. RT-PCR studies conducted by Vanhée-Brossollet *et al.*[4] had earlier highlighted this possibility in HIV-1-infected patients. Isolated 3′ LTR regions from different HIV-1 proviral DNA demonstrated promoter activity when oriented inversely to a reporter gene, suggesting possible antisense promoter activity in this segment. Deletion mutants and point mutation experiments further revealed the importance of an upstream stimulatory factor (USF) and the NF-κB binding sites for the regulation of this antisense promoter activity.[9–11] Other experiments have also concluded that an unexplored Sp1 binding site is important for antisense transcription. Furthermore, the viral Tat transactivator was observed to negatively act on LTR-derived antisense promoter activity. The negative function of Tat on HIV-1 antisense transcription, however, needs to be considered cautiously, as the experiments conducted to demonstrate Tat's downregulatory role depended on transfection experiments of plasmid constructs. It could thus be speculated that Tat activation of sense transcription in these constructs could lead to abundant transcription with antisense properties to the reporter gene and therefore inhibit the translation of the reporter mRNA. More recent data support this possibility and rather suggest a positive role of Tat in HIV-1 antisense transcription.[8]

These reports have provided strong evidence for the existence of antisense transcription in HIV-1 and FIV. However, due to the initial RT-PCR artefacts that generated false positives in these analyses and due to the difficulty in detecting antisense transcripts and the potentially encoded protein, the existence of antisense transcription in retroviruses was initially dismissed.

DISCOVERY OF HTLV-1 ANTISENSE TRANSCRIPTS

The study of antisense expression in HTLV-1 has provided more conclusive proof of the existence of antisense transcription in retroviruses. Early studies demonstrated that antisense transcripts existed in HTLV-1 and likely permitted the synthesis of a new viral protein with unknown characteristics.[1] The authors of this study provided interesting evidence for the existence of antisense transcripts through northern blot analysis of RNA extracted from an HTLV-1-infected cell line, demonstrating the presence of a 2.5 kb signal detected by different RNA probes extending through the 3′ portion of HTLV-1 proviral DNA. More than a decade later, an important finding more conclusively readdressed the existence of these transcripts in HTLV-1; Gaudray *et al.*[5] provided the first evidence of a protein termed HTLV-1 bZIP factor (HBZ) produced from an antisense transcript of a retrovirus. This protein was first identified by conducting a yeast two-hybrid assay against a cDNA library derived from an HTLV-1-infected cell line and using the CREB-2 transcription factor as bait. The identification of this new viral protein also led to the initial suggestion that HBZ downregulates HTLV-1 viral expression by formation of heterodimers with CREB-2, an important transcription factor mediating the Tax-dependent transactivation of the HTLV-1 LTR.

This important finding quickly led to a number of studies aimed at examining the transcript itself. The initial positioning of the HBZ ORF showed that it was located between the *env* gene and the last exon of the *tax* transcript (Fig. 1). Based on the positioning of the ORF and using an optimized RT-PCR approach, we detected such antisense transcripts in all tested HTLV-1-infected cell lines and ATL cells.[12] These results were also

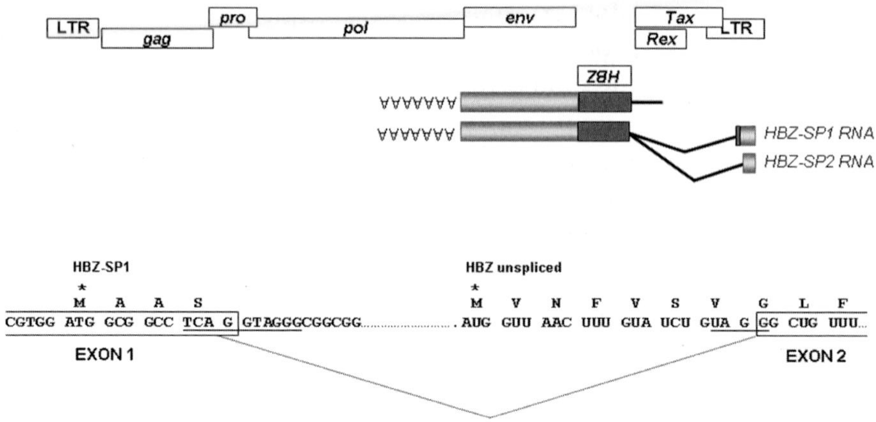

Fig. 1. Antisense transcripts in HTLV-1 and the produced HBZ isoforms. Antisense transcripts from the HTLV-1 virus are initiated from the 3′ LTR and are then alternatively spliced, producing a major spliced form (termed HBZ-SP1) that is responsible for encoding the most abundant HBZ isoform. In both spliced forms, transcripts are initiated at a similar position in the 3′ LTR (which constitutes exon 1) and are spliced to exon 2, which contains the main ORF region as well as the 3′ UTR. Alternatively, an unspliced transcript has been suggested that would be initiated downstream of the 3′ LTR. All of these transcripts are believed to terminate through a common polyA signal. The resulting spliced transcript replaces seven amino acids of the presumed isoform from an unspliced transcript with four new amino acids, all of which are localized in the 3′ LTR-derived exon 1. In the top panel, sense proviral genes are illustrated and compared to the position of the HBZ gene. For a clearer presentation, HTLV-1 accessory genes ($p12^{I}$, $p30^{II}$ and $p13^{II}$) are not indicated. For the spliced transcripts, the dark gray boxes represent the coding segment of HBZ while the lighter gray boxes are non-coding regions. In the bottom panel, nucleotidic sequences bordering the splice donor and splice acceptor sequence (underlined) are indicated and the corresponding amino acids are indicated above for each HBZ isoform. An asterisk pinpoints the corresponding methionine initiation codon for both HBZ-SP1 and unspliced HBZ. Boxes highlight the 3′ and 5′ borders of exon 1 and exon 2, respectively, of the HBZ-SP1 transcript.

confirmed by other teams using different approaches.[13,14] Further analyses have permitted the generation of an accurate characterization of the HTLV-1 antisense transcript. Indeed, RACE experiments have revealed that transcription initiation sites were all located in the 3′ LTR, precisely in the R and U5 regions. Importantly, these experiments showed that the antisense transcript was spliced and produced a major spliced form with the ATG initiation codon located in exon 1 in the 3′ LTR segment (Fig. 1). Further, results from our group indicated the possibility that antisense transcripts also exist as a minor spliced form, although no in-frame ATG initiation codon was present in the exon 1 sequence.[12] Focus was also given to the identification of the 3′ end of the transcript. In agreement with the initially suggested polyA signal,[1] we and others have confirmed its usage for 3′ processing of the HBZ transcript and addition of the polyA tail.[12,14] A GU-rich region has also been observed in the proximity of the polyA tail addition site. A potentially different 3′ end next to a less conserved polyA signal has, however, been suggested.[13]

REGULATION OF THE EXPRESSION OF HBZ

Early on, the mechanism of regulation of the antisense transcript in HTLV-1 was addressed and the existence of a potential TATA box located upstream of the ORF was suggested.[1] Using the HTLV-1 3′ LTR cloned upstream of the CAT reporter gene in the antisense orientation, as for HIV-1 antisense transcription, these authors provided evidence for a downregulatory effect of the Tax transactivator on antisense transcription. Again, however, a potential antisense effect by Tax in the plasmid context was a source of concern regarding these results.

The mechanism governing the regulation of antisense transcription is likely different from that of sense transcription. In fact, like HIV-1, the HTLV-1 3′ LTR is constantly hypomethylated (as opposed to the 5′ LTR), which might be used to argue for a more constitutive expression of HBZ mRNA.[15] In addition, antisense transcription appears to be low and the promoter strength might impact on this level as well as the likely interference coming from sense transcription. Only a limited number of studies have addressed HBZ gene regulation since the discovery of the structure of the HBZ transcript. Due to the actual localization of CAP sites, recent reports have all agreed that HBZ expression depends on a TATA-less promoter, thereby resulting in a multitude of transcription initiation sites.[12,14,16] A recent study has demonstrated that, like HIV-1 antisense transcription,[10] the constitutive expression of HBZ requires the presence of Sp1 binding sites positioned upstream of the CAP sites.[16] Using a 5′ end deleted HTLV-1 proviral DNA construct with the luciferase reporter gene in exon 2 of the HBZ gene, in T cells, we have also shown that known T-cell activators can induce, although modestly, the level of expression of HBZ.[17] Further studies have addressed the impact of Tax on HBZ expression and yielded opposite conclusions to those initially reported by Larocca *et al.*[1] Indeed in both transient and stable transfection experiments, Tax was shown to positively modulate HBZ expression through typical Tax responsive element 1 present in the 3′ LTR[16,17] (Fig. 2). Our

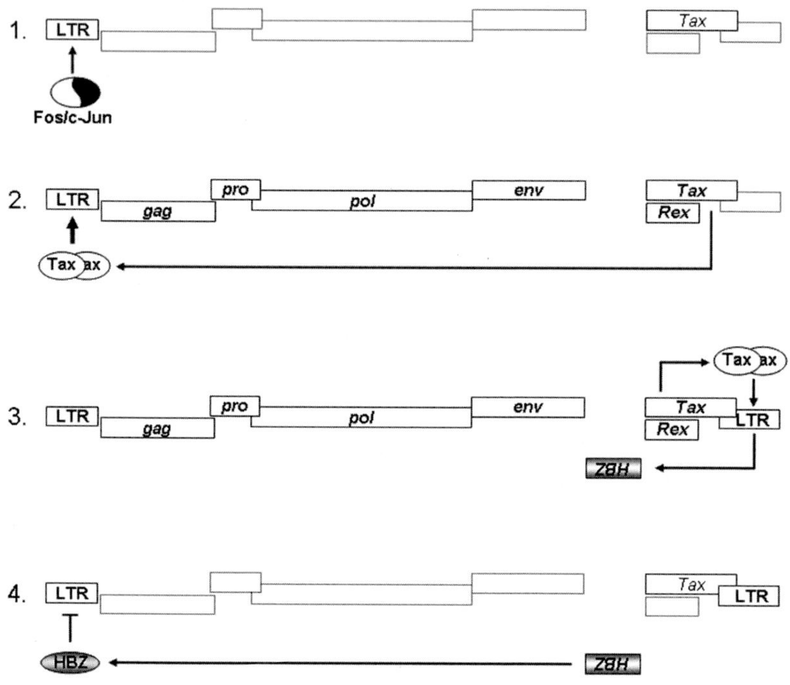

Fig. 2. Mutual regulation between Tax and HBZ. Previous experiments argue that Tax and HBZ reciprocally modulate their expression. This bidirectional modulation maintains Tax and HBZ levels in equilibrium, which might be needed to maintain cell proliferation and help infected cells escape from an immune response. In our model, we initially consider that many factors including the AP-1 complex (Fos/c-Jun) positively mediate viral expression at an early phase.[1] When a sufficient amount of Tax is produced, through the TRE1 elements, Tax strongly upregulates expression of all viral genes, including HBZ.[2,3] The presence of HBZ then downmodulates Tax- and AP-1-dependent viral gene expression to low levels, likely sufficient to maintain a limited amount of Tax only.[4] Again, HTLV-1 accessory genes have been omitted for clarity.

results also lend credence to a chromatin context effect on the extent of this Tax-mediated upregulation.

Being an antisense transcript, the HTLV-1 HBZ transcript is likely subjected to subtle regulatory mechanisms both at the transcriptional and post-transcriptional levels. Numerous studies have reported that the unspliced transcript could be detected in infected cells, albeit at lower levels than the spliced transcript.[13,16,18] The unspliced transcript recently identified by Yoshida *et al.*[16] reveals that CAP sites would be located 900 nucleotides downstream of the formerly identified CAP sites driving spliced transcript expression. In this study, an initiator sequence and a downstream promoter element is part of this new promoter region. Although endogenous RT priming derived RT-PCR artefacts need to be taken into account when studying antisense transcription, these studies reveal a particularly complex pattern of expression of HBZ, which could culminate in the production of various HBZ isoforms (see below).

The potential importance of post-transcriptional regulation in the production of the HBZ protein should also be considered. Splicing not only drives the expression of HBZ isoforms differing in their amino terminus but it also limits potential RNA interference with certain spliced sense transcripts, including Tax mRNA. Such a limited antisense effect might indeed allow both Tax and HBZ to coexist in infected cells and modulate different cellular genes as well as their own production. Further studies will be needed to identify potential splicing regulatory elements in the transcripts and the factors modulating the splicing pattern, possibly including the HTLV-1 Rex protein. Our report also suggested that an alternative spliced transcript could occur (although its encoding potential has not been demonstrated in infected cells) and this observation is directly relevant to potential post-transcriptional regulation. A final issue that needs to be considered pertains to the presence of the 1.45-kb-long 3′ untranslated region (UTR) segment. This sequence might modulate transcriptional stability and should also be analyzed to assess its importance in HBZ expression. Recently, the expression pattern of HBZ in infected cells was determined and it was shown that HBZ transcripts were different from other viral transcripts in that they appeared later than multi-spliced sense transcripts but were maintained at stable levels post-infection.[19] Transcriptional and post-transcriptional regulatory mechanisms might be acting to regulate the levels of antisense transcripts. All these potential regulatory mechanisms need to be more intensively investigated.

CHARACTERIZATION OF THE VARIOUS ISOFORMS OF THE HTLV-1 HBZ PROTEIN AND CELLULAR LOCALIZATION

The first studies on antisense-encoded proteins focused on the HIV-1 antisense ASP protein.[4,20] These studies have, however, been difficult as the ASP protein is not easily detected in infected and transfected cells. Although the existence of this ORF is largely dismissed, the important level of conservation at the amino acid level between different HIV-1 isolates and the recently established detection of antisense transcripts as well as a downstream polyA signal strongly argue in favor of the coding potential of this transcript *in vivo*.

The discovery of HBZ in HTLV-1 by Gaudray *et al.*[5] further indicates the possible existence of antisense transcript-encoded proteins in retroviruses. Although functional analyses of HBZ were initially carried out through a single isoform (derived from unspliced transcripts), complete characterization of these transcripts and their spliced nature argued for a more prominent isoform (Fig. 1). In fact, although the existence of the unspliced transcript-derived isoform is still possible, this protein has not yet been detected in any of the tested infected cell lines.[13] However, differences between most studied isoforms are subtle, and in fact involve amino end differences accounting for seven amino acids from the unspliced transcript-derived isoform swapped for four amino acids from the spliced transcript. This limited change in the amino acid composition of these two isoforms has therefore had little impact on the results from the initial functional studies in that both are capable of inhibiting Tax-mediated HTLV-1 LTR activation and Jun-dependent transactivation potential.[12,16] A recent study has, however, presented data showing that the HBZ isoform from the unspliced transcript is less potent in inhibiting Tax-mediated HTLV-1 LTR-driven expression than HBZ-SP1 (the spliced transcript-derived isoform), which likely results from lower production level and/or greater instability of this isoform.[16]

All studies agree with the fact that HBZ, regardless of the tested isoform, is mostly (if not exclusively) nuclear and present in the form of speckles.[5,13,21,22] In all cell lines tested, either through transfection experiments or the study of infected cell lines, HBZ has an unequivocally nuclear localization. Co-localization experiments further refined the nature of the speckle structures as not being Cajal bodies, splicing factor compartments or promyelocytic leukemia oncoprotein bodies.[21,22] We found that HBZ depended on three domains for its nuclear localization corresponding to basic-rich domains[21] (Fig. 3). Two out of three nuclear localization signals were necessary for proper nuclear localization. However, other studies have shown that nuclear localization did present subtle differences depending on the tested isoform.[13,22] In fact, the HBZ isoform initially characterized and produced from the unspliced transcript did localize to the nucleus as speckle structures, although it retained a perinucleolar staining pattern.[21] The spliced transcript-derived HBZ isoform was again nuclear but also presented an important nucleolar localization.[13,22] It will therefore be very crucial to determine how this moderate change in amino acid between isoforms can alter its nuclear distribution. As already mentioned, the difference in stability between the two isoforms could explain this difference of nuclear localization. However, post-translational modifications could also be the basis of this change in nuclear localization of HBZ.

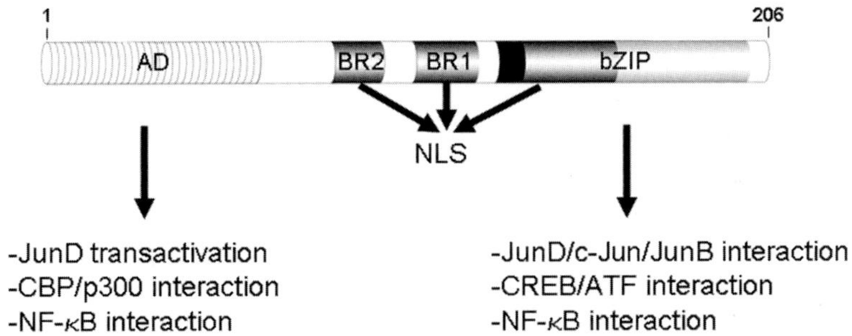

Fig. 3. The various domains of HBZ. The structure of the HBZ protein consists of several different domains. From the amino terminus, these are the activation domain (AD), two basic-rich regions (BR2 and BR1), the modulatory domain (MD), and the bZIP domain. These various regions are involved in different functional aspects as well as cellular localization.

STRUCTURE OF THE HBZ PROTEIN AND ITS INTERACTION PARTNERS

Aside from its basic domains required for nuclear localization, several domains in HBZ are important for its presumed viral function and alteration of cell growth (Fig. 3). The most studied domain remains the basic leucine zipper (bZIP) motif encompassing a typical leucine zipper adjacent to a basic domain reminiscent of a DNA binding-domain. Early studies by Gaudray *et al.*[5] demonstrated by *in vitro* assays and immunoprecipitation experiments in HTLV-1-infected cells that this domain was crucial for its interaction with members of the activating transcription factor/cyclic AMP-responsive element (CRE)-binding protein (ATF/CREB) family, which are also characterized by the presence of a bZIP domain. This interaction blocks the binding of CREB-2 and other related transcription factors to the HTLV-1 LTR and thereby abolishes Tax-dependent activation of promoter activity. These *in vitro* studies have been confirmed *in vivo* through standard chromatin immunoprecipitation (ChIP) experiments, which highlighted the displacement of CREB from the HTLV-1 LTR upon HBZ expression.[23] These experiments have also expanded the list of known HBZ-interacting ATF/CREB family members to include CREM-Ia and ATF-1.

The most well studied interaction with HBZ involves the Jun family members. These proteins are known to interact with members of the Fos family via their shared ZIP motifs and, as heterodimers, they can act upon cellular and viral gene expression by binding to AP-1 elements. As substantial homology existed between the HBZ and Fos family members, it soon became apparent that HBZ could interact with Jun family members. This was indeed the case, as illustrated by several studies.[24–26] However, outcomes were different depending on the Jun interacting partner. While HBZ interaction with c-Jun and JunB results in inhibition of transactivation through a complex bound to AP-1 sites, HBZ-JunD heterodimers instead lead to an increase in their transactivational potential. This augmented transactivation potential is thought to be mediated by a more optimal positioning of the heterodimer binding to cellular promoters and thereby more efficient interaction with nearby transcriptional factors or cofactors.[27,28] On the other hand, the blocking of the transactivating potential of both JunB and c-Jun has been inferred to occur by two different mechanisms depending on the HBZ isoform. Indeed, these cellular factors seem to be selectively sequestered by HBZ-SP1 in specific nuclear bodies, corresponding to transcriptionally inactive sites.[22] However, for c-Jun inactivation, Shimotohno's group highlighted the involvement of the 26S proteasome and specific degradation of c-Jun in a ubiquitination-independent manner, which would involve bridging of the proteasome with c-Jun by HBZ itself.[29] On the other hand, HBZ-SP1 fails to promote the degradation of c-Jun because its interaction with the proteasome is less efficient than that of HBZ.[29]

Additional interaction partners have been highlighted more recently. Through ChIP assays, we provided evidence that CREB binding protein (CBP) and p300 interact with HBZ, a complex which might also be responsible for the inhibition of Tax-dependent upregulation of viral gene expression.[30] These interactions were mapped to the KIX domain of these histone acetyltransferases, which is typically known to interact with CREB and Tax. Further experiments also revealed that this interaction was taking place via two LXXLL-like motifs located in the activation domain of HBZ. Sequestration of CBP and p300 through this interaction might be another mechanism by which Tax-mediated HTLV-1 LTR could be inhibited by HBZ expression. Another recent publication highlighted that the NF-κB subunit p65 might again be a target of HBZ and that its interaction would depend on the activation domain and the bZIP motif.[31] HBZ-dependent repression of the NF-κB complex was shown to have an impact on genes, which are normally modulated by this transcription factor.

As depicted in Fig. 3, additional domains are present in the HBZ protein and are typical of bZIP transcription factors. The N-terminus activation domain is responsible for the activation of JunD transactivational capacity.[26,28] Although comparable in sequence to DNA binding domains of c-Fos and other members of this family of transcription factor,[27] the basic region of HBZ bZIP is nonetheless extremely inefficient at contributing to DNA binding of heterodimers involving CREB-1, CREB-2, c-Jun or Jun-B partners to their target binding sites.[5,23,24] A region adjacent to the DNA binding domain that we have termed MD (modulatory domain) has important implications for the transactivational capacity of HBZ in the context of JunD activation.[27] This region is mainly composed of charged amino acids and has been shown to be different from the similarly positioned and charged amino acid-rich c-Fos region but equally capable of increasing the JunD activation potential.

FUNCTIONAL ASPECT OF HBZ IN RELATION TO REPLICATION OF HTLV-1

Initial studies highlighted the important effect of HBZ on HTLV-1 replication by virtue of its capacity to inhibit the Tax-mediated activation of HTLV-1 transcription.[5] HBZ over-expression in cells transfected with HTLV-1 proviral DNA has further reinforced the notion that HBZ was indeed a suppressor of Tax-mediated promoter activation.[23] As pointed out above, this inhibition likely results from the HBZ-dependent quenching of important transcription factors and histone acetyltransferase from Tax. Basal transcription is also believed to be similarly affected by HBZ and the mechanism of action would then involve heterodimer formation with c-Jun, which counter-represses its transactivational potential. However, studies also looked at the impact of suppressing HBZ expression or blocking HBZ translation in HTLV-1-infected cell lines.[32,33] Paradoxically, however, these approaches have led to no (or very limited) impact on HTLV-1 replication in cell culture, which would have been expected to trigger an increase in virion production. On the other hand, experiments *in vivo* have demonstrated that HTLV-1 persists more efficiently in rabbits infected with an HBZ-expressing HTLV-1 virus as compared to HBZ-deficient viruses. These conclusions were drawn from measurements of proviral DNA copy number and antibody titers. The authors then suggested that this was possibly linked to a hyperproduction of HTLV-1 viruses in HBZ-deficient virions, leading to its elimination through an ensuing strong antiviral immune response. More recent data using the *in vivo* rabbit model have further reinforced the notion that HBZ expression is maintained constitutively over time (unlike other HTLV-1 viral genes) and further correlates with proviral load.[19] Differences between *in vitro* and *in vivo* data with respect to the impact of HBZ on HTLV-1 replication are striking. One might nonetheless argue that HBZ controls the extent of viral production and that this control allows HTLV-1 to persist in infected individuals and avoid deleterious immune response. It is likely that in that scenario, the balance between Tax and HBZ levels is very crucial and that, as pointed out above, both modulate each other's expression. Thus, the interplay between both proteins is likely very subtle and depends on an adequate balance which shifts viral production from one direction to the other depending on the infected cell, the disease status and/or the phase of the replicative cycle.

THE IMPACT OF HBZ ON T-CELL PROLIFERATION AND TRANSFORMATION

Studies have thus far hinted at a more repressing function of HBZ by blocking the transactivational potency of various transcription factors such as CREB-2, c-Jun, Jun-B and NF-κB, which are generally involved in cell proliferation. These effects of HBZ might suggest that HBZ expression impacts negatively on T-cell proliferation and even culminates in cell death and/or cell-cycle arrest. However, it should be kept in mind that HBZ also has a more positive outcome on the JunD transactivational potential. This point is important since ATL cells are known to contain constitutively activated AP-1 complex, which in occurrence is mainly constituted of the JunD family member, as determined by supershift assay.[34] Observations favored a possible impact of HBZ on ATL development, specifically on ATL maintenance. It should first be highlighted that in ATL cells, the occurrence of 5′ LTR methylation and/or deletion is frequent, thereby leading to the absence of viral gene expression including the Tax protein.[35] It has been postulated that this renders ATL cells less likely to be targeted by the immune response through lower expression of viral protein. However, as opposed to the 5′ LTR, the 3′ region of the proviral DNA, including the 3′ LTR, is not only intact but is also hypomethylated.[15,36] This explains why HBZ expression has been conclusively detected in almost all tested ATL cells, including freshly isolated ATL samples.[13,14,18] This is not the case for Tax expression, and it could be speculated that this protein might rather be involved in the first step of transformation of ATL cells while HBZ would act as a maintenance factor in these cells. A recent report has further demonstrated that ATL cells harboring 5′ LTR-deleted proviral DNA retain the capacity to express HBZ.[35] As no possible sense-derived viral gene expression can be suggested for preserving HBZ expression, it is thus apparent that HBZ plays a direct role in ATL cell viability.[37,38] An interesting study has suggested such a possible role for HBZ in the maintenance of ATL cell proliferation. Kulmann *et al.*[28] have presented data showing that HBZ, through its heterodimerization with JunD, acts positively on the expression of the human telomerase reverse transcriptase gene. Importantly, telomerase activity has been linked to development and progression of ATL.[39]

Another intriguing possibility that has been suggested by Matsuoka's team[14] implies a positive role in T-cell proliferation for the HBZ transcript itself. This hypothesis is supported by an observed hyperproliferative effect by HBZ transcripts which have been mutated in their ATG initiation codon. It is worth noting that the unspliced transcript does not seem to have such an impact. This model would therefore argue for a possible bimodal function of HBZ whereby the HBZ protein could act upon different cell function/viral gene expression while the HBZ transcript would positively enhance cell proliferation by a yet-to-be-defined mechanism. A more recent version of these studies knocked out HBZ expression by lentiviral-mediated transduction of HBZ specific shRNA and resulted not only in the reduction of cell proliferation of HTLV-1-transformed cells in culture but also had an impact on size and tissue infiltration of these tumor cells upon inoculation of NOD/SCID$^{\gamma chain-/-}$ mice.[32] However, in these experiments, it cannot be distinguished whether the effect on cell proliferation *in vivo* and *in vitro* was due to a decrease in HBZ at the protein or mRNA level.

Recent experiments have provided interesting information as to the process of viral gene expression in different infected cells.[18] By real-time RT-PCR, quantification of HBZ and Tax mRNA levels was evaluated, indicating that primary ATL cells expressed high levels of HBZ but low levels of Tax mRNA. This pattern was distinguishable from that of infected cells, showing low HBZ and high Tax mRNA levels.

OTHER RETROVIRUSES AND ANTISENSE TRANSCRIPTION

As pointed out at the beginning of this chapter, the existence of ORFs located in the antisense strand of FIV and HIV-1 has been strongly suggested, although more studies are needed to achieve full detection of these proteins in infected cells. Comparison of HTLV-1 HBZ ORF with other simian and human retroviruses that belong to the PTLV (primate T-leukemia virus) family provides convincing evidence that STLV-1, STLV-2,

STLV-3, HTLV-2, HTLV-3 and HTLV-4 viruses are also encoding for an antisense strand-derived protein. Our initial studies showed that STLV-1 proviral DNA sequences were conserved in the regions harboring the splice acceptor, splice donor, ATG initiation codon and the rest of the HTLV-1 HBZ ORF.[12] However, recent studies have pointed to the possible existence of antisense strand ORF in other PTLV members, such as HTLV2, HTLV-3 and HTLV-4[40–42] (Fig. 4). In a recent report, antisense transcripts were detected in HTLV-3-infected cells.[42] If these transcripts do indeed culminate in the translation of the identified ORF, it is not certain that these new viral proteins will possess the same properties as HBZ. More studies will be necessary to further assess the existence of

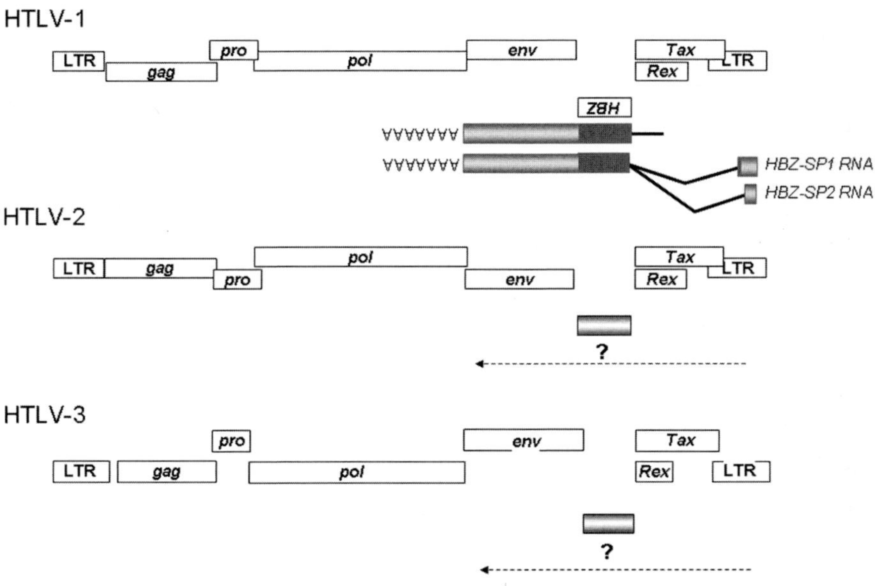

Fig. 4. Existence of antisense ORF in other HTLV viruses. In HTLV-2 and HTLV-3, ORFs are located in the antisense strand of their proviral DNA at a similar position to that of HBZ in HTLV-1. The potential for these ORFs to encode for a protein has been suggested, although studies are needed to further address this issue. For comparison, all three proviral DNAs are presented and the position of the HBZ gene can be compared to the presumed antisense ORF in HTLV-2 and HTLV-3. The arrows below both of these proviral DNAs highlight the potential transcripts that would permit protein synthesis. The existence of splicing in antisense transcription of HTLV-2 and HTLV-3 also needs to be considered.

antisense transcripts in these PTLVs. In addition, these studies should refine the possible link between ATL development and HBZ expression by analyzing the possible antisense encoded protein from the non-leukemia inducing HTLV-2 virus.

PERSPECTIVES

Since its discovery in 2002, substantial data have accumulated pertaining to the expression, regulation and function of HBZ. The discovery of this retrovirally encoded protein is unprecedented and opens up new encoding potential for retroviruses. In this regard, and based on an already published report,[40] the analysis of the antisense proviral DNA strand will ascertain whether other human (and non-human) retroviruses have evolved to allow viral gene expression using the antisense strand, regardless of the important constraints due to size which exist in the retroviral genome and the complexity that the positioning of such genes and their expression could impose on viral gene regulation. Initial analyses of other human retroviruses suggest that this is the case. It will be important to determine if HTLV-2, which is known to be non-leukemia inducing, would lead to the production of an antisense protein with differences in features from HBZ which would justify the differences in pathogenesis between HTLV-1 and HTLV-2. Studies will also be needed to truly assess the encoding potential of antisense strands from lentiviruses such as HIV-1 and FIV, both of which have formerly been suggested to contain an ORF embedded in the antisense strand. It should be stressed that the limited number of studies on HIV-1 antisense transcripts have demonstrated significant similarities in terms of HIV-1's transcriptional regulation with the HBZ gene (modulation by Tat and T-cell activators and the importance of the Sp1 binding site). All of these studies should provide a more complete picture of retroviral replication. Based on the experiments devoted to HBZ in *in vivo* models and in samples from HTLV-1-infected patients, it will be crucial to assess how HBZ could be involved in ATL development, if at all. The implication of the HBZ protein or its transcript will require the use of additional animal models with ATL-like symptoms. In addition, an important focus should be given to other HTLV-1-associated disorders such as

HTLV-1-associated myelopathy. It must be tested whether HBZ expression could also impact on the level of HIV-1 replication in co-infected individuals. In all these considerations, it will also be important to assess how HBZ is expressed in the context of other cells. Importantly, a recent report has demonstrated that cell-free HTLV-1 could infect dendritic cells.[43] Taking all of these perspectives into consideration, much important data have been gathered on the role played by HBZ in both replication and ATL, but further studies are needed to clearly demonstrate the implication of HBZ in ATL development. These studies might eventually allow researchers and clinicians to consider HBZ as an important new candidate for antiviral therapy.

ACKNOWLEDGMENTS

This work was supported by a grant from CanFAR (Canadian Foundation for AIDS Research) and by the Cancer Research Society Inc. (B. Barbeau), by institutional grants from the Centre National de la Recherche Scientifique (CNRS) and the Université Montpellier 1 (C. Devaux and J.-M. Mesnard), and by grants from the Association pour la Recherche sur le Cancer (ARC), the Agence Nationale de Recherche sur le SIDA (ANRS), the Fondation Recherche Médicale (FRM, Comité Languedoc), and the Ligue Contre le Cancer (Comité de l'Hérault) (J.-M. Mesnard).

REFERENCES

1. Larocca D, Chao LA, Seto MH, Brunck TK. (1989) Human T-cell leukemia virus minus strand transcription in infected T-cells. *Bioch Biophys Res Comm* **163**: 1006–1013.
2. Miller RH. (1988) Human immunodeficiency virus may encode a novel protein on the genomic DNA plus strand. *Science* **239**: 1420–1422.
3. Michael NL, Vahey MT, d'Arcy L, *et al.* (1994) Negative-strand RNA transcripts are produced in human immunodeficiency virus type 1-infected cells and patients by a novel promoter downregulated by Tat. *J Virol* **68**: 979–987.

4. Vanhée-Brossollet C, Thoreau H, Serpente N, *et al.* (1995) A natural antisense RNA derived from the HIV-1 *env* gene encodes a protein which is recognized by circulating antibodies of HIV+ individuals. *Virology* **206**: 196–202.
5. Gaudray G, Gachon F, Basbous J, *et al.* (2002) The complementary strand of HTLV-1 RNA genome encodes a bZIP transcription factor that down-regulates the viral transcription. *J Virol* **76**: 12813–12822.
6. Briquet S, Richardson J, Vanhée-Brossollet C, Vaquero C (2001) Natural antisense transcripts are detected in different cell lines and tissues of cats infected with feline immunodeficiency virus. *Gene* **267**: 157–164.
7. Bukrinsky MI, Etkin AI, Ivanovsky DI. (1990) Plus strand of the HIV provirus DNA is expressed at early stages of infection. *AIDS Res Hum Retrovir* **6**: 425–426.
8. Landry S, Halin M, Lefort S, *et al.* (2007) Detection, characterization and regulation of antisense transcripts in HIV-1. *Retrovirology* **4**: 71.
9. Ludwig LB, Ambrus JL, Krawczyk KA, *et al.* (2006) Human immunodeficiency virus-type 1 LTR DNA contains an intrinsic gene producing antisense RNA and protein products. *Retrovirology* **3**: 80.
10. Peeters A, Lambert PF, Deacon NJ. (1996) A fourth Sp1 site in the human immunodeficiency virus type 1 long terminal repeat is essential for negative-sense transcription. *J Virol* **70**: 6665–6672.
11. Bentley K, Deacon N, Sonza S, *et al.* (2004) Mutational analysis of the HIV-1 LTR as a promoter of negative sense transcription. *Arch Virol* **149**: 2277–2294.
12. Cavanagh M-H, Landry S, Audet B, *et al.* (2006) HTLV-I antisense transcripts initiating in the 3′ LTR are alternatively spliced and polyadenylated. *Retrovirology* **3**: 15.
13. Murata K, Hayashibara T, Sugahara K, *et al.* (2006) A novel alternative splicing isoform of human T-cell leukemia virus type 1 bZIP factor (HBZ-SI) targets distinct subnuclear localization. *J Virol* **80**: 2495–2505.
14. Satou Y, Yasunaga J-I, Yoshida M, Matsuoka M (2006) HTLV-I basic leucine zipper factor gene mRNA supports proliferation of adult T cell leukemia cells. *Proc Natl Acad Sci USA* **103**: 720–725.
15. Koiwa T, Hamano-Usami A, Ishida T, *et al.* (2002) 5′-long terminal repeat-selective CpG methylation of latent human T-cell leukemia virus type I provirus *in vitro* and *in vivo*. *J Virol* **76**: 9389–9397.

16. Yoshida M, Satou Y, Yasunaga J, *et al.* (2008) Transcriptional control of spliced and unspliced HTLV-1 bZIP factor gene. *J Virol* **82**: 9359–9368.
17. Landry S, Halin M, Vargas A, *et al.* (2009) Upregulation of human T-cell leukemia virus type 1 antisense transcription by the viral Tax protein. *J Virol* **83**: 2048–2054.
18. Usui T, Yanagihara K, Tsukasaki K, *et al.* (2008) Characteristic expression of HTLV-1 basic zipper factor (HBZ) transcripts in HTLV-1 provirus-positive cells. *Retrovirology* **5**: 34.
19. Li M, Kesic M, Yin H, *et al.* (2009) Kinetic analysis of human T-cell leukemia virus type 1 gene expression in cell culture and infected animals. *J Virol* **83**: 3788–3799.
20. Briquet S, Vaquero C. (2002) Immunolocalization studies of an antisense protein in HIV-1-infected cells and viral particles. *Virology* **292**: 177–184.
21. Hivin P, Frédéric M, Arpin-André C, *et al.* (2005) Nuclear localization of HTLV-I bZIP factor (HBZ) is mediated by three distinct motifs. *J Cell Sci* **118**: 1355–1362.
22. Hivin P, Basbous J, Raymond F, *et al.* (2007) The HBZ-SP1 isoform of human T-cell leukemia virus type I represses JunB activity by sequestration into nuclear bodies. *Retrovirology* **4**: 14.
23. Lemasson I, Lewis MR, Polakowski N, *et al.* (2007) Human T-cell leukemia virus type 1 (HTLV-1) bZIP protein interacts with the cellular transcription factor CREB to inhibit HTLV-1 transcription. *J Virol* **81**: 1543–1553.
24. Basbous J, Arpin C, Gaudray G, *et al.* (2003) HBZ factor of HTLV-I dimerizes with transcription factors JunB and c-Jun and modulates their transcriptional activity. *J Biol Chem* **278**: 43620–43627.
25. Matsumoto J, Ohshima T, Isono O, Shimotohno K (2005) HTLV-1 HBZ suppresses AP-1 activity by impairing both the DNA-binding ability and the stability of c-Jun protein. *Oncogene* **24**: 1001–1010.
26. Thébault S, Basbous J, Hivin P, *et al.* (2004) HBZ interacts with JunD and stimulates its transcriptional activity. *FEBS Lett* **562**: 165–170.
27. Hivin P, Arpin-André C, Clerc I, *et al.* (2006) A modified version of a Fos-associated cluster in HBZ affects Jun transcriptional potency. *Nucl Acid Res* **34**: 2761–2772.
28. Kuhlmann AS, Villaudy J, Gazzolo L, *et al.* (2007) HTLV-1 HBZ cooperates with JunD to enhance transcription of the human telomerase reverse transcriptase gene (hTERT). *Retrovirology* **4**: 92.

29. Isono O, Ohshima T, Saeki Y, *et al.* (2008) Human T-cell leukemia virus type 1 HBZ protein bypasses the targeting function of ubiquitination. *J Biol Chem* **283**: 34273–34282.
30. Clerc I, Polakowski N, Andre-Arpin C, *et al.* (2008) An interaction between the human T cell leukemia virus type 1 basic leucine zipper factor (HBZ) and the KIX domain of p300/CBP contributes to the down-regulation of Tax-dependent viral transcription by HBZ. *J Biol Chem* **283**: 23903–23913.
31. Zhao T, Yasunaga JI, Satou Y, *et al.* (2008) Human T-cell leukemia virus type 1 bZIP factor selectively suppresses the classical pathway of NF-kappaB. *Blood* **113**: 2755–2764.
32. Arnold J, Zimmerman B, Li M, *et al.* (2008) Human T-cell leukemia virus type-1 antisense-encoded gene, HBZ, promotes T-lymphocyte proliferation. *Blood* **112**: 3788–3797.
33. Arnold J, Yamamoto B, Li M, *et al.* (2006) Enhancement of infectivity and persistence *in vivo* by HBZ, a natural antisense coded protein of HTLV-1. *Blood* **107**: 3976–3982.
34. Mori N, Fujii M, Iwai K, *et al.* (2000) Constitutive activation of transcription factor AP-1 in primary adult T-cell leukemia cells. *Blood* **95**: 3915–3921.
35. Miyazaki M, Yasunaga J, Taniguchi Y, *et al.* (2007) Preferential selection of human T-cell leukemia virus type 1 provirus lacking the 5′ long terminal repeat during oncogenesis. *J Virol* **81**: 5714–5723.
36. Taniguchi Y, Nosaka K, Yasunaga J, *et al.* (2005) Silencing of human T-cell leukemia virus type I gene transcription by epigenetic mechanisms. *Retrovirology* **2**: 64.
37. Matsuoka M, Jeang KT. (2007) Human T-cell leukaemia virus type 1 (HTLV-1) infectivity and cellular transformation. *Nat Rev Cancer* **7**: 270–280.
38. Mesnard JM, Barbeau B, Devaux C. (2006) HBZ, a new important player in the mystery of adult-T-cell leukemia. *Blood* **108**: 3979–3982.
39. Uchida N, Otsuka T, Arima F, *et al.* (1999) Correlation of telomerase activity with development and progression of adult T-cell leukemia. *Leuk Res* **23**: 311–316.
40. Switzer WM, Qari SH, Wolfe ND, *et al.* (2006) Ancient origin and molecular features of the novel human T-lymphotropic virus type 3 revealed by complete genome analysis. *J Virol* **80**: 7427–7438.

41. Calattini S, Chevalier SA, Duprez R, *et al.* (2006) Human T-cell lymphotropic virus type 3: Complete nucleotide sequence and characterization of the human Tax3 protein. *J Virol* **80**: 9876–9888.
42. Chevalier SA, Ko NL, Calattini S, *et al.* (2008) Construction and characterization of a human T-cell lymphotropic virus type 3 infectious molecular clone. *J Virol* **82**: 6747–6752.
43. Jones KS, Petrow-Sadowski C, Huang YK, *et al.* (2008) Cell-free HTLV-1 infects dendritic cells leading to transmission and transformation of CD4(+) T cells. *Nat Med* **14**: 429–436.

CHAPTER 5

Regulation of HTLV-1 Transcription by Viral and Cellular Proteins

Nicholas Polakowski*,† and Isabelle Lemasson*,‡

SUMMARY

Human T-cell leukemia virus type 1 (HTLV-1) is a complex retrovirus that primarily infects CD4 positive T cells and causes pathological effects in a small percentage of infected individuals. One of the major diseases caused by the virus is an untreatable and often fatal form of leukemia known as adult T-cell leukemia (ATL). Presentation of ATL frequently occurs following several decades of persistent viral infection. This phenomenon is believed to be, in part, derived from acute regulation of HTLV-1 transcription. A subset of viral proteins and numerous cellular factors contribute to this process from the level of recruitment of the RNA polymerase II core transcription machinery to modification of chromatin structure. Based on the increasing diversity of interactions between viral and cellular factors involved in regulating HTLV-1 transcription, it is becoming apparent that levels of transcription are dictated by a variety of

*East Carolina University, Brody School of Medicine, 600 Moye Boulevard, Greenville, NC, 27834, USA.

†polakowskin@ecu.edu.

‡lemassoni@ecu.edu.

protein complexes capable of associating with the viral promoter. Cellular events triggering the formation of distinct regulatory complexes on the viral promoter is an important subject of current and future research.

INTRODUCTION

Human T-cell leukemia virus type 1 (HTLV-1) is a complex retrovirus that primarily infects CD4 positive T cells. Approximately 20 million people worldwide are infected with the virus, and a small percentage of this population will develop pathologies associated with infection. HTLV-1 is the etiologic agent of adult T-cell leukemia or lymphoma (ATL)[1,2] and tropical spastic paraparesis/HTLV-1-associated myelopathy (TSP/HAM).[3,4] ATL is an aggressive form of cancer, while TSP/HAM is a neurodegenerative disorder caused by demyelination of nerve cells that is believed to derive from an autoimmune response activated by the virus. In addition to these diseases, strong evidence suggests that HTLV-1 infection can lead to uveitis, and weaker epidemiologic links exist for other autoimmune disorders and increased susceptibility to infectious disease.

Following infection, the HTLV-1 genome becomes integrated into a host-cell chromosome. Although initially believed to be random,[5] integration is now known to favor active regions of the genome located close to CpG islands or at transcriptional start sites.[6-8] However, integration into actual transcriptional units is low.[8] The proviral genome contains elements common to other retroviruses, including the *gag*, *pol*, and *env* genes and identical 5′ and 3′ flanking sequences called long terminal repeats (LTRs) [Fig. 1(a)]. An additional component of the provirus, known as the *pX* region, is located primarily between the *env* gene and the 3′ LTR. This region encodes proteins that are required for viral propagation as well as proteins with regulatory and accessory functions [Tax, Rex, p30, p12, p13 and p21; Fig. 1(a)]. The 5′ LTR regulates expression of all viral genes with the exception of the antisense protein, HTLV-1 bZIP factor (HBZ). Expression of this protein is regulated by sequences positioned in the 3′ LTR. (In this chapter, HTLV-1 transcription refers to transcription originating from the 5′ LTR unless otherwise specified.)

The LTRs are subdivided into three regions, U3 (353 bp), R (228 bp) and U5 (176 bp). As the start site of transcription is located at the U3/R junction, U3 harbors the major elements of the viral promoter. Accessory elements are located in R and U5, but are not discussed in this chapter. The fundamental regulatory elements in U3 consist of a core promoter, containing a consensus TATA box, and Tax-responsive elements 1 and 2 (TxRE1 and TxRE2) that are involved in transcriptional activation through the viral protein Tax.

TxRE1 is collectively three semi-conserved 21-bp enhancer elements that are also called viral cAMP response elements (vCREs). With respect to the transcription start site, these elements are centered at −242 (the distal vCRE), −193 (the central vCRE) and −94 (the proximal vCRE). From 5′ to 3′, each vCRE is subdivided into domains A, B and C [Fig. 1(b)]. The central B domain is identical to the first five nucleotides of the consensus cellular CRE (TGACGTCA) and is consequently responsible for recruitment of multiple ATF/CREB members to the HTLV-1 promoter (see below).

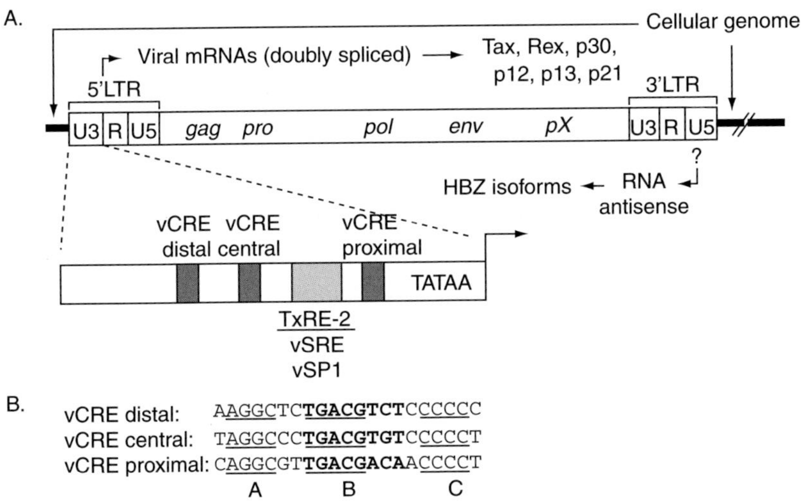

Fig. 1. Schematic representation of the integrated HTLV-1 provirus. (a) The 5′ LTR controls transcription of *gag*, *pol*, *env* and *pX* (encoding the proteins Tax, Rex, p30, p12 and p21) genes, while the 3′ LTR controls transcription of the antisense protein (HBZ). U3 is enlarged to show the main regulatory elements in the promoter. (b) The vCRE sequences are divided into domains A, B and C. Sequences identical to the first five nucleotides of the cellular CRE are underlined in domain B. Sequences important for Tax-binding are underlined in domains A and C.

This sequence is additionally recognized by several other factors.[9] Domains A and C respectively contain G- and C-rich sequences that facilitate binding of the viral transcription factor Tax to the vCREs. TxRE2 is centered at −140 between the proximal and central vCREs. Elements in TxRE2 modulate basal and Tax-dependent transcription.

In addition to having a fundamental role in the expression of viral proteins, HTLV-1 transcription is a mechanism for replication of the viral genome. Given these important functions in the viral life cycle, numerous cellular factors regulate basal HTLV-1 transcription and participate in transcriptional activation through Tax. Emerging evidence indicates that Tax-independent and -dependent transcription is not modulated through the formation of two unique complexes at the viral promoter, but rather involves effects of a variety of distinct complexes. This possibility is indicative of an adaptation that allows HTLV-1 to fine-tune levels of proviral transcription according to cellular metabolism and the environment of the infected T cell. Such a mechanism may mediate persistent and latent infection, allowing for long-term survival of the infected T cell in the peripheral blood. Finally, the site of integration may influence levels of transcription,[8] providing additional complexity to HTLV-1 transcriptional regulation.

HTLV-1 BASAL TRANSCRIPTION

HTLV-1 basal transcription is required for initial expression of the Tax protein, which is the viral transcription factor that mediates activation of HTLV-1 transcription. Tax has not been detected in infectious viral particles or as an HTLV-1 component transmitted through virological synapses by cell-to-cell contact.[10,11] Therefore, cellular and not viral factors initiate HTLV-1 basal transcription following integration or viral latency.

Regulation of Basal Transcription by AP-1

AP-1 transcription factors stimulate basal HTLV-1 transcription through interactions with core vCRE sequences, which are similar to the consensus AP-1 recognition sequence TGAGTCA[12,13] [Fig. 2(a)]. AP-1 is archetypically composed of a heterodimer of c-Jun and c-Fos, but also constitutes specific dimer combinations of other Jun and Fos members, including

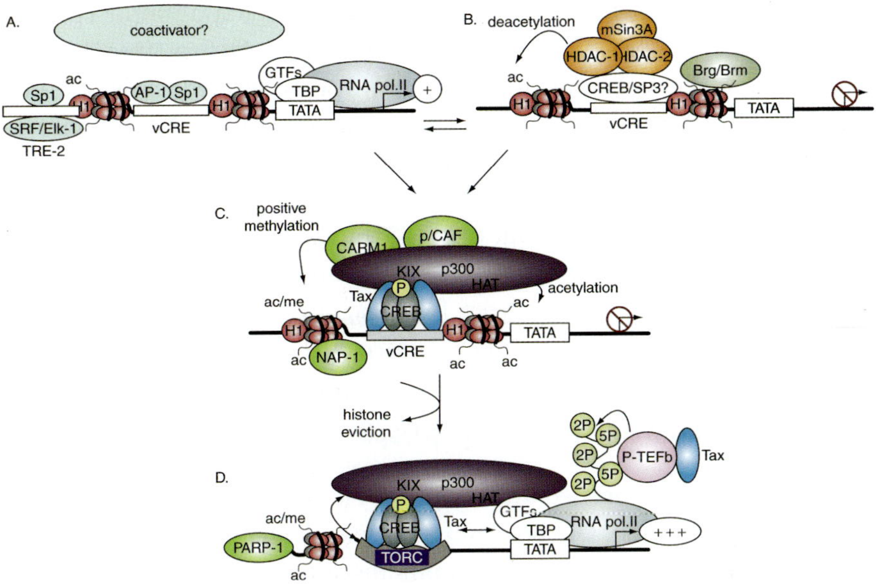

Fig. 2. Model of events leading to Tax-dependent transcription. Only the proximal vCRE and TxRE-2 are represented for simplicity. (a) Cellular factors regulate HTLV-1 basal transcription. (b) HTLV-1 transcription is silenced by repressor complexes bound to the viral promoter. (c) Binding of the Tax/CREB complex to the promoter leads to recruitment of p300/CBP and CARM-1 and removal of repressor complexes. p300/CBP and CARM-1 modify histones tails of promoter nucleosomes. (d) NAP-1 and RNA polymerase II remove nucleosomes from the promoter and downstream DNA. Tax binds to P-TEFb and stimulates the phosphorylation of serine 2, activating transcriptional elongation. Some proteins in the schematic are shown disconnected from Tax even though they interact with Tax. Double-headed arrows indicate interactions.

c-Jun, JunB, JunD, c-Fos, FosB, Fra-1 and Fra-2. Each of these proteins contains a basic leucine zipper (bZIP) domain.[14] The leucine zipper is composed of clusters of three to five heptad repeats. Leucines at the first position of each repeat form hydrophobic interactions with corresponding leucines on the partnering bZIP domain, stabilizing the interaction. Charged residues at the fifth and seventh positions in each repeat mediate specificity of the interaction between bZIP domains. Interacting leucine zippers form a coiled-coil structure that allows the adjacent basic regions to bind stably in the major groove of the DNA. This mechanism of DNA-binding is also important for activation of HTLV-1 transcription by Tax through ATF/CREB bZIP factors (see below). Interestingly, HTLV-1 transformed T cells express elevated levels of c-Jun, JunB, JunD, c-Fos and Fra-1 mRNA, and concomitantly exhibit high constitutive AP-1 activity.[15–18] The ATF/CREB factor, ATF-2, also serves as an AP-1 subunit.[19] Despite its lack of involvement in Tax-dependent activation of HTLV-1 transcription,[20] ATF-2 binds efficiently to the vCRE sequence *in vitro*.[21] Using chromatin immunoprecipitation (ChIP) assays, c-Fos, c-Jun and ATF-2 have been found at the viral promoter within a heterogeneous population of HTLV-1 infected T cells.[22]

Regulation of Basal Transcription by Sp1/Sp3

The ubiquitously expressed transcription factors, Sp1 and Sp3, regulate transcription through GC-box and GT-box elements that, respectively, represent their high- and low-affinity binding sites.[23] The HTLV-1 promoter contains two Sp1 binding sites: one located within the proximal vCRE and the other, designated vSp1, located within TxRE-2 [Fig. 1(a)].[24–27] Although Sp1 exhibits greater affinity for the vSp1 element, mutation of either site elicits a dramatic reduction in transcription, suggesting that both sites cooperate in regulating HTLV-1 transcription[26,28,29] [Fig. 2(a)]. The role of Sp1 in basal transcription is supported by results from ChIP and transfection assays, demonstrating that Sp1 is bound to chromosomally integrated HTLV-1 promoters in uninfected cells, and over-expression of Sp1 increases HTLV-1 transcription in the absence of Tax.[28,29] Mechanistically, Sp1 competes with CREB for binding to the proximal vCRE in the absence of Tax.[26,29] Surprisingly, in the presence of Tax, Sp1 is associated with a complex consisting of Tax, CREB and the proximal vCRE (described below) *in vitro* and slightly enhances Tax-dependent transcription.[29] *In vivo* formation of this complex is supported by ChIP data, demonstrating that the level of Sp1 bound to chromosomally integrated HTLV-1 promoters in uninfected cells is not altered in response to expression of Tax.[28] Interestingly, Sp3 is also found at the viral promoter in HTLV-1-infected T cells.[29] In contrast to Sp1, Sp3 represses basal and Tax-dependent transcription.[29] This effect is not surprising considering that Sp3 represses Sp1-mediated transactivation from promoters containing two or more Sp1 binding sites.[23]

Regulation of Basal Transcription by SRF/ELK-1

SRF controls transcription of "immediate early genes" in response to growth factor or mitotic stimuli and was recently shown to participate in HTLV-1 basal transcription [Fig. 2(a)]. This protein belongs to a family of transcription factors characterized by the presence of a specific DNA-binding/dimerization domain known as a MADS box.[30] The MADS box permits SRF to form homodimers which, in turn, associate with DNA through consensus elements called CArG boxes. This domain additionally mediates interactions with cofactors, such as ternary complex transcription factors (TCFs). The Ets domain transcription factor, Elk-1, is a TCF member that serves as an important cofactor for SRF. Indeed, serum response elements (SREs) in cellular promoters are defined by the presence of a CArG box and one or more adjacent Ets sites.[31] Interestingly, the HTLV-1 promoter contains a "viral" SRE (vSRE) that extends from −116 to −157 and, therefore, overlaps TxRE-2.[32] Within the vSRE, a sequence resembling the prototypical CArG box found in the c-fos promoter is centered at approximately −120.[33] Two upstream Ets sites complete the vSRE. While SRF and Elk-1 bind the vSRE independently *in vitro*, together they form a ternary complex with this element. In correlation with these properties, each factor augments basal transcription, and the combination of both factors produces an additive transcriptional effect. Results from ChIP assays confirm that both SRF and Elk-1 associate with the viral promoter in HTLV-1-infected cells.[32,33] Interestingly, Tax is known to bind to SRF and Elk-1, and through these interactions, Tax increases cellular gene transcription.[34,35] However, a role for the SRF/Elk-1 complex in Tax-dependent HTLV-1 transcription has not been reported.

ACTIVATION OF HTLV-1 TRANSCRIPTION BY THE VIRAL PROTEIN TAX

Once sufficient levels of Tax are obtained through HTLV-1 basal transcription, the viral transcription factor mediates formation of promoter complexes that activate HTLV-1 transcription. This process involves interactions between Tax and numerous cellular factors. Because many of these interactions involve overlapping regions of Tax,[36] it is likely that a variety of distinct Tax-containing complexes with diverse transcriptional regulator properties separately dictate levels of activated HTLV-1 transcription. The general events of Tax-dependent transcription are represented in Figs. 2(c) and 2(d).

Tax Forms a Complex with ATF/CREB Members on the vCREs

Activation of HTLV-1 transcription by Tax originates from an interaction between the viral protein and specific cellular bZIP transcription factors in the ATF/CREB family. Tax associates directly with CREB (CREB-1), CREM isoforms, CREB-2 (ATF4, TAXREB67) and ATF-1 (reviewed in Ref. 37), and many of these factors, in addition to Tax, are present at the viral promoter in HTLV-1-infected T-cell lines.[22] Additional ATF/CREB members have recently been added to this pool, including X-box binding protein 1 (XBP-1) and LZIP.[38,39] However, the interaction between Tax and CREB is currently believed to be central to transcriptional activation. This premise is supported by biochemical evidence indicating that Tax selectively associates with the vCRE as a complex with CREB.[20,40] Furthermore, results from reporter assays have shown that over-expression of a dominant negative mutant of CREB severely compromises Tax-dependent transcriptional activation from the LTR.[38]

Tax Binds the A and C Domains of the vCRE

The Tax/CREB complex appears to function as a single unit to confer its stable association with the HTLV-1 promoter and to activate transcription. As a homodimer, Tax contacts the basic region of CREB in solution, which serves to enhance CREB dimerization and, therefore, CREB DNA-binding activity.[41–43] Formation of Tax/CREB/vCRE complexes involves binding of CREB to the core cellular CRE-like sequence in the major groove of the DNA and contacts between Tax and the flanking 5′ G- and 3′ C-rich sequences in the minor groove of the DNA.[44–48] In the absence of Tax, the CREB/vCRE complex is highly unstable,[20,49] and very little CREB is detected at the LTR in cells lacking Tax.[50,51] Conversely, Tax alone is unable to bind the viral CRE.

Tax and CREB Recruit the Coactivators p300 and CBP to the HTLV-1 Promoter

Once bound to the HTLV-1 promoter, the Tax/CREB complex recruits the cellular coactivator p300 or its paralog CBP (p300/CBP). These proteins are intimately involved in regulation of gene expression (reviewed in Ref. 52). Owing to their ability to interact with more than 200 transcriptional regulators, p300 and CBP are recruited to numerous cellular promoters where they facilitate formation of preinitiation complexes. This effect involves their coactivator function, which is defined by their ability to bridge DNA-bound activators with the core transcription machinery, and to act as scaffolds for recruitment of other transcriptional regulators to these promoters. In addition, p300 and CBP are histone acetyltransferases (HATs) that target histones and transcription factors for acetylation (see below).

Initial evidence indicated that p300 and CBP were brought to the HTLV-1 promoter solely through direct interactions with Tax. Tax was found to contact the coactivator C/H1, KIX and CR2 domains, with each interaction contributing to viral transcription.[53–58] However, binding of Tax to the KIX domain appeared singularly critical for recruitment of p300/CBP to the HTLV-1 promoter.[56,59] This model diverged from the mechanism of coactivator-recruitment by CREB/cellular CRE complexes in which phosphorylation of CREB at S133 (pCREB) within its kinase inducible domain (KID $\rightarrow$ pKID) leads to direct binding of pKID to the KIX domain. Indeed, a vCRE-bound complex composed of Tax and solely the bZIP domain of CREB, which lacks KID, was sufficient for integration of the KIX domain into the complex.[53,54] Substituting the bZIP domain of CREB with that of ATF-1, CREM or CREB-2 produced similar quaternary complexes.[56,60,61] Additionally, Tax-dependent activation of transcription from the HTLV-1 promoter was achieved by over-expression of solely the bZIP domain of CREB or CREM in cells defective for transcriptional activation through CREB.[54,60] Together, these data support a model in which Tax circumvented the normal requirement for CREB phosphorylation for promoter-recruitment of p300/CBP.

Despite these findings, recent biochemical studies show that phosphorylation of CREB stabilizes the quaternary complex, which is, in part, attributed to the ability of the KIX domain to accommodate both Tax and pCREB simultaneously. Structurally, the KIX domain constitutes a hydrophobic groove with two distinct transcription factor-binding surfaces, each of which is differentially contacted by Tax and pCREB.[62] Cooperative binding of Tax and pCREB to the KIX domain dramatically enhances recruitment of p300 and CBP to vCRE sequences.[63] Interestingly, phosphorylation of CREB also stabilizes the association of Tax with the quaternary complex, which is believed to involve the formation of multiple protein–protein interactions in the complex.[64] In correlation with these effects, Tax stimulates CREB phosphorylation at S133 in HEK293, HeLa and murine fibroblast cells, and elevated pCREB levels are detected in HTLV-1-infected T-cell lines compared to uninfected T-cell lines.[64–66]

At the HTLV-1 promoter, p300 and CBP elicit multiple activities that contribute to activation of viral transcription. In *in vitro* transcription assays using "naked" DNA templates (DNA not assembled into nucleosomes), CBP, but not p300, functions as a coactivator for HTLV-1 transcription.[67] This activity is attributed to the N-terminal region of the protein that is involved in interactions with the core RNA polymerase II transcription machinery and contains the Tax-targeted C/H1 and KIX domains.[68] The HAT activity of both proteins is important in the context of the native environment of an infected T-cell nucleus, in which the provirus adopts a chromatin structure. The fundamental unit of chromatin, the nucleosome, is composed of DNA coiled around a protein octamer with two of each of the four core histones: H2A, H2B, H3 and H4.[69] *In vitro* assembly of naked DNA templates into nucleosomes represses HTLV-1 transcription.[67,70] This effect is countered by p300- or CBP-mediated acetylation of lysine residues on nucleosomal histones. Target lysine residues are concentrated in the N-terminal regions of the histones that protrude from the nucleosome and are referred to as histone tails. In the context of *in vitro* transcriptional activation from the vCREs, acetylation of the histone tails "opens" the chromatin structure, which allows the core RNA polymerase II transcription machinery access to the DNA.[67,70] Histone tails that are acetylated and/or contain other post-translational modifications also serve as docking sites for specific transcriptional regulators. This phenomenon, known as the histone code,[71] has not been evaluated in the context of HTLV-1 transcription. Interestingly, histone H1 is associated with the viral promoter *in vivo*.[50] This protein is a physiological component of chromatin that contributes to transcriptional repression by facilitating chromatin compaction. Biochemical evidence indicates that histone H1-containing chromatin restricts the ability of p300 to acetylate core histones.[72] However, at the HTLV-1 promoter, Tax prevents this effect.[72] Finally, additional results from *in vitro* transcription assays using chromatin templates indicate that p300 targets a non-histone protein for acetylation, and this effect is important for Tax-dependent transcriptional activation.[59] The identity of this substrate remains unknown.

The Role of TORC/CRTC Coactivators in Tax-Dependent HTLV-1 Transcription

In 2003, a new CREB coactivator family was discovered, called transducer of regulated CREB activity (TORC) or CREB regulated transcription coactivator (CRTC).[73,74] There are currently three known members of the TORC family: TORC1, TORC2 and TORC3. TORCs are expressed at low levels in all tissues. However, TORC1 is particularly abundant in certain regions of the brain, and TORC2 and TORC3 are predominantly found in T and B lymphocytes.[73] Although these proteins do not share extensive homology, they each harbor a conserved N-terminal coiled-coil domain that is believed to promote their formation into tetramers.[73] This domain additionally contacts CREB through its bZIP domain, which specifically requires arginine 314 found in CREB and related proteins.[73] As a subset of cellular promoters regulated by CREB, TORCs facilitate transcriptional activation by enhancing the interaction between CREB and the TAF(II)130 subunit of TFIID.[73] Although this mode of activation is frequently independent of CREB S133 phosphorylation,[73] results from recent studies demonstrate that TORC2 and p300/CBP cooperate in activating transcription from certain target promoters.[75,76] In this context, cAMP signaling is required, and recruitment of p300/CBP to these promoters occurs through a direct interaction between the CREB-associated TORC and the coactivators.[75,76]

Results from luciferase reporter assays have shown that all three members of the TORC family are involved in Tax-dependent HTLV-1 transcription.[77,78] As expected, TORC-responsive sequences localize to the vCREs. Like CREB, Tax interacts directly with TORC coiled-coil domains, suggesting that interactions of both Tax and CREB with this domain are involved in full transcriptional activation through these coactivators.[77,78] Interestingly, combined over-expression of p300 and individual TORCs produces an additive effect on HTLV-1 transcription in luciferase reporter assays.[78] Therefore, Tax potentially utilizes multiple mechanisms to activate and modulate transcription from the HTLV-1 promoter that involve the recruitment of either p300/CBP via the KIX domain, TORCs via the coiled-coil domain, or both coactivators. It is also possible that TORCs directly facilitate recruitment of p300/CBP to the viral promoter.[75,76] The environment and state of the HTLV-1-infected T cell are likely to dictate which mechanism of Tax-dependent HTLV-1 transcription is utilized.

For TORCs to function at the viral promoter they must first be localized to the nucleus. While TORC3 appears to be constitutively nuclear,[79] a key mode of TORC1 and TORC2 regulation involves their cytoplasmic-nuclear shuttling.[79,80] These proteins are phosphorylated at serine 171 by salt inducible kinase 2 (SIK2) and other members of the AMP kinase family,[79,81,82] resulting in their sequestration in the cytoplasm in complex with 14-3-3.[79] Certain signal transduction pathways promote dephosphorylation or altered phosphorylation of these cofactors, allowing them to translocate to the nucleus. The effect of Tax on the cellular localization of TORCs has not been addressed. It is possible that one or more pathways that promote TORC1- and TORC2-entry into the nucleus are activated by Tax. TORC1 and TORC2 translocate to the nucleus in response to signaling cascades induced by calcium,[80] cAMP[79–81] or activation of mitogen-activated/extracellular signal-regulated kinase 1 (MEKK1).[83] Interestingly, Tax is known to bind to, and activate, MEKK1,[84] and Tax and p12 (also encoded in the *pX* region of HTLV-1) augment calcium-induced signaling.[85,86] In addition, Tax stimulates CREB phosphorylation, mimicking effects of cAMP.[64–66] It is of interest to test whether these effects of Tax ultimately enhance HTLV-1 transcription through the functions of TORCs.

Tax Recruits the Coactivator P/CAF to the HTLV-1 Promoter

The p300/CBP associated factor (P/CAF) is an additional coactivator that plays a role in Tax-dependent transcription. P/CAF is part of several multi-protein complexes with co-activator functions and HAT activity.[87] P/CAF itself contains a HAT domain, and this enzymatic activity is essential for its role in activating transcription from cellular promoters.[87] P/CAF interacts directly with the C-terminal region of Tax (aa 245–353) and, consequently, associates with the Tax/CREB/vCRE complex and the quaternary complex containing p300 *in vitro*.[88,89] However, in HTLV-1-infected cells, P/CAF is not detected at the viral promoter by ChIP analysis.[22] This discrepancy may lie in the inability of ChIP antibodies to bind P/CAF when it is integrated into a protein complex on the viral promoter. In reporter assays, P/CAF stimulates Tax-dependent transcription from the HTLV-1 promoter, and the Tax mutant, M47 (L319R, L320S0), which is defective for HTLV-1 transcriptional activation and is unable to interact with this coactivator.[88,89] These results indicate that P/CAF plays an important role in Tax-dependent activation of HTLV-1 transcription. Despite containing a HAT domain, P/CAF activates HTLV-1 transcription in a HAT-independent manner, as determined using P/CAF mutants defective for HAT activity.[89,90] This observation is intriguing considering recent evidence that P/CAF mutants defective for HAT activity are sequestered in the cytoplasm.[91] Given these findings, it is possible that Tax is involved in translocating P/CAF into the nucleus, or alternatively, regulation of HTLV-1 transcription by P/CAF may occur through an indirect mechanism. Further work will provide an answer to this important question.

The Role of Chromatin Remodeling Complexes at HTLV-1 Promoter

Tax binds directly to components of ATP-dependent chromatin remodeling complex, switching defective/sucrose non-fermenting (SWI/SNF).

Through ATP hydrolysis, SWI/SNF remodeling activity alters DNA–histone interactions to mobilize nucleosomes.[92] Such events provide both positive and negative effects on expression of cellular genes. Human cells contain two distinct SWI/SNF complexes, designated BAF and PBAF. For BAF, ATPase activity is derived from BRG1 or BRM, while for PBAF, BRG1 exclusively serves as the catalytic subunit. In addition to these core components, the complexes also contain BRG1/BRM associated factors (BAFs), some of which are required for catalytic activity of the complexes, and others of which mediate recruitment of SWI/SNF to specific loci.[92] Tax has been found to bind to BRG1 and BAF-53, -57 and -155.[93]

The role of BRG1 in Tax-dependent HTLV-1 transcription remains controversial.[93] Results from the initial characterization of the Tax–BRG1 interaction indicated that BRG1 is essential for transcriptional activation by Tax. In lieu of these findings, a more recent study concluded that BRG1 is dispensable for Tax-dependent HTLV-1 transcription.[94] In the absence of Tax, BRG1 and BRM are associated with the viral promoter in uninfected cells carrying chromosomally integrated copies of an HTLV-1 LTR-*luciferase* gene construct.[50] However, formation of a transcriptionally competent Tax complex at these integrated sites leads to displacement of BRG1 and BRM from the DNA. These results suggest that the function of BRG1 and BRM in HTLV-1 transcription corresponds to maintenance of the HTLV-1 provirus in an assembled chromatin state in the absence of Tax.

Nucleosome-Displacement from the HTLV-1 Promoter

Tax-dependent activation of HTLV-1 transcription is associated with a reduction in the concentration of nucleosomes and linker histone H1 at the viral promoter and within the transcribed region.[50] Use of the transcription-defective Tax mutant, M47, initially correlated this effect with recruitment of RNA polymerase II to the promoter. Similar to wild-type Tax, M47 bound to the viral promoter concomitantly with CREB and p300 in uninfected cells carrying chromosomally integrated copies of an HTLV-1 LTR-*luciferase* gene. However, unlike wild-type Tax, M47 was unable to promote nucleosome eviction and failed to deliver the polymerase to the promoter.[50] Results from a recent study demonstrate that the histone chaperone nucleosome assembly protein 1 (NAP-1) and p300 can induce nucleosome eviction from the LTR *in vitro*.[95] This effect requires acetyl CoA, but is independent of ATP hydrolysis. Although NAP-1 is known to be involved in chromatin assembly, a recent study demonstrates that, in combination with the ATP-dependent chromatin remodeling factors, the CHD remodelers, NAP-1 stimulates nucleosome disassembly.[96] It is of interest to determine whether these proteins are involved in nucleosome displacement from the provirus in HTLV-1-infected T cells.

Regulation of Tax-Dependent Transcription by PARP-1

A role for the poly(ADP-ribose) polymerase-1 (PARP-1) in the regulation of HTLV-1 transcription has been reported.[97] PARP-1 is an abundant and ubiquitous nuclear enzyme that uses NAD^+ to ADP-ribosylate proteins. This protein is involved in DNA repair, transcriptional regulation, energy metabolism and apoptosis.[98] The role of PARP-1 in HTLV-1 transcription was initially characterized using partially purified protein fractions to reconstitute Tax-dependent transcription *in vitro*.[97] In this system, PARP-1 proved essential for transcriptional activation though Tax, an effect that was independent of ADP-ribosylase activity. These results were corroborated by studies in which Tax failed to activate transcription from the viral promoter in mouse embryonic fibroblasts deficient for PARP expression. Importantly, restoration of PARP-1 expression in these cells restored the transactivation function of Tax.

The function of PARP-1 in HTLV-1 transcription is not fully defined. In biochemical assays, PARP-1 alone binds the vCRE sequence, and in combination with Tax and CREB, it forms a complex on this element.[99] Given that naked templates were used for *in vitro* transcription, these results suggest that PARP-1 functions as a coactivator. However, it is possible that PARP-1 additionally contributes to HTLV-1 transcription in a chromatin context. Indeed, PARP-1 complexes associate with nucleosomes and displace histone H1 from chromatin.[98] It is possible that these functions underlie the loss of histone H1 from the viral promoter during Tax-dependent transcriptional activation.

Tax Recruits Methyltransferases to the Viral Promoter

In addition to acetylation, methylation of histone H3 lysine and arginine residues is detected at the HTLV-1 promoter. Methylation of histones is associated with the activation of gene expression or, alternatively, the repression and silencing of gene expression, depending on which amino acids are targeted for modification. Di- and tri-methylation of H3 K4 as well as arginine methylation on H3 are generally associated with active transcription.[100] In HTLV-1 infected T cells, H3 K4 di- and tri-methylation are detected at the viral promoter and throughout the provirus.[22] These histone marks are recognized by factors with diverse functions, such as chromatin remodeling, RNA splicing, DNA repair and gene activation.[100] Given the high level of HTLV-1 transcription in infected cells, H3 K4 methylation may facilitate recruitment of cellular factors involved in gene activation to the viral promoter. The methyltransferase targeting H3 K4 at the HTLV-1 promoter has not been identified, but it is well characterized that MLL, Set1, Set7/Set9 and SMYD3 perform this enzymatic modification.[100]

In addition to lysine modifications, results from ChIP assays revealed methylation of histone H3 arginine residues (R2, R17 and R26) at the HTLV-1 promoter in infected cells.[101] Coactivator-associated arginine methyltransferase (CARM1, also called PRMT4) was identified as the methyltranferase catalyzing these modifications and was found to be recruited to the viral promoter via a direct interaction with Tax. As histone H3 methylation by CARM1 (notably at R27) is correlated with gene activation, over-expression of catalytically active CARM1 increases Tax-dependent transcription.[101] The inverse effect occurs in response to siRNA-mediated reduction of CARM1 or inhibition of methyltransferase activity by adenosine-2,3-dialdehyde.[101,102] At the estrogen responsive promoter pS2, CARM1 methylates H3 R17 on histones pre-acetylated by CBP at H3 K18.[103] It would be informative to determine whether such crosstalk occurs at the viral promoter and whether there is an ordered recruitment of these enzymes to the promoter. Importantly, CARM1 also modulates activity of p300 and CBP through direct methylation of the coactivators.[104–106] Whether such events are involved in regulating HTLV-1 transcription remains to be explored.

Post-Translational Modification of Tax Modulates Transcription

Tax is known to be phosphorylated, ubiquitinated, sumoylated and acetylated. Phosphorylation of Tax at serines 300 or 301 is required for Tax translocation to "Tax speckled structures" in the nucleus. These regions contain SC35, a splicing factor, and are hot spots for nascent RNA synthesis.[107] Within Tax speckled structures, the viral protein co-localizes with CREB, RNA polymerase II, TBP and p300/CBP.[107,108] Mutation of serine 300 or 301 results in a diffuse pattern of Tax in the nucleus and abolishes Tax-dependent activation of transcription from the HTLV-1 promoter.[109] Additional phosphorylation at threonine 48, 184, and 215 and serine 336 are also detected on Tax. While mutation of threonine 184 or serine 336 does not influence activation of HTLV-1 transcription by Tax, phosphorylation of threonine 48 and 215 negatively regulate this function.[110]

Tax is additionally ubiquitinated, sumoylated and acetylated at specific lysine residues. Ubiquitination is detected on lysines 189, 197, 263, 280 and 284,[111,112] and sumoylation is detected on lysines 280 and 284.[112,113] It is unclear whether ubiquitination of Tax affects its ability to activate HTLV-1 transcription. This modification has been reported to inhibit HTLV-1 transcription or to have no effect on the transactivation function of Tax.[111,112,114] Sumoylation of Tax does not affect the ability of the viral protein to activate HTLV-1 transcription.[112] Finally, acetylation of Tax occurs on lysine 346, but requires prior phosphorylation of serine 300 or 302. Surprisingly, mutation of this residue does not affect Tax activity.[115]

Tax Interacts with General Transcription Factors

In addition to integrating cellular activator and coactivator interactions at the viral promoter, Tax interacts with specific core transcription factors. These include TFIIA,[116] TBP[117] and TAFII28.[118] It is likely that contacts with these and other core factors by Tax, CREB, p300 and other regulators contribute to synergistic activation of HTLV-1 transcription.

Tax Recruits the Elongation Factor, P-TEFb, to the Viral Promoter

Results from recent studies extend the role of Tax beyond that of mediating formation of the preinitiation complex. As the HTLV-1 core promoter contains a consensus TATA box, the final stages of preinitiation involve the binding of TBP to this element, which is generally the limiting step in initiation of transcription. In the normal context of HTLV-1 transcription, complexes containing Tax and a variety of cellular factors recruit TBP to the promoter. Interestingly, Tax is able to enhance transcription from synthetic promoters to which TBP is artificially tethered, signifying a possible role for the viral protein in the elongation step of transcription.[38]

Positive transcription elongation factor (P-TEFb) is implicated in post-transcription initiation effects of Tax. P-TEFb stabilizes the elongating RNA polymerase II complex, which is necessary for productive transcriptional elongation.[119] Consistent with Cdk9 and cyclin T1 constituting the fundamental subunits of P-TEFb, elongation effects involve Cdk9-mediated phosphorylation of negative elongation factors and, importantly, phosphorylation of Ser2 in the heptad repeat of the RNA polymerase II C-terminal domain (CTD). P-TEFb is found in two complexes in cells, an inactive complex containing the inhibitory subunits HEXIM1 and 7SK snRNA, and an active complex containing Brd4. Brd4 interacts directly with cyclin T1 and, through its bromodomain, recruits P-TEFb to promoters containing acetylated histones. In the context of HTLV-1 transcription, P-TEFb is present at the viral promoter and within the transcribed portion of the HTLV-1 genome *in vivo.*[120] A functional role for this factor in Tax-dependent HTLV-1 transcription was demonstrated by observations that both the inhibition of Cdk9 kinase activity or siRNA-mediated knock-down of Cdk9 repress transcription.[120] Results from biochemical studies indicated that Tax recruited P-TEFb to the viral promoter. Through this interaction, Tax was found to stimulate autophosphorylation of Cdk9, which represses its kinase activity. Consequently, Tax inhibited CTD Ser2 phosphoylation by Cdk9. A model was proposed in which Tax positioned P-TEFb on the HTLV-1 promoter in an inactive state in order to prevent anomalous firing of transcription prior to stable formation of the preinitiation complex. Subsequent uncharacterized events were hypothesized to mediate dephosphorylation of Cdk9, leading to activated transcriptional elongation.[120] Results from a more recent study demonstrated that recruitment of P-TEFb to the viral promoter occurs through a direct interaction between Tax and cyclin T1. Tax was found to selectively target the region of cyclin T1 contacted by Brd4 and, therefore, compete with Brd4 for cyclin-binding. In contrast to results from the initial biochemical studies, Tax increased CTD Ser2 phosphoylation *in vivo*.[121] Future work should clarify the effects of Tax on P-TEFb activity at the viral promoter.

Transcription Originating from the 3′ LTR

Analysis of retroviruses other than HTLV-1 have implied that transcription from the 3′ LTR is blocked when the 5′ LTR is transcriptionally active.[122] One mechanism of silencing is believed to be caused by the elongating RNA polymerase traversing the 3′ U3 region (transcriptional overlap interference[122]). Another mechanism is proposed, involving formation of a transcriptionally permissive chromatin state at the 5′ LTR and inducing a repressive chromatin structure at a 3′ LTR.[123] The discovery of the antisense RNA of HTLV-1[124] and the existence of the protein encoded by this RNA, HBZ,[125] demonstrated that transcription is initiated from the 3′ LTR and extends toward the *pX* region. Additional studies showed that transcription from the 3′ LTR is bidirectional, and RNA synthesis also extends from 3′ LTR into the 3′ flanking cellular DNA.[126] Consistent with these findings, Tax, CREB, p300, CBP, RNA polymerase II and TAFII250 are associated with the 3′ LTR in HTLV-1 infected T cells. Furthermore, histones within this region carry marks for active transcription, such as histone H3 and H4 acetylation and histone H3 K4 di- and tri-methylation. In correlation with the binding of factors to the 3′ LTR, Tax and CREB enhance expression of HBZ through the vCREs.[127,128] HBZ expression is also modulated by three Sp1 binding sites within the U5/R region.[127] Finally, recent evidence suggests that the integration site of the provirus influences the level of HBZ expression,[128] suggesting that the *HBZ* gene promoter is expanded to include regulatory elements in the cellular sequence flanking the 3′ LTR.

The HDAC-3-containing NCoR complex appears to preferentially regulate transcription from the 3′ LTR.[126] Results from ChIP assays demonstrate that, in comparison to the 5′ LTR, the 3′ LTR exhibits selective enrichment of both HDAC-3 and NCoR. Treatment with HDAC inhibitors augments transcription from the 3′ LTR, indicating that the NCoR complex modulates levels of transcription originating from this region. It is intriguing that different HDAC complexes are associated with the 5′and 3′ LTRs. One explanation for this phenomenon may involve recruitment of these complexes by factors bound to unique regions of the proviral genome, which are located outside the LTRs.

REPRESSION OF HTLV-1 TRANSCRIPTION

HTLV-1 infection, which was thought to be latent, is in fact a persistent active infection.[129] The viral protein Tax is highly immunogenic. Cytotoxic T-lymphocytes directed against Tax play a critical role in limiting proviral load. Several regulatory mechanisms are established to restrict HTLV-1 transcription and, therefore, expression of Tax. Additional post-transcriptional repression events mediated by the viral proteins p30 and Rex are reviewed in Ref. 130.

Repression of HTLV-1 Transcription by Methylation of the Viral Promoter

DNA methylation is an epigenetic mechanism of transcriptional repression which involves modification of the cytosine base in CpG dinucleotide sequences that, when present at high frequencies on the DNA, constitute CpG islands. DNA methylation is catalyzed by a variety of DNA dimethyltransferases.[131] The HTLV-1 provirus contains multiple CpG islands that are located in regions corresponding to 5′-LTR-*gag* (base pairs 1–1,360), *pol* (base pairs 3,876–4,509), *env* (base pairs 5,648–6,166), *env-pX* (base pairs 6,446–7,561), and *pX*-3′-LTR (base pairs 8,212–9,045).[132] A correlation between DNA methylation and repression of HTLV-1 transcription was first defined through artificial methylation of the viral promoter.[133] It has since been shown that the 5′ LTR in ATL cells is frequently hypermethylated, and the 3′ LTR is hypomethylated.[134] These findings are consistent with the fact that expression of Tax is often repressed in these cells, while in cells from HTLV-1 carriers, in which Tax expression is more prevalent, the 5′ LTR is less methylated.[132] In support of this observation, results from a recent study analyzing ATL cells found that complete methylation of the 5′ LTR is required to fully silence viral transcription.[135] Interestingly, methylation of the provirus initiates in internal sequences such as *gag*, *pol* and *env* and, over time, propagates to the 5′ LTR and *pX* region,[132] implying that the level of HTLV-1 transcription undergoes a gradual reduction during infection. Despite the repressive role of DNA methylation, additional mechanisms serve to inhibit HTLV-1 transcription. Indeed, ATL cells have been identified in which the *tax* gene is repressed but the viral promoter is only partially methylated.[132,135]

The exclusive link between DNA methylation and repression of HTLV-1 transcription was challenged by Ego *et al.*, who showed that Tax is capable of activating transcription from a heavily methylated viral promoter.[136] Interestingly, this process involved an interaction between Tax and methyl-CpG-binding domain protein 2 (MDB2) rather than CREB. MDB2 normally occupies repressor complexes, but it is also known to activate transcription.[137] When associated with the 5′ LTR, MDB2 recruits Tax, which binds directly to the C-terminal coiled-coil domain of MDB2. Transcriptional activation through the MDB2/Tax complex requires both proteins and is associated with acetylation of histone H3. The interaction between MDB2 and Tax appears to be specific, as another methyl-CpG binding protein MeCP2 does not interact with Tax. The authors of this study suggest that, in the context of a provirus harboring a heavily methylated 5′ LTR, Tax is expressed from an internal promoter located within the *env* gene.[138]

Repression of HTLV-1 Transcription by HDAC Complexes

Histone deacetylases (HDACs) are a family of enzymes that remove acetyl groups from modified lysine residues on histones, transcription factors, and other transcriptional regulatory proteins.[139] Deacetylation of histone tails is associated with chromatin condensation and repression of gene expression. Mammalian HDACs are grouped into four classes (I–IV) according to sequence homology with their yeast counterparts. Members of class I HDACs, including HDAC-1, -2, and -3, contribute to repression of HTLV-1 transcription. HDAC-1 and -2 are highly related proteins that interact to form the catalytic component of three multi-protein repressor complexes, designated Sin3, Mi-2/NuRD and CoREST. HDAC-3 represents the catalytic unit of the NCoR/SMRT complex.[139]

The role of repressor complexes at the HTLV-1 promoter was brought to light by results from luciferase reporter assays showing that overexpression of HDAC-1 represses HTLV-1 transcription.[140] This finding was confirmed and extended by experiments using HTLV-1-infected T cells. Treatment of these cells with HDAC inhibitors elevated their viral RNA levels and produced parallel increases in histone H4 acetylation at the viral promoter.[22] These effects were correlated with the presence of HDAC-1, -2 and -3 at the proviral LTRs.[22] Interestingly, additional results from ChIP assays showed that the Sin3 complex preferentially regulates the 5′ LTR, while the NCoR complex is associated with the 3′ LTR.[126] Given the high expression of Tax in the HTLV-1-infected T cells tested, levels of HDAC-1 and 2 at the 5′ LTR are significantly lower than at the *c-myb* promoter, which is repressed in these cells.[126] Indeed, binding of Tax to the viral promoter is mutually exclusive from binding of HDACs,[51,126] suggesting that a dynamic competition exists between transcriptional activators and repressors for occupancy of the promoter [Fig. 2(b)].

Many of the mechanistic details regarding exchange of HDAC complexes with Tax-containing complexes at the viral promoter remain elusive. Interestingly, Tax binds directly to HDAC-1, -2 and -3.[51,140,141] For HDAC-1, this interaction is mediated through the N-terminal domain of the enzyme.[140] It is believed that, during activation of HTLV-1 transcription, a subset of Tax displaces HDAC complexes from the viral promoter, while a separate subset of Tax is responsible for formation of the preinitiation complex on the promoter. Because HDAC complexes do not bind to DNA, specific promoter-bound factors anchor these complexes to the DNA. CREB was proposed to fulfill this role based on the requirement of the vCREs in HDAC-1-mediated repression and on results showing low-level binding of CREB to the viral promoter in cells lacking Tax.[51,126] It is possible, however, that a factor other than CREB anchors HDAC complexes to the vCREs. Indeed, the CREB/vCRE complex is biochemically unstable,[20,49] and Tax expression leads to substantial enrichment of CREB at the viral promoter *in vivo*.[50,51] Given that Sp1 recruits HDACs to several cellular promoters (reviewed in Ref. 23) and binds the proximal vCRE,[29] this factor (or Sp3) may play a key role in recruiting HDAC complexes to the viral promoter.

Interestingly, in an effort to eliminate cells carrying the integrated HTLV-1 provirus, a clinical trial was conducted to evaluate use of the HDAC inhibitor valproate (VPA) in treating patients with TSP/HAM.[142] In this study, VPA, when administered at pharmacologically relevant doses (1–3 mM), effectively decreased proviral loads in all HTLV-1-infected patients. This result suggests that reactivation of the viral expression by VPA leads to elimination of the cells expressing Tax by the immune system.

Repression of HTLV-1 Transcription by the Viral Protein HBZ

The HTLV-1-encoded protein, HTLV-1 bZIP factor (HBZ), antagonizes the role of Tax in HTLV-1 transcription (Fig. 3). As its name implies, this protein contains a bZIP domain that mediates formation of heterodimers with the cellular bZIP factors ATF1, CREB, CREB-2, CREM, c-Jun, JunB and JunD.[143–146] The overall effect of these interactions is to prevent the cellular factors from associating with their cognate DNA-binding sites and facilitating transcriptional activation. Indeed, the amino acid sequence of

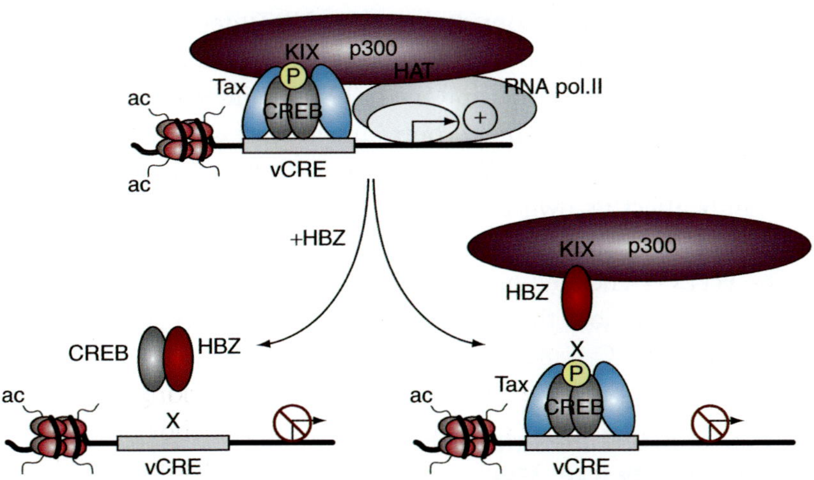

Fig. 3. Model of repression of Tax-dependent transcription by HBZ. Formation of the Tax-CREB-p300/CBP complex on the vCRE is disrupted by the formation of HBZ-CREB heterodimers or by binding of HBZ to the KIX domain of p300/CBP.

the basic region in HBZ diverges from that of corresponding regions in other bZIP factors, and current experimental evidence suggests that HBZ does not bind DNA directly. With respect to these observations, HBZ represses basal HTLV-1 transcription through its interaction with c-Jun, and it represses Tax-dependent transcription, in part, by inhibiting formation of the Tax/CREB/vCRE complex and a similar ternary complex containing CREB-2.[125,143,146] The mechanism underlying this effect involves sequestration of CREB by HBZ, but not disruption of the formed ternary complex.

In addition to its C-terminal bZIP domain, HBZ contains an N-terminal domain capable of binding p300/CBP.[147] Two LXXLL motifs (L denotes leucine and X denotes any amino acid) in this region directly contact the KIX domain of the coactivators. The LXXLL motif is found in several cellular transcription factors where it similarly binds the KIX domain.[148] Although HBZ additionally interacts with the coactivator HAT and C/H3 domains, stable binding to p300/CBP is conferred through its interaction with the KIX domain.[147] It is through this interaction that HBZ exhibits additional repressive effects on HTLV-1 transcription. *In vitro*, HBZ competes with Tax for binding to the KIX domain and is capable of displacing the KIX domain from a Tax/CREB/vCRE complex. The latter result is striking considering that HBZ does not disrupt the Tax/CREB/vCRE complex once it is formed.[146] Differential mutational analysis of the LXXLL motifs and the bZIP domain of HBZ indicate that both regions of the viral protein contribute to repression of HTLV-1 transcription. Repression of HTLV-1 transcription by HBZ is not expected to significantly affect its own expression, as key elements in the HBZ promoter are distinct from those in the 5′ LTR promoter.[127]

The HTLV-1-encoded protein p30 is also implicated in repression of HTLV-1 transcription through competition with Tax for binding to p300/CBP.[149]

Repression of HTLV-1 Transcription by Cellular bZIP Factors (C/EBP, ICER, ATFx)

Like HBZ, certain cellular factors contribute to repression of HTLV-1 transcription by modulating recruitment of CREB or related bZIP factors to the viral promoter. The CCAAT/enhancer-binding proteins (C/EBP) are a family of cellular bZIP transcription factors whose members include C/EBPα, C/EBPβ, C/EBPγ, C/EBPδ, C/EBPε, and C/EBPζ (CHOP). Each C/EBP protein carries an N-terminal activation domain and, through its leucine zipper, forms heterodimers with CREB-2.[19] Interestingly, while C/EBPβ, C/EBPδ and C/EBPε increase basal levels of HTLV-1 transcription, C/EBPα, C/EBPβ and C/EBPζ repress Tax-dependent transcription.[150–152] Although not fully characterized, the mechanism of transcriptional repression appears to depend on which C/EBP member is mediating repression. Specifically, C/EBPα and C/EBPζ repress HTLV-1 transcription by binding to CREB-2,[150,152] and C/EBPβ inhibits transcription by interacting with Tax.[151] Additional details regarding these mechanisms of repression remain to be elucidated. Surprisingly, these factors cooperate with Tax in activating expression of certain cellular genes,[153] suggesting that C/EBP-containing complexes recruited to cellular promoters are distinct from those that associate with the HTLV-1 promoter.

The bZIP factor ATFx is a member of the ATF/CREB family that was identified as a Tax-binding partner in a two-hybrid screen. ATFx binds the vCRE, but unlike other ATF/CREB members, ATFx represses Tax-dependent transcription.[154] The mechanism of inactivation of viral transcription by ATFx is currently unclear. Interestingly, ATFx forms heterodimers with C/EBPα and C/EBPγ.[19] It is therefore possible that, under certain cellular conditions, complexes containing ATFx, C/EBP members and Tax form complexes on the viral promoter that repress transcription. In turn, as over-expression of CREB or CREB-2 alleviates effects of ATFx,[154] these complexes may be replaced with complexes composed of Tax and alternative ATF/CREB members that activate HTLV-1 transcription.

The inducible cAMP early repressor (ICER) is a potent transcriptional repressor, as this protein does not contain an activation domain. ICER inhibits Tax-dependent transcription[155,156] by binding to the vCREs.[156] Whether ICER inhibits transcription as a homodimer or as a heterodimer with ATF/CREB members has not been reported. Since ICER is activated in PBMCs after mitogenic stimulation, it is hypothesized that ICER inhibits HTLV-1 transcription during this period.

Negative Feedback Regulation of HTLV-1 Transcription by Tax

The histone methyltransferase SUV39H1 is implicated in repression of HTLV-1 transcription. SUV39H1 methylates H3 K9, a modification found in heterochromatin and involved in repressing expression of euchromatic genes.[157] Interestingly, SUV39H1-mediated repression of HTLV-1 transcription is reported to involve Tax, as the viral protein augments SUV39H1 expression and is proposed to directly recruit the methyltransferase to the viral promoter[158] (Fig. 4).

B-cell leukemia 3 (Bcl-3) is a nuclear member of the IκB (inhibitor of NF-κB) family that either promotes or inhibits NF-κB target gene expression.[159] Bcl-3 is overexpressed in HTLV-1 infected T cells, and studies demonstrate that Tax activates expression of Bcl-3.[160–162] This process occurs through activation of the NF-κB pathway.[161] Interestingly, two groups reported that Bcl-3 represses Tax-dependent transcription in a vCRE-dependent manner[160,161] (Fig. 4). However, the precise mechanism of repression remains debatable. Results from one study indicate that Bcl-3 integrates into a Tax/CREB/TORC3 complex on the vCREs through an interaction between its ankyrin motifs and TORC3.[160] In this complex,

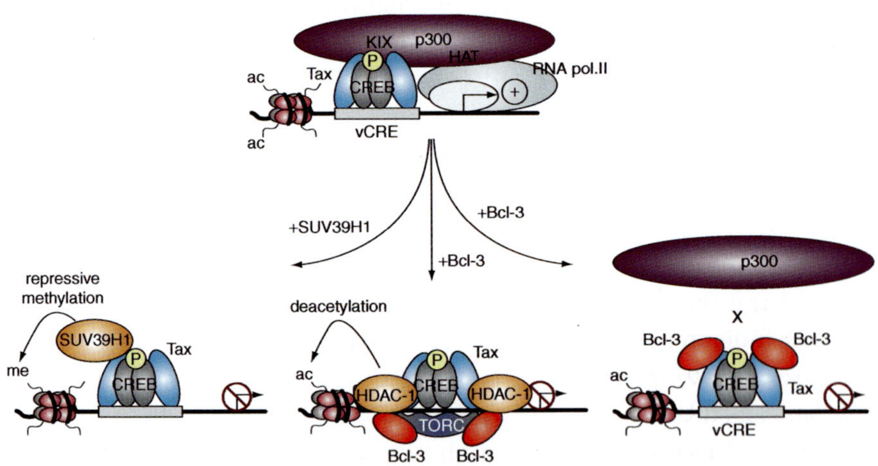

Fig. 4. Model of Tax-dependent repression by Bcl-3 and SUV39H1. The Tax-CREB-p300/CBP complex on the vCRE is transformed into a repressor complex by association of SUV39H1 or Bcl-3 with the complex.

Bcl-3 recruits HDAC-1 to the viral promoter, thereby abrogating histone acetylation and inhibiting transcription. In contrast, results from a separate study indicate that Bcl-3 is incorporated into a Tax/CREB complex on the vCREs through an interaction between its ankyrin motifs and Tax.[161] In this context, p300 is unable to associate with the promoter-bound complex. In activated T cells, over-expression of Bcl-3 leads to increased T cell survival, while Bcl-3-deficient T cells die abnormally fast.[159] Therefore, it is proposed that over-expression of Bcl-3 in HTLV-1 infected T cells enhances survival of these cells and concomitantly leads to a reduction in expression of viral proteins, facilitating evasion of the immune response to infection.

Although these negative feedback mechanisms through Tax are intriguing, additional studies are needed to define the signals that initiate them. Indeed, *de novo* Tax expression in cells activates transcription from the LTR, and viral RNA is abundant in HTLV-1-infected T cells that chronically express high levels of Tax. As activation of HTLV-1 transcription by Tax is mainly analyzed using heterogeneous cell populations, it is possible that the repressive effects of Tax are cell cycle-dependent.

REFERENCES

1. Hinuma Y, Nagata K, Hanaoka M, *et al.* (1981) Adult T-cell leukemia: Antigen in an ATL cell line and detection of antibodies to the antigen in human sera. *Proc Natl Acad Sci USA* **78**: 6476–6480.
2. Poiesz BJ, Ruscetti FW, Gazdar AF, *et al.* (1980) Detection and isolation of type C retrovirus particle from fresh and cultured lymphocytes of a patient with cutaneous T-cell lymphoma. *Proc Natl Acad Sci USA* **77**: 7415–7419.
3. Gessain A, Bari F, Vernant JC, *et al.* (1985) Antibodies to human T-lymphotropic virus type I in patients with tropical spastic paraparesis. *Lancet* **2**: 407–410.
4. Osame M, Usuku K, Izumo S, *et al.* (1986) HTLV-1 associated myelopathy: A new clinical entity. *Lancet* **1**: 1031–1032.
5. Seiki M, Eddy R, Shows TB, Yoshida M. (1984) Nonspecific integration of the HTLV provirus genome into adult T-cell leukaemia cells. *Nature* **309**: 640–642.

6. Leclercq I, Mortreux F, Cavrois M, *et al.* (2000) Host sequences flanking the human T-cell leukemia virus type 1 provirus *in vivo. J Virol* **74**: 2305–2312.
7. Doi K, Wu X, Taniguchi Y, *et al.* (2005) Preferential selection of human T-cell leukemia virus type I provirus integration sites in leukemic versus carrier states. *Blood* **106**: 1048–1053.
8. Meekings KN, Leipzig J, Bushman FD, *et al.* (2008) HTLV-1 integration into transcriptionally active genomic regions is associated with proviral expression and with HAM/TSP. *PLoS Pathog* **4**: e1000027.
9. Grant C, Barmak K, Alefantis T, *et al.* (2002) Human T cell leukemia virus type I and neurologic disease: Events in bone marrow, peripheral blood, and central nervous system during normal immune surveillance and neuroinflammation. *J Cell Physiol* **190**: 133–159.
10. Majorovits E, Nejmeddine M, Tanak Y, *et al.* (2008) Human T-lymphotropic virus-1 visualized at the virological synapse by electron tomography. *PLoS ONE* **3**: e2251.
11. Nejmeddine M, Barnard AL, Tanaka Y, *et al.* (2005) Human T-lymphotropic virus, type 1, Tax protein triggers microtubule reorientation in the virological synapse. *J Biol Chem* **280**: 29653–29660.
12. Fujii M, Tsuchiya H, Meng B, Seiki M. (1995) c-Jun, c-Fos and their family members activate the transcription mediated by three 21-bp repetitive sequences in the HTLV-I long terminal repeat. *Intervirology* **38**: 221–228.
13. Jeang KT, Chiu R, Santos E, Kim SG. (1991) Induction of the HTLV-I LTR by Jun occurs through the Tax-responsive 21-bp elements. *Virology* **181**: 218–227.
14. Vinson C, Acharya A, Taparowsky EJ. (2006) Deciphering B-ZIP transcription factor interactions *in vitro* and *in vivo. Biochim Biophys Acta* **1759**: 4–12.
15. Fujii M, Niki T, Matsuda T, *et al.* (1991) HTLV-I Tax induces expression of various immediate early serum response genes. *Oncogene* **6**: 1023–1029.
16. Hooper WC, Rudolph DL, Lairmore MD, Lal RB. (1991) Constitutive expression of c-jun and jun-B in cell lines infected with human T-lymphotropic virus types I and II. *Biochem Biophys Res Commun* **181**: 976–980.
17. Mori N, Fujii M, Iwai K, *et al.* (2000) Constitutive activation of transcription factor AP-1 in primary adult T-cell leukemia cells. *Blood* **95**: 3915–3921.

18. Iwai K, Mori N, Oie M, *et al.* (2001) Human T-cell leukemia virus type 1 Tax protein activates transcription through AP-1 site by inducing DNA binding activity in T cells. *Virology* **279**: 38–46.
19. Newman JR, Keating AE. (2003) Comprehensive identification of human bZIP interactions with coiled-coil arrays. *Science* **300**: 2097–2101.
20. Brauweiler A, Garl P, Franklin AA, *et al.* (1995) A molecular mechanism for HTLV-I latency and Tax transactivation. *J Biol Chem* **270**: 12814–12822.
21. Franklin A, Kubik MF, Uittenbogaard MN, *et al.* (1993) Transactivation by the human T-cell leukemia virus Tax protein is mediated through enhanced binding of activating transcription factor-2 (ATF-2) ATF-2 response and cAMP element-binding protein (CREB). *J Biol Chem* **268**: 21225–21231.
22. Lemasson I, Polakowski N, Laybourn PJ, Nyborg JK. (2002) Transcription factor binding and histone modifications on the integrated proviral promoter in HTLV-I-infected T-cells. *J Biol Chem* **277**: 49459–49465.
23. Wierstra I. (2008) Sp1: Emerging roles — beyond constitutive activation of TATA-less housekeeping genes. *Biochem Biophys Res Commun* **372**: 1–13.
24. Nyborg JK, Matthews M-AH, Yucel J, *et al.* (1990) Interaction of host cell proteins with the human T-cell leukemia virus type I transcriptional control region. II. A comprehensive map of protein binding sites facilitates construction of a simple chimeric promoter responsive to the viral Tax2 gene product. *J Biol Chem* **265**: 8237–8242.
25. Gégonne A, Bosselut R, Bailly R-A, Ghysdael J. (1993) Synergistic activation of the HTLV-I LTR ets-responsive region by transcription factors ets1 and Sp1. *EMBO J* **12**: 1169–1178.
26. Barnhart MK, Connor LM, Marriott SJ. (1997) Function of the human T-cell leukemia virus type 1 21-base-pair repeats in basal transcription. *J Virol* **71**: 337–344.
27. Wessner R, Yao J, Wigdahl B. (1997) Sp family members preferentially interact with the promoter proximal repeat within the HTLV-I enhancer. *Leukemia* **11**: 10–13.
28. Livengood JA, Nyborg JK. (2004) The high-affinity Sp1 binding site in the HTLV-1 promoter contributes to Tax-independent basal expression. *Nucl Acid Res* **32**: 2829–2837.
29. Yao J, Grant C, Harhaj E, *et al.* (2006) Regulation of human T-cell leukemia virus type 1 gene expression by Sp1 and Sp3 interaction with TRE-1 repeat III. *DNA Cell Biol* **25**: 262–276.

30. Niu Z, Li A, Zhang SX, Schwartz RJ. (2007) Serum response factor micromanaging cardiogenesis. *Curr Opin Cell Biol* **19**: 618–627.
31. Buchwalter G, Gross C, Wasylyk B. (2004) Ets ternary complex transcription factors. *Gene* **324**: 1–14.
32. Winter HY, Dayaram T, Marriott SJ. (2007) Activation of the human T-cell leukemia virus type 1 long terminal repeat by the ternary complex factor Elk-1. *J Virol* **81**: 13075–13081.
33. Wycuff DR, Yanites HL, Marriott SJ. (2004) Identification of a functional serum response element in the HTLV-I LTR. *Virology* **324**: 540–553.
34. Shuh M, Derse D. (2000) Ternary complex factors and cofactors are essential for human T-cell leukemia virus type 1 Tax transactivation of the serum response element. *J Virol* **74**: 11394–11397.
35. Winter HY, Marriott SJ. (2007) Human T-cell leukemia virus type 1 Tax enhances serum response factor DNA binding and alters site selection. *J Virol* **81**: 6089–6098.
36. Kashanchi F, Brady JN. (2005) Transcriptional and post-transcriptional gene regulation of HTLV-1. *Oncogene* **24**: 5938–5951.
37. Boxus M, Twizere JC, Legros S, *et al.* (2008) The HTLV-1 Tax interactome. *Retrovirology* **5**: 76.
38. Ching YP, Chun AC, Chin KT, *et al.* (2004) Specific TATAA and bZIP requirements suggest that HTLV-I Tax has transcriptional activity subsequent to the assembly of an initiation complex. *Retrovirology* **1**: 18.
39. Ku SC, Lee J, Lau J, *et al.* (2008) XBP-1, a novel human T-lymphotropic virus type 1 (HTLV-1) Tax binding protein, activates HTLV-1 basal and Tax-activated transcription. *J Virol* **82**: 4343–4353.
40. Zhao LJ, Giam CZ. (1992) Human T-cell lymphotropic virus type I (HTLV-I) transcriptional activator, Tax, enhances CREB binding to HTLV-I 21-base-pair repeats by protein–protein interaction. *Proc Natl Acad Sci USA* **89**: 7070–7074.
41. Tie F, Adya N, Greene WC, Giam CZ. (1996) Interaction of the human T-lymphotropic virus type 1 Tax dimer with CREB and the viral 21-base-pair repeat. *J Virol* **70**: 8368–8374.
42. Wagner S, Green MR. (1993) HTLV-I Tax protein stimulation of DNA binding of bZIP proteins by enhancing dimerization. *Science* **262**: 395–399.
43. Baranger A, Palmer CR, Hamm MK, *et al.* (1995) Mechanism of DNA binding enhancement by the HTLV-I transactivator Tax. *Nature* **376**: 606–608.

44. Kimzey AL, Dynan WS. (1998) Specific regions of contact between human T-cell leukemia virus type I Tax protein and DNA identified by photocross-linking. *J Biol Chem* **273**: 13768–13775.
45. Lenzmeier BA, Giebler HA, Nyborg JK. (1998) Human T-cell leukemia virus type 1 Tax requires direct access to DNA for recruitment of CREB binding protein to the viral promoter. *Mol Cell Biol* **18**: 721–731.
46. Lundblad R, Kwok RP, Laurance ME, *et al.* (1998) The human T-cell leukemia virus-1 transcriptional activator Tax enhances cAMP-responsive element-binding protein (CREB) binding activity through interactions with the DNA minor groove. *J Biol Chem* **273**: 19251–19259.
47. Kimzey AL, Dynan WS. (1999) Identification of a human T-cell leukemia virus type I Tax peptide in contact with DNA. *J Biol Chem* **274**: 34226–34232.
48. Lenzmeier BA, Baird EE, Dervan PB, Nyborg JK. (1999) The Tax protein–DNA interaction is essential for HTLV-I transactivation *in vitro*. *J Mol Biol* **291**: 731–744.
49. Yin MJ, Gaynor RB. (1996) Complex formation between CREB and Tax enhances the binding affinity of CREB for the human T-cell leukemia virus type 1 21-base-pair repeats. *Mol Cell Biol* **16**: 3156–3168.
50. Lemasson I, Polakowski NJ, Laybourn PJ, Nyborg JK. (2006) Tax-dependent displacement of nucleosomes during transcriptional activation of human T-cell leukemia virus type 1. *J Biol Chem* **281**: 13075–13082.
51. Lu H, Pise-Masison CA, Linton R, *et al.* (2004) Tax relieves transcriptional repression by promoting histone deacetylase 1 release from the human T-cell leukemia virus type 1 long terminal repeat. *J Virol* **78**: 6735–6743.
52. Goodman RH, Smolik S. (2000) CBP/p300 in cell growth, transformation, and development. *Genes Dev* **14**: 1553–1577.
53. Kwok RP, Laurance ME, Lundblad JR, *et al.* (1996) Control of cAMP-regulated enhancers by the viral transactivator Tax through CREB and the co-activator CBP. *Nature* **380**: 642–646.
54. Giebler HA, Loring JE, Van Orden K, *et al.* (1997) Anchoring of CREB binding protein to the human T-cell leukemia virus type 1 promoter: A molecular mechanism of Tax transactivation. *Mol Cell Biol* **17**: 5156–5164.
55. Yan JP, Garrus JE, Giebler HA, *et al.* (1998) Molecular interactions between the coactivator CBP and the human T-cell leukemia virus Tax protein. *J Mol Biol* **281**: 395–400.

56. Harrod R, Tang Y, Nicot C, *et al.* (1998) An exposed KID-like domain in human T-cell lymphotropic virus type 1 Tax is responsible for the recruitment of coactivators CBP/p300. *Mol Cell Biol* **18**: 5052–5061.
57. Lemasson I, Nyborg JK. (2001) Human T-cell leukemia virus type I Tax repression of p73beta is mediated through competition for the C/H1 domain of CBP. *J Biol Chem* **276**: 15720–15727.
58. Scoggin KE, Ulloa A, Nyborg JK. (2001) The oncoprotein Tax binds the SRC-1-interacting domain of CBP/p300 to mediate transcriptional activation. *Mol Cell Biol* **21**: 5520–5530.
59. Georges SA, Giebler HA, Cole PA, *et al.* (2003) Tax recruitment of CBP/p300, via the KIX domain, reveals a potent requirement for acetyltransferase activity that is chromatin-dependent and histone tail-independent. *Mol Cell Biol* **23**: 3392–3404.
60. Laurance ME, Kwok RP, Huang MS, *et al.* (1997) Differential activation of viral and cellular promoters by human T-cell lymphotropic virus-1 Tax and cAMP-responsive element modulator isoforms. *J Biol Chem* **272**: 2646–2651.
61. Gachon F, Thebault S, Peleraux A, *et al.* (2000) Molecular interactions involved in the transactivation of the human T-cell leukemia virus type 1 promoter mediated by Tax and CREB-2 (ATF-4). *Mol Cell Biol* **20**: 3470–3481.
62. Ramirez JA, Nyborg JK. (2007) Molecular characterization of HTLV-1 Tax interaction with the KIX domain of CBP/p300. *J Mol Biol* **372**: 958–969.
63. Geiger TR, Sharma N, Kim YM, Nyborg JK. (2008) The human T-cell leukemia virus type 1 Tax protein confers CBP/p300 recruitment and transcriptional activation properties to phosphorylated CREB. *Mol Cell Biol* **28**: 1383–1392.
64. Kim YM, Ramirez JA, Mick JE, *et al.* (2007) Molecular characterization of the Tax-containing HTLV-1 enhancer complex reveals a prominent role for CREB phosphorylation in Tax transactivation. *J Biol Chem* **282**: 18750–18757.
65. Trevisan R, Daprai L, Acquasaliente L, *et al.* (2004) Relevance of CREB phosphorylation in the anti-apoptotic function of human T-lymphotropic virus type 1 Tax protein in serum-deprived murine fibroblasts. *Exp Cell Res* **299**: 57–67.
66. Trevisan R, Daprai L, Paloschi L, *et al.* (2006) Antiapoptotic effect of human T-cell leukemia virus type 1 Tax protein correlates with its CREB transcriptional activity. *Exp Cell Res* **312**: 1390–1400.

67. Lu H, Pise-Masison CA, Fletcher TM, *et al.* (2002) Acetylation of nucleosomal histones by p300 facilitates transcription from Tax-responsive human T-cell leukemia virus type 1 chromatin template. *Mol Cell Biol* **22**: 4450–4462.
68. Kashanchi F, Duvall JF, Kwok RP, *et al.* (1998) The coactivator CBP stimulates human T-cell lymphotrophic virus type I Tax transactivation *in vitro*. *J Biol Chem* **273**: 34646–34652.
69. Luge K, Mader AW, Richmond RK, *et al.* (1997) Crystal structure of the nucleosome core particle at 2.8 Å resolution. *Nature* **389**: 251–260.
70. Georges SA, Kraus WL, Luger K, *et al.* (2002) p300-mediated Tax transactivation from recombinant chromatin: histone tail deletion mimics coactivator function. *Mol Cell Biol* **22**: 127–137.
71. Jenuwein T, Allis CD. (2001) Translating the histone code. *Science* **293**: 1074–1080.
72. Konesky L, Nyborg JK, Laybour PJ. (2006) Tax abolishes histone H1 repression of p300 acetyltransferase activity at the human T-cell leukemia virus type 1 promoter. *J Virol* **80**: 10542–10553.
73. Conkright MD, Canettieri G, Screaton R, *et al.* (2003) TORCs: Transducers of regulated CREB activity. *Mol Cell* **12**: 413–423.
74. Iourgenko V, Zhang W, Mickanin C, *et al.* (2003) Identification of a family of cAMP response element-binding protein coactivators by genome-scale functional analysis in mammalian cells. *Proc Natl Acad Sci USA* **100**: 12147–12152.
75. Ravnskjaer K, Kester H, Li Y, *et al.* (2007) Cooperative interactions between CBP and TORC2 confer selectivity to CREB target gene expression. *EMBO J* **26**: 2880–2889.
76. Xu W, Kasper LH, Lerac S, *et al.* (2007) Individual CREB-target genes dictate usage of distinct cAMP-responsive coactivation mechanisms. *EMBO J* **26**: 2890–2903.
77. Koga H, Ohshima T, Shimotohno K. (2004) Enhanced activation of Tax-dependent transcription of human T-cell leukemia virus type I (HTLV-I) long terminal repeat by TORC3. *J Biol Chem* **279**: 52978–52983.
78. Siu YT, Chin KT, Siu KL, *et al.* (2006) TORC1 and TORC2 coactivators are required for Tax activation of the human T-cell leukemia virus type 1 long terminal repeats. *J Virol* **80**: 7052–7059.

79. Screaton RA, Conkright MD, Katoh Y, *et al.* (2004) The CREB coactivator TORC2 functions as a calcium- and cAMP-sensitive coincidence detector. *Cell* **119**: 61–74.
80. Bittinger MA, McWhinnie E, Meltzer J, *et al.* (2004) Activation of cAMP response element-mediated gene expression by regulated nuclear transport of TORC proteins. *Curr Biol* **14**: 2156–2161.
81. Koo SH, Flechner L, Qi L, *et al.* (2005) The CREB coactivator TORC2 is a key regulator of fasting glucose metabolism. *Nature* **437**: 1109–1111.
82. Takemori H, Kajimura J, Okamoto M. (2007) TORC-SIK cascade regulates CREB activity through the basic leucine zipper domain. *FEBS J* **274**: 3202–3209.
83. Siu YT, Ching YP, Jin DY. (2008) Activation of TORC1 transcriptional coactivator through MEKK1-induced phosphorylation. *Mol Biol Cell* **19**: 4750–4761.
84. Yin MJ, Christerson LB, Yamamoto Y, *et al.* (1998) HTLV-I Tax protein binds to MEKK1 to stimulate IkappaB kinase activity and NF-kappaB activation. *Cell* **93**: 875–884.
85. Good L, Maggirwar SB, Harhaj EW, Sun SC. (1997) Constitutive dephosphorylation and activation of a member of the nuclear factor of activated T cells, NF-AT1, in Tax-expressing and type I human T-cell leukemia virus-infected human T cells. *J Biol Chem* **272**: 1425–1428.
86. Ding W, Albrecht B, Kelley RE, *et al.* (2002) Human T-cell lymphotropic virus type 1 p12(I) expression increases cytoplasmic calcium to enhance the activation of nuclear factor of activated T cells. *J Virol* **76**: 10374–10382.
87. Nagy Z, Tora L. (2007) Distinct GCN5/PCAF-containing complexes function as co-activators and are involved in transcription factor and global histone acetylation. *Oncogene* **26**: 5341–5357.
88. Harrod R, Kuo YL, Tang Y, *et al.* (2000) p300 and p300/cAMP-responsive element-binding protein associated factor interact with human T-cell lymphotropic virus type-1 Tax in a multi-histone acetyltransferase/activator-enhancer complex. *J Biol Chem* **275**: 11852–11857.
89. Jiang H, Lu H, Schiltz RL, *et al.* (1999) PCAF interacts with Tax and stimulates Tax transactivation in a histone acetyltransferase-independent manner. *Mol Cell Biol* **19**: 8136–8145.

90. Okada M, Jeang KT. (2002) Differential requirements for activation of integrated and transiently transfected human T-cell leukemia virus type 1 long terminal repeat. *J Virol* **76**: 12564–12573.
91. Blanco-Garcia N, Asensio-Juan E, de la Cruz X, Martinez-Balbas MA. (2009) Autoacetylation regulates P/CAF nuclear localization. *J Biol Chem* **284**: 1343–1352.
92. Roberts CW, Orkin SH. (2004) The SWI/SNF complex — chromatin and cancer. *Nat Rev Cancer* **4**: 133–142.
93. Wu K, Bottazzi E, de la Fuente C, *et al.* (2004) Protein profile of Tax-associated complexes. *J Biol Chem* **279**: 495–508.
94. Zhang L, Liu M, Merling R, Giam CZ. (2006) Versatile reporter systems show that transactivation by human T-cell leukemia virus type 1 Tax occurs independently of chromatin remodeling factor BRG1. *J Virol* **80**: 7459–7468.
95. Sharma N, Nyborg JK. (2008) The coactivators CBP/p300 and the histone chaperone NAP1 promote transcription-independent nucleosome eviction at the HTLV-1 promoter. *Proc Natl Acad Sci USA* **105**: 7959–7963.
96. Walfridsson J, Khorosjutina O, Matikainen P, *et al.* (2007) A genome-wide role for CHD remodelling factors and Nap1 in nucleosome disassembly. *EMBO J* **26**: 2868–2879.
97. Anderson MG, Scoggin KE, Simbulan-Rosenthal CM, Steadman JA. (2000) Identification of poly(ADP-ribose) polymerase as a transcriptional coactivator of the human T-cell leukemia virus type 1 Tax protein. *J Virol* **74**: 2169–2177.
98. Kraus WL. (2008) Transcriptional control by PARP-1: Chromatin modulation, enhancer-binding, coregulation, and insulation. *Curr Opin Cell Biol* **20**: 294–302.
99. Zhang Z, Hildebrandt EF, Simbulan-Rosenthal CM, Anderson MG. (2002) Sequence-specific binding of poly(ADP-ribose) polymerase-1 to the human T cell leukemia virus type-I tax responsive element. *Virology* **296**: 107–116.
100. Shukla A, Chaurasia P, Bhaumik SR. (2009) Histone methylation and ubiquitination with their cross-talk and roles in gene expression and stability. *Cell Mol Life Sci* **66**: 1419–1433.
101. Jeong SJ, Lu H, Cho WK, *et al.* (2006) Coactivator-associated arginine methyltransferase 1 enhances transcriptional activity of the human T-cell

lymphotropic virus type 1 long terminal repeat through direct interaction with Tax. *J Virol* **80**: 10036–10044.

102. Dasgupta A, Jung KJ, Jeong J, Brady JN. (2008) Inhibition of methyltransferases results in induction of g2/m checkpoint and programmed cell death in human T-lymphotropic virus type 1-transformed cells. *J Virol* **82**: 49–59.
103. Daujat S, Bauer UM, Shah V, *et al.* (2002) Crosstalk between CARM1 methylation and CBP acetylation on histone H3. *Curr Biol* **12**: 2090–2097.
104. Xu W, Chen H, Du K, *et al.* (2001) A transcriptional switch mediated by cofactor methylation. *Science* **294**: 2507–2511.
105. Chevillard-Briet M, Trouche D, Vandel L. (2002) Control of CBP co-activating activity by arginine methylation. *EMBO J* **21**: 5457–5466.
106. Zika E, Fauquier L, Vandel L, Ting JP. (2005) Interplay among coactivator-associated arginine methyltransferase 1, CBP, and CIITA in IFN-gamma-inducible MHC-II gene expression. *Proc Natl Acad Sci USA* **102**: 16321–16326.
107. Semmes OJ, Jeang KT. (1996) Localization of human T-cell leukemia virus type 1 Tax to subnuclear compartments that overlap with interchromatin speckles. *J Virol* **70**: 6347–6357.
108. Bex F, Yin MJ, Burny A, Gaynor RB. (1998) Differential transcriptional activation by human T-cell leukemia virus type 1 Tax mutants is mediated by distinct interactions with CREB binding protein and p300. *Mol Cell Biol* **18**: 2392–2405.
109. Bex F, Murphy K, Wattiez R, *et al.* (1999) Phosphorylation of the human T-cell leukemia virus type 1 transactivator Tax on adjacent serine residues is critical for Tax activation. *J Virol* **73**: 738–745.
110. Durkin SS, Ward MD, Fryrear KA, Semmes OJ. (2006) Site-specific phosphorylation differentiates active from inactive forms of the human T-cell leukemia virus type 1 Tax oncoprotein. *J Biol Chem* **281**: 31705–31712.
111. Chiari E, Lamsoul I, Lodewick J, *et al.* (2004) Stable ubiquitination of human T-cell leukemia virus type 1 Tax is required for proteasome binding. *J Virol* **78**: 11823–11832.
112. Nasr R, Chiari E, El-Sabban M, *et al.* (2006) Tax ubiquitylation and sumoylation control critical cytoplasmic and nuclear steps of NF-kappaB activation. *Blood* **107**: 4021–4029.
113. Lamsoul I, Lodewick J, Lebrun S, *et al.* (2005) Exclusive ubiquitination and sumoylation on overlapping lysine residues mediate NF-kappaB activation

by the human T-cell leukemia virus Tax oncoprotein. *Mol Cell Biol* **25**: 10391–10406.

114. Peloponese JM, Jr., Iha H, Yedavalli VR, *et al.* (2004) Ubiquitination of human T-cell leukemia virus type 1 Tax modulates its activity. *J Virol* **78**: 11686–11695.
115. Lodewick J, Lamsoul I, Polania A, *et al.* (2009) Acetylation of the human T-cell leukemia virus type 1 Tax oncoprotein by p300 promotes activation of the NF-kappaB pathway. *Virology* **386**: 68–78.
116. Clemens KE, Piras G, Radonovich MF, *et al.* (1996) Interaction of the human T-cell lymphotropic virus type 1 Tax transactivator with transcription factor IIA. *Mol Cell Biol* **16**: 4656–4664.
117. Caron C, Rousset R, Beraud C, *et al.* (1993) Functional and biochemical interaction of the HTLV-I Tax1 transactivator with TBP. *EMBO J* **12**: 4269–4278.
118. Caron C, Mengus G, Dubrowskaya V, *et al.* (1997) Human TAF(II)28 interacts with the human T cell leukemia virus type I Tax transactivator and promotes its transcriptional activity. *Proc Natl Acad Sci USA* **94**: 3662–3667.
119. Bres V, Yoh SM, Jones KA. (2008) The multi-tasking P-TEFb complex. *Curr Opin Cell Biol* **20**: 334–340.
120. Zhou M, Lu H, Park H, *et al.* (2006) Tax interacts with P-TEFb in a novel manner to stimulate human T-lymphotropic virus type 1 transcription. *J Virol* **80** 4781–4791.
121. Cho WK, Zhou M, Jang MK, *et al.* (2007) Modulation of the Brd4/P-TEFb interaction by the human T-lymphotropic virus type 1 Tax protein. *J Virol* **81**: 11179–11186.
122. Cullen BR, Lomedico PT, Ju G. (1984) Transcriptional interference in avian retroviruses — implications for the promoter insertion model of leukaemogenesis. *Nature* **307**: 241–245.
123. Emerman M, Temin HM. (1984) Genes with promoters in retrovirus vectors can be independently suppressed by an epigenetic mechanism. *Cell* **39**: 449–467.
124. Larocca D, Chao LA, Seto MH, Brunck TK. (1989) Human T-cell leukemia virus minus strand transcription in infected T-cells. *Biochem Biophys Res Commun* **163**: 1006–1013.
125. Gaudray G, Gachon F, Basbous J, *et al.* (2002) The complementary strand of the human T-cell leukemia virus type 1 RNA genome encodes a bZIP

transcription factor that down-regulates viral transcription. *J Virol* **76**: 12813–12822.

126. Lemasson I, Polakowski N, Laybourn PJ, Nyborg JK. (2004) Transcription regulatory complexes bind the human T-cell leukemia virus 5′ and 3′ long terminal repeats to control gene expression. *Mol Cell Biol* **24**: 6117–6126.
127. Yoshida M, Satou Y, Yasunaga J, *et al.* (2008) Transcriptional control of spliced and unspliced human T-cell leukemia virus type 1 bZIP factor (HBZ) gene. *J Virol* **82**: 9359–9368.
128. Landry S, Halin M, Vargas A, *et al.* (2009) Upregulation of human T-cell leukemia virus type 1 antisense transcription by the viral Tax protein. *J Virol* **83**: 2048–2054.
129. Asquith B, Bangham CR. (2008) How does HTLV-I persist despite a strong cell-mediated immune response? *Trends Immunol* **29**: 4–11.
130. Baydoun HH, Bellon M, Nicot C. (2008) HTLV-1 Yin and Yang: Rex and p30 master regulators of viral mRNA trafficking. *AIDS Rev* **10**: 195–204.
131. Prokhortchouk E, Defossez PA. (2008) The cell biology of DNA methylation in mammals. *Biochim Biophys Acta* **1783**: 2167–2173.
132. Taniguchi Y, Nosaka K, Yasunaga J, *et al.* (2005) Silencing of human T-cell leukemia virus type I gene transcription by epigenetic mechanisms. *Retrovirology* **2**: 64.
133. Saggioro D, Forino M, Chieco-Bianchi L. (1991) Transcriptional block of HTLV-I LTR by sequence-specific methylation. *Virology* **182**: 68–75.
134. Koiwa T, Hamano-Usami A, Ishida T, *et al.* (2002) 5′-long terminal repeat-selective CpG methylation of latent human T-cell leukemia virus type 1 provirus *in vitro* and *in vivo*. *J Virol* **76**: 9389–9397.
135. Takeda S, Maeda M, Morikawa S, *et al.* (2004) Genetic and epigenetic inactivation of *tax* gene in adult T-cell leukemia cells. *Int J Cancer* **109**: 559–567.
136. Ego T, Tanaka Y, Shimotohno K. (2005) Interaction of HTLV-1 Tax and methyl-CpG-binding domain 2 positively regulates the gene expression from the hypermethylated LTR. *Oncogene* **24**: 1914–1923.
137. D'Alessio AC, Szyf M. (2006) Epigenetic tete-a-tete: The bilateral relationship between chromatin modifications and DNA methylation. *Biochem Cell Biol* **84**: 463–476.
138. Nosaka T, Ariumi Y, Sakurai M, *et al.* (1993) Novel internal promoter/enhancer of HTLV-I for Tax expression. *Nucl Acid Res* **21**: 5124–5129.

139. Yang XJ, Seto E. (2008) The Rpd3/Hda1 family of lysine deacetylases: From bacteria and yeast to mice and men. *Nat Rev Mol Cell Biol* **9**: 206–218.
140. Ego T, Ariumi Y, Shimotohno K. (2002) The interaction of HTLV-1 Tax with HDAC1 negatively regulates the viral gene expression. *Oncogene* **21**: 7241–7246.
141. Villanueva R, Iglesias AH, Camelo S, *et al.* (2006) Histone deacetylase 3 represses HTLV-1 Tax transcription. *Oncol Rep* **16**: 581–585.
142. Lezin A, Gillet N, Olindo S, *et al.* (2007) Histone deacetylase mediated transcriptional activation reduces proviral loads in HTLV-1 associated myelopathy/tropical spastic paraparesis patients. *Blood* **110**: 3722–3728.
143. Basbous J, Arpin C, Gaudray G, *et al.* (2003) The HBZ factor of human T-cell leukemia virus type I dimerizes with transcription factors JunB and c-Jun and modulates their transcriptional activity. *J Biol Chem* **278**: 43620–46327.
144. Matsumoto J, Oshima T, Isono O, Shimotohno K. (2005) HTLV-1 HBZ suppresses AP-1 activity by impairing both the DNA-binding ability and the stability of c-Jun protein. *Oncogene* **24**: 1001–1010.
145. Thebault S, Basbous J, Hivin P, *et al.* (2004) HBZ interacts with JunD and stimulates its transcriptional activity. *FEBS Lett* **562**: 165–170.
146. Lemasson I, Lewis MR, Polakowski N, *et al.* (2007) Human T-cell leukemia virus type 1 (HTLV-1) bZIP protein interacts with the cellular transcription factor CREB to inhibit HTLV-1 transcription. *J Virol* **81**: 1543–1553.
147. Clerc I, Polakowski N, Andre-Arpin C, *et al.* (2008) An interaction between the human T cell leukemia virus type 1 basic leucine zipper factor (HBZ) and the KIX domain of p300/CBP contributes to the down-regulation of Tax-dependent viral transcription by HBZ. *J Biol Chem* **283**: 23903–23913.
148. Plevin MJ, Mills MM, Ikura M. (2005) The LxxLL motif: A multifunctional binding sequence in transcriptional regulation. *Trends Biochem Sci* **30**: 66–69.
149. Zhang W, Nisbet JW, Albrecht B, *et al.* (2001) Human T-lymphotropic virus type 1 p30(II) regulates gene transcription by binding CREB binding protein/p300. *J Virol* **75**: 9885–9895.

150. Gachon F, Gaudray G, Thebault S, *et al.* (2001) The cAMP response element binding protein-2 (CREB-2) can interact with the C/EBP-homologous protein (CHOP). *FEBS Lett* **502**: 57–62.
151. Hivin P, Gaudray G, Devaux C, Mesnard JM. (2004) Interaction between C/EBPbeta and Tax down-regulates human T-cell leukemia virus type I transcription. *Virology* **318**: 556–565.
152. Grant C, Nonnemacher M, Jain P, *et al.* (2006) CCAAT/enhancer-binding proteins modulate human T cell leukemia virus type 1 long terminal repeat activation. *Virology* **348**: 354–369.
153. Tsukada J, Misago M, Serino Y, *et al.* (1997) Human T-cell leukemia virus type I Tax transactivates the promoter of human prointerleukin-1beta gene through association with two transcription factors, nuclear factor-interleukin-6 and Spi-1. *Blood* **90**: 3142–3153.
154. Forgacs E, Gupta SK, Kerry JA, Semmes OJ. (2005) The bZIP transcription factor ATFx binds human T-cell leukemia virus type 1 (HTLV-1) Tax and represses HTLV-1 long terminal repeat-mediated transcription. *J Virol* **79**: 6932–6939.
155. Bodor J, Spetz AL, Strominger JL, Habener JF. (1996) cAMP inducibility of transcriptional repressor ICER in developing and mature human T lymphocytes. *Proc Natl Acad Sci USA* **93**: 3536–3541.
156. Newbound GC, O'Rourke JP, Collins ND, *et al.* (2000) Repression of Tax-mediated human t-lymphotropic virus type 1 transcription by inducible cAMP early repressor (ICER) protein in peripheral blood mononuclear cells. *J Med Virol* **62**: 286–292.
157. Krauss V. (2008) Glimpses of evolution: Heterochromatic histone H3K9 methyltransferases left its marks behind. *Genetica* **133**: 93–106.
158. Kamoi K, Yamamoto K, Misawa A, *et al.* (2006) SUV39H1 interacts with HTLV-1 Tax and abrogates Tax transactivation of HTLV-1 LTR. *Retrovirology* **3**: 5.
159. Palmer S, Chen YH. (2008) Bcl-3, a multifaceted modulator of NF-kappaB-mediated gene transcription. *Immunol Res* **42**: 210–218.
160. Hishiki T, Ohshima T, Ego T, Shimotohno K. (2007) BCL3 acts as a negative regulator of transcription from the human T-cell leukemia virus type 1 long terminal repeat through interactions with TORC3. *J Biol Chem* **282**: 28335–28343.

161. Kim YM, Sharma N, Nyborg JK. (2008) The proto-oncogene Bcl3, induced by Tax, represses Tax-mediated transcription via p300 displacement from the human T-cell leukemia virus type 1 promoter. *J Virol* **82**: 11939–11947.
162. Ng PW, Iha H, Iwanaga Y, *et al.* (2001) Genome-wide expression changes induced by HTLV-1 Tax: Evidence for MLK-3 mixed lineage kinase involvement in Tax-mediated NF-kappaB activation. *Oncogene* **20**: 4484–4496.

CHAPTER 6

Cellular Factors Involved in HIV-1 RNA Transport

Claire A. Williams[†], Andrew M. L. Lever[†] and Truus E. M. Abbink[*,†]

SUMMARY

HIV-1 assembly and genomic RNA encapsidation have been intensively studied for many years. Many details of the interaction between the RNA packaging signal and the nucleocapsid (NC) domain of the structural Gag protein are now understood. However, there are still many unknowns regarding the spatial and temporal control of the RNA packaging process. It is generally assumed that cellular mRNAs are complexed with RNA-binding proteins during or shortly after transcription. These ribonucleoprotein complexes (RNPs) are subsequently trafficked out of the nucleus into the cytoplasm, where they can be translated at various locations, stored in stress granules during stress responses or destroyed by various RNA degradation machineries. It is assumed that the HIV-1 genomic RNA (gRNA) is also subject to these processes. The composition

*Corresponding author.

†Department of Medicine, University of Cambridge, Addenbrooke's Hospital, Level 5, Box 157, Hills Road, Cambridge, CB2 0QQ, United Kingdom. E-mail: gema2@cam.ac.uk.

of the genomic RNP may thus determine the fate of the viral RNA. Indeed, it has been shown that altering the expression level of certain host proteins can affect HIV-1 translation, gRNA localization or particle assembly, implying that these proteins are important for efficient viral replication. Consequently, there is an increasing interest in targeting these host factors as an additional approach to antiviral treatment.

INTRODUCTION

The assembly of HIV-1 particles involves the packaging (and possibly dimerization) of gRNA into viral cores composed of Gag and Gag-Pol proteins. The particles carry two copies of the RNA genome that are non-covalently linked at the 5′ termini, forming an RNA dimer structure. The 5′ untranslated region (5′ UTR) contains an RNA packaging signal ψ that is recognized by the NC domain of Gag. This interaction results in both RNA and protein conformational changes, leading to a step-wise multimerization of the structural proteins that appear as differently sized assembly intermediates.[1] In addition, Gag multimerization depends on interactions between its capsid (CA) domains and its interaction with cellular membranes through its N-terminal myristoylation motif in the matrix (MA) domain.[2] The C-terminal p6 domain of Gag has been identified as critical for late stages of assembly: budding and release. Virus particles assemble at the plasma membrane and their budding depends on members of several cellular protein families that are subverted by the virus to fulfil this important role.[3,4]

The 5′ UTR of HIV-1 gRNA is highly conserved and regulates various steps of the viral life cycle. Both gRNA dimerization and packaging are regulated by sequence and structural motifs in the 5′ UTR. Other steps of the viral life cycle, including RNA and protein expression, splicing and reverse transcription depend on various motifs present in the leader as well. In order to coordinate these various functions in time and space, the gRNA may sequentially adopt several different RNA conformations that influence protein recruitment to the RNP complex. In particular, viral RNA transport through the nuclear pore to the cytoplasm and to the particle assembly point is expected to be significantly affected by RNA rearrangements.[5] The gRNA not only serves as an mRNA to produce the structural proteins Gag and Gag-Pol but is also packaged into new virus particles. These competing processes need coordination, because a certain level of Gag and Gag-Pol is required for virion particle formation. Finally, RNA dimer formation may occur prior to, or at the time of, particle assembly, possibly at the site where the gRNA and Gag interact.[6,7] These processes and their tight regulation are extremely important for the viral life cycle and constitute potential targets for antiviral drugs.

In this chapter, we briefly outline HIV-1 RNA processing. We describe recent insights into cellular co-factors involved in these processes, the potential for short interfering RNA (siRNA) library screens to identify additional factors and for developing new antiviral drugs targeting such factors.

HIV-1 RNAs are transcribed from the 5′ long terminal repeat (LTR) promoter region (Fig. 1, step 1). This initially results in the production of fully spliced 2kb viral mRNA transcripts that migrate into the cytoplasm

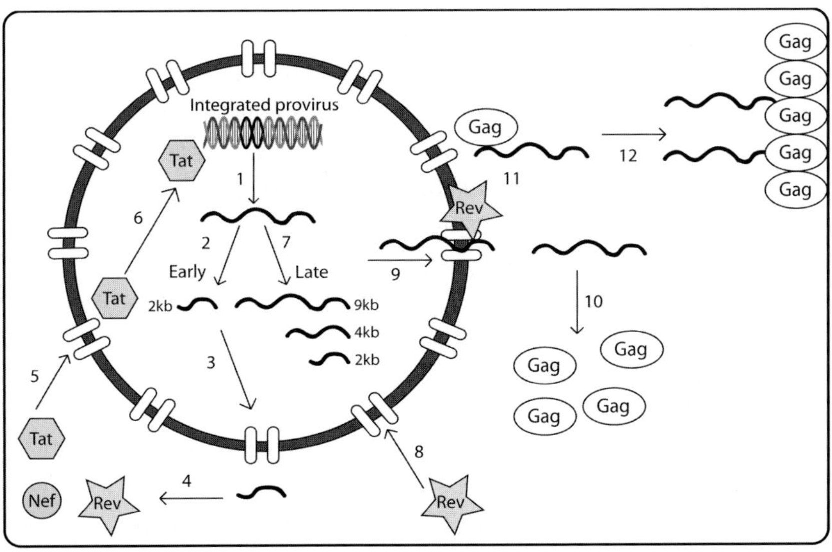

Fig. 1. Overview of HIV-1 RNA biology. Viral RNAs are transcribed, capped, polyadenylated, spliced (or not), transported to the cytoplasm, translated, and/or packaged (as monomers or dimers). These processes are likely to be coordinated by various RNA sequences and structural motifs as well as viral and cellular factors. The numbers indicate the different steps of viral gene expression and particle assembly and are explained in the text.

and are translated to produce Tat, Rev and Nef (Fig. 1, steps 2–4). The Tat protein enters the nucleus where it interacts with the trans-activation response (TAR) hairpin, which is present at the 5′ end of all nascent viral transcripts (Fig. 1, steps 5 and 6).[8] Tat recruits Cyclin T1 and Cyclin-dependent kinase (CDK) 9 to the TAR hairpin, allowing CDK9 to phosphorylate the C-terminal domain of RNA polymerase II. This phosphorylation increases the transcription efficiency several hundred-fold above the basal level.

Besides the 2kb class of mRNAs, HIV-1 expresses a 4kb class of incompletely spliced RNAs and the 9kb full length or gRNA which is unspliced (Fig. 1, step 7). The former class of mRNAs encode the Env proteins and the accessory proteins Vpr, Vpu and Vif, whilst the latter encode the Gag and Gag-Pol polyproteins. During HIV-1 infection, more than 40 differently spliced mRNAs are produced, which allows for sophisticated control of viral gene expression.[9,10] The inefficient splicing of the HIV-1 gRNA has been attributed to imperfect splice acceptor sites and non-canonical branch point sequences.[11–16] Furthermore, splice enhancers and silencers throughout the viral genome have been implicated in splicing regulation.[17–20]

In general, intron-containing mRNAs are recognized by commitment factors and retained in the nucleus to prevent expression of abnormal proteins.[21] HIV-1 overcomes this block by use of the Rev protein, which enters the nucleus following translation (Fig. 1, step 8).[22,23] The inefficiency of HIV-1 gRNA splicing allows Rev to bind the Rev-responsive element (RRE) present within the *env* region of the gRNA and the 4kb class of mRNAs. It is generally accepted that Rev binding to the RPE-containing RNAs prevents splicing of these RNAs to the 2kb class because of efficient transport of the RNAs to the Exportin-1 pore.[24] Rev binds to the cellular nuclear export protein Exportin-1 or CRM1.[25] The RRE/Rev/Exportin-1 complex passes through nuclear pores into the cytoplasm (Fig. 1, step 9), where the RPE-containing RNAs are released from the transport complex.[26]

The gRNA may then be translated (Fig. 1, step 10). In addition, gRNA is packaged into assembling virions (Fig. 1, steps 11 and 12). The selection of gRNA from a much larger pool of cellular and spliced viral RNA is specific and efficient. The process seems to be dependent on ongoing translation of the gRNA.[27,28] Also, translation inhibition studies have shown that HIV-1 gRNA exists in a single population capable of both translation and packaging.[29] The switch mechanism from translation to packaging is unknown. A change in the gRNA structure has been suggested, but there is no experimental data to support this hypothesis.[30–33] An alternative mechanism may exist that reroutes the genomic HIV-1 RNA from the translation machinery to the encapsidation machinery. One option is that the translation machinery itself is inhibited during virus infection, for instance, as a consequence of cell cycle arrest.[34,35] At the same time, the build-up of viral proteins may shift the competition between translation and packaging toward the latter process. Indeed, high levels of Gag or Rev inhibit HIV-1 translation by binding to the Ψ packaging signal in gRNA.[33,36] This 120-nucleotide sequence resides in the gRNA 5′UTR, immediately upstream of the Gag start codon and is important in the selective packaging of gRNA.[37–41] An interaction between the gRNA and the NC domain of Gag is central to this process.[40,42–44] Ψ alone is unable to support the packaging of a heterologous RNA[45]; sequences within the entire 5′UTR and the Gag open reading frame contribute to RNA packaging.[46] As a consequence of Gag binding, the structure of Ψ is changed.[47]

The initial interaction between Gag and the viral genome has been suggested to occur in a perinuclear location (Fig. 1, step 11).[7] The exact journey of the Gag-RNA complex from this perinuclear region to the assembly site at the plasma membrane remains a mystery (Fig. 1, step 12), although microtubules and actin filaments have been implicated.[48] A role for the Ψ packaging signal was suggested to be important for the kinetics of the Gag-gRNA trajectory.[7] Others have implicated the NC domain of Gag in the localization of the protein.[49,50] Wild-type Gag localizes to punctae in the cytoplasm and to the plasma membrane. The punctate pattern possibly reflects the localization of Gag at the multivesicular structures that are thought to be important for HIV-1 assembly and budding.[51] Mutation of the NC domain was found to significantly shift Gag localization from punctae to being diffusely spread throughout the cytoplasm,[49] but the reason is unclear. Analysis of Gag mutants in which the NC domain is swapped for a leucine zipper (LZ) domain may shed some light on this.[52] Gag-LZ chimeras are able to multimerize and assemble into virus-like particles (VLPs), but these contain no detectable gRNA.

THE CELLULAR SUPPORT

The HIV-1 gRNA serves as a substrate for splicing, polyadenylation, nuclear export, translation and packaging into virus particles. The gRNA should not be considered as "naked" and is generally assumed to complex with various host proteins involved in cellular RNA metabolism. The composition of the RNP may change during the different stages of viral replication and this may, as a consequence, determine the fate of the viral RNA. Indeed, nuclear events (alluded to as nuclear history) have been implicated in determining the fate of an RNA once it reaches the cytoplasm.[53]

Apart from nuclear export and the RNA packaging steps, in which viral proteins are a prerequisite, post-transcriptional processing of the viral RNA is heavily dependent on cellular factors. These interactions can coordinate sometimes competing processes in RNA biology. We will describe various host proteins that have been shown to be important for HIV-1 RNA biology, focusing on proteins involved in post-splicing reactions (Table 1). The proteins and their roles in HIV-1 replication are grouped according to their endogenous functions where possible. For more information on host factors involved in HIV-1 transcription and RNA splicing the reader is referred to recent reviews.[54,55]

HIV-1 and RNA Helicases

HIV-1 does not encode its own RNA helicase and has been proposed to subvert cellular RNA helicases for its own benefit.[80] RNA helicases comprise a ubiquitously expressed protein family that is conserved in species as diverse as bacteria and humans. Traditionally, they are described as enzymes that utilize the energy provided from the hydrolysis of NTPs to unwind RNA duplexes, but their functions are actually much broader than this. RNA helicases are capable of disrupting RNA–RNA and RNA–DNA base pairing, displacing proteins from RNA, annealing RNA strands, and stabilizing RNA structures. Given their range of functions, it is not surprising that RNA helicases function in all aspects of RNA metabolism and are often described as RNA chaperones.[81,82]

Table 1. Host Proteins Implicated in HIV-1 RNA Biology

Host Protein	Proposed Function(s)	Reference
DDX1	Maintaining the spatial localization of Rev	56
DDX3	Required for Rev-mediated nuclear export	57
RHA	Role in gene expression and virion assembly	58–60
UPF1	Possible role in viral translation	61
DDX24	Involved in translation and assembly	62
hnRNP A1	Involved in RNA splicing and Rev-mediated nuclear export	63
hnRNP-A2	Involved in RNA splicing and gRNA trafficking	64, 65
hnRNP-E1	Involved in translation	66
hnRNP-E2	Involved in RNA splicing and gene expression	66
Sam68	Involved in Rev-mediated nuclear export and translation	67, 68
hRIP	Role in cytoplasmic release of Rev-exported RNA	69
Purα	Role in transcription and Rev-mediated nuclear export	70, 71
Staufen	Involved in gRNA transport, translation and assembly	72–76
ABCE1	Role in late stage of assembly and release	77
KIF4	Role in cytoplasmic transport of assembly intermediates	78
SOCS1	Role in intracellular trafficking and stability of Gag	79

DEAD box RNA helicase 1 (DDX1) has several proposed functions including roles in transcription, 3′ processing of pre-mRNA, and DNA repair. It has been implicated in playing a role in several human cancers.[83–87] The DDX1 interaction with Rev was shown in yeast two-hybrid systems, immunoprecipitation and pull-down assays.[55] It binds to an N-terminal region of Rev known as the nuclear diffusion inhibitory signal (NIS). The NIS region is responsible for the predominantly nuclear localization of Rev, which contains both nuclear export and targeting signals.[88] Knocking down DDX1 by RNA interference (RNAi) leads to a loss of this nuclear predominance and a change in the splicing pattern of RPE-containing mRNA. Low levels of DDX1 inhibit HIV-1 protein production, whilst DDX1 over-expression increases viral replication.[56] Although DDX1 appears capable of binding RPE-containing RNA, it is not yet known if this is important during the course of an HIV-1 infection. Even though it may not act by directly binding to viral RNA, DDX1 is clearly important for the nuclear export of viral RNA via its effects on Rev subcellular distribution.

Despite being CD4-negative, astrocytes can sustain low-level infection by HIV-1. This infection is very different from that in macrophages and lymphocytes, being characterized by over-expression of completely spliced viral RNA at the expense of unspliced and partially spliced species.[89] The altered splicing pattern correlates with an abnormal distribution of Rev, the majority being found in the cytoplasm, rather than in the nucleus.[90] This aberrant distribution mirrors that seen by Fang *et al.* when they decreased DDX1 expression in cells.[56] Over-expressing DDX1 in astrocytes restores the nuclear predominance of Rev localization and increases the level of HIV-1 replication.[91] It was concluded that low nuclear DDX1 restricts Rev function in astrocytes.

A second RNA helicase involved in Rev function is DDX3. Over-expression of DDX3 increases expression from Rev-dependent reporter plasmids.[57] Whilst knock-down of DDX3 does not affect cytoplasmic levels of spliced RNAs, cytoplasmic levels of RPE-containing RNAs are decreased. These results strongly implicate DDX3 in Rev-mediated export of incompletely spliced HIV-1 viral transcripts (Fig. 1, step 9). DDX3 is a nucleocytoplasmic shuttling protein that localizes to nuclear pores and interacts with both Exportin-1 and Rev. The interaction with Exportin-1 was not dependent upon the DDX3 nuclear export signal, suggesting that it is more than simply cargo for Exportin-1. The importance of DDX3 for HIV-1 replication was shown when over-expression of DDX3 increased viral replication in a T-cell line, and knock-down by RNAi inhibited it. The latter finding was confirmed by others.[92] Knocking down DDX3 did not have any obvious adverse effect on cell cycle nor did it induce apoptosis, highlighting DDX3 as a potential antiviral target.

The precise mechanism of action of DDX3 is unknown. It has been hypothesized that it may function similarly to Dbp5p, a yeast DEAD box helicase essential for mRNA export from the nucleus.[93] Dbp5p associates with the cytoplasmic face of nuclear pore complexes and shuttles between the nucleus and cytoplasm.[94,95] It is proposed to be responsible for the release of mRNA from export receptors once they have passed through the nuclear pore complex into the cytoplasm.[96] Study of the Balbiani ring mRNA in the larval salivary glands of *Chironomus tentans* (*C. tentans*) suggests that the mRNA and associated proteins undergo a large conformational change during translocation through the nuclear pore.[5] Work suggesting that the homologue of Dbp5p in *C. tentans* associates with mRNA cotranscriptionally and remains associated throughout nuclear export has led to the speculation that Dbp5p may also be involved in the RNP conformational changes seen during nuclear export.[97] DDX3 may fulfil similar functions to Dbp5p, but in the Rev/RRE/Exportin-1-mediated transport of gRNA.

Another well-investigated RNA helicase in the HIV-1 life cycle is RNA helicase A (RHA or DDX9). This protein has been implicated in several steps of the HIV-1 life cycle. As well as the usual helicase domains, RHA contains two double-stranded RNA-binding domains, the second of which is capable of binding the stem of the TAR hairpin in HIV-1 transcripts. Over-expression of RHA leads to enhanced viral transcription, possibly due to the recruitment of RNA polymerase II by RHA, or as a result of unwinding of the transcript.[58] RHA also associates with the RRE *in vivo* in a Rev-independent manner. Over-expression of RHA increases the ratio of unspliced and singly spliced mRNAs to multiply spliced mRNAs. Therefore, an additional proposed role of RHA is to release mRNA transcripts from spliceosomes prior to the completion of splicing.[59] RHA has been shown to be packaged into HIV-1 virions, either through an interaction with viral RNA or the Gag polyprotein.[60] Its knock-down using siRNA leads to a decrease in infectious virion production, suggesting that it functions either as a chaperone during the assembly or maturation of virions, or perhaps in promoting reverse transcription in newly infected cells.

Finally, RHA has been shown to play a role in the translation of mRNAs with highly structured 5′ UTRs.[98] Such mRNAs, for example those of retroviruses, require scanning of a highly structured sequence by initiating ribosomal subunits. Unpublished data in a recent review claimed that knock-down of RHA inhibits the translation of HIV-1 Gag, providing another potential role for the multi-functional RHA in HIV-1 replication.[99]

A fourth RNA helicase implicated in HIV-1 RNA biology is UPF1. UPF1 is an evolutionarily conserved cytoplasmic RNA helicase that plays a central role in nonsense-mediated mRNA decay (NMD).[100] This mRNA surveillance system leads to the degradation of mRNAs which contain a premature termination codon, preventing the production of truncated proteins.[101] Additionally, UPF1 is proposed to play roles in Staufen-mediated mRNA decay, histone mRNA degradation, splicing, DNA replication, and possibly translation.[61,102–105] UPF1 was detected in the RNP containing Staufen, Gag and the HIV-1 gRNA. Confocal microscopy showed diffuse cytoplasmic staining of both Gag and UPF1 in transfected cells. It was calculated that 10–20 percent of the UPF1 signal co-localized with Gag, especially at the cell periphery. RNAi-mediated knock-down of UPF1 reduced the levels of viral RNA and Gag resulting in a significant decrease in virus production.[61] The infectivity of the virus particles was unfortunately not measured, so a specific effect on RNA packaging was not addressed. Over-expression of an siRNA-resistant UPF1 gene (UPF1 rescue) rescued this phenotype only partially. Gag protein expression levels were restored to 80 percent compared to non-silencing controls, whereas the genomic and subgenomic viral RNA levels were restored to only 20 percent. This difference points towards a role for UPF1 in HIV-1 RNA translation, which needs to be further investigated. In addition, it suggests that any effect of UPF1 knock-down on viral RNA levels may be indirect. RNAi-mediated knock-down of UPF1 was previously shown to influence the expression of many genes.[106] Over-expression of UPF1 elevated all viral RNA levels and Gag protein levels. Screening a set of UPF1 mutant proteins identified the ATPase activity and the zinc finger domain of the protein as important for these effects. More work is required to define the precise role of UPF1 in HIV-1 replication, including its potential role in viral RNA stability, and to rule out possible indirect effects of RNAi-mediated UPF1 knock-down on HIV-1 RNA biology.[105,106]

A fifth RNA helicase has recently been suggested to play a role in HIV-1 replication, although its mechanism of action remains unclear. DDX24 has been proposed to play a role in viral packaging because its knock-down leads to a decrease in infectious particle production, despite an increase in viral protein production.[62] RNA incorporation into virions produced from DDX24 knock-down cells appears to be diminished, without affecting RNA dimerization levels. In co-immunoprecipitation studies, DDX24 interacts with both Gag and Rev. DDX24 appears to affect Rev function, as knock-down of the protein increases Rev-mediated nuclear export of HIV-1 RNA. This observation potentially explains the increase in viral protein production. Secondly, DDX24 knock-down appears to specifically inhibit the packaging of RNA transcripts that have been exported by the Rev-dependent pathway. Accumulating evidence suggests that Rev can enhance encapsidation of RNA into virions,[107] and it is possible that DDX24 may assist in this process. Alternatively, DDX24 may function in regulating the switch from gRNA translation to encapsidation. As DDX24 is predominantly a nuclear protein, it may contribute to the "nuclear history" of the viral RNA and its fate in the cytoplasm.

HIV-1 and hnRNP Proteins

A second class of proteins important for HIV-1 replication is the family of heterogeneous nuclear ribonucleoproteins (hnRNPs). These compose a diverse family of proteins that were originally identified because of their association with hnRNA. A characteristic of this class is the presence of one or more RNA-binding domains.[55] In addition, the proteins contain auxiliary domains that have been proposed to function in protein–protein interactions.[108] A vast number of isoforms of hnRNP proteins are generated by alternative splicing and post-translational modification. hnRNP proteins fulfil various post-transcriptional, gene-regulatory functions, such as splicing, RNA transport and translation.[109] Some, but not all, hnRNPs can shuttle in and out of the nucleus.[110,111] The shuttling class of hnRNPs generally associates with mRNA in the nucleus and accompanies mRNAs to the cytoplasm, where they modulate mRNA localization, translational efficiency and stability.

To date, several hnRNP family members have been shown to be involved in HIV-1 RNA biology. hnRNP-A1 and -A2 have been implicated in HIV-1 splicing, binding to several sequences in the HIV-1 RNA. hnRNP-A1 has also been found to enhance Rev-mediated nuclear export of gRNA by interacting with the INS-1 element in the Gag open reading frame.[63] Others confirmed the role of hnRNP A1 in Gag production, but could not detect a change in gRNA localization upon RNAi-mediated knock-down of the host protein.[65] hnRNP-A2 has been shown to bind to two hnRNP-A2 responsive elements, A2RE-1 and A2RE-2, that are present in the Gag and Vpr coding sequences, respectively.[64] Consequently, the gRNA contains both A2RE-1 and -2 motifs, whereas the Vif and Vpr mRNAs contain A2RE-2, and Tat mRNA contains a 5′ truncated A2RE-2.

The role of these motifs in RNA transport was initially addressed in murine oligodendrocytes. The cells were injected with truncated fluorescent HIV-1 transcripts, and RNA transport through the cells was analyzed. Although this showed involvement of the two A2RE motifs in RNA transport, the system was far from physiological for HIV-1 replication and did not address the effect of Rev-mediated nuclear export on hnRNP-A2 involvement in HIV-1 gRNA transport.

Mutational analyses of the motifs in a more physiological T-cell line indicated that A2RE-2 is more important than A2RE-1 in HIV-1 replication.[64] Inactivation of A2RE-2 did not affect early gene expression events, such as splicing, Rev-mediated nuclear export and translation. However, a late defect in the localization of HIV-1 gRNA, Gag and Vpr proteins was observed. Wild-type gRNA, Gag and Vpr proteins localize to punctate speckles throughout the cytoplasm. The A2RE-2-inactivated gRNA accumulated in and around the nucleus, both in HeLa and Cos7 cells, and Gag and Vpr proteins derived from the A2RE-2-mutated genome were diffuse in the cytoplasm. It was not investigated whether the gRNA accumulation was of nuclear or cytoplasmic origin. In a second set of experiments, the same group described that significant knock-down of hnRNP-A2 protein levels (>80 percent) resulted in the accumulation of the gRNA in the vicinity of the perinuclear microtubule-organizing center (MTOC). The gRNA was transported to the MTOC at a late stage of virus production, which likely depended on dynamin- and/or dynein-mediated movement. Inhibition of Rev-mediated nuclear export did not prevent the perinuclear association of gRNA upon hnRNP-A2 knock-down, indicating that the gRNA targeting the MTOC region was derived from the cytoplasm.

Surprisingly, the RNA packaging levels are differentially affected when either hnRNP-A2 is knocked down or the A2RE-2 motif is inactivated. Upon A2RE-2 mutagenesis, RNA packaging is decreased with a concomitant decrease in particle infectivity, whereas hnRNP-A2 knock-down results in a reproducible 40–60 percent increase in virion-associated RNA levels but a ~50 percent drop in infectivity per particle. It is not easy to reconcile the different phenotypes seen with alternative methods of disrupting the hnRNP-A2–gRNA interaction. Plausibly, hnRNP-A2 binding activity to HIV-1 gRNA is required for proper gRNA trafficking and optimal production of infectious particles. In both hnRNP-A2 knock-down and A2RE-2 mutagenesis assays, the typical late punctate pattern of Gag proteins and gRNA localization is lost. This is consistent with the hnRNP-A2 protein interacting with the gRNA and keeping it on track along a packaging pathway. This track may involve microtubules, since hnRNP-A2 has been found to bind to microtubule adaptor proteins in neurons.[112]

Recently, the effect of hnRNP-A2 on HIV-1 particle production and infectivity was analyzed by a second group.[113] In their assay system, which analyzed HIV-1 NL4-3 virus production in HEK 293 cells, virus production and infectivity increased after efficient knock-down of hnRNP-A2, whereas over-expression of hnRNP-A2 resulted in a significant drop in virus production and infectivity. The discrepancy between the results from both groups could be explained by a difference in the HIV-1 strain (NL4-3 versus HxBru), siRNAs or cell type (HEK 293 versus HeLa) and further analysis of the involvement of hnRNP-A2 in HIV-1 infection in T cells is needed. Neither paper describing the siRNA-mediated knock-down of hnRNP-A2 in virus producer cells found any significant changes in HIV-1 splicing. This is in disagreement with several papers that support a role for this host protein in HIV-1 splicing, suggesting that for HIV-1 splicing, hnRNP-A2 is redundant (reviewed in Refs. 54 and 114).

Recently, hnRNP-A2 has been shown to interact with hnRNP-E1 in neuronal cells.[115] hnRNP-E1 was shown to inhibit translation of A2RE-containing mRNAs in an hnRNP-A2-dependent manner, which may implicate hnRNP-E1 in translational repression during microtubule-mediated transport of A2RE RNAs. A role for hnRNP-E1 has also been identified in HIV-1 gene expression.[66] Together with hnRNP-E2, the protein was identified as binding to an exonic splicing silencer sequence (ESS3a) that is present in the 3′ end of all viral RNAs. Over-expression of hnRNP-E1 inhibited translation of the viral mRNAs, whereas lowering hnRNP-E1 or hnRNP-E2 levels by RNAi enhanced viral translation. Steady-state HIV-1 RNA levels, nuclear export of RPE-containing RNAs and splicing were not greatly affected by modulation of the hnRNP-E proteins, and the small difference (approximately two-fold) observed in RNA levels could not account for the significantly higher protein expression (approximately five-fold). This small difference in RNA level is probably caused by a difference in RNA stability, since subgenomic viral RNAs that are not spliced are similarly affected, suggestive of a non-specific effect. Modulation of hnRNP-E1 and -E2 expression levels resulted in similar effects on translation from subgenomic HIV-1 mRNAs that lacked the ESS3 sequence. Additional hnRNP-E binding motifs may exist in the HIV-1 genome, but they were not sought. It would be interesting to determine if the hnRNP-E1 role in HIV-1 production is dependent on hnRNP-A2.

hnRNP-E1 has been detected in virus particles, suggesting that besides its role in translation, it may be involved in gRNA packaging.[116] The hnRNP-E proteins have been implicated in the switch from poliovirus RNA translation to replication (which is arguably analogous to the HIV-1 gRNA switch from translation to packaging).[117,118] It is tempting to speculate that the hnRNP-E proteins might be involved in the switch from translation to packaging of HIV-1 gRNA. However, there is no substantive evidence for this as yet.

HIV-1 and Sam68

A requirement for correct 3′ UTR end formation was previously shown in Rev-mediated nuclear export.[119] Subgenomic HIV-1 constructs in which the polyadenylation site was replaced with a hammerhead ribozyme cleavage site produced transcripts that were not exported to the cytoplasm, despite being correctly cleaved. More recently, the level of cleavage and polyadenylation of the HIV-1 RNAs was shown to be influenced by the host protein Sam68 (Src-associated substrate in mitosis of 68kD).[120] Over-expression of Sam68 increases polyadenylation of HIV-1 RNAs, suggesting a role for the protein in proper 3′ end formation. Sam68 is a nuclear protein that contains a high affinity RNA binding domain, a C-terminal nuclear targeting signal, and SH3-domain interaction motifs. It has been implicated in cell differentiation, RNA metabolism and T-cell signaling.[121–123] The protein preferentially binds to RNAs with UAAA, AUUU or polyU sequences.[121]

Previously, Sam68 was shown to bind to the RRE motif *in vitro* and in 293T cells.[67] In addition, the authors showed that Sam68 could partially substitute and synergize with Rev in RRE-mediated gene expression. Furthermore, Sam68 over-expression in 293T cells enhanced

RRE-mediated RNA transport to the cytoplasm with a concomitant increase in protein expression.[124] In agreement with this, decreasing Sam68 expression by antisense RNA treatment significantly inhibited HIV-1 NL4-3 replication in the Jurkat and CEM T-cell lines.[125] This was attributed to inhibition of nuclear export via Exportin-1 affecting RRE-dependent gene expression, which was confirmed when Sam68 knock-down with siRNAs resulted in a decrease in RPE-containing mRNAs in the cytoplasm.[126] Unfortunately, polyadenylation levels were not determined for the HIV-1 RNAs in the Sam68 knock-down experiments.

A C-terminal deletion of the Sam68 protein (Sam68ΔC) destroys its function in HIV-1 production in a dominant negative fashion. The deletion mutant lacks the nuclear targeting signal, resulting in a predominantly cytoplasmic localization. There is some controversy in the field over the mechanism behind this dominant negative effect. It has been suggested that the Sam68ΔC protein induces stress granules that suppress translation of at least the HIV-1 Nef mRNA.[127] The authors established that Nef (but neither Tat nor Rev) mRNAs were targeted to the Sam68ΔC-induced stress granules. This sequestration suppressed Nef translation and could explain an inhibition of HIV-1 replication in T cells by the Sam68ΔC mutant. However, it cannot explain the reduced virus particle production in 293T cells, for which Nef is dispensable. Whether gRNA localizes to stress granules upon Sam68ΔC expression has not been addressed.

Others have claimed that there is an inhibitory effect of the Sam68ΔC mutant on the loading of Env mRNAs onto the translation machinery.[68] The Sam68ΔC mutant did not affect viral polyadenylation or polyA-tail length, suggesting that the earlier finding of enhancement of polyadenylation by Sam68 over-expression is not relevant for its positive effect on Rev-dependent gene expression. The translational defect may have been caused by a reduced association of the polyA-binding protein 1 (PABP1) with HIV-1 RNAs. Further research is required to determine how Sam68 influences PABP1 association with the viral RNA. Overall, Sam68, in analogy with Rev, plays versatile roles in cellular and HIV-1 gene expression that reach beyond alternative splicing and nuclear export, reviewed in Ref. 128. For further details, the reader is referred to two recent reviews.[129,130]

HIV-1 and hRIP

Like DDX3, the human Rev interacting protein (hRIP) was identified in two independent yeast two-hybrid analyses searching for cellular co-factors of Rev.[131,132] The protein localizes predominantly to the nucleus and shares many sequence motifs with proteins of the nucleoporin family.[132] When endogenous hRIP is inhibited by a dominant negative mutant, Rev-mediated export of RPE-containing RNA is specifically decreased. In this situation, *in situ* hybridization shows that these RNAs accumulate at the nuclear periphery rather than entering the cytoplasm.[69] In support of this finding, knock-down of hRIP with RNAi leads to the same aberrant pattern of RPE-containing RNA localization. In addition, hRIP siRNA treatment of a human T-cell line and monocyte-derived macrophages leads to significantly decreased levels of viral replication. This phenotype can be rescued by the expression of an siRNA-resistant hRIP (hRIP rescue), making an off-target effect of the siRNA unlikely.[133] These results clearly demonstrate the importance of hRIP for proper Rev function and viral replication. Given the *in situ* hybridization results, hRIP is presumably involved in the release of RNA from a perinuclear location into the cytoplasm. It may be involved in the disassembly of the Rev–RNA complexes reaching the cytoplasm. Alternatively, hRIP may facilitate the subsequent transport of the viral RNA to the translation or assembly machinery. Its exact mechanism remains to be determined.

HIV-1 and Purα

Purα is an evolutionarily conserved nucleic acid binding protein that preferentially binds single-stranded DNA. It shows a specific affinity for GGN repeat sequences and binds several regulatory proteins.[134] Purα has been implicated in mRNA transport to sites of translation, transcription regulation, cell cycle regulation, translation regulation, and DNA replication.[135–140] This multi-functional protein can also interact with both Tat and Rev. Purα was shown to bind Tat in an RNA-dependent manner.[141,142] Further work showed an interaction between Purα and the TAR hairpin. Over-expression of Purα increases the expression of an LTR-driven reporter construct, dependent on an intact TAR hairpin, suggesting a role for Purα in HIV-1 gene expression.[70]

With regards to translation regulation, Purα appears to enhance the expression of Rev-dependent transcripts.[71] When over-expressed, it increases the expression of RPE-containing luciferase constructs in the presence of Rev, but has little effect alone. RNAi-mediated knock-down of Purα significantly affects Rev-dependent protein expression. Pull-down assays demonstrated an interaction between Rev and Purα. Finally, Purα is capable of binding the RRE in the presence and absence of Rev. The mechanistic action of Purα in Rev function is unclear and its broad range of cellular functions does not help narrow down the possibilities. Given its association with Staufen-containing transport granules, one could envisage that Purα might be involved in the cytoplasmic transport of RPE-containing RNAs to the site of translation.

HIV-1 and Staufen

Staufen is a protein with four double-stranded RNA-binding domains and a C-terminal tubulin-binding domain. It associates with polysomes and the rough endoplasmic reticulum.[143] It has also been shown to be involved in mRNA transport, mRNA decay and translational control.[74,102,144] Staufen is selectively packaged into HIV-1, HIV-2 and Moloney murine leukemia virus particles.[72] Staufen has also been found to associate with telomerase in the nucleolus.[145]

Interactions between Staufen and HIV-1 gRNA have been demonstrated in cells and virus particles. In virions, there is a correlation between the level of gRNA and Staufen. When Staufen is over-expressed, gRNA packaging is increased three-fold, but the infectivity of the resultant virions is reduced. In agreement with Staufen's cellular role, it has been speculated to be involved in gRNA trafficking (and consequently packaging).[72] Knock-down of Staufen by siRNA decreases infectious particle production. The combination of these findings led the authors to conclude that Staufen was present in viral RNP complexes in the cytoplasm of infected cells, where it may be involved in their transport or in recruiting gRNA.[73] Recently, Staufen levels were shown to affect stress granule formation.[146] Therefore, the effects of Staufen knock-down and over-expression on HIV-1 assembly may be indirect.

Despite the observation that Staufen is not packaged into virus-like particles (VLPs) generated by the expression of Gag alone, Staufen seems to interact with Gag.[72] Co-immunoprecipitation and bioluminescence resonance energy transfer (BRET) studies show that this interaction requires the NC domain of Gag, with the zinc finger domains being especially important, and the C-terminal and dsRNA-binding domains of Staufen. This interaction is probably RNA-independent. Further BRET analysis indicates that Staufen has a role in the multimerization of Gag polyproteins in living cells[75] and that the N terminal of the protein is important for this function, even though this region is not thought to interact with Gag.[76] Examination of cellular fractions indicated that this likely occurs specifically in membrane-associated complexes, which have been shown previously to be the primary site of Gag multimerization. Both over-expression and knock-down of Staufen increased Gag self-association and VLP formation. It would appear that the level of Staufen in producer cells is important for particle formation, but what specific role it is playing in assembly remains unclear.[75] Finally, a role for Staufen in HIV-1 translation has also been proposed.[74,147]

HIV-1 and ABCE1

ATP binding cassette protein family E (ABCE1, also known as HP68 and RLI) is an RNase L inhibitor.[148] It is involved in translation initiation,[149] and biogenesis and nuclear export of ribosomes.[150] It was identified as having a role in HIV-1 assembly in a cell-free system that allows for the separation of co-translational and post-translational steps in assembly.[151] ABCE1 specifically bound to assembly intermediates rather than fully assembled particles.[152] An interaction between Gag and ABCE1 was shown in a chronically infected T-cell line and in transfected cells, with the two proteins co-localizing in confocal microscopy studies. The NC domain of Gag is crucial for the interaction with the host protein, which is apparently RNA-independent.[153] Depletion of ABCE1 from the cell-free system did not affect total Gag protein levels but did significantly decrease the production of complete particles. A recent study suggests that ABCE1 remains associated with Gag assembly intermediates until the onset of virion release; the host protein remains in the cytoplasm upon viral budding. When HIV-1 protease is inactivated, ABCE1 and assembling particles remain associated for an increased period of time and budding is delayed.[77] These results demonstrate an important role for ABCE1 during HIV-1 assembly and release, possibly as a chaperone for Gag in the presence or absence of gRNA.

An earlier study suggested that over-expression of ABCE1 causes an increase in viral RNA and protein levels, and thus in viral replication.[154] This increase was likely caused by a decrease in interferon-induced RNase L activity due to ABCE1's endogenous activity. This suggests that ABCE1 may be acting both directly and indirectly in the viral life cycle to enhance viral replication. Interestingly, this study also showed that the expression of ABCE1 increases over the course of an HIV-1 infection, suggesting a potential mechanism for HIV-1 to inhibit antiviral defences.

HIV-1 Assembly and the Cytoskeleton

The initial contact between Gag and the gRNA has been suggested to take place in the perinuclear region,[7] whilst virion assembly takes place at the plasma membrane.[3] The route taken by Gag and the gRNA from a perinuclear site to the plasma membrane has not been elucidated. It is currently an area of debate whether the viral RNA and Gag are transported to the plasma membrane as a complex, or whether they interact once they reach the plasma membrane. Either way, it is unlikely that the complexes diffuse to the plasma membrane, since they are too big. Transport on microtubules has been suggested to be involved.[48,155] Microtubules are part of the cytoskeleton, which fulfils both structural and transport functions in eukaryotic cells. The intermediate filaments are responsible for the cell's structural stability, whereas the actin filaments and the microtubules function as transport systems for macromolecular cargos, including RNPs and vesicular structures. Transport on both systems occurs in a polarized fashion, depending on the polymerization of the filament's monomers at the growing (plus) end. As a consequence, plus end movement directs cargos toward the cell periphery, whereas minus end movement is centripetal. Myosin proteins perform the slow movement on actin bundles, and kinesin and dyenin proteins are required for the rapid transport along microtubules. Kinesins move in the plus end direction, whereas dynein moves in the opposite direction. The MTOC is located at the minus end of microtubules and resides perinuclearly. Besides its role in microtubular transport, this structure/organelle is responsible for the mitotic spindle function during cell division. Movements on both actin and microtubules depend on the ATPase activity of the motor proteins; for further details see Ref. 48.

The involvement of actin bundles and microtubules in HIV-1 assembly has been suggested, but controversies exist. Nuclear actin has been shown to be important for Rev-mediated RNA export from the nucleus.[156,157] Others have investigated the effect of actin-destabilizing drugs on virus assembly in T cells and found contradictory effects.[158] Upon treatment with mycalolide B (an actin-depolymerizer), Gag and Env were not targeted to the plasma membrane but to cytoplasmic aggregates that were shown to contain membranous vesicles. As a consequence, the release of virus particles was greatly impaired. In contrast, others have analyzed pseudotyped VLP production in primary macrophages in the presence of the actin inhibitor cytochalasin D.[3] They showed that Gag lost its diffuse intracellular localization upon treatment, whereas it was still targeted to the plasma membrane. Assembly site formation, Gag polyprotein processing and VLP release were not altered. It was therefore suggested that the diffuse intracellular Gag localization was due to endocytosis of released particles, which would depend on an intact actin network. Indeed, inhibition of endocytosis resulted in a similar change in a codon-optimized Gag–GFP protein. Not only do these data indicate the plasma membrane as the assembly site for HIV-1 in macrophages, but they also seem to solve the controversy over the endosome membranes as virus assembly sites in these cells. Unfortunately, VLP infectivity and gRNA localization were not analyzed in the cytochalasin D-treated cells.

There are several pieces of evidence suggesting that the microtubule network may be involved in the cytoplasmic transport of HIV-1 gRNA and viral assembly intermediates.

Firstly, it was shown that Gag localization was altered in Cos-1 cells by knocking down KIF4 expression.[78] KIF4 is a kinesin involved in cytokinesis that can bind Gag proteins of HIV-1 and other retroviruses.[159,160] Upon inhibition of the endogenous KIF4 function by either RNAi-mediated knock-down or by over-expression of a truncated, dominant negative form of the protein, intracellular Gag levels decreased, probably through reduced Gag protein stability. In addition, Gag was shown to build up in a perinuclear site close to the MTOC and did not traffic efficiently to the punctae. This perinuclear site was not surrounded by a nuclear membrane, and was not equivalent to aggresomes, multivesicular bodies or lysosomes. It was established that the perinuclear site contained an E2-sumoylating enzyme, Ubc9. Inactivation of the HIV-1 Gag sumoylation site affects the infectivity of the virus particles without affecting particle production. These data suggest a role for KIF4 in Gag transport and suggest that Gag transport is mediated along microtubules. Indeed, treatment of T cells with nocodazole, a compound that inhibits microtubule polymerization, prevents formation of the viral assembly site and reduces cell-to-cell transmission of HIV-1 at the virological synapse.[155] On the other hand, nocodazole treatment does not affect release of virus particles in the absence of a virological synapse.[3] This could indicate that the nocodazole-resistant microtubules are involved in the transport of Gag and possibly the HIV-1 assembly intermediates. Similarly, these stable microtubules are important for cellular transport of the infecting core after cell entry.[161]

Another protein involved in Gag trafficking to the plasma membrane is suppressor of cytokine signaling 1 (SOCS1). SOCS1 expression is induced upon HIV-1 infection (possibly by Nef) and was shown to stimulate the late steps of HIV-1 replication, which are cytoplasmic trafficking, Gag stability and virus production.[79,162] The infectivity of the virus particles was not affected, suggesting that SOCS1 did not affect RNA packaging levels significantly. SOCS1 binds the MA and NC domains of Gag, independently of gRNA, and was shown to co-localize with Gag in punctae from the perinuclear regions to the plasma membrane. Reducing SOCS1 expression by siRNAs reduces trafficking and assembly of HIV-1 Gag, resulting in perinuclear aggregates that are subjected to lysosomal degradation. These observations suggest a direct role for the host protein in HIV-1 particle production. The increased level of SOCS1 expression seen during HIV-1 infection may prevent this aberrant Gag trafficking pathway. The importance of SOCS1 was confirmed in infected Jurkat T-cell lines and primary T cells. Together, these data suggest that SOCS1 is a crucial factor that is subverted by HIV-1 for its own benefit.

Interestingly, a recent paper implicated dynein in localizing gRNA, Gag and the LAMP1 (lysosome associated membrane protein 1) marker for late endosomal vesicles to possibly a similar perinuclear region.[50] Inhibition of this trajectory by RNAi-mediated knock-down of the dynein heavy chain (DHC) increased virus particle release and intracellular Gag levels. These data suggest that dynein reroutes HIV-1 assembly products from the cell periphery inwards to the perinuclear region and thereby inhibits particle release. Unfortunately, this function was not addressed in T cells. The perinuclear locations of Gag observed in KIF4, SOCS1 and DHC knock-down experiments appear to be different, although all may affect Gag protein stability. It is of importance to determine if this Gag degradation is a functional part of the viral life cycle or merely a degradative path for excess or misassembled/misfolded proteins.

HIV-1 AND siRNA LIBRARY SCREENS

RNAi is a recently developed, powerful tool that can be used to identify additional cellular proteins required for optimal HIV-1 replication. A genome wide siRNA screen, hereafter referred to as the Brass screen, identified various cellular factors necessary for HIV-1 infection.[163] A modified HeLa cell line was transfected with over 21,000 pools of siRNAs and subsequently infected with HIV-1 (strain III-B). The siRNA pools were assessed for their effect on infectious particle production. This approach identified 273 cellular factors with possible roles in the HIV-1 life cycle. Only 36 of these proteins have previously been reported to interact with HIV-1.

A second genome wide siRNA library screen (König screen) was carried out with the aim of identifying host factors involved specifically in the early stages of HIV-1 replication.[164] 293T cells were transfected with siRNAs against ~20,000 human genes and subsequently infected with a pseudotyped HIV-1, expressing firefly luciferase. Luciferase activity was measured to determine early infection events. This screen identified 295 proteins that may be important for the viral life cycle.

Most recently, a third genome wide screen (Zhou screen) was performed in a CD4-expressing HeLa cell line.[165] Cells were transfected with around 20,000 siRNA pools and subsequently infected with HIV-1 (HXB2 strain). The impact of siRNA treatment on Tat-mediated transactivation of transcription was measured at 48 hrs and 96 hrs post-infection along with an assessment of infectious virus production. It was shown that 224 genes were important for HIV-1 replication.

It is interesting to compare the results of these screens and examine which cellular proteins have been identified in them. For example, the Brass and König screens are in agreement for 13 genes. This low level of overlap is perhaps unsurprising considering that the screens were carried out to identify proteins in late and early replication steps, respectively. Surprisingly, only 15 genes were identified in both the Brass and the Zhou screens despite similar experimental protocols. There are many possible reasons for the differences, including different approaches to designing the siRNAs, different viral strains used, and different measures of viral replication. A more detailed comparison of the three screens has recently been published.[166] Regarding the cellular factors discussed in this review, DDX3 was identified by the Brass and the Zhou screen, Sam68 was positive in the König screen, and Purα in the Brass screen. The fact that numerous other proteins were not positive suggests either that functional redundancy exists for some of the factors or proteins knocked down were essential to cell viability and so ruled out on those grounds. Alternatively, the siRNAs may not have successfully knocked down their target. It is worth considering the fact that due to the technical challenge of transfecting T cells, all of the screens were carried out in non-physiological cell lines. It will be interesting to see how reproducible the positive results from these screens are in different cell lines. These experiments highlight both the power of siRNA technology and the difficulties in interpreting the results from large scale screens.

ANTIVIRALS TARGETING CELLULAR HOST FACTORS INVOLVED IN HIV-1 RNA METABOLISM

Given the lack of a potential HIV-1 vaccine on the horizon, the continued development of antiviral drugs is of paramount importance. The majority of currently available drugs target viral proteins. The problem with such drugs is the development of viral drug resistance.[167] A possible solution to prevent this is to target cellular proteins required for optimal HIV-1 infection or their interactions with the virus. The obvious difficulty with this type of drug is the potential for cellular toxicity. The ideal target would be a cellular factor that is redundant for the cell but essential for viral replication. Alternatively, it may be possible to target the interaction between the virus and cellular factor without altering the protein's endogenous function. The recently introduced anti-retroviral drug maraviroc targets the interaction between the virus and the cellular chemokine receptor CCR5.[168] Several other cellular factors required for HIV-1 replication have been proposed as potential drug targets, including APOBEC3G and LEDGF.[169]

Two groups have recently published papers exploring the possibility of targeting DDX3 to inhibit HIV-1 replication. Yedavalli *et al.*[170] screened a panel of ring-expanded nucleosides (RENs) for activity against DDX3 function and HIV-1 replication. RENs have previously been shown to be active against the helicases of several viruses.[171] Two compounds were identified that inhibited DDX3 activity in an *in vitro* RNA unwinding assay and reduced HIV-1 replication in T cells and macrophages *in vitro.* Neither of the compounds was toxic in cells or in mice at the concentrations needed to inhibit viral replication. It remains to be determined whether it is the activity of the RENs against DDX3 that is responsible for the effects seen on viral replication.

Meanwhile, another group took a computer-based pharmacophore modeling approach using the X-ray crystallographic structure of DDX3 complexed with AMP.[172] They identified the structural features important for the interaction between the helicase and AMP, and screened databases of commercially available compounds that could be expected to interact with the DDX3 nucleotide-binding pocket. The compounds determined to have the most suitable interaction with the helicase were tested in *in vitro* systems. One compound was identified which inhibited the ATPase activity of DDX3 and the replication of HIV-1 in cell culture without evidence of cytotoxicity. Although these studies provide early evidence that inhibiting DDX3 activity is not detrimental to cell viability, it remains to be seen what the long-term effects will be. The endogenous role of DDX3 is currently unclear and even controversial. The protein has been suggested to function in interferon signaling and to modulate translation.[173–175] Moreover, DDX3 may be specifically required for the translation of mRNAs with a complex, highly structured 5′ UTR.[176] Finally, DDX3 has been identified as a component of functional spliceosomes and is associated with a number of human cancers.[177,178] Inhibition of the DDX3 ATPase activity may therefore affect various cellular processes.

CONCLUDING REMARKS

Although much of the research discussed in this chapter is still "work in progress", it is obvious that the recent development of siRNA screens, analysis of protein–protein interactions in live cells and live cell imaging are impressive and powerful tools. These techniques will, in future, inform us about the dynamic composition of gRNA-containing RNPs during their transport from the nucleus to the assembly site. This knowledge will allow us to design much-needed new antiviral drugs. At the moment, the most promising candidate discussed here is DDX3.[170,172] Although still in their infancy, the DDX3-targeting compounds have the potential to target other pathological processes that rely on DDX3 function. These include hepatitis C virus infection and tumorigenesis,[178,179] although the potentially divergent role of DDX3 in cancers of different tissues would need addressing. In addition, development of powerful genetic, biochemical and cell biological techniques will give us insight into physiological trafficking routes of cellular RNA in uninfected cells. The RNP composition on cellular RNAs is still enigmatic. As has been shown in the past, for example with RNA splicing, virological studies have the potential to reveal information about cellular processes.

ACKNOWLEDGMENTS

The authors thank Natasha Lee for critical reading of the chapter. We thank Neil Bennett for fruitful discussions. The limited scope of the chapter did not allow us to cite all references; we do apologize to those whose work we may have inadvertently failed to mention.

REFERENCES

1. Lingappa JR, Thielen BK. (2009) Assembly of immature HIV-1 capsids using a cell-free system. *Methods Mol Biol* **485**: 185–195.
2. Li H, Dou J, Ding L, Spearman P. (2007) Myristoylation is required for human immunodeficiency virus type 1 Gag–Gag multimerization in mammalian cells. *J Virol* **81**: 12899–12910.
3. Jouvenet N, Neil SJ, Bess C, *et al.* (2006) Plasma membrane is the site of productive HIV-1 particle assembly. *PLoS Biol* **4**: e435.
4. Carlton JG, Martin-serrano J. (2009) The ESCRT machinery: New functions in viral and cellular biology. *Biochem Soc Trans* **37**: 195–199.
5. Mehlin H, Daneholt B, Skoglund U. (1992) Translocation of a specific premessenger ribonucleoprotein particle through the nuclear pore studied with electron microscope tomography. *Cell* **69**: 605–613.
6. Greatorex J. (2004) The retroviral RNA dimer linkage: Different structures may reflect different roles. *Retrovirology* **1**: 22.
7. Poole E, Strappe P, Mok HP, *et al.* (2005) HIV-1 Gag-RNA interaction occurs at a perinuclear/centrosomal site: Analysis by confocal microscopy and FRET. *Traffic* **6**: 741–755.
8. Stevens M, De Clercq E, Balzarini J. (2006) The regulation of HIV-1 transcription: Molecular targets for chemotherapeutic intervention. *Med Res Rev* **26**: 595–625.
9. Purcell DF, Martin MA. (1993) Alternative splicing of human immunodeficiency virus type 1 mRNA modulates viral protein expression, replication, and infectivity. *J Virol* **67**: 6365–6378.
10. Schwartz S, Felber BK, Benko DM, *et al.* (1990) Cloning and functional analysis of multiply spliced mRNA species of human immunodeficiency virus type 1. *J Virol* **64**: 2519–2529.
11. Rabson AB, Graves BJ. (1997) Synthesis and processing of viral RNA. In: *Retroviruses*, JM Coffin, SH Hughes & HE Varmus (eds.), pp. 205–262. Cold Spring Harbor, New York, USA.
12. Dyhr-Mikkelsen H, Kjems J. (1995) Inefficient spliceosome assembly and abnormal branch site selection in splicing of an HIV-1 transcript *in vitro*. *J Biol Chem* **270**: 24060–24066.
13. Jacquenet S, Ropers D, Bilodeau PS, *et al.* (2001) Conserved stem-loop structures in the HIV-1 RNA region containing the A3 3′ splice site and its

cis-regulatory element: Possible involvement in RNA splicing. *Nucl Acids Res* **29**: 464–478.

14. O'Reilly MM, McNally MT, Beemon KL. (1995) Two strong 5′ splice sites and competing, suboptimal 3′ splice sites involved in alternative splicing of human immunodeficiency virus type 1 RNA. *Virology* **213**: 373–385.
15. Si Z, Amendt BA, Stoltzfus CM. (1997) Splicing efficiency of human immunodeficiency virus type 1 tat RNA is determined by both a suboptimal 3′ splice site and a 10 nucleotide exon splicing silencer element located within tat exon 2. *Nucl Acids Res* **25**: 861–867.
16. Suh D, Seguin B, Atkinson S, *et al.* (2003) Mapping of determinants required for the function of the HIV-1 env nuclear retention sequence. *Virology* **310**: 85–99.
17. Amendt BA, Si ZH, Stoltzfus CM. (1995) Presence of exon splicing silencers within human immunodeficiency virus type 1 tat exon 2 and tat-rev exon 3: Evidence for inhibition mediated by cellular factors. *Mol Cell Biol* **15**: 6480.
18. Tange TO, Damgaard CK, Guth S, *et al.* (2001) The hnRNP A1 protein regulates HIV-1 tat splicing via a novel intron silencer element. *EMBO J* **20**: 5748–5758.
19. Domsic JK, Wang Y, Mayeda A, *et al.* (2003) Human immunodeficiency virus type 1 hnRNP A/B-dependent exonic splicing silencer ESSV antagonizes binding of U2AF65 to viral polypyrimidine tracts. *Mol Cell Biol* **23**: 8762–8772.
20. Zahler AM, Damgaard CK, Kjems J, Caputi M. (2004) SC35 and heterogeneous nuclear ribonucleoprotein A/B proteins bind to a juxtaposed exonic splicing enhancer/exonic splicing silencer element to regulate HIV-1 tat exon 2 splicing. *J Biol Chem* **279**: 10077–10084.
21. Izaurralde E, Mattaj IW. (1995) RNA export. *Cell* **81**: 153–159.
22. Zapp ML, Green MR. (1989) Sequence-specific RNA binding by the HIV-1 Rev protein. *Nature* **342**: 714–716.
23. Fischer U, Meyer S, Teufel M, *et al.* (1994) Evidence that HIV-1 Rev directly promotes the nuclear export of unspliced RNA. *EMBO J* **13**: 4105–4112.
24. Iacampo S, Cochrane A. (1996) Human immunodeficiency virus type 1 Rev function requires continued synthesis of its target mRNA. *J Virol* **70**: 8332–8339.

25. Askjaer P, Jensen TH, Nilsson J, *et al.* (1998) The specificity of the CRM1-Rev nuclear export signal interaction is mediated by RanGTP. *J Biol Chem* **273**: 33414–33422.
26. Hope TJ. (1999) The ins and outs of HIV Rev. *Arch Biochem Biophys* **365**: 186–191.
27. Dorman N, Lever A. (2000) Comparison of viral genomic RNA sorting mechanisms in human immunodeficiency virus type 1 (HIV-1), HIV-2, and Moloney murine leukemia virus. *J Virol* **74**: 11413–11417.
28. Poon DT, Chertova EN, Ott DE. (2002) Human immunodeficiency virus type 1 preferentially encapsidates genomic RNAs that encode Pr55(Gag): Functional linkage between translation and RNA packaging. *Virology* **293**: 368–378.
29. Butsch M, Boris-Lawrie K. (2000) Translation is not required to generate virion precursor RNA in human immunodeficiency virus type 1-infected T cells. *J Virol* **74**: 11531–11537.
30. Abbink TE, Berkhout B. (2003) A novel long distance base-pairing interaction in human immunodeficiency virus type 1 RNA occludes the Gag start codon. *J Biol Chem* **278**: 11601–11611.
31. Abbink TE, Ooms M, Haasnoot PC, Berkhout B. (2005) The HIV-1 leader RNA conformational switch regulates RNA dimerization but does not regulate mRNA translation. *Biochemistry* **44**: 9058–9066.
32. Ooms M, Huthoff H, Russell R, *et al.* (2004) A riboswitch regulates RNA dimerization and packaging in human immunodeficiency virus type 1 virions. *J Virol* **78**: 10814–10819.
33. Anderson EC, Lever AM. (2006) Human immunodeficiency virus type 1 Gag polyprotein modulates its own translation. *J Virol* **80**: 10478–10486.
34. Goh WC, Rogel ME, Kinsey CM, *et al.* (1998) HIV-1 Vpr increases viral expression by manipulation of the cell cycle: a mechanism for selection of Vpr *in vivo. Nat Med* **4**: 65–71.
35. Gummuluru S, Emerman M. (1999) Cell cycle- and Vpr-mediated regulation of human immunodeficiency virus type 1 expression in primary and transformed T-cell lines. *J Virol* **73**: 5422–5430.
36. Groom HCT, Anderson EC, Dangerfield JA, Lever AML. (2009) Rev regulates translation of human immunodeficiency virus type 1 RNAs. *J Gen Virol* **90**: 1141–1147.

37. Lever A, Gottlinger H, Haseltine W, Sodroski J. (1989) Identification of a sequence required for efficient packaging of human immunodeficiency virus type 1 RNA into virions. *J Virol* **63**: 4085–4087.
38. Aldovini A, Young RA. (1990) Mutations of RNA and protein sequences involved in human immunodeficiency virus type 1 packaging result in production of noninfectious virus. *J Virol* **64**: 1920–1926.
39. Clavel F, Orenstein JM. (1990) A mutant of human immunodeficiency virus with reduced RNA packaging and abnormal particle morphology. *J Virol* **64**: 5230–5234.
40. Darlix JL, Gabus C, Nugeyre MT, *et al.* (1990) Cis elements and trans-acting factors involved in the RNA dimerization of the human immunodeficiency virus HIV-1. *J Mol Biol* **216**: 689–699.
41. Hayashi T, Shioda T, Iwakura Y, Shibuta H. (1992) RNA packaging signal of human immunodeficiency virus type 1. *Virology* **188**: 590–599.
42. Sakaguchi K, Zambrano N, Baldwin ET, *et al.* (1993) Identification of a binding site for the human immunodeficiency virus type 1 nucleocapsid protein. *Proc Natl Acad Sci USA* **90**: 5219–5223.
43. Damgaard CK, Dyhr-Mikkelsen H, Kjems J. (1998) Mapping the RNA binding sites for human immunodeficiency virus type-1 gag and NC proteins within the complete HIV-1 and -2 untranslated leader regions. *Nucl Acids Res* **26**: 3667–3676.
44. Freed EO. (1998) HIV-1 gag proteins: Diverse functions in the virus life cycle. *Virology* **251**: 1–15.
45. Berkowitz RD, Hammarskjold ML, Helga-Maria C, *et al.* (1995) 5′ regions of HIV-1 RNAs are not sufficient for encapsidation: Implications for the HIV-1 packaging signal. *Virology* **212**: 718–723.
46. Houzet L, Paillart JC, Smagulova F, *et al.* (2007) HIV controls the selective packaging of genomic, spliced viral and cellular RNAs into virions through different mechanisms. *Nucl Acids Res* **35**: 2695–2704.
47. Zeffman A, Hassard S, Varani G, Lever A. (2000) The major HIV-1 packaging signal is an extended bulged stem loop whose structure is altered on interaction with the Gag polyprotein. *J Mol Biol* **297**: 877–893.
48. Naghavi MH, Goff SP. (2007) Retroviral proteins that interact with the host cell cytoskeleton. *Curr Opin Immunol* **19**: 402–407.

49. Grigorov B, Decimo D, Smagulova F, *et al.* (2007) Intracellular HIV-1 Gag localization is impaired by mutations in the nucleocapsid zinc fingers. *Retrovirology* **4**: 54.
50. Lehmann M, Milev M, Abrahamyan L, *et al.* (2009) Intracellular transport of human immunodeficiency virus type 1 genomic RNA and viral production are dependent on dynein motor function and late endosome positioning. *J Biol Chem* **284**: 14572–14585.
51. Perlman M, Resh MD. (2006) Identification of an intracellular trafficking and assembly pathway for HIV-1 Gag. *Traffic* **7**: 731–745.
52. Crist RM, Datta SA, Stephen AG, *et al.* (2009) Assembly properties of human immunodeficiency virus type 1 Gag-leucine zipper chimeras: Implications for retrovirus assembly. *J Virol* **83**: 2216–2225.
53. Swanson CM, Puffer BA, Ahmad KM, *et al.* (2004) Retroviral mRNA nuclear export elements regulate protein function and virion assembly. *EMBO J* **23**: 2632–2640.
54. Cochrane AW, McNally MT, Mouland AJ. (2006) The retrovirus RNA trafficking granule: From birth to maturity. *Retrovirology* **3**: 18.
55. Dreyfuss G, Kim VN, Kataoka N. (2002) Messenger-RNA-binding proteins and the messages they carry. *Nat Rev Mol Cell Biol* **3**: 195–205.
56. Fang J, Kubota S, Yang B, *et al.* (2004) A DEAD box protein facilitates HIV-1 replication as a cellular co-factor of Rev. *Virology* **330**: 471–480.
57. Yedavalli VS, Neuveut C, Chi YH, *et al.* (2004) Requirement of DDX3 DEAD box RNA helicase for HIV-1 Rev-RRE export function. *Cell* **119**: 381–392.
58. Fujii R, Okamoto M, Aratani S, *et al.* (2001) A role of RNA helicase A in cis-acting transactivation response element-mediated transcriptional regulation of human immunodeficiency virus type 1. *J Biol Chem* **276**: 5445–5451.
59. Li J, Tang H, Mullen TM, *et al.* (1999) A role for RNA helicase A in post-transcriptional regulation of HIV type 1. *Proc Natl Acad Sci USA* **96**: 709–714.
60. Roy BB, Hu J, Guo X, *et al.* (2006) Association of RNA helicase A with human immunodeficiency virus type 1 particles. *J Biol Chem* **281**: 12625–12635.
61. Ajamian L, Abrahamyan L, Milev M, *et al.* (2008) Unexpected roles for UPF1 in HIV-1 RNA metabolism and translation. *RNA* **14**: 914–927.

62. Ma J, Rong L, Zhou Y, *et al.* (2008) The requirement of the DEAD-box protein DDX24 for the packaging of human immunodeficiency virus type 1 RNA. *Virology* **375**: 253–264.
63. Najera I, Krieg M, Karn J. (1999) Synergistic stimulation of HIV-1 Rev-dependent export of unspliced mRNA to the cytoplasm by hnRNP A1. *J Mol Biol* **285**: 1951–1964.
64. Beriault V, Clement JF, Levesque K, *et al.* (2004) A late role for the association of hnRNP A2 with the HIV-1 hnRNP A2 response elements in genomic RNA, Gag, and Vpr localization. *J Biol Chem* **279**: 44141–44153.
65. Levesque K, Halvorsen M, Abrahamyan L, *et al.* (2006) Trafficking of HIV-1 RNA is mediated by heterogeneous nuclear ribonucleoprotein A2 expression and impacts on viral assembly. *Traffic* **7**: 1177–1193.
66. Woolaway K, Asai K, Emili A, Cochrane A. (2007) hnRNP E1 and E2 have distinct roles in modulating HIV-1 gene expression. *Retrovirology* **4**: 28.
67. Reddy TR, Xu W, Mau JK, *et al.* (1999) Inhibition of HIV replication by dominant negative mutants of Sam68, a functional homolog of HIV-1 Rev. *Nat Med* **5**: 635–642.
68. Marsh K, Soros V, Cochrane A. (2008) Selective translational repression of HIV-1 RNA by Sam68DeltaC occurs by altering PABP1 binding to unspliced viral RNA. *Retrovirology* **5**: 97.
69. Sanchez-Velar N, Udofia EB, Yu Z, Zapp ML. (2004) hRIP, a cellular cofactor for Rev function, promotes release of HIV RNAs from the perinuclear region. *Gene Dev* **18**: 23–34.
70. Chepenik LG, Tretiakova AP, Krachmarov CP, *et al.* (1998) The single-stranded DNA binding protein, Pur-alpha, binds HIV-1 TAR RNA and activates HIV-1 transcription. *Gene* **210**: 37–44.
71. Kaminski R, Darbinian N, Sawaya BE, *et al.* (2008) Puralpha as a cellular co-factor of Rev/RRE-mediated expression of HIV-1 intron-containing mRNA. *J Cell Biochem* **103**: 1231–1245.
72. Mouland AJ, Mercier J, Luo M, *et al.* (2000) The double-stranded RNA-binding protein Staufen is incorporated in human immunodeficiency virus type 1: Evidence for a role in genomic RNA encapsidation. *J Virol* **74**: 5441–5451.
73. Chatel-Chaix L, Clement JF, Martel C, *et al.* (2004) Identification of Staufen in the human immunodeficiency virus type 1 Gag ribonucleoprotein complex and a role in generating infectious viral particles. *Mol Cell Biol* **24**: 2637–2648.

74. Dugre-Brisson S, Elvira G, Boulay K, *et al.* (2005) Interaction of Staufen1 with the 5′ end of mRNA facilitates translation of these RNAs. *Nucl Acids Res* **33**: 4797–4812.
75. Chatel-Chaix L, Abrahamyan L, Frechina C, *et al.* (2007) The host protein Staufen1 participates in human immunodeficiency virus type 1 assembly in live cells by influencing pr55Gag multimerization. *J Virol* **81**: 6216–6230.
76. Chatel-Chaix L, Boulay K, Mouland AJ, Desgroseillers L. (2008) The host protein Staufen1 interacts with the Pr55Gag zinc fingers and regulates HIV-1 assembly via its N-terminus. *Retrovirology* **5**: 41.
77. Dooher JE, Schneider BL, Reed JC, Lingappa JR. (2007) Host ABCE1 is at plasma membrane HIV assembly sites and its dissociation from Gag is linked to subsequent events of virus production. *Traffic* **8**: 195–211.
78. Martinez NW, Xue X, Berro RG, *et al.* (2008) Kinesin KIF4 regulates intracellular trafficking and stability of the human immunodeficiency virus type 1 Gag polyprotein. *J Virol* **82**: 9937–9950.
79. Ryo A, Tsurutani N, Ohba K, *et al.* (2008) SOCS1 is an inducible host factor during HIV-1 infection and regulates the intracellular trafficking and stability of HIV-1 Gag. *Proc Natl Acad Sci USA* **105**: 294–299.
80. Jeang KT, Yedavalli V. (2006) Role of RNA helicases in HIV-1 replication. *Nucl Acids Res* **34**: 4198–4205.
81. Bleichert F, Baserga SJ. (2007) The long unwinding road of RNA helicases. *Mol Cell* **27**: 339–352.
82. Rajkowitsch L, Chen D, Stampfl S, *et al.* (2007) RNA chaperones, RNA annealers and RNA helicases. *RNA Biol* **4**: 118–130.
83. Abdelhaleem M. (2004) Do human RNA helicases have a role in cancer? *Biochim Biophys Acta* **1704**: 37–46.
84. Li L, Monckton EA, Godbout R. (2008) A role for DEAD box 1 at DNA double-strand breaks. *Mol Cell Biol* **28**: 6413–6425.
85. Ishaq M, Ma L, Wu X, *et al.* (2009) The DEAD-box RNA helicase DDX1 interacts with RelA and enhances nuclear factor kappaB-mediated transcription. *J Cell Biochem* **106**: 296–305.
86. Bleoo S, Sun X, Hendzel MJ, *et al.* (2001) Association of human DEAD box protein DDX1 with a cleavage stimulation factor involved in 3′-end processing of pre-MRNA. *Mol Biol Cell* **12**: 3046–3059.

87. Sunden Y, Semba S, Suzuki T, *et al.* (2007) DDX1 promotes proliferation of the JC virus through transactivation of its promoter. *Microbiol Immunol* **51**: 339–347.
88. Kubota S, Pomerantz RJ. (2000) The nuclear function of the nuclear diffusion inhibitory signal of human immunodeficiency virus type 1: Critical roles in dominant nuclear localization and intracellular stability. *J Hum Virol* **3**: 173–181.
89. Tornatore C, Meyers K, Atwood W, *et al.* (1994) Temporal patterns of human immunodeficiency virus type 1 transcripts in human fetal astrocytes. *J Virol* **68**: 93–102.
90. Neumann M, Afonina E, Ceccherini-Silberstein F, *et al.* (2001) Nucleocytoplasmic transport in human astrocytes: Decreased nuclear uptake of the HIV Rev shuttle protein. *J Cell Sci* **114**: 1717–1729.
91. Fang J, Acheampong E, Dave R, *et al.* (2005) The RNA helicase DDX1 is involved in restricted HIV-1 Rev function in human astrocytes. *Virology* **336**: 299–307.
92. Ishaq M, Hu J, Wu X, *et al.* (2008) Knockdown of cellular RNA helicase DDX3 by short hairpin RNAs suppresses HIV-1 viral replication without inducing apoptosis. *Mol Biotechnol* **39**: 231–238.
93. Snay-Hodge CA, Colot HV, Goldstein AL, Cole CN. (1998) Dbp5p/Rat8p is a yeast nuclear pore-associated DEAD-box protein essential for RNA export. *EMBO J* **17**: 2663–2676.
94. Schmitt C, von Kobbe C, Bachi A, *et al.* (1999) Dbp5, a DEAD-box protein required for mRNA export, is recruited to the cytoplasmic fibrils of nuclear pore complex via a conserved interaction with CAN/Nup159p. *EMBO J* **18**: 4332–4347.
95. Hodge CA, Colot HV, Stafford P, Cole CN. (1999) Rat8p/Dbp5p is a shuttling transport factor that interacts with Rat7p/Nup159p and Gle1p and suppresses the mRNA export defect of xpo1–1 cells. *EMBO J* **18**: 5778–5788.
96. Lund MK, Guthrie C. (2005) The DEAD-box protein Dbp5p is required to dissociate Mex67p from exported mRNPs at the nuclear rim. *Mol Cell* **20**: 645–651.
97. Zhao J, Jin SB, Bjorkroth B, *et al.* (2002) The mRNA export factor Dbp5 is associated with Balbiani ring mRNP from gene to cytoplasm. *EMBO J* **21**: 1177–1187.

98. Hartman TR, Qian S, Bolinger C, *et al.* (2006) RNA helicase A is necessary for translation of selected messenger RNAs. *Nat Struct Mol Biol* **13**: 509–516.
99. Bolinger C, Boris-Lawrie K. (2009) Mechanisms employed by retroviruses to exploit host factors for translational control of a complex proteome. *Retrovirology* **6**: 8.
100. Sun X, Perlick HA, Dietz HC, Maquat LE. (1998) A mutated human homologue to yeast Upf1 protein has a dominant-negative effect on the decay of nonsense-containing mRNAs in mammalian cells. *Proc Natl Acad Sci USA* **95**: 10009–10014.
101. Akimitsu N. (2008) Messenger RNA surveillance systems monitoring proper translation termination. *J Biochem* **143**: 1–8.
102. Kim YK, Furic L, Desgroseillers L, Maquat LE. (2005) Mammalian Staufen1 recruits Upf1 to specific mRNA 3′UTRs so as to elicit mRNA decay. *Cell* **120**: 195–208.
103. Kaygun H, Marzluff WF. (2005) Regulated degradation of replication-dependent histone mRNAs requires both ATR and Upf1. *Nat Struct Mol Biol* **12**: 794–800.
104. Mendell JT, ap Rhys CM, Dietz HC. (2002) Separable roles for rent1/hUpf1 in altered splicing and decay of nonsense transcripts. *Science* **298**: 419–422.
105. Azzalin CM, Lingner J. (2006) The human RNA surveillance factor UPF1 is required for S phase progression and genome stability. *Curr Biol* **16**: 433–439.
106. Mendell JT, Sharifi NA, Meyers JL, *et al.* (2004) Nonsense surveillance regulates expression of diverse classes of mammalian transcripts and mutes genomic noise. *Nat Genet* **36**: 1073–1078.
107. Brandt S, Blissenbach M, Grewe B, *et al.* (2007) Rev proteins of human and simian immunodeficiency virus enhance RNA encapsidation. *PLoS Pathog* **3**: e54.
108. Krecic AM, Swanson MS. (1999) hnRNP complexes: Composition, structure, and function. *Curr Opin Cell Biol* **11**: 363–371.
109. Shyu AB, Wilkinson MF. (2000) The double lives of shuttling mRNA binding proteins. *Cell* **102**: 135–138.
110. Pinol-Roma S, Dreyfuss G. (1992) Shuttling of pre-mRNA binding proteins between nucleus and cytoplasm. *Nature* **355**: 730–732.

111. Visa N, Izaurralde E, Ferreira J, *et al.* (1996) A nuclear cap-binding complex binds Balbiani ring pre-mRNA cotranscriptionally and accompanies the ribonucleoprotein particle during nuclear export. *J Cell Biol* **133**: 5–14.
112. Kosturko LD, Maggipinto MJ, D'Sa C, *et al.* (2005) The microtubule-associated protein tumor overexpressed gene binds to the RNA trafficking protein heterogeneous nuclear ribonucleoprotein A2. *Mol Biol Cell* **16**: 1938–1947.
113. Jablonski JA, Caputi M. (2009) Role of cellular RNA processing factors in human immunodeficiency virus type 1 mRNA metabolism, replication, and infectivity. *J Virol* **83**: 981–992.
114. Stoltzfus CM, Madsen JM. (2006) Role of viral splicing elements and cellular RNA binding proteins in regulation of HIV-1 alternative RNA splicing. *Curr HIV Res* **4**: 43–55.
115. Kosturko LD, Maggipinto MJ, Korza G, *et al.* (2006) Heterogeneous nuclear ribonucleoprotein (hnRNP) E1 binds to hnRNP A2 and inhibits translation of A2 response element mRNAs. *Mol Biol Cell* **17**: 3521–3533.
116. Chertova E, Chertov O, Coren LV, *et al.* (2006) Proteomic and biochemical analysis of purified human immunodeficiency virus type 1 produced from infected monocyte-derived macrophages. *J Virol* **80**: 9039–9052.
117. Perera R, Daijogo S, Walter BL, *et al.* (2007) Cellular protein modification by poliovirus: The two faces of poly(rC)-binding protein. *J Virol* **81**: 8919–8932.
118. Schwartz M, Chen J, Janda M, *et al.* (2002) A positive-strand RNA virus replication complex parallels form and function of retrovirus capsids. *Mol Cell* **9**: 505–514.
119. Huang Y, Carmichael GG. (1996) Role of polyadenylation in nucleocytoplasmic transport of mRNA. *Mol Cell Biol* **16**: 1534–1542.
120. McLaren M, Asai K, Cochrane A. (2004) A novel function for Sam68: Enhancement of HIV-1 RNA 3′ end processing. *RNA* **10**: 1119–1129.
121. Tremblay GA, Richard S. (2006) mRNAs associated with the Sam68 RNA binding protein. *RNA Biol* **3**: 90–93.
122. Najib S, Rodriguez-Bano J, Rios MJ, *et al.* (2005) Sam68 is tyrosine phosphorylated and recruited to signalling in peripheral blood mononuclear cells from HIV infected patients. *Clin Exp Immunol* **141**: 518–525.

123. Najib S, Martin-Romero C, Gonzalez-Yanes C, Sanchez-Margalet V. (2005) Role of Sam68 as an adaptor protein in signal transduction. *Cell Mol Life Sci* **62**: 36–43.
124. Reddy TR, Tang H, Xu W, Wong-Staal F. (2000) Sam68, RNA helicase A and Tap cooperate in the post-transcriptional regulation of human immunodeficiency virus and type D retroviral mRNA. *Oncogene* **19**: 3570–3575.
125. Li J, Liu Y, Kim BO, He JJ. (2002) Direct participation of Sam68, the 68-kilodalton Src-associated protein in mitosis, in the CRM1-mediated Rev nuclear export pathway. *J Virol* **76**: 8374–8382.
126. Modem S, Badri KR, Holland TC, Reddy TR. (2005) Sam68 is absolutely required for Rev function and HIV-1 production. *Nucl Acids Res* **33**: 873–879.
127. Henao-Mejia J, Liu Y, Park IW, *et al.* (2009) Suppression of HIV-1 Nef translation by Sam68 mutant-induced stress granules and Nef mRNA sequestration. *Mol Cell* **33**: 87–96.
128. Groom HCT, Anderson EC, Lever AML. (2009) Rev: Beyond nuclear export. *J Gen Virology* **90**: 1303–1318.
129. Suhasini M, Reddy TR. (2009) Cellular proteins and HIV-1 Rev function. *Curr HIV Res* **7**: 91–100.
130. Cochrane AW. (2009) Inhibition of HIV-1 gene expression by Sam68deltaC: Multiple targets but a common mechanism? *Retrovirology* **6**: 22.
131. Bogerd HP, Fridell RA, Madore S, Cullen BR. (1995) Identification of a novel cellular cofactor for the Rev/Rex class of retroviral regulatory proteins. *Cell* **82**: 485–494.
132. Fritz CC, Zapp ML, Green MR. (1995) A human nucleoporin-like protein that specifically interacts with HIV Rev. *Nature* **376**: 530–533.
133. Yu Z, Sanchez-Velar N, Catrina IE, *et al.* (2005) The cellular HIV-1 Rev cofactor hRIP is required for viral replication. *Proc Natl Acad Sci USA* **102**: 4027–4032.
134. Bergemann AD, Johnson EM. (1992) The HeLa Pur factor binds single-stranded DNA at a specific element conserved in gene flanking regions and origins of DNA replication. *Mol Cell Biol* **12**: 1257–1265.
135. Johnson EM, Kinoshita Y, Weinreb DB, *et al.* (2006) Role of Pur alpha in targeting mRNA to sites of translation in hippocampal neuronal dendrites. *J Neurosci Res* **83**: 929–943.

136. Haas S, Thatikunta P, Steplewski A, *et al.* (1995) A 39-kD DNA-binding protein from mouse brain stimulates transcription of myelin basic protein gene in oligodendrocytic cells. *J Cell Biol* **130**: 1171–1179.
137. Chen NN, Khalili K. (1995) Transcriptional regulation of human JC polyomavirus promoters by cellular proteins YB-1 and Pur alpha in glial cells. *J Virol* **69**: 5843–5848.
138. Stacey DW, Hitomi M, Kanovsky M, *et al.* (1999) Cell cycle arrest and morphological alterations following microinjection of NIH3T3 cells with Pur alpha. *Oncogene* **18**: 4254–4261.
139. Gallia GL, Darbinian N, Jaffe N, Khalili K. (2001) Single-stranded nucleic acid-binding protein, Pur alpha, interacts with RNA homologous to 18S ribosomal RNA and inhibits translation *in vitro*. *J Cell Biochem* **83**: 355–363.
140. Shimotai Y, Minami H, Saitoh Y, *et al.* (2006) A binding site for Pur alpha and Pur beta is structurally unstable and is required for replication *in vivo* from the rat aldolase B origin. *Biochem Biophys Res Commun* **340**: 517–525.
141. Krachmarov CP, Chepenik LG, Barr-Vagell S, *et al.* (1996) Activation of the JC virus Tat-responsive transcriptional control element by association of the Tat protein of human immunodeficiency virus 1 with cellular protein Pur alpha. *Proc Natl Acad Sci USA* **93**: 14112–14117.
142. Gallia GL, Darbinian N, Tretiakova A, *et al.* (1999) Association of HIV-1 Tat with the cellular protein, Puralpha, is mediated by RNA. *Proc Natl Acad Sci USA* **96**: 11572–11577.
143. Marion RM, Fortes P, Beloso A, *et al.* (1999) A human sequence homologue of Staufen is an RNA-binding protein that is associated with polysomes and localizes to the rough endoplasmic reticulum. *Mol Cell Biol* **19**: 2212–2219.
144. Kiebler MA, Hemraj I, Verkade P, *et al.* (1999) The mammalian staufen protein localizes to the somatodendritic domain of cultured hippocampal neurons: Implications for its involvement in mRNA transport. *J Neurosci* **19**: 288–297.
145. Le S, Sternglanz R, Greider CW. (2000) Identification of two RNA-binding proteins associated with human telomerase RNA. *Mol Biol Cell* **11**: 999–1010.
146. Thomas MG, Tosar LJ, Desbats MA, *et al.* (2009) Mammalian Staufen 1 is recruited to stress granules and impairs their assembly. *J Cell Sci* **122**: 563–573.

147. Parkin NT, Cohen EA, Darveau A, *et al.* (1988) Mutational analysis of the 5′ non-coding region of human immunodeficiency virus type 1: Effects of secondary structure on translation. *EMBO J* 7: 2831–2837.
148. Bisbal C, Martinand C, Silhol M, *et al.* (1995) Cloning and characterization of a RNAse L inhibitor: A new component of the interferon-regulated 2-5A pathway. *J Biol Chem* **270**: 13308–13317.
149. Chen ZQ, Dong J, Ishimura A, *et al.* (2006) The essential vertebrate ABCE1 protein interacts with eukaryotic initiation factors. *J Biol Chem* **281**: 7452–7457.
150. Kispal G, Sipos K, Lange H, *et al.* (2005) Biogenesis of cytosolic ribosomes requires the essential iron-sulphur protein Rli1p and mitochondria. *EMBO J* **24**: 589–598.
151. Lingappa JR, Hill RL, Wong ML, Hegde RS. (1997) A multistep, ATP-dependent pathway for assembly of human immunodeficiency virus capsids in a cell-free system. *J Cell Biol* **136**: 567–581.
152. Zimmerman C, Klein KC, Kiser PK, *et al.* (2002) Identification of a host protein essential for assembly of immature HIV-1 capsids. *Nature* **415**: 88–92.
153. Lingappa JR, Dooher JE, Newman MA, *et al.* (2006) Basic residues in the nucleocapsid domain of Gag are required for interaction of HIV-1 gag with ABCE1 (HP68), a cellular protein important for HIV-1 capsid assembly. *J Biol Chem* **281**: 3773–3784.
154. Martinand C, Montavon C, Salehzada T, *et al.* (1999) RNase L inhibitor is induced during human immunodeficiency virus type 1 infection and down regulates the 2-5A/RNase L pathway in human T cells. *J Virol* **73**: 290–296.
155. Jolly C, Mitar I, Sattentau QJ. (2007) Requirement for an intact T-cell actin and tubulin cytoskeleton for efficient assembly and spread of human immunodeficiency virus type 1. *J Virol* **81**: 5547–5560.
156. Kimura T, Hashimoto I, Yamamoto A, *et al.* (2000) Rev-dependent association of the intron-containing HIV-1 gag mRNA with the nuclear actin bundles and the inhibition of its nucleocytoplasmic transport by latrunculin-B. *Gene Cells* **5**: 289–307.
157. Hofmann W, Reichart B, Ewald A, *et al.* (2001) Cofactor requirements for nuclear export of Rev response element (RRE)- and constitutive transport element (CTE)-containing retroviral RNAs: An unexpected role for actin. *J Cell Biol* **152**: 895–910.

158. Sasaki H, Ozaki H, Karaki H, Nonomura Y. (2004) Actin filaments play an essential role for transport of nascent HIV-1 proteins in host cells. *Biochem Biophys Res Commun* **316**: 588–593.
159. Kim W, Tang Y, Okada Y, *et al.* (1998) Binding of murine leukemia virus Gag polyproteins to KIF4, a microtubule-based motor protein. *J Virol* **72**: 6898–6901.
160. Tang Y, Winkler U, Freed EO, *et al.* (1999) Cellular motor protein KIF-4 associates with retroviral Gag. *J Virol* **73**: 10508–10513.
161. Naghavi MH, Valente S, Hatziioannou T, *et al.* (2007) Moesin regulates stable microtubule formation and limits retroviral infection in cultured cells. *EMBO J* **26**: 41–52.
162. Qiao X, He B, Chiu A, *et al.* (2006) Human immunodeficiency virus 1 Nef suppresses CD40-dependent immunoglobulin class switching in bystander B cells. *Nat Immunol* **7**: 302–310.
163. Brass AL, Dykxhoorn DM, Benita Y, *et al.* (2008) Identification of host proteins required for HIV infection through a functional genomic screen. *Science* **319**: 921–926.
164. Konig R, Zhou Y, Elleder D, *et al.* (2008) Global analysis of host-pathogen interactions that regulate early-stage HIV-1 replication. *Cell* **135**: 49–60.
165. Zhou H, Xu M, Huang Q, *et al.* (2008) Genome-scale RNAi screen for host factors required for HIV replication. *Cell Host Microbe* **4**: 495–504.
166. Goff SP. (2008) Knockdown screens to knockout HIV-1. *Cell* **135**: 417–420.
167. Shafer RW, Schapiro JM. (2008) HIV-1 drug resistance mutations: An updated framework for the second decade of HAART. *AIDS Rev* **10**: 67–84.
168. Dolin R. (2008) A new class of anti-HIV therapy and new challenges. *N Engl J Med* **359**: 1509–1511.
169. Greene WC, Debyser Z, Ikeda Y, *et al.* (2008) Novel targets for HIV therapy. *Antiviral Res* **80**: 251–265.
170. Yedavalli VS, Zhang N, Cai H, *et al.* (2008) Ring expanded nucleoside analogues inhibit RNA helicase and intracellular human immunodeficiency virus type 1 replication. *J Med Chem* **51**: 5043–5051.
171. Zhang N, Chen HM, Koch V, *et al.* (2003) Ring-expanded ("fat") nucleoside and nucleotide analogues exhibit potent *in vitro* activity against flaviviridae NTPases/helicases, including those of the West Nile virus, hepatitis C virus, and Japanese encephalitis virus. *J Med Chem* **46**: 4149–4164.

172. Maga G, Falchi F, Garbelli A, *et al.* (2008) Pharmacophore modeling and molecular docking led to the discovery of inhibitors of human immunodeficiency virus-1 replication targeting the human cellular aspartic acid-glutamic acid-alanine-aspartic acid box polypeptide 3. *J Med Chem* **51**: 6635–6638.
173. Schroder M, Baran M, Bowie AG. (2008) Viral targeting of DEAD box protein 3 reveals its role in TBK1/IKKepsilon-mediated IRF activation. *EMBO J* **27**: 2147–2157.
174. Shih JW, Tsai TY, Chao CH, Wu Lee YH. (2008) Candidate tumor suppressor DDX3 RNA helicase specifically represses cap-dependent translation by acting as an eIF4E inhibitory protein. *Oncogene* **27**: 700–714.
175. Lee CS, Dias AP, Jedrychowski M, *et al.* (2008) Human DDX3 functions in translation and interacts with the translation initiation factor eIF3. *Nucl Acids Res* **36**: 4708–4718.
176. Lai MC, Lee YH, Tarn WY. (2008) The DEAD-box RNA helicase DDX3 associates with export messenger ribonucleoproteins as well as tip-associated protein and participates in translational control. *Mol Biol Cell* **19**: 3847–3858.
177. Zhou Z, Licklider LJ, Gygi SP, Reed R. (2002) Comprehensive proteomic analysis of the human spliceosome. *Nature* **419**: 182–185.
178. Botlagunta M, Vesuna F, Mironchik Y, *et al.* (2008) Oncogenic role of DDX3 in breast cancer biogenesis. *Oncogene* **27**: 3912–3922.
179. Ariumi Y, Kuroki M, Abe K, *et al.* (2007) DDX3 DEAD-box RNA helicase is required for hepatitis C virus RNA replication. *J Virol* **81**: 13922–13926.

CHAPTER 7

Integration Site Selection by Retroviruses and Retroviral Vectors

Claudia Cattoglio[†] and Fulvio Mavilio[*,‡]

INTRODUCTION

The integration process, i.e., the stable insertion of viral DNA into the host-cell genome, is one of the defining features of the life cycle of retroviruses. It is the result of an evolutionary strategy ensuring long-lasting viral gene expression and permanent transmission to the host-cell progeny. Diverse genomic DNA sequences can function as acceptors for retroviral integration, and only a weak sequence consensus exists at the viral insertion site.[1] Nevertheless, target-site selection is not random: recent large-scale surveys of retroviral integration in infected murine and human cells uncovered some genomic features systematically and specifically associated with retroviral insertions, and revealed that each

*Corresponding author.
†Laboratory of Gene Expression, Istituto Scientifico H. San Raffaele, Via Olgettina 58, 20132 Milano, Italy.
‡Department of Biomedical Sciences, University of Modena and Reggio Emilia, Via Gottardi 100, Modena, Italy. E-mail: fulvio.mavilio@unimore.it.

retrovirus has a unique, characteristic pattern of integration within mammalian genomes (reviewed in Ref. 2).

Despite intense research efforts, the molecular mechanisms determining retroviral target-site selection are still largely unknown; a deeper understanding of the subject would have a double impact, on basic retrovirology as well as on its clinical applications. The identification of viral and/or cellular players controlling integration preferences would provide new targets for anti-retroviral therapy, which would be especially valuable in the case of HIV. The gene therapy field in particular will benefit from a deeper knowledge of retroviral integration mechanisms. Replication-defective, integration-competent retroviral vectors have been extensively used to stably deliver therapeutic genes to a variety of cell types, and are still the only available tool to permanently integrate a therapeutic transgene with an efficiency compatible with clinical application. Vectors derived from the murine leukemia viruses (MLV) have been used in a variety of gene therapy protocols with remarkable success in the treatment of hematopoietic disorders such as severe combined immunodeficiencies (SCIDs). Although the oncogenic potential of murine retroviruses had been known for decades, retroviral vectors were considered relatively safe because integration was believed to be random, and the chance of accidentally disrupting or activating a gene was thought to be remote (10^{-7} per insertion[3]). This assumption was rapidly reassessed when a lymphoproliferative disorder was reported in a patient treated for X-linked SCID by MLV-transduced hematopoietic progenitors. The malignant T-cell clone carried a single copy of the therapeutic vector integrated at the 5′ of the LMO2 proto-oncogene, sufficiently close to the viral LTR enhancer to upregulate its expression.[4] At present, five out of 18 patients enrolled in two independent X-linked SCID clinical trials have experienced insertional oncogenesis, and in all but one case the malignant clone hosted an MLV insertion near the LMO2 gene.[5,6] Subsequent studies on the integration preferences of MLV-derived vectors partially explained the apparently high frequency of such events.[2] Understanding retroviral integration mechanisms has therefore become a major safety issue in the gene-therapy field, where comparative analyses of the integration patterns of different retroviruses should ideally identify the vector(s) with the least harmful profile(s).

This chapter summarizes our current knowledge of the genomic and viral determinants of integration-site selection by retroviruses and their vector counterparts in the human genome, with a focus on the Moloney MLV (Mo-MLV) gamma-retrovirus and the human immunodeficiency virus type 1 (HIV-1) lentivirus, for which in-depth studies are available. We start by listing a series of genomic features that can be positively or negatively correlated to retroviral integration, and continue by analyzing the insertion preferences genus by genus, choosing one retrovirus as representative of an entire family. There is a brief description of techniques for integration-site recovering and analysis. The last part of the chapter proposes models for possible mechanisms of integration targeting, and highlights implications and perspectives of target-site selection for both sides, the virus and the host cell.

GENOME-WIDE STUDIES OF RETROVIRAL INTEGRATION IN MAMMALIAN GENOMES

Before completion of genome sequencing projects, it was impossible to obtain an accurate global picture of retroviral integration events. Early studies using *in vitro* integration models identified several factors enhancing or reducing insertion efficiency, such as DNA bending induced by nucleosomal assembly, steric hindrance of DNA binding proteins, and DNA physical structure. However, target-site selection *in vivo* remained poorly understood, with early studies on Mo-MLV and avian sarcoma-leukosis viruses (ASLV) producing conflicting results. As soon as almost complete sequences were available for several vertebrate genomes, genome-wide approaches were used to analyze integration targeting in a statistically rigorous manner. Large-scale, high-throughput methods were designed to clone and sequence the junctions between proviral and host-cell DNA (see Box 1). More recently, massive parallel sequencing techniques have been adapted to integration studies, providing a 100-fold increase in throughput over the classical Sanger technology.[7–9] Once retrieved, the position of integration sites can be correlated with a series of general genomic feature and cell-type specific annotations. This has now been done in a variety of cell types, derived from different species (avian, murine, human and non-human primate primary cells and/or cell lines) after acute

Box 1. Methods for the isolation of integration sites from genomic DNA of infected cells.

To retrieve retroviral integration sites, cells are first infected with the retrovirus or the retroviral vector of interest; two days of culture after infection are sufficient to achieve complete integration. Long-term culture periods after transduction can result in clonal outgrowth and are only allowed when the interest is in selecting growth-promoting integrations. Infected cells are then harvested and genomic DNA is isolated. Different strategies are then available.

Inverse (I) PCR Genomic DNA extracted from infected cells is first cleaved with a restriction enzyme that cuts only once within the provirus (1), and self-ligated to form circles containing the vector-genome junction (2). Circles are then subjected to nested PCR with two couples of primers annealing on opposite strands of the proviral LTR (3). The PCR products containing the genomic sequence from the insertion nucleotide up to the nearest restriction enzyme recognition site are finally cloned and sequenced, using the classic Sanger procedure or massive sequencing (4). See Ref. 70.

Linker-mediated (LM) PCR DNA from infected cells is cleaved with one or several 4-bp cutter restriction enzyme(s), to generate fragments ranging from tens to a few thousands of base pairs. An incompatible restriction enzyme cutting inside the provirus is used to prevent linker ligation and amplification of the internal backbone (1). DNA is then ligated to DNA linkers with sticky ends and a nested PCR is performed, using a couple of primers binding to the viral LTR and another couple to the DNA linker. Depending on the primer used, either 3′ LTR- or 5′ LTR-genome junctions, or both, can be sequenced (2). PCR products are then cloned and sequenced (3). See Ref. 71.

(*Continued*)

Box 1. (*Continued*)

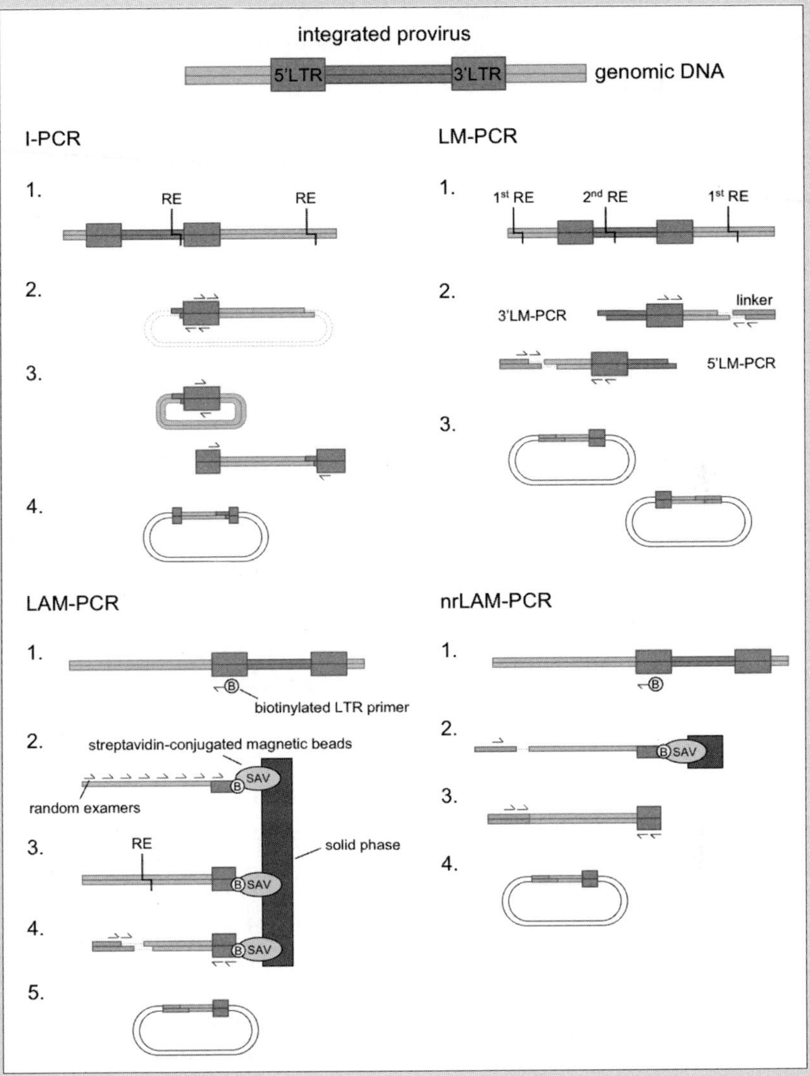

(*Continued*)

Box 1. (*Continued*)

Linear amplification-mediated (LAM) PCR Genomic DNA from infected cells is first subjected to a linear PCR by primer extension, using LTR-specific biotinylated primers (1). Amplified fragments are then enriched by magnetic tag selection of extension primers, and double-strand DNA is synthesized by random exanucleotide priming (2). After digestion with one or several 4-bp cutter restriction enzyme(s) (3), DNA linkers with sticky ends are ligated and nested PCR is performed essentially as in the LM-PCR (4-5). The approach is particularly suitable for low-efficiently infected cells. See Ref. 72.

Non-restrictive linear amplification-mediated (nrLAM) PCR This technique has been recently proposed as an advanced LAM-PCR approach to circumvent biases in integration-site recovery due to restriction enzyme digestion of genomic DNA. The single-stranded DNA originated by the first linear PCR (1) is directly ligated to a single-stranded DNA linker via T4 RNA-ligase (2), with no need for restriction digestion. The single-strand ligation product is then directly used for the nested PCR, as in LM-PCR and classic LAM-PCR approaches (3). PCR products are then cloned and sequenced. See Ref. 14.

infection with different retroviruses or retroviral vectors (among others: HIV-1, simian immunodeficiency virus, Mo-MLV, and ASLV; extensively reviewed by Bushman *et al.*[2]). Since no significant differences in terms of integration distribution were found between replication-competent retroviruses and vectors derived from them,[10] the former were soon replaced by the latter in insertion studies.

Primary DNA Sequence

One of the most obvious characteristics to be investigated for a role in integration-site selection is the primary DNA sequence at the target site.

While the integrase has strict sequence requirements for the viral DNA ends (the dinucleotide CA, invariably located 2 base pairs from both ends of the viral termini, and sequences up to 15 base pairs upstream of the CA), target-site sequences are very diverse. A recent study re-analyzing integration sites from HIV-1-, Mo-MLV-, ASLV- and SIV-infected cells found a statistically weak palindromic consensus, centered on the virus-specific duplicated target-site sequence.[1] The consensus is weakly conserved but distinguishable between different retroviruses, as later confirmed by other larger surveys[9,11] (Fig. 1). The same consensus is also found around insertion sites catalyzed by pre-integration complexes (PICs) in naked genomic DNA *in vitro*, suggesting that the observed preferences are due to the integration machinery itself rather than due to host factors. The most widely accepted hypothesis is that this consensus reflects a need for certain bases at certain positions to meet the spatial or energy requirements of the integration complex.

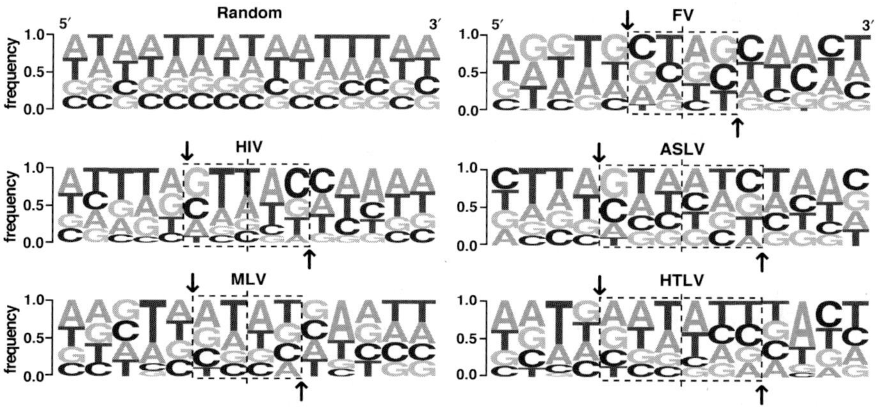

Fig. 1. Favored DNA sequences at retroviral integration sites. Retroviral insertion (specified for both strands by the vertical arrows) results in target-site duplication (4- to 6-base pair repeats, in dashed boxes). For each position around the integration site, the height of each base reflects its frequency at that site (from 0 to 100 percent). The base preference shows a weak palindromic pattern centered on the duplicated target sites (symmetry is marked by a dashed vertical line). Sequence logos were obtained by the enoLOGOS Web-based tool,[77] using the base composition reported in Ref. 20 as input.

General Genomic Features

Virtually any general genomic feature whose position is annotated along the genome can be explored to seek a correlation with retroviral integration. This does not necessarily require programming abilities, thanks to several publicly available genome browsers and/or databases that enable researchers to visualize, annotate and analyze entire genomes (see Ref. 12). for an extensive list of molecular biology databases). The position of integration sites is therefore easily correlated with the presence and the density of coding and non-coding transcription units, CpG islands, GC-content, centromeric regions and repetitive elements, fragile sites, sequences conserved across species, transcription factor binding sites, etc. A positive or negative association between retroviral insertions and the genomic feature of interest is calculated by comparing the integration frequency near that feature to the frequency expected from random integration. For example, considering the well-characterized RefSeq genes as the reference category for coding RNAs, around 30 percent of the human genome consists of transcription units; any retrovirus that targets genes at significantly higher frequency shows a preference for intragenic versus intergenic regions. Whatever the bias towards genes, a relevant question that often arises is whether one or another functional category of genes is favored by the integration; functional annotation tools based on the gene ontology categories are publicly available to tackle the issue (among others, the DAVID Bioinformatics Resources 2008[13]). Collections of computer-generated, random integration sites are normally used for statistical comparisons. However, using completely random sites can be misleading: the experimental procedure commonly used to map insertion sites entails the cleavage of genomic DNA with restriction enzymes (see Box 1), introducing a recovery bias. Recent studies showed that, depending on the restriction enzymes used, as much as one third of the human genome can be inaccessible to integration analysis.[9,14] Hence, the perfect random controls must be "matched" to have the same potential bias i.e., random sites are constrained to lie at the same distance from a restriction site as the experimental sites. Another critical issue is the number of integration sites needed to assign a trend to the data; low numbers and weak statistics may lead to unreproducible conclusions. This clearly depends on how much a given genomic feature is relevant to the integration-site selection operated by the retrovirus: the stronger the association, the fewer the insertion sites needed to detect it. The peculiar preference of Mo-MLV virus for transcription start sites (see Mo-MLV section) is so strong that a few hundred integrations are sufficient to detect it.

Cell-Specific Genomic Features

Many interesting genomic features to be associated with retroviral target-site selection are not universally defined, but vary between different cell types. Transcriptional activity, DNA accessibility (DNase I hypersensitive or resistant sites) and chromatin organization are all attributes that can dramatically change from one cellular context to another, since they reflect cell-specific transcriptional regulatory pathways. Public repositories of transcription profiles exist (e.g., the Gene Expression Omnibus-GEO[15]), which contain thousands of datasets derived from cell lines and primary cells of different species cultured under various conditions. Ad hoc expression arrays performed on the cell of interest are clearly the best choice, but a careful selection of deposited data can be a reasonable approximation. Genome-wide annotations of DNase I hypersensitive sites[16] and of a set of epigenetic modifications[17–19] are by contrast available for only a few cell types; therefore particular caution is needed if they are used to draw conclusions in a different cell context. The field is, however, in rapid expansion, and in a few years many more of such genome-wide, large-scale data sets will be available for a variety of cell types.

INTEGRATION PREFERENCES OF RETROVIRUSES AND RETROVIRAL VECTORS

Based on evolutionary relatedness, retroviruses are classified into seven genera (Alpha-, Beta-, Delta-, Epsilon- and Gamma-retroviridae, Spumaviridae and Lentiviridae). Except for Epsilon-retroviruses, at least one member of each family has been investigated for integration preferences, where the number of studies reflects the clinical relevance of that virus (HIV-1 and

Mo-MLV integration profiles are the most extensively characterized). These surveys revealed different patterns of favored and disfavored target sites for each retroviral family, suggesting differential involvement of viral and cellular players in each integration process. Interestingly, unsupervised clustering of different retroviruses on the basis of their integration preferences beautifully overlaps with phylogenetic trees based on the sequence similarity of their integrases, which, in turn, are in good agreement with traditional trees based on genomic sequences (Fig. 2).[20] This adds an exquisite evolutionary flavor to the strategy of target-site selection used by single retroviruses; indeed, the survival of a retrovirus cannot but depend on the site in the host-cell genome in which it integrates.

The integration preferences of a representative member for each retroviral family are discussed in detail below and summarized for the most commonly correlated genomic features in Table 1. Possible mechanisms underlying these preferences are discussed in the following section.

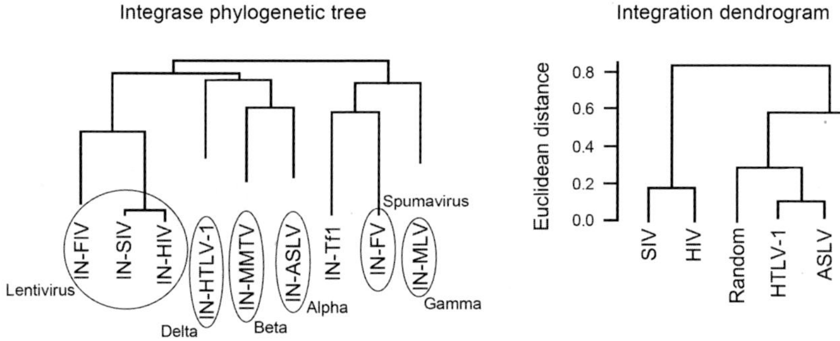

Fig. 2. Clustering of different retroviruses based on their integrase sequence and integration profile. The phylogenetic tree on the left is obtained from the alignment of the amino acid sequences of the integrase (IN) of eight different retroviruses, belonging to six different *genera* (circled), and of the yeast retrotransposon Tf1. The integration dendrogram on the right derives from an unsupervised hierarchical clustering (euclidean distance, average linkage) of the location of retroviral insertions with respect to 69 distinct genomic features. Retroviruses with a similar integration profile segregate together (HIV and SIV, FV and MLV, HTLV-1 and ASLV), and those with the less biased pattern (HTLV-1 and ASLV) segregate with random sequences. The integration tree is in good accordance with the integrase phylogenetic analysis, suggesting that the most closely related integrases direct integration into regions of the genome with similar features (Modified from Ref. 20).

Table 1. Favored or disfavored genomic features around retroviral integration sites. For each virus, the frequency of integration around the specified feature is calculated and divided by the expected frequency for random integration. The resulting number (fold-enrichment over random control sites) is reported in the table. A value of 1 indicates that the genomic feature is irrelevant to target-site selection; a value of less than 1 suggests that integration around that genomic feature is disfavored. Cell shades are indicative of the level of over- or under-representation with respect to controls (red stands for highly significant over-representation; light blue stands for under-representation). The source of the data (reference), the cell type and the number of integrations retrieved for each virus are indicated in the first two rows. ND: not determined.

	Fold Enrichment Over Random Control Sites					
	HIV-1	Mo-MLV	FV	ASLV	MMTV	HTLV-1
Reference(s)	Wang *et al.*[10]	Wu *et al.*[67] Lewinsky *et al.*[17] Cattoglio *et al.*[24]	Trobridge *et al.*[69] Nowrouzi *et al.*[70]	Narezkina *et al.*[71] Mitchell *et al.*[72]	Faschinger *et al.*[73] Brady *et al.*[65]	Derse *et al.*[14]
Cell Type (No. of Integrations)	Jurkat (40,569)	HeLa (903) HeLa (544) $CD34^+$ HSCs (1,030)	$CD34^+$ HSCs (1,821) Fibroblasts (1,008) 293 (628)	293-TVA (469) HeLa (226)	Hs578T (298) NMuMG (170)	HeLa (541)
RefSeq Genes	2	1.35	1	1.4	1	1.3
Gene Activity	1.3–1.5	1.3	1	1.3	1	ND
TSS (±5 kb)	<1	6–7	4	1.3	<1	2
CpG Islands (±1 kb)	<1	5	3.5	1.6	<1	1.3
DNaseI Hypersensitive Sites (±1 kb)	<1	13	6	1.8	1	1
G/C Content	<1	1.04	1.05	1	1	1
Gene Density (1Mb)	1.7	1.6	1.2	1.2	1	1.2

n n n n n n Level of representation with respect to control datasets (from white to dark grey indicates under- to over-representation).

Moloney Murine Leukemia Virus (Mo-MLV)

Mo-MLV is part of the Gamma-retrovirus family, historically known as Oncoretroviridae for their ability to induce tumors. Because they reach high expression levels in the hematopoietic system, Mo-MLV-based vectors carrying wild-type LTRs have been largely used in gene therapy for blood disorders. The severe adverse events reported in the X-SCID clinical trials led the scientific community to reconsider the safety of MLV-derived vectors and to look at their integration characteristics with a renewed interest. Mo-MLV integration profiling in a variety of mammalian cells revealed a modest preference for active genes, and a striking distribution around transcription start sites (TSSs), with ~20 percent of insertions landing 5kb upstream or downstream from the +1 position of any gene [Fig. 3(a)]. TSSs have therefore been considered as a major genomic determinant of Mo-MLV integration-site selection. On closer examination, however, it appears that the Mo-MLV bias for TSSs is simply one of the consequences of a more general preference of Mo-MLV PICs for genomic regions with a role in transcriptional regulation. In fact, regions flanking Mo-MLV integrations are particularly enriched in CpG islands [Fig. 3(b)], DNase I hypersensitive sites, transcription factor binding sites [TFBSs, Fig. 3(c)],[21–23] and phylogenetically conserved sequences (unpublished observation). The link between Mo-MLV integration and the host-cell transcriptional program is even more apparent when looking at the nature of the genes targeted by the virus: an extensive analysis of Mo-MLV insertion sites in human hematopoietic stem/progenitor cells (HSCs) showed that genes involved in the development and function of the hematopoietic and immune systems are targeted at high frequency,[21] suggesting that cell-specific gene expression programs are instrumental in directing Mo-MLV PICs to the genome. In accordance with this hypothesis, the analysis of TFBSs around Mo-MLV integrations in HSCs, T-cells, keratinocytes and HeLa cells identified cell-specific patterns (Ref. 23 and unpublished observations). Another striking feature of the Mo-MLV insertion profile is the high incidence of hot spots, i.e., genomic loci recurrently targeted by the virus; about 20 percent of over 1,000 Mo-MLV integrations in HSCs resulted clustered near proto-oncogenes and/or genes controlling hematopoietic and immune functions. More recently, massive parallel sequencing studies revealed an even higher frequency of hot spots and indicated that a high clustering level is a peculiar characteristic of MLV integration (unpublished observations). Recurrent integration around certain categories of genes explains in part the phenomenon of clonal dominance reported in long-term pre-clinical and clinical studies of Mo-MLV-transduced cells. In fact, analysis of the progeny of transduced HSCs in mice,[24] non-human primates[25] and humans[26,27] identified dominant hematopoietic clones that hosted Mo-MLV insertions near proto-oncogenes or other genes controlling cell growth, proliferation or signal transduction. The conclusion of these studies was that vector-induced deregulation of certain categories of genes confers some growth and/or survival advantage to transduced progenitors, resulting in their *in vivo* amplification. The high frequency of Mo-MLV hot spots with the same characteristics suggests, however, that clonal dominance is not entirely the result of *in vivo* selection, but it is also favored by the existence of highly preferred regions for retroviral integration that make clonal amplification more likely to occur.

Human Immunodeficiency Virus Type 1 (HIV-1)

HIV-1 is one of the several members of the Lentiviridae genus. Due to obvious therapeutic implications, the HIV-1 integration pattern was the first to be extensively characterized.[10] The most evident feature of the HIV-1 insertion profile is probably its preference for genes and gene-rich regions, with 60 to 80 percent of integration sites, depending on the target cell, landing within a transcription unit. Unlike Mo-MLV, HIV-1 integrations are evenly spread along the whole transcription unit, with no bias around TSSs [Fig. 3(a)]. Genes transcriptionally active at the time of infection are more likely to be targeted by HIV PICs. Local A/T-rich target DNA sequences are also favored, in accordance with the role proposed for the A/T-hook DNA-binding protein LEDGF/p75 in tethering HIV-1 PICs (see next section). HIV-1 integration is disfavored in genomic regions containing CpG islands [Fig. 3(b)], G/C-rich sequences, DNase I hypersensitive sites, and TFBSs [Fig. 3(c)], denoting a negative preference for transcriptional regulatory domains, in sharp contrast to Mo-MLV. Accordingly, a broad group of "housekeeping" genes, controlling cell cycle, metabolism and replication are found among HIV-1 integration targets.

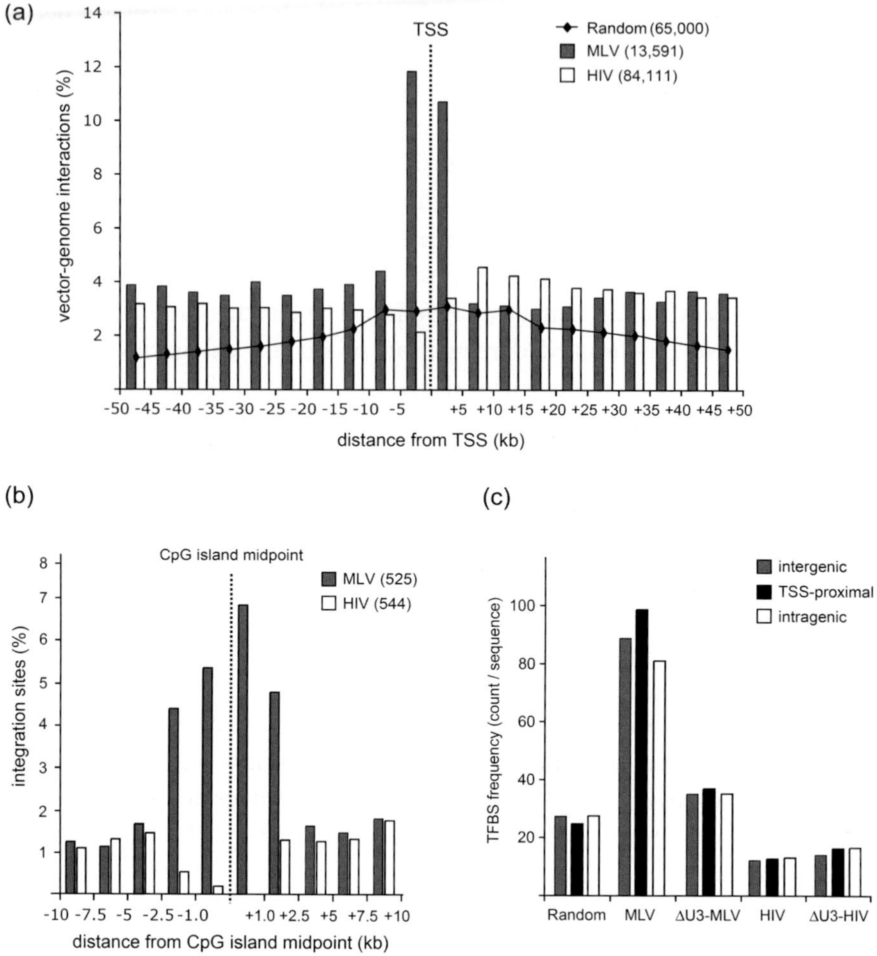

Fig. 3. Genomic features associated to Mo-MLV and HIV-1 target-site selection. (a) Distribution of Mo-MLV and HIV-1 integration sites with respect to the transcription start site (TSS) of target genes. The y-axis indicates the percentage of targeted genes the TSS of which is located at the distance specified on the x-axis from the insertion site. The line shows the profile of 65,000 computer-generated random sites.[78] MLV insertions are clearly clustered around TSSs, while HIV integrates along the whole transcription unit. The HIV dataset derives from Jurkat cells,[9] while the MLV dataset derives from primary human T cells (unpublished data). (b) Frequency of MLV and HIV integration near CpG islands. The y-axis represents the percentage of integration sites having at least one CpG island within the specified distance (from −10 to +10 kb). Genomic regions of 2.5 kb around MLV integrations are enriched in CpG islands, while regions flanking HIV insertions are CpG

Compared to MLV, there were fewer hot spots in HIV-1-infected HSCs, in genomic regions particularly enriched in HIV-preferred features, i.e., gene-dense regions and highly active genes.[21] A massive survey of more than 40,000 HIV integrations in a T-cell line has recently revealed associations with epigenetic modifications by computational analysis.[9] Strong positive correlation was seen with markers of transcriptionally active chromatin, such as mono-, di-, and tri-methylation of histone 3 lysine 4 (H3 K4), H3 K9/K14 acetylation, and H4 acetylation, while negative correlation was found with H3 K27 tri-methylation, marking heterochromatic regions. Although partially redundant with measures of gene density and chromatin structures, epigenetic modifications were shown to influence integration also independently of known genomic features. For gene therapy applications, the low propensity of HIV-1 to integrate in regulatory regions and the lack of a bias towards TSSs predict a better safety profile for the use of lentiviral versus gamma-retroviral vectors.

Foamy Virus (FV)

Foamy viruses, or Spumaviruses, are complex exogenous retroviruses mainly prevalent in non-human primates. FV vectors have been developed that possess broad host range, large packaging capacity and high transduction efficiency of human hematopoietic cells, making them promising substitutes for Mo-MLV vectors in the gene therapy of hematological disorders. Genome-wide profiling of FV integration sites, however, gave results that do not justify their use in the clinical setting.

---

island-depleted. Distances are calculated from the CpG island midpoint. Data are extrapolated from Ref. 22. (c) Influence of transcription factor binding sites (TFBSs) on MLV and HIV target-site selection. The plot represents the number of TFBSs found in genomic regions flanking (±1 kb) the insertion sites of random sequences and MLV/HIV vectors with wild-type or enhancer-less long terminal repeats (Δ). MLV integrates in genomic regions with a high content of TFBSs, independently of the presence of genes or TSSs; such preference is lost upon LTR deletion (ΔMLV). Conversely, the presence of TFBSs is detrimental to HIV integration, and no effect is seen after enhancer deletion (ΔHIV). Insertions are annotated as "TSS-proximal" when occurring within a distance of ±5 kb from the TSS of any gene, as "intragenic" when occurring into a gene at a distance of >5 kb from the TSS, and as "intergenic" in all other cases. Original data from Ref. 23.

The FV integration pattern effectively resembles that of Mo-MLV, but with an approximately two-fold reduction of over-represented features with respect to random controls (e.g., 20 percent of Mo-MLV insertions are found around TSSs compared to 10 percent of FV integration sites). Such a decrease is, however, not sufficient to predict a significant improvement in safety for FV-based vectors with respect to Mo-MLV constructs.

Avian Sarcoma and Leukosis Virus (ASLV)

ASLV is a member of the Alpha-retrovirus family whose natural hosts are chickens. However, pseudotyped viral particles can be produced that are able to infect mammalian cells. Insertion studies of this virus revealed only a weak, though still detectable, bias in favor of active genes, gene-dense regions, and genomic features associated with genes (e.g., slight enrichment compared to random sites for CpG islands, DNase I hypersensitive sites and regulatory regions). Integrations are evenly distributed along the whole transcription unit, with no preference for transcription start sites (TSSs). The nearly random ASLV insertion profile in mammals is encouraging the development of optimized vectors as safe transfer tools for gene-therapy applications.[28]

Mouse Mammary Tumor Virus (MMTV)

MMTV is a representative of the Beta-retrovirus genus for which a single large-scale mapping study of integration sites has been performed, looking at both murine and human mammary cell lines. MMTV displays the most random integration-site distribution among retroviruses to date, with no preference for active genes, TSSs, gene-dense regions, CpG islands, or DNase hypersensitive sites.[29,30]

Human T-Cell Leukemia Virus Type 1 (HTLV-1)

HTLV-1 is the only component of the Delta-retroviral family for which the integration profile has been evaluated to date. The virus integrates into the human genome with little but significant preference for TSSs, transcription units, promoters and gene-dense regions. No over-representation with respect to random sequences is observed near DNase I hypersensitive sites or CpG islands, and the GC-content of surrounding genomic regions is comparable to controls. A role of the target-cell transcriptional activity on target-site selection by HTLV-1 has not been described.

PROPOSED MECHANISMS FOR INTEGRATION SITE SELECTION

The genus-specific integration patterns of retroviruses imply a distinct molecular mechanism directing integration-site selection for each retroviral family. Target DNA accessibility can explain some common characteristics, like preference for active genes and avoidance of centromeric heterochromatin, but other peculiar features, like the MLV preference for promoter-proximal regions, require a different, more complex model.

Ty Retrotransposons: A Paradigm for Tethered Integration

Studies of retrotransposons in yeast provide an interesting model mechanism. Ty elements are well-studied retrotransposons that replicate through cycles of transcription, reverse transcription and integration similar to those of retroviruses, except for the fact that all steps occur inside a single cell. This lifestyle poses special problems. The yeast genome is 60 to 70 percent coding and a randomly integrating element is at high risk of committing suicide by insertional inactivation of host genes. Probably for this reason, Ty elements evolved strategies to actively select targets where insertion would not compromise host fitness. There appear to be at least three distinct mechanisms to avoid host genes, exemplified by the Ty1, Ty3 and Ty5 elements. Both Ty1 and Ty3 integrate at the 5′ end of genes transcribed by RNA polymerase III, in regions that tolerate insertions with no adverse consequences. The Ty3 element targets tRNA genes with extraordinary precision, inserting within a few base pairs of the TSS. This is probably mediated by local tethering of PICs to the TFIIIB component of the Pol-III basal transcription machinery.[31] The Ty1 element integrates less precisely, in a window of ~750 base pairs upstream of TSSs. The histone deacetylase Hos2, and the Trithorax-group protein Set3, both components of the Set3 complex, have been recently proposed to tether Ty1 to tRNA genes.[32] The Ty5 element shows different integration specificity, with 95 percent of insertions found either at telomeres or at silent mating loci; in this case, the heterochromatin protein silent information regulator 4 (Sirp4) is involved in specifying integration sites, through direct interaction with the Ty5-encoded integrase.[33,34] In all these cases, there is therefore evidence that Ty integration complexes are tethered to their preferred sites by interaction with specific cellular proteins, mediating integration at specific sites. It is reasonable to suppose that such a tethering mechanism might also operate for retroviruses, with a targeting strategy evolutionarily adapted to promote their specific requirements for propagation. As discussed above, intracellular Ty retrotransposons evolved to direct their integration outside transcription units, thus minimizing the risk of host gene perturbation. In contrast, acutely infecting retroviruses need to maximize the production of progeny virions, and integration in transcriptionally active regions may facilitate high-level transcription. The retroviral integration machinery probably evolved accordingly, allowing interactions with host nuclear proteins located in active chromatin regions.

Tethering Models for Retroviral Integration

Proof-of-principle that retroviral PICs can be tethered to integration sites by cellular interactors is provided by several *in vitro* studies performed with engineered integrases, fused to sequence-specific DNA binding domains. Such hybrid integrases are capable of targeted integration *in vitro*, demonstrating that tethering of integrase protein to target sites can constrain integration-site selection. Different combinations were tested, all with positive results. The HIV-1 integrase was fused to the DNA-binding domains of the λ phage repressor protein λR,[35] the *Escherichia coli* LexA repressor,[36] and the zinc finger proteins E2C[37] and zif268.[38] The ASLV integrase was also fused to the *E. coli* LexA DNA-binding domain.[39] These recombinant integrases directed integration *in vitro* near the binding site specific for the newly acquired DNA-binding domain. A certain level of efficiency was also observed *in vivo*, where the HIV integrase/E2C fusion protein was shown to re-direct integration into a unique E2C-binding site within the 5′ untranslated region of the erbB-2 gene on human chromosome 17, with a frequency 7- to 10-fold higher (from 0.15 to 1–1.5 percent) than that mediated by the wild-type integrase.[40] Off-target integration was still largely predominant, but the study proved the concept that tethering may mediate integration targeting, and that integrase-DNA interactions may be engineered to affect integration-site selection.

If tethering is indeed involved in retroviral integration-site selection, the main challenge becomes the identification of chromosomal ligands for the retroviral integration machinery and of their interacting counterparts within PICs. In principle, any viral or cellular component of the PIC could act as a binding partner in a tethering interaction. For the best-studied HIV-1, several cellular proteins have been isolated as physically bound to viral PICs, and for some of them the association occurs via direct interaction with the viral integrase. These include members of the DNA repair machinery (hRad1[41]), chromatin remodeling complexes (Ini-1,[42] EED[43]), and constitutive chromatin components, [HMGI(Y),[44] LEDGF/p75]. The PSIP1/LEDGF/p75 protein is the most studied and best characterized interactor of the HIV-1 integrase.

LEDGF/p75: A candidate for lentiviral integration tethering

Despite its name, the lens epithelium-derived growth factor (LEDGF/p75) is a ubiquitously expressed nuclear protein, tightly associated with chromatin throughout the cell cycle.[45] The protein was identified in affinity-based screenings for its tight binding to the HIV-1 integrase and was shown to stimulate its catalytic activity *in vitro*.[46–48] LEDGF/p75 is a member of the hepatoma-derived growth factor-related protein (HRP) family, characterized by a conserved N-terminal PWWP domain, found in a variety of nuclear proteins.[49,50] Of the six described HRP family members, only LEDGF/p75 and its highly homologous HRP2 contain a second conserved domain at the C-terminus, termed integrase-binding domain (IBD), which allows their interaction with different lentiviral integrases.[51] The PWWP domain, together with a nuclear localization signal and a double copy of an AT-hook DNA-binding domain, mediates LEDGF/p75 association with chromatin, with no apparent sequence specificity.[48,52] The cellular functions of LEDGF/p75 remain largely unknown, although a role in transcriptional activation was proposed soon after the protein was identified. The role it plays in HIV infectivity has been intensively studied. The most robust results came from studies in human cells with RNA interference knock-downs of LEDGF/p75[53–57] and in murine cells with homozygous gene-trap mutations in the LEDGF/p75 locus.[57,58] When LEDGF/p75 protein is depleted, the first effect is a re-localization of the integrase to the cell cytoplasm, with loss of chromosomal association and increased proteosomal degradation. A second, important consequence is an overall reduction of HIV-1 infectivity, due to a severe impairment in the integration process. Residual integration sites were analyzed, which revealed a decrease in the typical HIV preference for transcription units, and an increase in insertion near CpG islands and promoter regions, classical targets of other retroviruses. Integration did not become random, however, and transcribed genes were still favored. Therefore, it is still plausible that cell factors other than LEDGF/p75 participate in PIC tethering to chromosomes, although LEDGF/p75 remains a dominant cellular binding partner of the HIV-1 integrase, required for efficient integration and replication of the virus.

Overall, these studies suggested a model whereby one domain of LEDGF/p75 binds chromatin at active transcription units while the other one acts as a receptor for incoming PICs; by enhancing the DNA strand transfer activity of the integrase, LEDGF7p75 would then direct integration to a site nearby. Such a tethering model predicts that LEDGF/p75 should accumulate preferentially on active transcription units, but this has not been experimentally demonstrated so far. It is not even known how LEDGF/p75 might recognize active transcription units. Given the positive correlation between HIV-1 insertion sites and certain post-translational histone modifications (described in HIV-1 integration section), one possible model predicts that LEDGF/p75 recognizes active transcription units via the histone modifications that specifically mark them.

Transcription factors and gamma-retroviral integration tethering

Only a single cellular factor has been isolated to date as physically associated to the Mo-MLV PICs. This is the cellular protein BAF, identified as, and named after, the barrier-to-autointegration factor in Mo-MLV infection.[59] BAF is an inhibitor of suicide integration of the Mo-MLV provirus and promotes efficient intermolecular DNA recombination once a suitable chromosomal target is identified. Although essential for PIC integration activity, interaction with BAF alone does not explain the characteristic Mo-MLV preference for regulatory regions (TSSs, CpG islands, DNase I hypersensitive sites), and in general for regions with a high content of TFBSs, independently of their location with respect to genes and TSSs.[23] The MLV integrase is essential in mediating these preferences, since an HIV-1 vector packaged with a Mo-MLV integrase gains Mo-MLV predilection for TSSs, CpG islands[22] and TFBS-rich regions.[23] Interestingly, the preference for TFBS-rich regions is dramatically reduced when the U3 transcriptional enhancer is deleted from the Mo-MLV LTRs [Fig. 3(a)].[23] These results indicate that the integrase and the U3 enhancer are the major viral determinants of Mo-MLV selection of regulatory regions in the genome. The model that follows is that cellular transcription factors binding the Mo-MLV U3 enhancer cooperate with the integrase in directing PICs towards regions actively engaged by the transcriptional machinery. Accordingly, some of the TFBSs enriched around Mo-MLV integrations are consensus motifs for transcription factors already known to bind the U3 enhancer and drive proviral expression after integration (e.g., members of the ETS family and the bivalent YY1 transcription factor). The existence of cell-specific and non-specific clusters of TFBS around MLV integration sites suggests that MLV PICs interact with general components of the enhancer-binding complexes (e.g., co-regulators, chromatin remodeling or mediator complexes) rather than with specific transcription factors or transcription factor families.[23] Biochemical support to this hypothesis is provided by recent data indicating that the MLV integrase may interact directly with chromatin remodeling, DNA repair and transcription factors.[60] Tethering of PICs to transcription factories, where promoters and regulatory regions are relocated by cell-specific mechanisms, may also easily explain the Mo-MLV high frequency of integration hot spots and preferred targeting of genes involved in cell-specific pathways. Cooperation between transcription factors and integrase may be seen as an evolutionary derivative of the mechanisms by which yeast retrotransposons target their integration to specific genomic regions. As discussed before, a specific domain of the Ty 3 retrotransposase directs integration of the retrotransposon to promoters transcribed by the DNA polymerase III, by tethering to Pol III-specific transcription factors. This domain is lacking in the retroviral integrases, which are otherwise related to retrotransposases, and may have been functionally replaced by the association with transcription factors bound to LTR elements. As a result, MLV PICs are able to target a large collection of Pol II-specific, rather than a few Pol III-specific, regulatory elements throughout the genome.

Tethering models for other retroviruses

The nearly random distribution of ASLV, HTLV-1 and MMTV integration sites has two plausible explanations. One possibility is that the integration machinery of these viruses does not interact, or interacts only weakly, with cellular factors bound near or within genes; alternatively, PICs are tethered to the host genome by proteins ubiquitously distributed along chromosomes, independently from genes and regulatory regions. It is also possible that the specific host factor(s) responsible for PIC tethering are not expressed in the cell context used for the integration studies. In the case of ASLV, for instance, PICs could interact with chicken host factors that have poor or no homologues in mammalian cells, leading to reduced target specificity. However, an analysis of ASLV integrations performed in chicken embryo fibroblasts tends to disfavor this hypothesis.[61]

The FV integration pattern is similar to, although less pronounced than, that of the Mo-MLV, suggesting that analogous mechanisms mediate the target-site selection of the two viruses. Recent findings on the trafficking of FV PICs in the nucleus could explain this sort of "attenuated" insertion profile. A short domain within the C-terminus of the structural FV Gag protein efficiently binds H2A/H2B core histones and mediates tethering of PICs to host chromatin prior to integration.[62] Due to the even distribution of H2A/H2B in the genome, this docking is not restricted to discrete chromosomal location and could, in theory, attenuate other specific preferences dictated by the viral integrase and/or LTRs.

IMPLICATIONS AND PERSPECTIVES OF INTEGRATION SITE SELECTION

Integration-Site Selection and Viral Evolution

The choice of the integration site has a major impact on the survival and fitness of a retrovirus, by determining the overall efficiency of proviral gene expression. It is therefore reasonable that each viral family has evolved a molecular strategy to direct integration in order to maximize the likelihood of propagating to target cells.

Gamma-retroviruses, and arguably Spumaviruses, may have evolved a mechanism coupling target-site selection to gene regulation to take advantage of nearby cellular promoters and/or enhancers to activate their own expression. Conversely, integration of viral LTR enhancers in the proximity of cell-specific growth regulators increases the chance of clonal expansion or transformation by insertional gene activation, possibly resulting in expansion of infected cells and indefinite viral propagation.

Lentiviruses have evidently evolved a different integration strategy, to target open chromatin regions while minimizing interference with the cell transcriptional machinery. This is expected to promote maximum production of daughter virions during the limited lifespan of infected cells in the phase of active replication. On the other hand, integration into active genes, but at a distance from promoters and regulatory regions, may be more permissive for the latent phase of the viral life cycle, if one thinks of latency as a deliberate HIV survival strategy, which is not necessarily the case.[63] *In vitro* latency models, in which silent HIV proviruses are re-activated by treatment with tumor-necrosis factor-α and then profiled for their integrations, suggest that transcriptional latency may also derive from integration in a "silencing" genomic environment (centromeric heterochromatin, long intergenic regions, or very highly expressed domains).[64] The relationship between HIV latency and integration sites remains, uncertain, however.

It is less obvious how the nearly random integration patterns of Alpha-, Beta-, and Delta-retroviruses are favorable for their survival. In this case, the host–virus interaction may have evolved to reduce damage to the host, or simply not evolved in any specific direction.

Integration-Site Selection and Implications for Gene Therapy

The covalent integration of viral DNA into the host-cell genome carries an intrinsic mutagenic potential, which is further exacerbated by the integration profile and/or some structural properties of certain retroviruses. This has an obvious impact in the gene therapy field, where the main interest is to design the most efficient gene transfer vectors with the lowest possible genotoxic potential.

One of the main risks associated with retroviral gene transfer is the transcriptional deregulation of host genes by viral LTRs and, depending on the type of gene affected, its possible effects on the host-cell biology. The propensity of Mo-MLV to target regulatory regions, the strength of its LTRs in a variety of cell types, and the frequency and characteristics of its integration hot spots give Mo-MLV-based vectors a relatively high genotoxic risk. A self-inactivating, U3-deleted design is predicted to improve the safety profile of Mo-MLV-based vectors, since it results in a reduced targeting of regulatory regions[23] and a lower chance of transcriptional interference. Because they rely on internal promoters for transgene expression, self-inactivating vectors can still transactivate nearby genes,[65] though an accurate choice of promoters/enhancers and insulating elements should significantly reduce this risk.[66,67] Still, the Mo-MLV integrase remains an unwanted protagonist of the virus tendency to target potentially dangerous regions of the genome. The use of HIV-derived vectors should decrease the chances of insertional gene activation by generally reducing integration in the proximity of regulatory enhancers and promoters. However, since transactivation can operate thousands of base pairs from the target gene transcription start site,[65,68] again suitable promoters and/or insulators should be included in the vector design.

Another genotoxic risk exists associated with the insertion of viral elements in regions other than promoters, e.g., functional splicing and polyadenylation signals. The presence of splice donor and acceptor sites in the retroviral backbone was demonstrated to enhance viral vector titers and transgene expression. However, upon integration within a transcription unit, the viral splicing signals may function as alternative donors or acceptors for those of the target gene. The result may be aberrant splicing, leading to truncated or otherwise mutated products, with potentially altered function. Insertion of an active polyadenylation signal can have similar consequences, producing either premature transcript termination of a host gene (strong polyA signals) or read-through transcription from a proviral promoter (weak polyA signals). The highly disruptive potential of viral post-transcriptional regulatory elements is testified by the integration profile of evolutionarily fixed retroviral elements in the human genome. Human endogenous retroviruses (HERVs), i.e., dead retroviral elements that account for about 8 percent of the entire human genome, are preferentially located outside transcription units, away from gene-rich regions and regulatory elements. When inside a transcription unit, HERVs are found in opposite transcriptional orientation relative to the host gene, so that their splice sites and/or polyadenylation signal cannot disrupt gene transcription. Integration profiling of a resurrected HERV showed, instead, a preference for genomic regions involved in active transcription, with no bias for sense or antisense orientation with respect to target genes.[30] The observation indicates that integrations leading to the insertion of strong splice sites and polyadenylation signals inside transcription units are indeed deleterious to host viability, and, in the case of HERVs accumulating in the human germ line, have been removed by purifying selection in the host cell. Evidence for negative selection of cells harboring certain integration events has emerged also in the follow-up of clinical gene therapy studies.[69] Given the extremely high frequency of integration within transcription units, lentiviral vectors are particularly susceptible to the risk of disrupting host mRNA synthesis. To date, the issue has not been given the attention it deserves, but dedicated studies are necessary, since the results could undermine the idea of self-inactivating lentiviral vectors as the perfect tools for gene therapy.

The clinical use of viral vectors derived from retroviruses other than MLV and HIV, with a theoretically less harmful integration profile (e.g., ASLV, MMTV), is under evaluation, but the technology available for these vectors is currently far less sophisticated and advanced than that developed in years of vectorology for gamma-retroviral and lentiviral vectors.

REFERENCES

1. Wu X, Li Y, Crise B, *et al.* (2005) Weak palindromic consensus sequences are a common feature found at the integration target sites of many retroviruses. *J Virol* **79**: 5211–5214.
2. Bushman F, Lewinski M, Ciuffi A, *et al.* (2005) Genome-wide analysis of retroviral DNA integration. *Nat Rev Microbiol* **3**: 848–858.
3. Stocking C, Bergholz U, Friel J, *et al.* (1993) Distinct classes of factor-independent mutants can be isolated after retroviral mutagenesis of a human myeloid stem cell line. *Growth Factors* **8**: 197–209.
4. Hacein-Bey-Abina S, Von Kalle C, Schmidt M, *et al.* (2003) LMO2–associated clonal T cell proliferation in two patients after gene therapy for SCID-X1. *Science* **302**: 415–419.
5. Hacein-Bey-Abina S, Garrigue A, Wang GP, *et al.* (2008) Insertional oncogenesis in 4 patients after retrovirus-mediated gene therapy of SCID-X1. *J Clin Invest* **118**: 3132–3142.
6. Howe SJ, Mansour MR, Schwarzwaelder K, *et al.* (2008) Insertional mutagenesis combined with acquired somatic mutations causes leukemogenesis following gene therapy of SCID-X1 patients. *J Clin Invest* **118**: 3143–3150.
7. Margulies M, Egholm M, Altman WE, *et al.* (2005) Genome sequencing in microfabricated high-density picolitre reactors. *Nature* **437**: 376–380.
8. Wang GP, Garrigue A, Ciuffi A, *et al.* (2008) DNA bar coding and pyrosequencing to analyze adverse events in therapeutic gene transfer. *Nucleic Acids Res* **36**: e49.
9. Wang GP, Ciuffi A, Leipzig J, *et al.* (2007) HIV integration site selection: Analysis by massively parallel pyrosequencing reveals association with epigenetic modifications. *Genome Res* **17**: 1186–1194.
10. Schroder AR, Shinn P, Chen H, *et al.* (2002) HIV-1 integration in the human genome favors active genes and local hotspots. *Cell* **110**: 521–529.
11. Berry C, Hannenhalli S, Leipzig J, *et al.* (2006) Selection of target sites for mobile DNA integration in the human genome. *PLoS Comput Biol* **2**: e157.
12. Galperin MY (2008) The Molecular Biology Database Collection: 2008 update. *Nucleic Acids Res* **36**: D2–D4.
13. Dennis G, Jr., Sherman BT, Hosack DA, *et al.* (2003) DAVID: Database for Annotation, Visualization, and Integrated Discovery. *Genome Biol* **4**: P3.

14. Gabriel R, Eckenberg R, Paruzynski A, *et al.* (2009) Comprehensive genomic access to vector integration in clinical gene therapy. *Nat Med* **15**: 1431–1436.
15. http://www.ncbi.nlm.nih.gov/projects/geo/index.cgi Gene Expression Omnibus, provided by the National Center for Biotechnology Information.
16. Crawford GE, Holt IE, Whittle J, *et al.* (2006) Genome-wide mapping of DNase hypersensitive sites using massively parallel signature sequencing (MPSS). *Genome Res* **16**: 123–131.
17. Wang Z, Zang C, Rosenfeld JA, *et al.* (2008) Combinatorial patterns of histone acetylations and methylations in the human genome. *Nat Genet* **40**: 897–903.
18. Barski A, Cuddapah S, Cui K, *et al.* (2007) High-resolution profiling of histone methylations in the human genome. *Cell* **129**: 823–837.
19. Cui K, Zang C, Roh TY, *et al.* (2009) Chromatin signatures in multipotent human hematopoietic stem cells indicate the fate of bivalent genes during differentiation. *Cell Stem Cell* **4**: 80–93.
20. Derse D, Crise B, Li Y, *et al.* (2007) Human T-cell leukemia virus type 1 integration target sites in the human genome: Comparison with those of other retroviruses. *J Virol* **81**: 6731–6741.
21. Cattoglio C, Facchini G, Sartori D, *et al.* (2007) Hot spots of retroviral integration in human CD34$^+$ hematopoietic cells. *Blood* **110**: 1770–1778.
22. Lewinski MK, Yamashita M, Emerman M, *et al.* (2006) Retroviral DNA integration: Viral and cellular determinants of target-site selection. *PLoS Pathog* **2**: e60.
23. Felice B, Cattoglio C, Cittaro D, *et al.* (2009) Transcription factor binding sites are genetic determinants of retroviral integration in the human genome. *PLoS ONE* **4**: e4571.
24. Kustikova O, Fehse B, Modlich U, *et al.* (2005) Clonal dominance of hematopoietic stem cells triggered by retroviral gene marking. *Science* **308**: 1171–1174.
25. Calmels B, Ferguson C, Laukkanen MO, *et al.* (2005) Recurrent retroviral vector integration at the Mds1/Evi1 locus in nonhuman primate hematopoietic cells. *Blood* **106**: 2530–2533.
26. Ott MG, Schmidt M, Schwarzwaelder K, *et al.* (2006) Correction of X-linked chronic granulomatous disease by gene therapy, augmented by insertional activation of MDS1–EVI1, PRDM16 or SETBP1. *Nat Med* **12**: 401–409.

27. Schwarzwaelder K, Howe SJ, Schmidt M, *et al.* (2007) Gammaretrovirus-mediated correction of SCID-X1 is associated with skewed vector integration site distribution in vivo. *J Clin Invest* **117**: 2241–2249.
28. Hu J, Renaud G, Gomes TJ, *et al.* (2008) Reduced genotoxicity of avian sarcoma leukosis virus vectors in rhesus long-term repopulating cells compared to standard murine retrovirus vectors. *Mol Ther* **16**: 1617–1623.
29. Faschinger A, Rouault F, Sollner J, *et al.* (2008) Mouse mammary tumor virus integration site selection in human and mouse genomes. *J Virol* **82**: 1360–1367.
30. Brady T, Lee YN, Ronen K, *et al.* (2009) Integration target site selection by a resurrected human endogenous retrovirus. *Genes Dev* **23**: 633–642.
31. Kirchner J, Connolly CM, Sandmeyer SB (1995) Requirement of RNA polymerase III transcription factors for in vitro position-specific integration of a retroviruslike element. *Science* **267**: 1488–1491.
32. Mou Z, Kenny AE, Curcio MJ (2006) Hos2 and Set3 promote integration of Ty1 retrotransposons at tRNA genes in Saccharomyces cerevisiae. *Genetics* **172**: 2157–2167.
33. Zhu Y, Dai J, Fuerst PG, *et al.* (2003) Controlling integration specificity of a yeast retrotransposon. *Proc Natl Acad Sci USA* **100**: 5891–5895.
34. Zhu Y, Zou S, Wright DA, *et al.* (1999) Tagging chromatin with retrotransposons: Target specificity of the Saccharomyces Ty5 retrotransposon changes with the chromosomal localization of Sir3p and Sir4p. *Genes Dev* **13**: 2738–2749.
35. Bushman FD (1994) Tethering human immunodeficiency virus 1 integrase to a DNA site directs integration to nearby sequences. *Proc Natl Acad Sci USA* **91**: 9233–9237.
36. Goulaouic H, Chow SA (1996) Directed integration of viral DNA mediated by fusion proteins consisting of human immunodeficiency virus type 1 integrase and Escherichia coli LexA protein. *J Virol* **70**: 37–46.
37. Tan W, Zhu K, Segal DJ, *et al.* (2004) Fusion proteins consisting of human immunodeficiency virus type 1 integrase and the designed polydactyl zinc finger protein E2C direct integration of viral DNA into specific sites. *J Virol* **78**: 1301–1313.
38. Bushman FD, Miller MD (1997) Tethering human immunodeficiency virus type 1 preintegration complexes to target DNA promotes integration at nearby sites. *J Virol* **71**: 458–464.

39. Katz RA, Merkel G, Skalka AM (1996) Targeting of retroviral integrase by fusion to a heterologous DNA binding domain: In vitro activities and incorporation of a fusion protein into viral particles. *Virology* **217**: 178–190.
40. Tan W, Dong Z, Wilkinson TA, *et al.* (2006) Human immunodeficiency virus type 1 incorporated with fusion proteins consisting of integrase and the designed polydactyl zinc finger protein E2C can bias integration of viral DNA into a predetermined chromosomal region in human cells. *J Virol* **80**: 1939–1948.
41. Mulder LC, Chakrabarti LA, Muesing MA (2002) Interaction of HIV-1 integrase with DNA repair protein hRad18. *J Biol Chem* **277**: 27489–27493.
42. Kalpana GV, Marmon S, Wang W, *et al.* (1994) Binding and stimulation of HIV-1 integrase by a human homolog of yeast transcription factor SNF5. *Science* **266**: 2002–2006.
43. Violot S, Hong SS, Rakotobe D, *et al.* (2003) The human polycomb group EED protein interacts with the integrase of human immunodeficiency virus type 1. *J Virol* **77**: 12507–12522.
44. Li L, Yoder K, Hansen MS, *et al.* (2000) Retroviral cDNA integration: Stimulation by HMG I family proteins. *J Virol* **74**: 10965–10974.
45. Engelman A, Cherepanov P (2008) The lentiviral integrase binding protein LEDGF/p75 and HIV-1 replication. *PLoS Pathog* **4**: e1000046.
46. Cherepanov P, Maertens G, Proost P, *et al.* (2003) HIV-1 integrase forms stable tetramers and associates with LEDGF/p75 protein in human cells. *J Biol Chem* **278**: 372–381.
47. Emiliani S, Mousnier A, Busschots K, *et al.* (2005) Integrase mutants defective for interaction with LEDGF/p75 are impaired in chromosome tethering and HIV-1 replication. *J Biol Chem* **280**: 25517–25523.
48. Turlure F, Maertens G, Rahman S, *et al.* (2006) A tripartite DNA-binding element, comprised of the nuclear localization signal and two AT-hook motifs, mediates the association of LEDGF/p75 with chromatin in vivo. *Nucleic Acids Res* **34**: 1653–1675.
49. Qiu C, Sawada K, Zhang X, *et al.* (2002) The PWWP domain of mammalian DNA methyltransferase Dnmt3b defines a new family of DNA-binding folds. *Nat Struct Biol* **9**: 217–224.
50. Stec I, Nagl SB, van Ommen GJ, *et al.* (2000) The PWWP domain: A potential protein-protein interaction domain in nuclear proteins influencing differentiation? *FEBS Lett* **473**: 1–5.

51. Cherepanov P, Devroe E, Silver PA, *et al.* (2004) Identification of an evolutionarily conserved domain in human lens epithelium-derived growth factor/transcriptional co-activator p75 (LEDGF/p75) that binds HIV-1 integrase. *J Biol Chem* **279**: 48883–48892.
52. Llano M, Vanegas M, Hutchins N, *et al.* (2006) Identification and characterization of the chromatin-binding domains of the HIV-1 integrase interactor LEDGF/p75. *J Mol Biol* **360**: 760–773.
53. Ciuffi A, Llano M, Poeschla E, *et al.* (2005) A role for LEDGF/p75 in targeting HIV DNA integration. *Nat Med* **11**: 1287–1289.
54. Llano M, Delgado S, Vanegas M, *et al.* (2004) Lens epithelium-derived growth factor/p75 prevents proteasomal degradation of HIV-1 integrase. *J Biol Chem* **279**: 55570–55577.
55. Llano M, Vanegas M, Fregoso O, *et al.* (2004) LEDGF/p75 determines cellular trafficking of diverse lentiviral but not murine oncoretroviral integrase proteins and is a component of functional lentiviral preintegration complexes. *J Virol* **78**: 9524–9537.
56. Maertens G, Cherepanov P, Pluymers W, *et al.* (2003) LEDGF/p75 is essential for nuclear and chromosomal targeting of HIV-1 integrase in human cells. *J Biol Chem* **278**: 33528–33539.
57. Marshall HM, Ronen K, Berry C, *et al.* (2007) Role of PSIP1/LEDGF/p75 in lentiviral infectivity and integration targeting. *PLoS ONE* **2**: e1340.
58. Sutherland HG, Newton K, Brownstein DG, *et al.* (2006) Disruption of Ledgf/Psip1 results in perinatal mortality and homeotic skeletal transformations. *Mol Cell Biol* **26**: 7201–7210.
59. Lee MS, Craigie R (1994) Protection of retroviral DNA from autointegration: Involvement of a cellular factor. *Proc Natl Acad Sci USA* **91**: 9823–9827.
60. Studamire B, Goff SP (2008) Host proteins interacting with the Moloney murine leukemia virus integrase: Multiple transcriptional regulators and chromatin binding factors. *Retrovirology* **5**: 48.
61. Barr SD, Leipzig J, Shinn P, *et al.* (2005) Integration targeting by avian sarcoma-leukosis virus and human immunodeficiency virus in the chicken genome. *J Virol* **79**: 12035–12044.
62. Tobaly-Tapiero J, Bittoun P, Lehmann-Che J, *et al.* (2008) Chromatin tethering of incoming foamy virus by the structural Gag protein. *Traffic* **9**: 1717–1727.
63. Persaud D, Zhou Y, Siliciano JM, *et al.* (2003) Latency in human immunodeficiency virus type 1 infection: No easy answers. *J Virol* **77**: 1659–1665.

64. Lewinski MK, Bisgrove D, Shinn P, *et al.* (2005) Genome-wide analysis of chromosomal features repressing human immunodeficiency virus transcription. *J Virol* **79**: 6610–6619.
65. Maruggi G, Porcellini S, Facchini G, *et al.* (2009) Transcriptional enhancers induce insertional gene deregulation independently from the vector type and design. *Mol Ther* **17**: 851–856.
66. Nienhuis AW, Dunbar CE, Sorrentino BP (2006) Genotoxicity of retroviral integration in hematopoietic cells. *Mol Ther* **13**: 1031–1049.
67. Baum C, Schambach A, Bohne J, *et al.* (2006) Retrovirus vectors: Toward the plentivirus? *Mol Ther* **13**: 1050–1063.
68. Hargrove PW, Kepes S, Hanawa H, *et al.* (2008) Globin lentiviral vector insertions can perturb the expression of endogenous genes in beta-thalassemic hematopoietic cells. *Mol Ther* **16**: 525–533.
69. Recchia A, Bonini C, Magnani Z, *et al.* (2006) Retroviral vector integration deregulates gene expression but has no consequence on the biology and function of transplanted T cells. *Proc Natl Acad Sci USA* **103**: 1457–1462.
70. Li J, Shen H, Himmel KL, *et al.* (1999) Leukaemia disease genes: Large-scale cloning and pathway predictions. *Nat Genet* **23**: 348–353.
71. Wu X, Li Y, Crise B, *et al.* (2003) Transcription start regions in the human genome are favored targets for MLV integration. *Science* **300**: 1749–1751.
72. Schmidt M, Zickler P, Hoffmann G, *et al.* (2002) Polyclonal long-term repopulating stem cell clones in a primate model. *Blood* **100**: 2737–2743.
73. Trobridge GD, Miller DG, Jacobs MA, *et al.* (2006) Foamy virus vector integration sites in normal human cells. *Proc Natl Acad Sci USA* **103**: 1498–1503.
74. Nowrouzi A, Dittrich M, Klanke C, *et al.* (2006) Genome-wide mapping of foamy virus vector integrations into a human cell line. *J Gen Virol* **87**: 1339–1347.
75. Narezkina A, Taganov KD, Litwin S, *et al.* (2004) Genome-wide analyses of avian sarcoma virus integration sites. *J Virol* **78**: 11656–11663.
76. Mitchell RS, Beitzel BF, Schroder AR, *et al.* (2004) Retroviral DNA integration: ASLV, HIV, and MLV show distinct target site preferences. *PLoS Biol* **2**: E234.
77. Workman CT, Yin Y, Corcoran DL, *et al.* (2005) enoLOGOS: A versatile web tool for energy normalized sequence logos. *Nucleic Acids Res* **33**: W389–W392.
78. Hematti P, Hong BK, Ferguson C, *et al.* (2004) Distinct genomic integration of MLV and SIV vectors in primate hematopoietic stem and progenitor cells. *PLoS Biol* **2**: e423.

CHAPTER 8

Syncytins in Normal and Pathological Placentas

André Malassiné[†,‡,§], Christian Lavialle[¶,||], Jean Louis Frendo[†,‡,§,**], Anne Dupressoir[§,¶,||] and Danièle Evain-Brion[*,†,‡,§]

INTRODUCTION

During the course of evolution, the genomes of vertebrates have been invaded by many different transposable elements that today constitute nearly half of the DNA of mammals. Five to ten percent of mammalian genomes are occupied by a class of elements named endogenous retroviruses (ERVs), which are the remnants of past infections by ancestral retroviruses (for a review, see Ref. 1). In the case of humans, the numerous human endogenous retroviruses (HERVs) identified have been classified into several families, such as HERV-K and HERV-W, based on the amino

[*] Corresponding author: daniele.evain-brion@parisdescartes.fr.
[†] INSERM, U767, 4 avenue de l'Observatoire, 75006, Paris, France.
[‡] Université Paris Descartes, Paris, France.
[§] PremUP, Fondation, 75006, Paris, France.
[¶] Unité des Rétrovirus Endogènes et Éléments Rétroïdes des Eucaryotes Supérieurs, Centre National de la Recherche Scientifique, Unité Mixte de Recherche 8122, Institut Gustave Roussy, 39 rue Camille Desmoulins, 94805 Villejuif, France.
[||] Université Paris-Sud XI, 91405 Orsay, France.
[**] CNRS, Paris, France.

acid specificity of the tRNA used to prime reverse transcription. Most of the currently identified ERVs usually display a simple genetic structure close to that of the integrated proviral form of exogenous retroviruses, with the three basic *gag*, *pol* and *env* genes flanked by two long terminal repeats (LTRs)[2]. The *env* gene encodes the envelope glycoprotein (Env) that interacts with a specific receptor and mediates fusion of the viral envelope with the plasma membrane of the target cells.

Occasionally, retroviruses have infected cells belonging to the germline. In that case, if the provirus persists in a germ cell until fertilization, it will be found in every cell of the offspring and be transmitted "vertically" to the descendants in a Mendelian fashion. The persistence of a newly integrated provirus in a population will depend on its impact on host fitness. Many functional endogenous proviruses have probably been lost during evolution due to deleterious effects of their insertions (negative selection). The ERVs that have been successfully "endogenized" are subjected to random genetic drift, which in the absence of any selection pressure leads to progressive inactivation of the proviruses and disruption of proviral genes, ultimately beyond recognition.

Unlike in other species such as the mouse, where many endogenous complete proviruses that are still active have been identified, no infectious HERV has been detected to date. However, some human HERVs still contain at least one of the open reading frames (ORFs) for the retroviral proteins. Screening of the human genome has identified about 70 HERV copies with one full-length or almost full-length ORF (reviewed in Ref. 2). Among the identified HERV ORFs, 18 ORFs for full-length Env proteins belonging to 11 distinct HERV families have been identified.[3,4] It is worth noting that these 18 *env* genes are transcribed at various levels in humans, several of them displaying a placenta-specific expression pattern.[4,5]

ENDOGENOUS RETROVIRUS GENE EXPRESSION IN PLACENTAS: DISCOVERY OF SYNCYTINS

In several instances, a few ERV genes were found to have been co-opted by their host in a kind of "symbiotic" relationship to contribute new physiological functions through the recruitment of viral protein-coding sequences. The most remarkable example of this discovery is the recruitment of endogenous *env* genes for a physiological role in the placenta of eutherian mammals. The involvement of ERVs in this organ had been long suspected because of the observation by electron microscopy, since the early 1970s, in both human and animal placental tissues, of retroviral particles, later followed by the detection of reverse transcriptase activity and retroviral proteins (reviewed in Refs. 6–8). Moreover, specific features of the retroviral Env proteins are consistent with functions that they might be expected to fulfill if they were expressed in placental tissues. One of these is their capacity to induce cell–cell fusion, resulting in the formation of multinucleated syncytia — a direct consequence of an activity critical for virus entry, i.e., the fusion of the viral envelope with the plasma membrane of the infected cell. In a number of mammalian species, the placenta is one of the few tissues where cell fusion takes place, leading to the formation of a syncytial layer: the syncytiotrophoblast, which covers the outer surface of the chorionic epithelium. Another feature of a number of Env proteins is their immunosuppressive activity carried by a small domain of their transmembrane subunit,[9] which could contribute to the immunosuppressive properties of the placenta. Expression of Env proteins in the placenta could also have a protective effect against exogenous retroviruses through receptor interference.

In humans, two *env* genes carried by endogenous proviruses, belonging to the HERV-W and HERV-FRD families, and designated *syncytin-1* (*syncytin/ERVWE1*) and *syncytin-2* (*HERV-FRD*), respectively, have been identified for which convincing evidence of their involvement in placental physiology could be provided.[10–13] *Syncytin* genes have been highly conserved in evolution, from the time when the respective HERV-W and HERV-FRD retroviruses carrying them entered the primate lineage 20 and 40 million years ago, respectively, and they display remarkably few polymorphisms in the extant human population, both facts providing strong evidence for purifying selection.[10,14,15] Both the human *syncytin-1* and *-2* proteins were found to be highly fusogenic in *ex vivo* cell–cell fusion assays, but the cells prone to fusion in each case being different suggested different receptor usage.[10]

Although it entered the primate lineage 20 million years ago, the human HERV-W provirus harboring *syncytin-1* as its *env* gene still displays the recognizable hallmarks of integrated retroviral sequences with two LTRs and *gag*, *pol* and *env* genes, although the *gag* and *pol* ORFs are disrupted.[11,16] The locus maps to chromosome 7q21–q22.[13] The structure of the 3.1 kb mRNA encoding *syncytin-1* is consistent with conservation of the promoter activity of the ancestral HERV-W retrovirus 5′ LTR, as well as of the splicing pattern of the subgenomic *env* mRNA and termination within the 3′ LTR.[16] Transcriptional regulation of the *syncytin-1* gene has been extensively investigated. Besides LTR regulatory elements, an upstream cellular enhancer was identified.[16–20] Interestingly, this included the first reported[21] functional binding sites for the placenta-specific transcription factor glial-cell missing-1 (GCM-1; also known as GCMa) and showed that GCM-1 does regulate *syncytin-1* expression, in agreement with the overlapping expression profile of the two genes in placental tissues (see below).

In contrast to *syncytin-1*, transcriptional regulation of the *syncytin-2* gene has not yet been investigated. The 40 million-year-old human HERV-FRD provirus containing the *syncytin-2* gene is still recognizable with its two LTRs flanking residual *gag* and *pol* disrupted sequences followed by the intact ORF of the *syncytin-2 env* gene. The HERV-FRD locus is located at chromosomal band 6p24.1.[10] A major spliced transcript of about 3.5 kb can be deduced from the GenBank reference mRNA sequence and EST cDNA clones (accession number NM_207582 and, for example, DA863475, respectively). As for *syncytin-1*, transcription again starts within the 5′ LTR and splicing occurs between the conserved donor and acceptor sites of the ancestral subgenomic *env* mRNA.

The properties of human *syncytin* proteins can be inferred from the well-known features of the envelope glycoprotein from simple infectious retroviruses, such as murine leukemia virus (MLV) (for reviews, see Refs. 22 and 23. Both the *syncytin-1* and *syncytin-2* proteins contain 538 amino acids. They exhibit a signal peptide of 20 or 16 amino acids, respectively, a furin cleavage site that fits the consensus (RNKR or RVKR, respectively) followed by a hydrophobic stretch consistent with a fusion peptide, disulphide bridge-forming CX2C and CX6CC motifs, heptad repeat sequences and a transmembrane anchorage domain.[10,24–26] Both the receptors for *syncytin-1* and *-2* have been found. That of *syncytin-1* was rapidly identified, simply by screening for already known retrovirus receptors,[12] as a sodium-dependent neutral amino acid transporter with broad substrate specificity for small neutral amino acids including L-alanine, L-glutamine, L-leucine and L-glycine, variously designated as ASCT2/ATB°/SLC1A5. ASCT2 had previously been demonstrated to be a receptor for ("type D") betaretroviruses such as MPMV or the RD114/ simian type D retrovirus, and denoted RDR. The receptor for *syncytin-2* was identified through systematic screening of a human X hamster radiation hybrid panel of 93 clones (GenBridge4) for susceptibility to infection by a retroviral pseudotype displaying *syncytin-2* at its surface.[27] The gene was mapped to chromosome 1p34.2 and identified as encoding a multipass transmembrane protein, named major facilitator superfamily domain containing 2 (MFSD2) that belongs to a large family of putative carbohydrate transporters conserved in evolution.

Another human endogenous *env* gene with an almost full-length ORF first found to be expressed at a high level in the placenta was that carried by ERV3 (HERV-R), a single copy HERV mapping to chromosome 7q11.2.[28] However, several arguments suggest that this gene either is not essential or is redundant.[4,15,29] Other Env proteins, two contained in tandemly integrated HERV-V proviruses and one in a HERV-P(b) provirus, were found to be expressed in human placenta,[3,5,30] but their role, in placental physiology, if any, remains to be established.

The question comes to mind as to whether ERVs could have also been diverted in other species for *a syncytin*-like function in the placenta. Taking advantage of the knowledge of the mouse genome sequence, two novel full-length ERV *env* genes were discovered, each present as a single copy and specifically expressed in mouse placenta.[31] In contrast to the human genes, the murine *syncytin-A* and *-B* genes are not framed by sequences from the ancestral retroviruses that carried and inserted them into the mouse genome. Akin to the human *syncytin-1* gene, the *syncytin-B*, but not the *syncytin-A* gene, appears to be regulated by GCM1, consistent with their coexpression in the same syncytial cell layer (Ref. 32; see below). Both the *syncytin-A* and *-B* proteins display the classical structure of full-length retroviral Env proteins and are highly fusogenic in *ex vivo* transfection assays, but the fact that the target cells prone to fusion in each case are different indicates that they likely bind different cell-surface receptors. Orthologues of both the *syncytin-A* and *-B* genes were identified in the genome of all the members of the Muroidea superfamily investigated (rat, hamster, vole and gerbil), but not in that of evolutionarily more distant rodents (squirrel, woodchuck and guinea pig).[31] Importantly, the muroid *syncytin* genes are not orthologous to the primate *syncytin* genes and are phylogenetically unrelated, illustrating a striking example of the phenomenon of convergent evolution.

Besides primates and muroids, involvement of ERV Env proteins has also been investigated in a species from a third branch of eutherian mammals, the ruminants, namely the sheep, whose genome harbors multiple endogenous Jaagsiekte sheep retrovirus (enJSRV) *env* genes.[33] However, no evidence could be obtained for induction of syncytia formation by any of the enJSRV *env* genes cloned to date. Rather, later results obtained by loss-of-function experiments *in utero* indicate that enJSRV Env plays a part in the development of the peri-implantation ovine conceptus, although the cellular and molecular mechanisms that mediate the biological effects of enJSRVs Env within cells remain unknown.[33]

Undoubtedly, further investigation in other branches of the eutherian tree should allow us to assess whether similar independent recruitment of ERVs in phylogenetically remote species is the rule and whether the decade-old theory of a cause-effect relationship between this and the emergence of a placenta in mammals holds true.[7,8]

SYNCYTIN EXPRESSION AND LOCALIZATION

Human Placenta: Trophoblast Differentiation

During placental development in humans, the definitive structure of the placenta appears very early, by day 21 of pregnancy (for a review, see Ref. 34). Following blastocyst adhesion, extremely rapid trophoblast proliferation begins and trophoblast cells undergo cell fusion to immediately form the multinucleated syncytiotrophoblast, which invades the maternal uterine stroma. By day 21 after ovulation, the chorionic villus — the definitive functional and structural unit of the placenta — is already established (Fig. 1). The trophoblast differentiates along two pathways: the extravillous and the villous trophoblast. In the case of the extravillous phenotype, the cytotrophoblast cells of the anchoring villi that are in contact with the uterine wall proliferate, detach from the basement membrane,

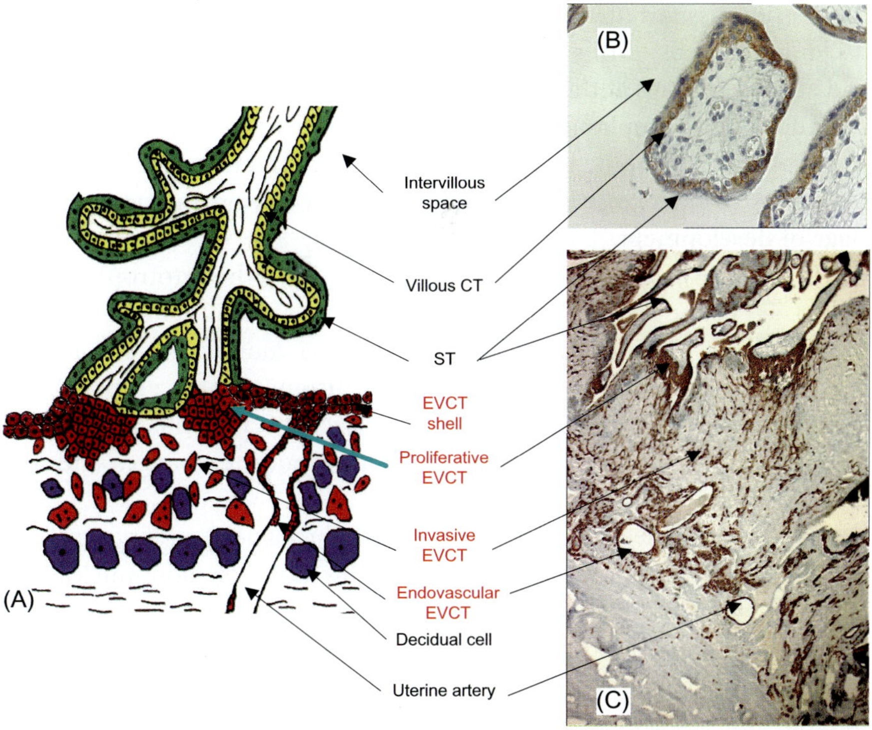

Fig. 1. Diagram of the utero-placental interface in the first trimester pregnancy. The chorionic villi are floating in the intervillous space (Panels A and B). In these villi, cytotrophoblastic cells (CT) differentiate by fusion to generate the syncytiotrophoblast (ST, Panels A, B, C). In the anchoring villi, the extravillous cytotrophoblastic cells (EVCT) in contact with the endometrium proliferate and invade the decidua and the lumen of associated spiral uterine arteries (Panels A and C). EVCT forms a continuous shell at the level of the decidua and forms plugs in the spiral arteries (Panel A). Panels B and C: cytokeratin07 immunostaining of villous and extravillous trophoblastic cells.

and aggregate into multilayered columns of non-polarized cells that rapidly invade the uterus. This trophoblastic invasion is confined to the endometrium, the first third of the myometrium, and the associated spiral arterioles. The extravillous trophoblastic cells scattered throughout the decidua and myometrium differentiate into multinucleated placental giant cells [Figs. 1(A) and 1(C)]. These giant cells appear to arise from cell fusion of extravillous trophoblasts. Thc extravillous trophoblast not only invades the uterine tissue but also forms a continuous shell at the level of the decidua. The cells of this shell anchor the placenta to the maternal tissue and also form plugs at the tips of the utero-placental arteries. These plugs of endovascular trophoblasts prevent, until the 12th week of gestation, access of maternal blood to the intervillous space and therefore protect the conceptus from excessively high oxygen levels during this very critical stage of development.[35]

In the case of the villous phenotype [Fig. 1(B)], the cytotrophoblastic cells (CTs) of the floating villi (in the intervillous space) remain attached to the villous basement membrane, forming a monolayer of epithelial cells, which proliferate and differentiate, by fusion, to form a syncytiotrophoblast (ST) that covers the entire surface of the villi. The ST is multifunctional and primarily engaged in absorption, exchanges, and specific hormonal functions.

The multinucleated ST is regenerated throughout pregnancy by a continuous turnover process including proliferation of mononuclear CT followed by the induction of early stages of apoptosis and fusion of these CT into ST. Apoptotic progression in the ST leads to the accumulation of condensed nuclei into syncytial knots and the shedding of these aggregates into the intervillous spaces. ST formation can be reproduced *in vitro* using different models. Choriocarcinoma cells, i.e., BeWo cells, are able to fuse in the presence of cAMP to form a multinucleated syncytium. However, the last step of differentiation, the gathering of nuclei into a central cluster, is missing. Purified villous CTs, cultured on plastic dishes, aggregate and fuse to form the multinucleated ST [Fig. 3(C)] with pregnancy specific hormonal production (i.e., hCG, progesterone).[36]

Syncytin-1

By *in situ* hybridization on term placental sections, HERV-W family mRNAs were initially detected only in the ST of the villous trophoblast.[13] However, depending on the specificities of antibodies and/or the various staining methods used, there is little consensus regarding *syncytin-1* localization in the villous trophoblast. Using immunohistochemistry, Blond *et al.*[12] localized the protein mainly in the ST, but also in the CTs of 13-week villi. Using a polyclonal antibody on normal second trimester and term placentas, Lee *et al.*[37] localized the *syncytin-1* immunostaining near the basal membrane of the ST. Yu *et al.*[21] published data showing a uniform staining of the whole ST in both term and preterm placenta, while Smallwood *et al.*[38] localized the staining in both the ST and CTs in first trimester placenta. Interestingly, *in vitro* isolated primary CTs from early and term placentas express the *syncytin-1* transcripts.[39] More recently according to Muir,[40] in all first and second trimester villous tissues examined, *syncytin-1* is not confined to the ST, but is also detected strongly in the underlying CTs. Using two different monoclonal antibodies, we have confirmed this double localization, as illustrated in Fig. 2(A).

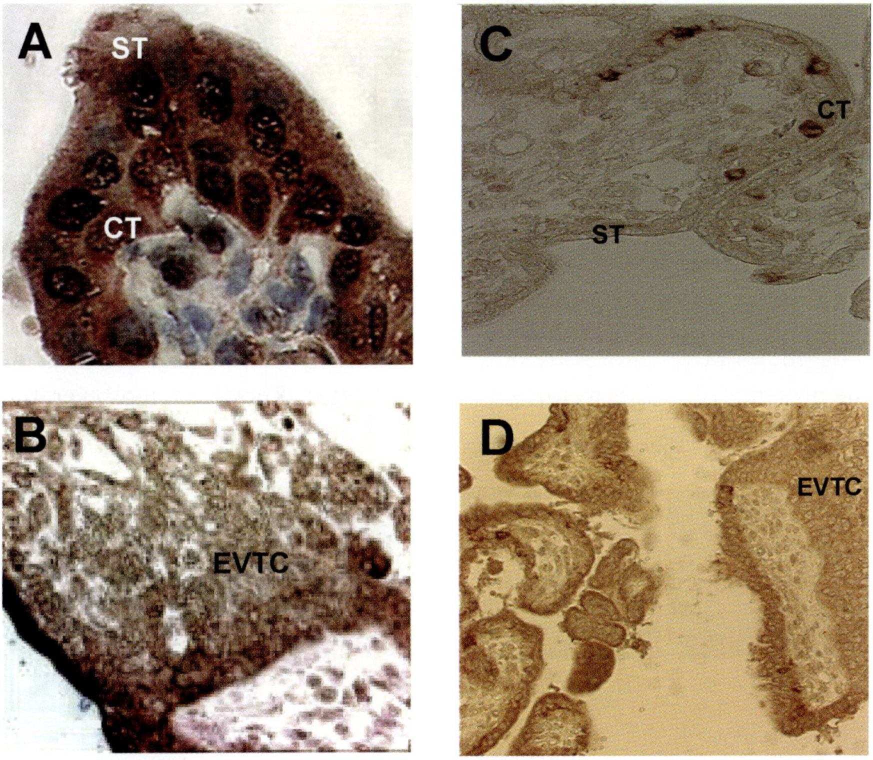

Fig. 2. Syncytin-1 and syncytin-2 localization in first trimester placenta. Syncytin-1 is immunodetected in the syncytiotrophoblast (ST), in the cytotrophoblastic cells (CT) of the villous trophoblast (Panel A) and in all the cell types of the extravillous phenotype lineage (Panel B, with permission of Elsevier). Syncytin-2 is immunodetected in some cytotrophoblastic cells of the floating placental villi (Panel C) and is not detected in the extravillous trophoblast (Panel D).

In contrast, *syncytin-1* immunolocalization in the extravillous trophoblast has been little studied. Using monoclonal and polyclonal antibodies, *syncytin-1* was immunolocalized in all the cell types of the extravillous phenotype lineage: CTs of the cell column, interstitial extravillous trophoblastic cells, multinucleated giant cells and endovascular trophoblastic cells [Fig. 2(B)]. In addition, the presence of *syncytin-1* transcripts and protein was demonstrated in cultured extravillous trophoblastic cells[41] [Fig. 3(A)]. Muir *et al.*[40] confirmed this study by demonstrating that

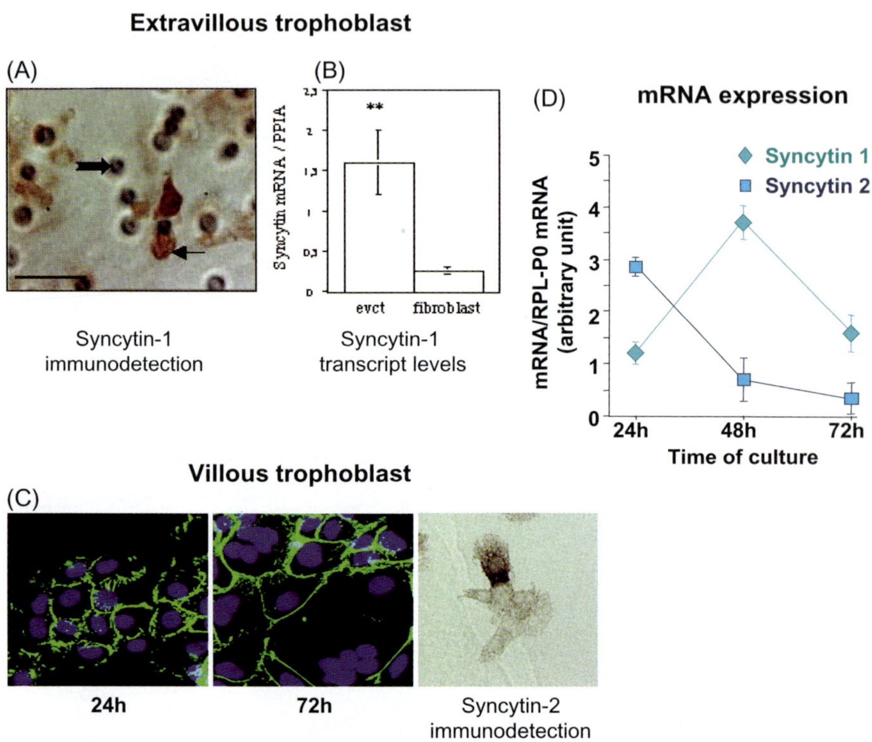

Fig. 3. Syncytin-1 and syncytin-2 expression in trophoblastic cells in culture. Panel A: Immunostaining with anti-syncytin-1 antibody shows a positive staining in extravillous trophoblastic cells (arrow) cultured for two days on matrigel transwell and migrating across the 8-μm diameter pores of the membrane (bold arrow). Panel B: Syncytin-1 mRNA expression in extravillous trophoblastic cells cultured for two days on matrigel and in placental fibroblasts cultured for the same period on plastic dishes (with permission of Elsevier).

syncytin-1 is widely expressed by various populations of normal human trophoblasts (villous and extravillous), as well as by ectopic trophoblasts in tubal pregnancy. *Syncytin-1* is also expressed in the choriocarcinoma cell lines JAR and JEG-3, which are used as models of villous and extravillous cells, respectively.

It appears now that, being expressed in all types of trophoblastic cells, *syncytin-1* can be considered as a specific marker of the human trophoblast.[41]

In 2003, Smallwood *et al.* were the first to study the temporal expression of *syncytin-1*.[38] The authors demonstrated that *syncytin-1* mRNA production is highest in term placenta but protein synthesis is lowest, compared with that in early pregnancy. Using TaqMan based RT-PCR assays, Okahara *et al.*[42] demonstrated that *syncytin-1* showed an almost constant expression throughout gestation. According to Chen *et al.*,[43] the level of *syncytin-1* mRNA expression increases significantly from the first trimester of pregnancy until 37 weeks of gestation, and then decreases.

Syncytin-1 induces the formation of syncytia upon interaction with the type D mammalian retrovirus receptor, RDR/ASCT2 (see above). According to Chen *et al.*,[44] the RDR/ASCT2 transcript levels are lower in second trimester and term placentas as compared to first trimester placentas. Using a polyclonal antibody, the type D mammalian virus receptor was localized in the various trophoblasts with the villous and extravillous phenotypes.[41] The co-localization of *syncytin-1* and its receptor in some trophoblastic cells that do not fuse (proliferative, intermediate and endovascular extravillous trophoblastic cells) suggests that *syncytin-1* and

---

Panel C: Morphological differentiation and syncytin-2 immunostaining during *in vitro* culture of isolated villous trophoblastic cells. After 24 hrs cytotrophoblastic cells (CT) are mainly aggregated and after 78 hrs a large amount of CTs have fused into syncytiotrophoblast (desmoplakin immonostaining and DAPI nuclear staining). Syncytin-2 immunostaining is observed in some aggregated CTs with a stronger staining at intercellular sites. Panel D: Syncytin-1 and syncytin-2 transcript variation during *in vitro* differentiation of villous trophoblastic cells. Syncytin-2 rapidly decreases during syncytiotrophoblast formation while syncytin-1 increases with cell aggregation and fusion and then slightly decreases.

its receptor appear to be required but are not sufficient for trophoblastic cell fusion. Recently, using another antibody, Hayward *et al.*[45] localized RDR/ASCT2 expression mainly in the cytotrophoblast compartment, the receptor being largely absent from the ST. Local variation in RDR/ASCT2 expression is not associated with the proliferative status of CTs, suggesting to these authors that the fusion of placental trophoblastic cells is not regulated by local or temporal variations in RDR/ASCT2 expression.

Syncytin-2

Using a highly specific monoclonal antibody, *syncytin-2* is only detected at the level of the CTs, which form a continuous single layer of cuboidal cells beneath the ST in the villi of the first trimester placenta.[46] Immunostaining is observed only in the cytoplasm of some CTs, but never in the ST nor in the mesenchymal core of the villi [Fig. 2(C)]. In the second trimester placenta [Figs. 4(A)]and 4(B)], *syncytin-2* immunostaining is present: (i) in the cytoplasm of CTs, and (ii) in their thin elongated cytoplasmic processes coming into contact with the ST and covering the villus basal lamina. At term (Fig. 5), immunostaining is detected at low magnification in a fraction of the flat CTs and extends into their thin cytoplasmic processes. Higher magnification shows the continuity of *syncytin-2* immunostaining between the cytoplasm surrounding the nuclei and that of the thin elongated cytoplasmic processes. Therefore, this localization highlights the modification of CT shape from cuboidal in early placentas to flat with cytoplasmic processes in term placentas.[47] Furthermore, *in vitro* detection of *syncytin-2* transcripts is restricted to villous CTs and decreases significantly with time in culture [Fig. 3(D)]. *In vitro* immunostaining is also observed in some aggregated CTs, with stronger staining at the intercellular boundaries [Fig. 3(C)]. Consistent with these results, *syncytin-2* transcripts were recently detected *in situ* only in CTs.[27,48] In contrast, using a polyclonal antibody, Chen *et al.*[26] observed a faint *syncytin-2* immunostaining in the ST of first trimester placental villi.

As already mentioned above, the human receptor for *syncytin-2*, MFSD2, has been recently identified.[27] *In situ* hybridization of human placental tissue using an MFSD2-specific probe provided evidence for receptor expression at the level of the ST. Taking into account the

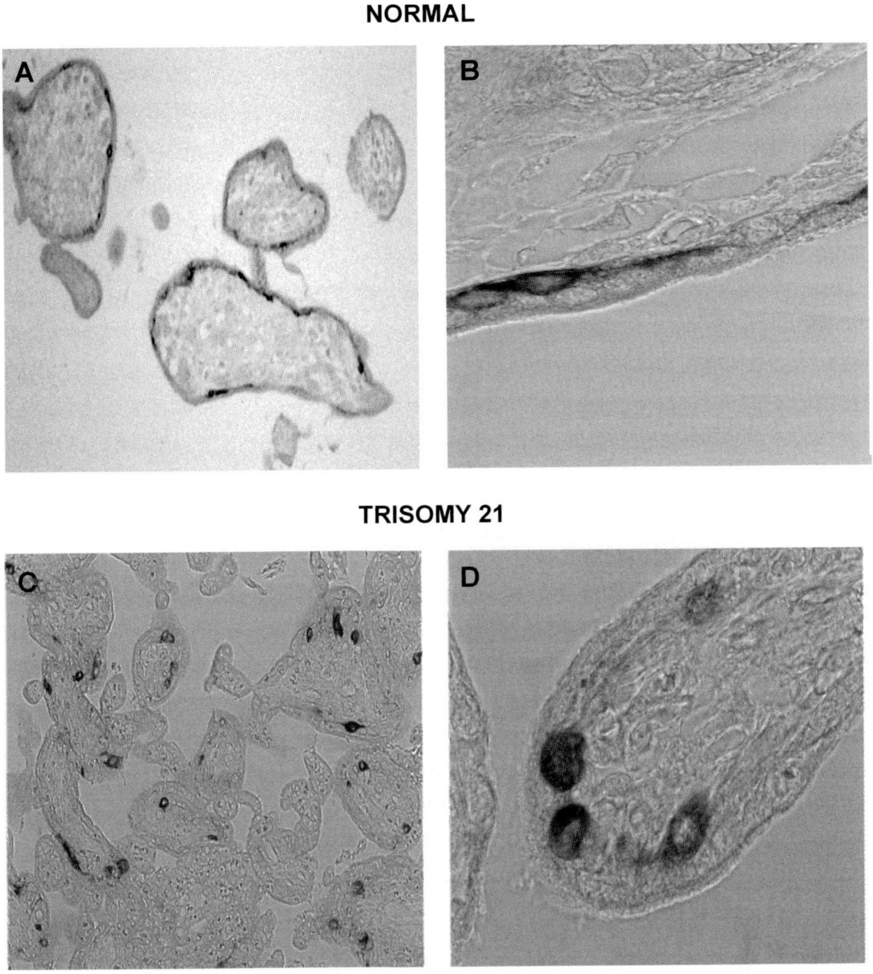

Fig. 4. Syncytin-2 localization in age-matched second trimester normal and trisomy 21-affected placenta. In normal placenta, immunostaining with anti-syncytin-2 antibody shows positive reactivity in a fraction of elongated cytotrophoblastic cells (Panel A). Syncytin-2 immunostaining is observed in the cytoplasm and in the thin cytoplasmic processes of the cytotrophoblastic cells (Panel B). In trisomy 21-affected placenta syncytin-2 is detected in some cuboidal cytotrophoblastic cells (Panels C and D).

difference in expression of *syncytin-2* and its receptor MFSD2 between placental cell types, a model can be proposed in which an oriented process of cell–cell fusion takes place with "in-fusion" of the mononucleated CT expressing *syncytin-2* into the ST expressing MFSD2.[27]

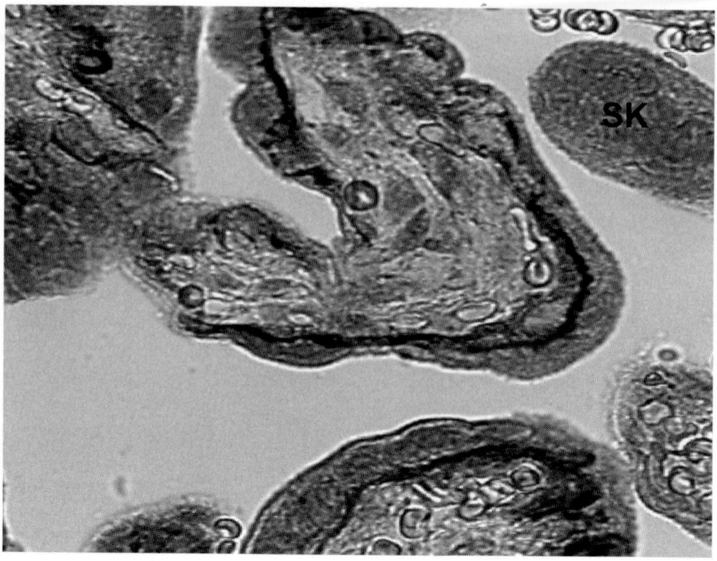

Fig. 5. *Syncytin-2* localization in preeclamptic term placenta. *Syncytin-2* is detected in the thin elongated cytoplasmic processes of the cytotrophoblastic cells of the villi. No staining is observed in the syncytial knots (SK).

There is no consensus regarding the temporal expression of *syncytin-2*. Using TaqMan-based RT-PCR, Okahara *et al.*[42] demonstrated that in contrast to *syncytin-1*, *syncytin-2* expression gradually decreases as pregnancy proceeds. Using immunoblotting with a polyclonal antibody, Chen *et al.*[26] observed a significant increase in the level of *syncytin-2* transcripts at the second and third trimester compared with the first trimester.

Syncytin-1 and *syncytin-2* expression during *in vitro* trophoblast differentiation

Using our well-established *in vitro* model of villous and extravillous CT isolation and culture, we investigated the levels of *syncytin-1* and *syncytin-2* transcripts in these cells. *Syncytin-1* is expressed at a higher level in villous as compared to extravillous CTs,[46] whereas *syncytin-2* is only expressed in villous CTs. Interestingly, during *in vitro* fusion and differentiation of villous CTs into ST, a striking difference in expression was observed between the two syncytin genes. While *syncytin-1* expression increases with cell aggregation and fusion, *syncytin-2* transcripts decrease upon fusion of CTs into ST [Fig. 3(D)].

Other Species

As described above, mice have independently acquired two ERV genes encoding the retroviral envelope proteins, *syncytin-A* and *-B*.[31] The *syncytin-A* and *-B* genes are specifically expressed in the mouse placenta and can trigger cell–cell fusion *in vitro* in several cell lines. *In vitro*, during the differentiation of trophoblast stem cells into ST, *syncytin-A* mRNA and protein expression is collinear with the fusion process and the protein is immunolocalized in the ST.[49] *In situ* hybridization of placenta cryosections further localizes the transcripts in the ST-containing labyrinthine zona.[31] Recently, a detailed *in situ* hybridization study revealed that *syncytin-A* mRNA specifically localizes to the ST-I layer of the labyrinth, which is closer to maternal blood, whereas *syncytin-B* mRNA is detected in the ST-II layer, which is closest to the fetal capillaries.[32]

The expanded earlier enJSRV *env* gene in sheep is another example of an *env* gene expressed in trophoblasts.[50] The enJSRV *env* is first expressed in the trophectoderm of the ovine placenta at day 12, when mononuclear trophectoderm cells begin proliferating in the process of blastocyst elongation. Hyaluronidase 2 (HYAL2), the receptor for the enJSRV *env* protein, is located within binucleate cells and "multinucleated syncytial plaques."[33]

THE FUNCTIONS OF SYNCYTINS

Considerable evidence has now been gathered to suggest that *syncytin-1* is involved in the multifactorial and dynamic process of trophoblastic cell fusion to form the syncytial layer of the villous placenta.[51] It has been demonstrated that transfection of a recombinant *syncytin-1* in a variety of cell types induces the formation of giant syncytia[12] and that fusion of trophoblastic choriocarcinoma cells (BeWo) can be inhibited by an anti-*syncytin-1* antiserum.[13] Furthermore forskolin induced syncytialization of BeWo cells causes a marked increase in *syncytin-1* expression that precedes syncytial formation. In primary culture of trophoblastic cells, Frendo *et al.*[39] showed that *syncytin-1* expression was upregulated after stimulating the fusion of primary CTs by a cAMP analog. Cell fusion and differentiation of these normal trophoblastic cells are markedly inhibited by specific *syncytin-1* antisense oligonucleotides. This was the first true evidence that *syncytin-1* may play a role in trophoblastic cell fusion *in vivo*. *Syncytin-1* binds to the type D retrovirus receptor ASCT2 and may use another neutral amino acid transporter (ASCT1) as an auxiliary receptor.[52] The hASCT2-binding domain of *syncytin-1* has been identified.[53] This domain contains several sub-domains that are poorly conserved, but a region of 18 residues containing the SDGGGX2DX2R conserved motif was proven to be essential for *syncytin-1*–hASCT2 interaction. Furthermore, conflicting results have been reported as to the presence or not of an inhibitory cytoplasmic domain, similar to the MLV R peptide, in *syncytin-1*. Some C-terminal truncation mutants were reported to activate fusogenicity,[24,54] whereas others were found to be inhibitory.[25] Further work is required to establish what role, if any, the cytoplasmic domain of syncytins may play in cell fusion regulation.

As demonstrated by *ex vivo* cell–cell fusion assays, *syncytin-2* is fusogenic.[10] However, this activity has not yet been demonstrated in normal trophoblastic cells. In the mouse, using a specific antibody or antisense oligonucleotides, Gong *et al.*[49] demonstrated that inhibition of *syncytin-A* leads to an obvious decrease in ST formation. The *syncytin-A* mediated fusion mechanism was recently analyzed: N-terminal heptad repeats of the protein are shown to be inhibitory to cell–cell fusion, whereas C-terminal heptad repeats seem to have no inhibitory activity.[55] In sheep, it has been demonstrated that injection *in utero* of morpholino antisense oligonucleotides that block enJSRV envelope protein production in the conceptus trophectoderm, induces a retarded trophectoderm outgrowth and binucleated cell differentiation, leading to pregnancy loss.[33]

The ability of syncytins to induce cell fusion may not be their only physiologic role. An anti-apoptotic function under certain conditions was also demonstrated for *syncytin-1*.[56,57]

Recently, using an allogeneic tumor rejection assay in the mouse, *syncytin-2* was found to be immunosuppressive, but — rather unexpectedly — *syncytin-1* was not.[58] As for the human syncytins, only one of the two mouse syncytins (*syncytin-B*) displays immunosuppressive activity.[58] Further scrutiny of the sequence of the immunosuppressive domain in the syncytins of other muroids showed a high degree of conservation, and importantly, in every instance, at least one of the two syncytins harbored an immunosuppressive function. Altogether, these data strongly suggest that immunosuppression is an essential function of syncytins since it has been conserved in both the muroid and the primate lineages in at least one member of each pair.[58] Understanding the involvement of *syncytin-A* and *-B* in the mouse placenta may provide valuable insights into the role in health and disease of the functionally analogous *syncytin-1* and *-2* genes in humans.

REGULATION OF SYNCYTIN EXPRESSION

One putative regulator of *syncytin-1* expression is the GCM-1 expanded earlier (on p. 4), which is required for placental development in the mouse. Interestingly, GCM-1 is able to activate *syncytin-1* mRNA expression via two GCM-1-binding sites upstream of the 5′ LTR of the *syncytin-1* gene in BeWo and JEG3 cells but not in HeLa cells.[21] In the villous trophoblast, *in situ* hybridization studies show that the GCM-1 mRNA is expressed in villous CTs. GCM-1 protein expression is seen mainly within the nuclei of a subset of CTs consistent with its role as a transcription factor.[59] The cAMP/protein kinase A (PKA) pathway was found to activate *syncytin-1* expression via regulation of GCM-1 activity.[60,61] Furthermore, Chuang *et al.* demonstrated that CBP-mediated GCM-1 acetylation underlies the activated cAMP/PKA signaling pathway that stimulates trophoblastic fusion and that the regulation of GCM activities is regulated by histone acetyl transferases (HATs) and histone deacetylases (HDACs).[62] These observations provide a mechanistic explanation for the stimulation of the fusion of cultured trophoblastic cells by the cAMP-inducer forskolin. Forskolin also activates the expression of *syncytin-2*,[26] but the molecular mechanisms involved are at present unknown. Interestingly, in the mouse, expression of GCM-1 is detected in the ST-II, but not the ST-I, layer, suggesting that it specifically activates the *syncytin-B* gene. Indeed, *syncytin-B* was found to be downregulated in GCM-1 mutant mice, whereas *syncytin-A* expression was unaltered.[32] Recently, bZIP-type transcription factors TORC1 (transducer of regulated CREB) and OASIS (old astrocyte specifically induced substance) were shown to bind and stimulate the promoter of GCM-1 in trophoblastic cells.[63] Over-expression of TORC1 or OASIS in choriocarcinoma cells led to placental cell fusion. Furthermore, inhibition of GCM-1 by siRNA and antisense oligonucleotide methods result in an increased rate of proliferation but prevent *de novo* ST formation in syncytially denuded floating villous explants.[59]

Because local ischemia and hypoxia are factors known to be involved in placental pathophysiology, the influence of hypoxia on *syncytin-1* and ASCT2 was analyzed with various models. It must be first pointed out that trophoblastic cell fusion and differentiation are inhibited by hypoxia.[64] Using BeWo cells, Knerr *et al.*[65] demonstrated that hypoxia decreased *syncytin-1* mRNA levels in cultured cells, whereas those of ASCT2 mRNA were not altered significantly. In isolated perfused cotyledons (a more physiological model) hypoxia also reduces *syncytin-1* expression but not that of its receptor. Kudo *et al.*[66] showed that low-oxygen conditions (2 percent) suppress fusion of choriocarcinoma cells together with the expression of *syncytin-1* and its receptor. These effects are rapidly reversed if the oxygen concentration is increased from 2 percent to 20 percent. These results suggest that under conditions of low ambient oxygen, dysregulation of *syncytin-1* expression may disturb the normal process of ST formation. Epigenetic factors such as the methylation process might regulate syncytin expression. Indeed CpG methylation plays a fundamental role in the transcriptional suppression of *syncytin-1* in non-placental tissues, and by contrast, demethylation of the *syncytin-1* promoter in trophoblast is a prerequisite for its expression and for differentiation of the multinucleated ST.[67] The re-methylation pattern observed during pregnancy is consistent with the significant decrease of *syncytin-1* observed by Chen *et al.*[43] between 37 and 40 weeks of gestation. Conversely, *syncytin-2* remains unmethylated throughout gestation, suggesting that the progressive decrease of expression from the first trimester until term[42] could be due to a modulation of the pool of transcription factors.

SYNCYTINS AND PLACENTAL PATHOLOGIES

Preeclampsia

Preeclampsia is a major and frequent complication of human pregnancy with serious maternal and fetal — intrauterine growth retardation (IUGR) — consequences.[68] Preeclampsia is diagnosed on the basis of hypertension (≥14/9) and significant proteinuria (≥0.30 g/24 hr) during the second trimester. Preeclampsia is specific to humans and is placental in origin. The pathophysiology of preeclampsia is poorly understood, but pathological studies and recent experimental data suggest the following: (i) poor placental perfusion due to defective invasion and remodeling of the uterine spiral arteries by the extravillous cytotrophoblasts during the first trimester; (ii) maternal factors of genetic and/or environmental origin; and (iii) placental ischemia and release of toxic ST fragments circulating in maternal blood, leading to maternal endothelial dysfunction and clinical signs. The time interval between the cause (defective vascular invasion during the first trimester) and the consequences (clinical signs during the second trimester) complicates the study of this disorder of placental origin.

As analyzed by immunohistochemistry and illustrated in Fig. 5, an increase in ST apoptosis is observed in placenta from preeclamptic women and trophoblast fragments are shed into the maternal blood in greater numbers.[69] In addition, changes in trophoblast turnover might occur with an increase in cell proliferation being correlated with hypoxia conditions and *syncytin-1* expression decrease.

Using *in situ* hybridization and immunohistochemistry with a polyclonal antibody, Lee *et al.*[37] were the first to demonstrate that *syncytin-1* expression is reduced in PE. Similarly in placental villi, *syncytin-1* mRNAs are lower in patients with PE or HELLP (hemolysis, elevated liver enzymes, low platelets) syndrome than in healthy control subjects.[70] In addition, at the protein level, using immunoblots, *syncytin-1* was also shown to be reduced in PE placentas by approximately 50 percent.[43] Recently, a quantitative analysis showed that the transcriptional level of four HERVs (*syncytin-1*, *syncytin-2*, HERV-H7/F and HERV-Fb) were lower in placentas obtained from pregnant women with pregnancy induced hypertension than in those from normotensive pregnant women.[48] However, no difference in *syncytin-2* localization was observed between preeclamptic placentas and gestational age-matched control placentas.[47] Finally, according to Langbein *et al.*[71] cultured trophoblastic cells from PE- and HELLP-associated IUGR pregnancies have an intrinsic impaired ability to fuse. All these results suggest that the significant reduction in syncytin expression in preeclamptic placenta might be related to hypoxia and abnormal trophoblast turnover. However, further investigations are required to elucidate the relationship between decrease in syncytin expression and the complex physiopathology of this disease.

Trisomy 21

Trisomy of chromosome 21 (T21), which causes the phenotype known as Down's syndrome, is the major known genetic cause of mental retardation and is found in around 1:800 live births. Little is known about placental development in this aneuploid condition despite the fact that the trophoblast carries the genetic abnormality. However, a defect in ST formation in T21-affected placentas is observed. Cultured cytotrophoblasts, isolated from T21-affected placentas, aggregate but fuse poorly or belatedly (Fig. 4).[72,73] This is in agreement with previous histological observations pointing to an increased percentage of two-layered trophoblasts in T21 placentas.[74,75] In addition, we demonstrated that this *in vitro* defect or delay in ST formation is characterized by a dramatic decrease in the synthesis of syncytiotrophoblastic pregnancy-associated hormones[76] and by the secretion of an hyperglycosylated human chorionic gonadotropin with low bioactivity.[77] This abnormal trophoblast fusion implicates at least in part over-expression of super oxide dismutase-1 (SOD-1).[78] In addition, we recently showed that during the second trimester of pregnancy, *syncytin-2* is immunolocalized in some cuboidal CTs in T21 placentas, whereas in normal placentas it is observed in flat CTs, extending into their cytoplasmic processes. These results highlight the abnormal trophoblast differentiation observed in trisomy 21-affected placentas.

CONCLUSION

Besides the profound impact that transposable elements have had on genome evolution and organization by imposing selection pressures on their hosts, it has become apparent that ERVs have contributed beneficial functions several times. In the case of mammals, this is best illustrated by the "domestication" of ERV *env* genes to serve as cellular fusogenic proteins, the syncytins, within the placenta of primates and rodents. However, challenging tasks lie ahead to delineate the precise role of each protein *in vivo*. In the mouse, knocking out each *syncytin* gene or knocking in appropriately designed mutant alleles should allow precise elucidation of *in vivo* function, the mechanism of action of their product, and their potential functional redundancy or complementation. Although valuable pieces of information can be gathered from the study of the mouse as a model organism easily amenable to experimentation, these conclusions cannot be directly extrapolated to humans since the rodent and primate *syncytin* genes are homologous, not orthologous, and the formation of syncytial structures follows different patterns.

We still have a lot to learn about *syncytin* function in normal and pathological human placenta. Indeed, trophoblast fusion is one essential and limiting step of the human trophoblast differentiation pathway and trophoblast turnover. It is a multifactorial and dynamic process that is finely regulated and still poorly understood. We still do not know the molecular mechanisms by which the fusogenic syncytins unite membranes. Fusion mediated by *syncytin-1* and *-2* is dependent on their binding to cell-surface receptors. It will be of interest to analyze the effect of syncytins on these receptors and their functions. Interestingly, the *syncytin-1* and -2 receptors are transporters that could be implicated in feto-placental transfers and exchanges. However, it appears that amino acid transporters act as regulators of intracellular signaling, leading to the concept of transceptors.[79] Therefore, syncytin binding to receptors with transceptor function could stimulate intracellular signaling involved in cell–cell fusion. This new concept remains to be investigated, as does the identification of the signaling pathways.

Studies of pathological placentas show that *syncytin-1* expression illustrates the abnormal trophoblast development observed during preeclampsia. Furthermore, *syncytin-2* illustrates the abnormal trophoblast differentiation observed in T21 affected pregnancies. The absence of a consensus regarding the temporal expression of syncytins requires further investigations. Moreover, genetic or epigenetic modulation of syncytin expression during pregnancy and in placental pathology remains to be investigated.

REFERENCES

1. Jern P, Coffin JM. (2008) Effects of retroviruses on host genome function. *Annu Rev Genet* **42**: 709–732.
2. de Parseval N, Heidmann T. (2005) Human endogenous retroviruses: From infectious elements to human genes. *Cytogenet Genome Res* **110**: 318–332.
3. Villesen P, Aagaard L, Wiuf C, Pedersen FS. (2004) Identification of endogenous retroviral reading frames in the human genome. *Retrovirology* **1**: 32.
4. de Parseval N, Lazar V, Casella JF, *et al.* (2003) Survey of human genes of retroviral origin: Identification and transcriptome of the genes with coding capacity for complete envelope proteins. *J Virol* **77**: 10414–10422.
5. Blaise S, de Parseval N, Heidmann T. (2005) Functional characterization of two newly identified human endogenous retrovirus coding envelope genes. *Retrovirology* **2**: 19.
6. Harris J. (1991) The evolution of placental mammals. *FEBS Lett* **295**: 3–4.
7. Harris J. (1998) Placental endogenous retrovirus (ERV): Structural, functional, and evolutionary significance. *Bioessays* **20**: 307–316.
8. Villareal L. (1997) On viruses, sex, and motherhood. *J Virol* **71**: 859–865.
9. Cianciolo GJ, Copeland TD, Oroszlan S, Snyderman R. (1985) Inhibition of lymphocyte proliferation by a synthetic peptide homologous to retroviral envelope proteins. *Science* **230**: 453–455.
10. Blaise S, de Parseval N, Benit L, Heidmann T. (2003) Genomewide screening for fusogenic human endogenous retrovirus envelopes identifies syncytin 2, a gene conserved on primate evolution. *Proc Natl Acad Sci USA* **100**: 13013–13018.
11. Blond J-L, Beseme F, Duret L, *et al.* (1999) Molecular characterization and placental expression of HERV-W, a new human endogenous retrovirus family. *J Virol* **73**: 1175–1185.

12. Blond J-L, Lavillette D, Cheynet V, *et al.* (2000) An envelope glycoprotein of the human endogenous retrovirus HERV-W is expressed in the human placenta and fuses cells expressing the type D mammalian retrovirus receptor. *J Virol* **74**: 3321–3329.
13. Mi S, Lee X, Li X-P, *et al.* (2000) Syncytin is a captive retroviral envelope protein involved in human placental morphogenesis. *Nature* **403**: 785–789.
14. Mallet F, Bouton O, Prudhomme S, *et al.* (2004) The endogenous retroviral locus ERVWE1 is a bonafide gene involved in hominoid placental physiology. *Proc Natl Acad Sci USA* **101**: 1731–1736.
15. de Parseval N, Diop G, Blaise S, *et al.* (2005) Comprehensive search for intra- and inter-specific sequence polymorphisms among coding envelope genes of retroviral origin found in the human genome: Genes and pseudogenes. *BMC Genomics* **6**: 117.
16. Prudhomme S, Oriol G, Mallet F. (2004) A retroviral promoter and a cellular enhancer define a bipartite element which controls env ERVWE1 placental expression. *J Virol* **78**: 12157–12168.
17. Lee WJ, Kwun HJ, Jang KL. (2003) Analysis of transcriptional regulatory sequences in the human endogenous retrovirus W long terminal repeat. *J Gen Virol* **84**: 2229–2235.
18. Cheng YH, Richardson BD, Hubert MA, Handwerger S. (2004) Isolation and characterization of the human syncytin gene promoter. *Biol Reprod* **70**: 694–701.
19. Cheng YH, Handwerger S. (2005) A placenta-specific enhancer of the human syncytin gene. *Biol Reprod* **73**: 500–509.
20. Lin C, Lin M, Chen H. (2005) Biochemical characterization of the human placental transcription factor GCMa/1. *Biochem Cell Biol* **83**: 188–195.
21. Yu C, Shen K, Lin M, *et al* (2002) GCMa regulates the syncytin-mediated trophoblastic fusion. *J Biol Chem* **277**: 50062–50068.
22. Hunter E. (1997) *Viral Entry and Receptors.* Cold Spring Harbor Lab. Press, Plainview, NY.
23. White JM, Delos SE, Brecher M, Schornberg K. (2008) Structures and mechanisms of viral membrane fusion proteins: Multiple variations on a common theme. *Crit Rev Biochem Mol Biol* **43**: 189–219.
24. Chang C, Chen PT, Chang GD, *et al.* (2004) Functional characterization of the placental fusogenic membrane protein syncytin. *Biol Reprod* **71**: 1956–1962.

25. Cheynet V, Ruggieri A, Oriol G, *et al.* (2005) Synthesis, assembly, and processing of the Env ERVWE1/syncytin human endogenous retroviral envelope. *J Virol* **79**: 5585–5593.
26. Chen CP, Chen LF, Yang SR, *et al.* (2008) Functional characterization of the human placental fusogenic membrane protein syncytin 2. *Biol Reprod* **79**: 815–823.
27. Esnault C, Priet S, Ribet D, *et al.* (2008) A placenta-specific receptor for the fusogenic, endogenous retrovirus-derived, human syncytin-2. *Proc Natl Acad Sci USA* **105**: 17532–17537.
28. Boyd MT, Bax CM, Bax BE, *et al.* (1993) The human endogenous retrovirus ERV-3 is upregulated in differentiating placental trophoblast cells. *Virology* **196**: 905–909.
29. de Parseval N, Heidmann T. (1998) Physiological knockout of the envelope gene of the single-copy ERV-3 human endogenous retrovirus in a fraction of the Caucasian population. *J Virol* **72**: 3442–3445.
30. Kjeldbjerg AL, Villesen P, Aagaard L, Pedersen FS. (2008) Gene conversion and purifying selection of a placenta-specific ERV-V envelope gene during simian evolution. *BMC Evol Biol* **8**: 266.
31. Dupressoir A, Marceau G, Vernochet C, *et al.* (2005) Syncytin-A and syncytin-B, two fusogenic placenta-specific murine envelope genes of retroviral origin conserved in Muridae. *Proc Natl Acad Sci USA* **102**: 725–730.
32. Simmons DG, Natale DR, Begay V, *et al.* (2008) Early patterning of the chorion leads to the trilaminar trophoblast cell structure in the placental labyrinth. *Development* **135**: 2083–2091.
33. Dunlap KA, Palmarini M, Varela M, *et al.* (2006) Endogenous retroviruses regulate periimplantation placental growth and differentiation. *Proc Natl Acad Sci USA* **103**: 14390–143905.
34. Benirschke K, Kaufmann P. (2000) *Pathology of the Human Placenta.* Springer-Verlag, New York.
35. Burton GJ. (2008) Oxygen, the Janus gas: Its effects on human placental development and function. *J Anat* **215**: 27–35.
36. Kliman H, Nestler J, Sermasi E, *et al.* (1986) Purification, characterization, and *in vitro* differentiation of cytotrophoblasts from human term placentae. *Endocrinology* **118**: 1567–1582.

37. Lee X, Keith JC, Jr., Stumm N, *et al.* (2001) Downregulation of placental syncytin expression and abnormal protein localization in pre-eclampsia. *Placenta* **22**: 808–812.
38. Smallwood A, Papageorghiou A, Nicolaides K, *et al.* (2003) Temporal regulation of the expression of syncytin (HERV-W), maternally imprinted PEG10, and SGCE in human placenta. *Biol Reprod* **69**: 286–293.
39. Frendo J-L, Olivier D, Cheynet V, *et al.* (2003) Direct involvement of HERV-W Env glycoprotein in human trophoblast cell fusion and differentiation. *Mol Cell Biol* **23**: 3566–3574.
40. Muir A, Lever AM, Moffett A. (2006) Human endogenous retrovirus-W envelope (syncytin) is expressed in both villous and extravillous trophoblast populations. *J Gen Virol* **87**: 2067–2071.
41. Malassine A, Handschuh K, Tsatsaris V, *et al.* (2005) Expression of HERV-W Env glycoprotein (syncytin) in the extravillous trophoblast of first trimester human placenta. *Placenta* **26**: 556–562.
42. Okahara G, Matsubara S, Oda T, *et al.* (2004) Expression analyses of human endogenous retroviruses (HERVs): Tissue-specific and developmental stage-dependent expression of HERVs. *Genomics* **84**: 982–990.
43. Chen CP, Wang KG, Chen CY, *et al.* (2006) Altered placental syncytin and its receptor ASCT2 expression in placental development and pre-eclampsia. *BJOG* **113**: 152–158.
44. Chen EH, Olson EN. (2005) Unveiling the mechanisms of cell–cell fusion. *Science* **308**: 369–373.
45. Hayward MD, Potgens AJ, Drewlo S, *et al.* (2007) Distribution of human endogenous retrovirus type W receptor in normal human villous placenta. *Pathology* **39**: 406–412.
46. Malassine A, Blaise S, Handschuh K, *et al.* (2007) Expression of the fusogenic HERV-FRD Env glycoprotein (syncytin 2) in human placenta is restricted to villous cytotrophoblastic cells. *Placenta* **28**: 185–191.
47. Malassine A, Frendo JL, Blaise S, *et al.* (2008) Human endogenous retrovirus-FRD envelope protein (syncytin 2) expression in normal and trisomy 21-affected placenta. *Retrovirology* **5**: 6.
48. Kudaka W, Oda T, Jinno Y, *et al.* (2008) Cellular localization of placenta-specific human endogenous retrovirus (HERV) transcripts and their possible implication in pregnancy-induced hypertension. *Placenta* **29**: 282–289.

49. Gong R, Huang L, Shi J, *et al.* (2007) Syncytin-A mediates the formation of syncytiotrophoblast involved in mouse placental development. *Cell Physiol Biochem* **20**: 517–526.
50. Dunlap KA, Palmarini M, Adelson DL, Spencer TE. (2005) Sheep endogenous betaretroviruses (enJSRVs) and the hyaluronidase 2 (HYAL2) receptor in the ovine uterus and conceptus. *Biol Reprod* **73**: 271–279.
51. Malassiné A, Frendo J-L, Evain-Brion D. (2010) Trisomy 21-affected placentas highlight prerequisite factors for human trophoblast fusion and differentiation. *Int J Dev Biol* **54**: 475–482.
52. Lavillette D, Marin M, Ruggieri A, *et al.* (2002) The envelope glycoprotein of human endogenous retrovirus type W uses a divergent family of amino acid transporters/cell surface receptors. *J Virol* **76**: 6442–6452.
53. Cheynet V, Oriol G, Mallet F. (2006) Identification of the hASCT2-binding domain of the Env ERVWE1/syncytin-1 fusogenic glycoprotein. *Retrovirology* **3**: 41.
54. Drewlo S, Leyting S, Kokozidou M, *et al.* (2006) C-Terminal truncations of syncytin-1 (ERVWE1 envelope) that increase its fusogenicity. *Biol Chem* **387**: 1113–1120.
55. Peng X, Pan J, Gong R, *et al.* (2007) Functional characterization of syncytin-A, a newly murine endogenous virus envelope protein. Implication for its fusion mechanism. *J Biol Chem* **282**: 381–389.
56. Knerr I, Schnare M, Hermann K, *et al.* (2007) Fusiogenic endogenous-retroviral syncytin-1 exerts anti-apoptotic functions in staurosporine-challenged CHO cells. *Apoptosis* **12**: 37–43.
57. Knerr I, Soder S, Licha E, *et al.* (2008) Response of HEK293 and CHO cells overexpressing fusiogenic syncytin-1 to mitochondrion-mediated apoptosis induced by antimycin A. *J Cell Biochem* **105**: 766–775.
58. Mangeney M, Renard M, Schlecht-Louf G, *et al.* (2007) Placental syncytins: genetic disjunction between the fusogenic and immunosuppressive activity of retroviral envelope proteins. *Proc Natl Acad Sci USA* **104**: 20534–20539.
59. Baczyk D, Drewlo S, Proctor L, *et al.* (2009) Glial cell missing-1 transcription factor is required for the differentiation of the human trophoblast. *Cell Death Differ* **16**: 719–727.
60. Chang CW, Chuang HC, Yu C, *et al.* (2005) Stimulation of GCMa transcriptional activity by cyclic AMP/protein kinase A signaling is attributed to CBP-mediated acetylation of GCMa. *Mol Cell Biol* **25**: 8401–8414.

61. Knerr I, Schubert SW, Wich C, *et al.* (2005) Stimulation of GCMa and syncytin via cAMP mediated PKA signaling in human trophoblastic cells under normoxic and hypoxic conditions. *FEBS Lett* **579**: 3991–3998.
62. Chuang HC, Chang CW, Chang GD, *et al.* (2008) Histone deacetylase 3 binds to and regulates the GCMa transcription factor. *Nucl Acid Res* **34**: 1459–1469.
63. Schubert SW, Abendroth A, Kilian K, *et al.* (2008) bZIP-Type transcription factors CREB and OASIS bind and stimulate the promoter of the mammalian transcription factor GCMa/Gcm1 in trophoblast cells. *Nucl Acid Res* **36**: 3834–3846.
64. Alsat E, Wyplosz P, Malassiné A, *et al.* (1996) Hypoxia impairs cell fusion and differentiation process in human cytotrophoblast, *in vitro*. *J Cell Physiol* **168**: 346–353.
65. Knerr I, Weigel C, Linnemann K, *et al.* (2003) Transcriptional effects of hypoxia on fusiogenic syncytin and its receptor ASCT2 in human cytotrophoblast BeWo cells and in *ex vivo* perfused placental cotyledons. *Am J Obstet Gynecol* **189**: 583–588.
66. Kudo Y, Boyd CA, Millo J, *et al.* (2003) Manipulation of CD98 expression affects both trophoblast cell fusion and amino acid transport activity during syncytialization of human placental BeWo cells. *J Physiol* **550**: 3–9.
67. Matouskova M, Blazkova J, Pajer P, *et al.* (2006) CpG methylation suppresses transcriptional activity of human syncytin-1 in non-placental tissues. *Exp Cell Res* **312**: 1011–1020.
68. Roberts JM, Cooper DW. (2001) Pathogenesis and genetics of pre-eclampsia. *Lancet* **357**: 53–56.
69. Mayhew TM. (2006) Allometric studies on growth and development of the human placenta: Growth of tissue compartments and diffusive conductances in relation to placental volume and fetal mass. *J Anat* **208**: 785–794.
70. Knerr I, Beinder E, Rascher W. (2002) Syncytin, a novel human endogenous retroviral gene in human placenta: Evidence for its dysregulation in preeclampsia and HELLP syndrome. *Am J Obstet Gynecol* **86**: 210–213.
71. Langbein M, Strick R, Strissel PL, *et al.* (2008) Impaired cytotrophoblast cell–cell fusion is associated with reduced syncytin and increased apoptosis in patients with placental dysfunction. *Mol Reprod Dev* **75**: 175–83.
72. Frendo J-L, Vidaud M, Guibourdenche J, *et al.* (2000) Defect of villous cytotrophoblast differentiation into syncytiotrophoblast in Down syndrome. *J Clin Endocrinol Metab* **85**: 3700–3707.

73. Massin N, Frendo J-L, Guibourdenche J, *et al.* (2001) Defect of syncytiotrophoblast formation and hCG expression in Down's syndrome. *Placenta* 22 (Suppl A): S93–S97.
74. Oberweiss D, Gillerot Y, Koulischer L, *et al.* (1983) Le placenta des Trisomie 21 dans le dernier trimestre de la gestation. *J Gynecol Obstet Biol Reprod* **12**: 345–349.
75. Roberts L, Sebire N, Fowler D, Nicolaides K. (2000) Histomorphological features of chorionic villi at 10–14 weeks of gestation in trisomic and chromosomally normal pregnancies. *Placenta* **21**: 678–683.
76. Pidoux G, Gerbaud P, Laurendeau I, *et al.* (2004) Large variability of trophoblast gene expression within and between human normal term placentae. *Placenta* **25**: 469–473.
77. Frendo J-L, Guibourdenche J, Pidoux G, *et al.* (2004) Trophoblast production of a weakly bioactive human chorionic gonadotropin in trisomy 21-affected pregnancy. *J Clin Met Endocrinol* **89**: 727–732.
78. Frendo J-L, Thérond P, Bird T, *et al.* (2001) Overexpression of copper zinc superoxide dismutase impairs human trophoblast cell fusion and differentiation. *Endocrinology* **142**: 3638–3648.
79. Hundal HS, Taylor PM. (2009) Amino acid transceptors: Gate keepers of nutrient exchange and regulators of nutrient signaling. *Am J Physiol Endocrinol Metab* **296**: E603–E613.s

CHAPTER 9

Vpu, Tetherin and Innate Immunity: Antiviral Restriction of Retroviral Particle Release

Stuart J. D. Neil*

SUMMARY

The role of the HIV-1 accessory genes in the pathogenesis of HIV/AIDS has become of increasing interest recently. The ability of these proteins to modulate aspects of both innate and adaptive immune responses to facilitate viral replication is an emerging theme in our understanding of the importance of the lentiviral accessory genes. The Vpu protein of HIV-1 has at least two roles in viral replication: to induce the degradation of the primary viral receptor CD4, and to promote the release of newly formed virus particles from infected cells. Recent advances have shed light on the biological basis for the latter function. Vpu targets an interferon-regulated host membrane protein, tetherin (CD317/BST2), that inhibits the release of nascent retroviral particles. In this chapter, I review our current understanding of the function of Vpu with particular reference to its targeting

*Department of Infectious Disease, King's College London School of Medicine, Guy's Hospital, London SE1 9RT, United Kingdom, Tel: +44 207 188 8279, E-mail: stuart.neil@kcl.ac.uk.

of tetherin. I also discuss tetherin-mediated restriction of other enveloped mammalian viruses and the possible strategies by which other viruses may target its activity. Finally, I address the potential implications of Vpu and tetherin interactions in human disease.

INTRODUCTION

Viral protein U (Vpu) is a 16 KDa membrane protein encoded by human immunodeficiency type 1 (HIV-1),[1,2] the simian immunodeficiency viruses that infect chimpanzees (SIVcpz), and their likely evolutionary precursors found in several Old World monkey species such as the greater spot-nosed monkey (SIVgsn), the mona monkey (SIVmon) and the moustached macaque (SIVmus),[3] but not any other primate immunodeficiency virus lineage or HIV-2. The protein is expressed from the same spliced mRNA species that also encode the viral envelope glycoprotein.[4] The requirement for Vpu in HIV-1 replication is cell-type dependent. Vpu is dispensable for virus growth in many transformed cell types[5–7] but Vpu(−) virus replication is impaired in primary $CD4^+$ T cells and severely inhibited in macrophages,[8] suggesting it is essential *in vivo*.

Vpu consists of an N-terminal transmembrane domain (Fig. 1), and a cytoplasmic tail that forms two alpha helices linked by a conserved DSGNES motif that can be phosphorylated by casein kinase II (CKII).[9,10] Evidence from reconstituted membranes suggests that Vpu may oligermerize into a pentamer to form a cation permeable ion channel.[11,12] In this regard, it has some structural similarities to other small ion channels encoded by diverse eukaryotic viruses, termed viroporins,[13] the prototypic member being the M2 proton channel of influenza viruses. Expression of Vpu in *Xenopus* oocytes has demonstrated that it can mediate ion flux across the plasma membrane and the TM domain is required for its known biological functions.[12,14] However, the relevance of Vpu's ion channel effect to its function *in vivo* is unclear, although small molecule ion channel inhibitors derived from amilorides do appear to have antiviral activity on HIV-1 replication.[15,16]

Vpu localizes predominantly to internal membrane structures, the trans-Golgi network (TGN) and endosomes as well as the endoplasmic reticulum,[17] although Vpu proteins from Clade C HIV-1 strains have a

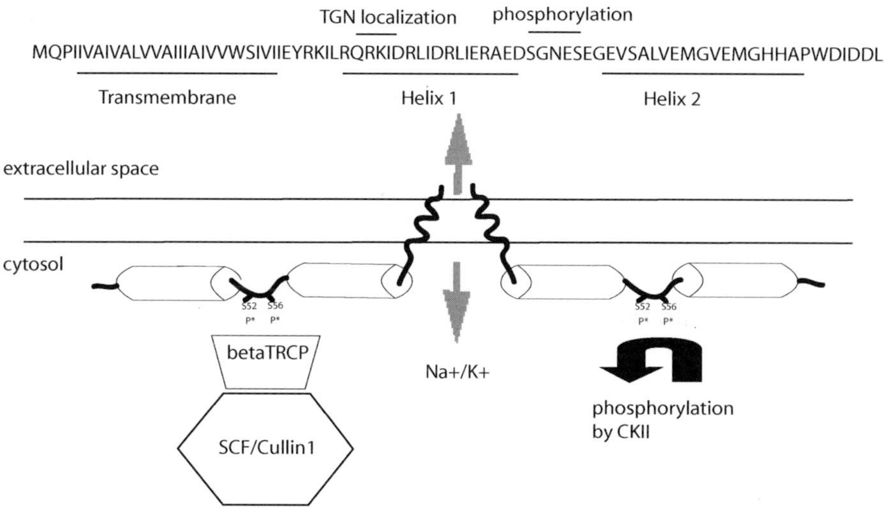

Fig. 1. A schematic of the Vpu protein of HIV-$1_{NL4.3}$. Vpu consists of an N-terminal transmembrane domain and a cytoplasmic tail that forms two alpha helices linked by casein kinase II (CKII) phospho-acceptor sites on serines 52 and 56. A small basic RKI motif may be involved in Vpu trafficking to the TGN.[74] The Vpu TM domains form a pentameric ion-permeable channel in the membrane and are predicted to be pitched at a 15° angle, with some or all of the cytoplasmic tail lying parallel to the membrane.[14] The cytoplasmic tail of Vpu binds to that of CD4, and phosphorylation dependent recruitment of βTRCP leads to CD4 dislocation and proteasomal degradation.

more prominent localization at the plasma membrane.[18,19] In the HIV-1 life cycle Vpu has two major functions. One is to downregulate the expression of the primary viral receptor, CD4 (described below). The other, until recently poorly characterized, is that Vpu is able to stimulate the release of new virus particles from infected cells. It is this activity of Vpu that forms the major focus of this chapter.

CD4 DOWNREGULATION

Vpu triggers a downregulation and degradation of CD4 in infected cells.[20] This is not a unique function in the HIV-1 genome as the accessory gene Nef also performs a similar role (reviewed in Ref. 21). However, the action of these proteins is spatially and mechanistically distinct: while Nef removes CD4 from the plasma membrane for lysozomal degradation, Vpu induces the degradation of newly synthesized CD4 at the endoplasmic reticulum.[20] Removal of cell surface CD4 prevents superinfection when the infected cell encounters new virus particles, and probably facilitates the formation of Env-driven virological synapses with new target cells. Why HIV-1 needs two distinct mechanisms for CD4 downregulation is unclear, especially given that the function in SIVs that lack Vpu genes is restricted to Nef alone. However, by specifically targeting CD4 before it leaves the ER, Vpu might facilitate the proper folding of nascent envelope trimers that could be compromised by premature interaction with CD4.[22]

The mechanism of CD4 downregulation has been reasonably well characterized over the last 15 years. The cytoplasmic tail of Vpu and that of CD4 interact.[23–26] Phosphorylation of the serine residues 52 and 56 in the DSGNES motif by CKII induces the recruitment of beta-TRCP to the Vpu cytoplasmic tail.[27] Beta-TRCP acts as an adaptor for the SCF/Cullin1 E3 ubquitin ligase through its F box domain.[27] CD4 then undergoes a form of ER-associated degradation,[28,29] a process by which the cell senses misfolded proteins in the ER and extracts them for proteasomal destruction.[30] CD4 molecules are physically dislocated from the ER membrane to the cytoplasm, where they are sequentially deglycosylated, polyubiquitylated and degraded by the proteasome.[28–30] This process resembles that used by the human cytomegalovirus proteins US2 and US11, which both hijack different cellular ERAD pathways to dislocate and destroy class I MHC molecules as part of that virus's immune evasion strategy. Indeed, Vpu has also been implicated in class I MHC downregulation. Vpu-mediated CD4 dislocation is sensitive to inhibition by dominant negative mutants of p97/cdc48/VCP,[31] an AAA-ATPase that drives the dislocation process. Several ERAD pathways in eukaryotes have been described and it is yet to be determined what other cellular factors are required for Vpu-mediated CD4 dislocation.

Vpu AND HIV-1 PARTICLE RELEASE

The "other function" of Vpu stemmed from the enigmatic observation that it was required for the efficient release of new viral particles from infected T cells.[2,32] This phenotype was independent of envelope/CD4 interactions and was variable amongst cell types tested: in several human and primate cell lines, the effects of Vpu on particle release were minor to non-existent, but in primary T cells and macrophages, Vpu is required for efficient particle production.[8] HIV-1 protein synthesis and the processing of the structural Gag molecules was unaffected in cells infected with Vpu(−) viruses.[33] However, the clues to how this phenotype arises were apparent from early electron micrographs of T-cell lines infected with Vpu-defective HIV-1; there appeared to be an accumulation of viral particles associated with cell surface and in intracellular vacuoles.[34,35] While little more evidence of the biochemical basis for this boost to virus release could be gleaned, a key early observation of this aspect of Vpu function was that the effects on CD4 and virus release were distinct. Vpu-mediated virus release was sensitive to brefeldin A, indicating that unlike the degradation of CD4, this process required the secretory pathway.[36] Also, mutants of Vpu lacking the conserved CKII phosphorylation site that were defective for CD4 down-regulation retained partial viral release function,[36] but the TM domain was an absolute requirement.[37]

Heinrich Gottlinger and colleagues did a seminal study of the effect of Vpu on viral release.[35] They showed that not only the release of HIV-1 particles, but also the release of virus-like particles derived from other lentiviruses and oncoretroviruses, could be stimulated from HeLa cells by co-expression of Vpu, indicating that Vpu was required in some cells for a process common to many diverse retroviruses. In the EM, mature HIV-1 particles with cone-shaped capsids were observed decorating cell surfaces and also within the lumens of vacuoles in Jurkat T cells infected with Vpu-defective HIV-1. This group had previously defined the HIV-1 "late-domain," a peptide sequence within HIV-1 Gag p6 required for assembly and release of mature virions with fully processed mature capsids.[38] These observations allowed them to define a difference between the failure of Vpu(−) viral particle release and the "stalk-mutant" immature viral particles of p6-defective mutants with electron-dense cores that remain contiguous with the cell membrane. In a prescient speculation, the authors further suggested that perhaps Vpu(−) virus release might be impeded by the expression of a cellular inhibitory activity for which Vpu was required to protect the nascent virion.

While two cellular factors, the tetratricopeptide protein UBP[39] and the Na+/K+ ion channel TASK-1[40] were isolated as putative Vpu-interacting proteins that might be required for optimal virus release, the conceptual advance in Vpu biology was initiated by the group of P. Spearman.[7] Inspired by the recent identification of APOBEC3G as the major target of the HIV-1 Vif protein,[41] they tested the hypothesis that Vpu might also overcome a dominant cellular inhibitor of HIV-1 replication. This they clearly demonstrated by showing that efficient HIV-1 particle release from African green monkey COS-7 cells that normally had no requirement for Vpu could become Vpu-dependent upon fusion with human cells that constitutively exhibited this phenotype. This result, like earlier similar studies with Vif,[42] was highly suggestive that Vpu was able to overcome a species-specific inhibitor of HIV-1 particle release in human cells.

THE PHENOTYPE OF Vpu-DEFECTIVE HIV-1 PARTICLE RELEASE

The notion of a restriction factor targeting retroviral assembly/release was particularly attractive because it allowed the division of cells into "permissive" and "restrictive" based on the viral requirement for Vpu. Interestingly, not all human cells were restrictive.[5] While HeLa cells exhibit a defect in Vpu-defective HIV-1 release of about 20-fold, human HOS, 293T and HT1080 did not. These easy to manipulate cells allowed for new comprehensive studies into the cell biological basis of the block to Vpu-defective virus release. In restrictive, but not permissive, cells, Vpu(−) virions could be visualized in both late endosomes and at the plasma membrane. PM localization of Gag proteins temporally preceded endosomal accumulations[5,43] and their appearance could be significantly inhibited by endocytic blockade by dominantly negative forms of Rab5a, Dynamin and EPS-15,[5] indicating that in the absence of Vpu, virions assembling at the PM were being taken into the cell by an endocytic process. However, endocytic blockade was not sufficient in itself to abolish the Vpu requirement; virion release was still impeded but now the virus accumulated in "sheets" on the cell surface. These virions contained mature capsid, suggesting they were mature particles physically separate from the cell. The direct proof of this came from the demonstration that these virions, unlike those bearing late domain defects, could be removed from the cell surface by protease treatment: assembly and ESCRT-dependent budding and membrane scission were unimpeded, but fully assembled Vpu(−) virions remained associated with the PM by a proteinaceaous tether, and were then subsequently endocytosed.[5] These tethers could sometimes be visualized linking mature viruses to the cell and each other by EM[6] (see Fig. 2). These data implied

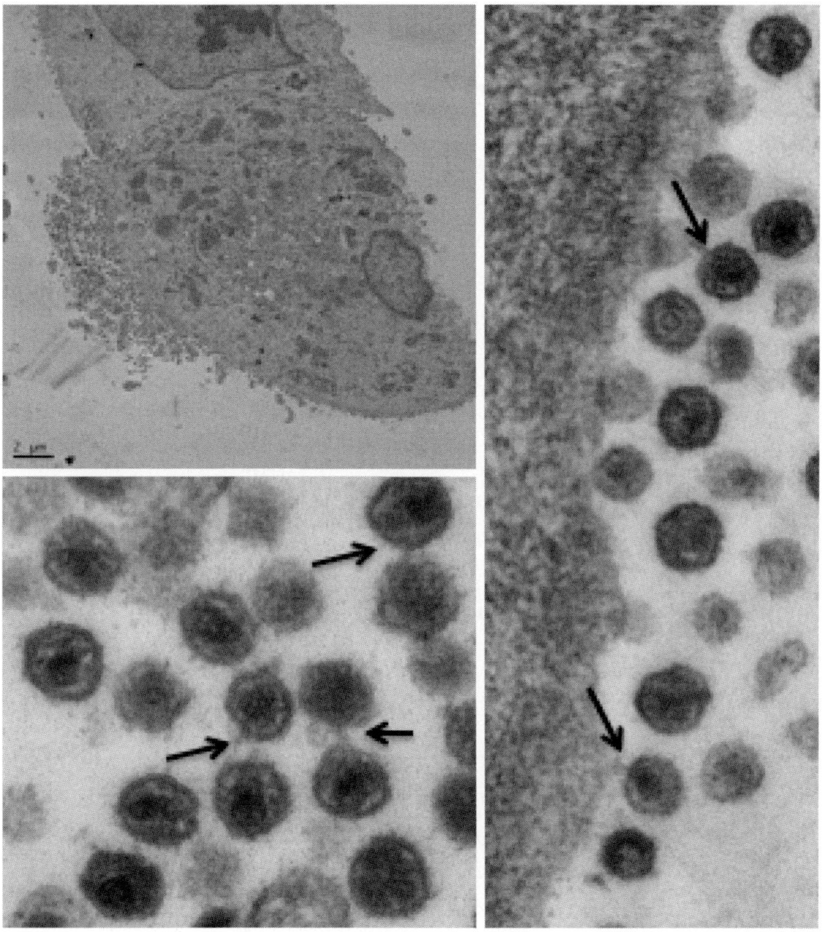

Fig. 2. Tethering of Vpu-deficient HIV-1 mutants on cell surfaces. Electron micrographs of HT1080 cells stably expressing tetherin infected with a Vpu-defective HIV-1 mutant. Massive accumulations of predominantly mature virions remain associated with the cell surface. Electron dense tethers (arrowed) can often be visualized between virion and the cell, or between virions themselves.

that the putative restriction factor itself might be a membrane or membrane-associated protein that could be incorporated into budding virions. An added knock-on implication of this work, which was formally shown shortly afterward, was that retroviral assembly is initiated and completed at the plasma membrane[44,45] and not at late endosomal compartments as had previously been postulated.

There were interesting parallels with the effects of type I interferon on HIV-1 release and the phenotype described above, particularly early studies showing IFN-treated macrophages and T cells accumulating virions in endosomes and on cell surfaces.[46–48] Since some tissue culture adapted isolates of HIV-1 harbor defects in one or more accessory genes, we looked at whether there was any relationship between type I IFN and Vpu function. Treatment of permissive human cells with IFNα induced a tethering phenotype analogous to that seen in HeLa cells and which could be abolished by Vpu expression.[6] Vpu-defective HIV-1 replication in primary peripheral blood mononuclear cells was severely limited by IFNα through a similar mechanism. Furthermore, this tethering activity was not limited to retroviral particles because the release of viral-like particles derived from the VP40 matrix protein of Ebola virus could also be restricted by IFNα treatment and rescued by Vpu.[6] Thus, the tethering activity likely represented a novel mechanism of innate immune inhibition that was not specific for retroviruses but in principle could inhibit the release of any enveloped viral particle that budded at the cell surface.

TETHERIN

The conclusions of the above studies suggested that an interferon-induced antiviral factor existed that could restrict enveloped virion release by tethering the nascent particles to the plasma membrane of the infected cell.[5,6] Furthermore, such a factor may well be a plasma membrane or membrane associated protein itself. This was conceptually an attractive way to inhibit viral spread relatively non-specifically. Comparative gene expression analyses of permissive and restrictive human cells versus genes induced by IFNα in permissive cells allowed a very small number of genes to be identified that fitted these criteria.[49] Amongst these was bone marrow stromal cell antigen 2 (BST2), also known as CD317, a membrane protein of hitherto unknown function. Expression of BST2 was shown to be both necessary and sufficient to induce a requirement for Vpu in HIV-1 particle release, and permissive cells stably expressing BST2 acquired all the cell biological attributes of HIV-1 release from restrictive cells (protease-sensitive tethering of mature viral particles at the PM and their accumulation in late endosomal compartments) and was given the new epithet "tetherin" to reflect this.[49] An independent study confirmed these data[50] after examining the results of a proteomic screen for targets of the immunomodulatory protein K5 from human herpesvirus 8.[51]

Tetherin is a small (181 amino acids) type II membrane protein of unusual topology[52] (Fig. 3). It consists of an N-terminal cytoplasmic tail, a transmembrane anchor, and an extracellular domain that includes a three extracellular cysteine residues, potential N-linked glycosylation sites, and a putative coiled-coil domain that may be involved in multimerization. Tetherin runs as a dimer on non-reducing SDS-PAGE gels through disulfide linkages between the extracellular cysteines.[53] The addition of oligosaccharides to tetherin results in the protein running as a variable-sized smear depending on the cell type. The most unusual feature of tetherin is that its C-terminal is processed to form a GPI anchor.[52] This feature is absolutely required for tetherin to restrict HIV-1 particle release.[49]

Tetherin is expressed on mature B cells and plasma cells, and in mice is a marker of plasmacytoid dendritic cells (PDCA-1).[54] In addition, it can be upregulated on myeloid cells and lymphocytes by various activatory stimuli and can be induced in many cell types by type I IFN.[54] Tetherin also corresponds to the tumor antigen HM1.24, expressed widely on multiple myeloma cells, and has been of interest in this regard as a target for cancer immunotherapy.[55,56] While its expression in the bone marrow stroma and on B cells has led to it being postulated to have a role in B cell development,[55,56] other than its effects on viral particle release, the only known immunological attribute of tetherin is to act as a ligand for leukocyte inhibitory receptor ILT7.[57]

Tetherin localizes to the PM and multiple membrane compartments including the trans-Golgi network (TGN).[54] In polarized epithelial cells, tetherin appears to localize predominantly to the apical surface, and, through an interaction with the BAR-RacGAP protein RICH2, plays a role in organizing the sub-apical F-actin network.[58] A tyrosine-based motif in the cytoplasmic tail of tetherin acts as a binding site for the adaptor proteins AP-1 and AP-2 and is required for the endocytic uptake of tetherin and delivery to the TGN.[59]

The topology and predicted structural features of tetherin suggest an attractive mode of action that would restrict the release of nascent virions. By

(a)

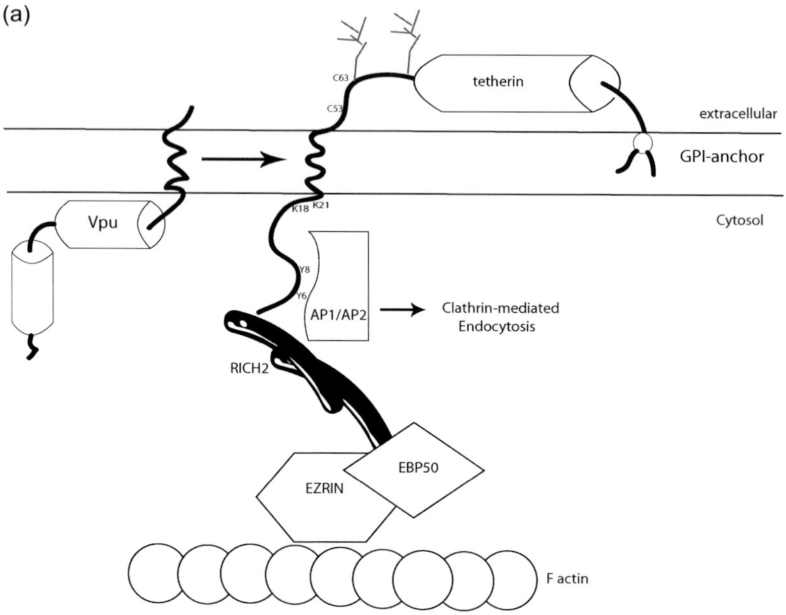

Fig. 3. The molecular and cellular biology of tetherin. (a) A schematic of the predicted topology of human tetherin. Tetherin is a type II membrane protein consisting of an N-terminal cytoplasmic tail, a transmembrane helix, an ectodomain comprising a dimerization domain with three conserved cysteine residues and a putative coiled-coil. The C-terminus of tetherin is processed to form a GPI anchor that is essential for virus restriction. The cytoplasmic tail has binding sites for the clathrin adaptor proteins AP1 and AP2,[59] and is linked to the sub-apical F-actin network in polarized cells via an interaction with the Bar-RacGAP containing protein RICH2.[58] (b, c) A tethering-model for restriction of virus release by tetherin.[5,49] Tetherin dimers become incorporated into assembling virions at the plasma membrane of infected cells. After scission of virus and cell membranes, tetherin molecules act as crosslinking bridges between the cells and mature particles, inhibiting virion release. Accumulations of virions can then be internalized and trafficked to late endosomes, possibly for destruction. (c) Potential configurations of tetherin crosslinks in which dimers are linked through disulfide and/or coiled-coil interactions to link virus and cell membranes.

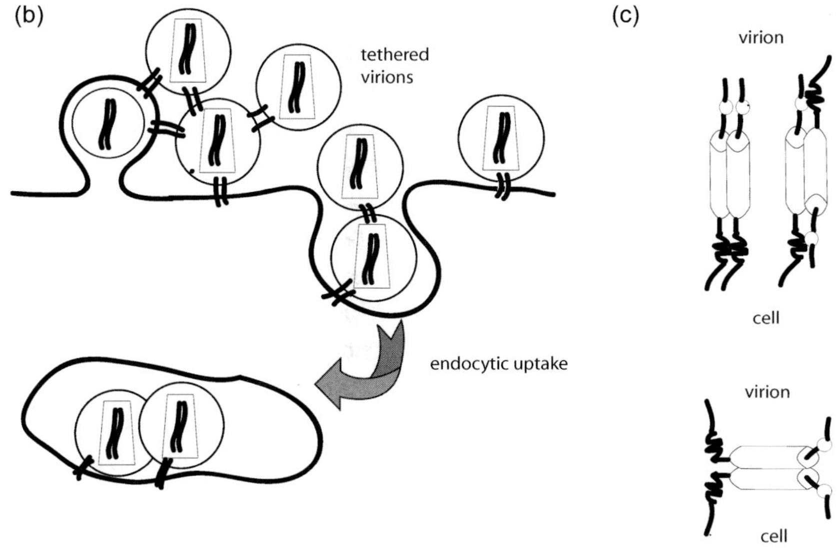

Fig. 3. (*Continued*)

having two membrane anchors, dimers and higher-order multimers of tetherin, linked either through disulfide bonds and/or coiled-coil interactions, could potentially span the membrane of the cell and the virus simultaneously.[49] In the EM electron dense tethers are sometimes visible between the cell and the virus, or between two virions.[6,49] The putative GPI anchor of tetherin is essential for viral restriction,[49] as is its disulfide-mediated dimerization and coiled-coil domain.[60,61] Since GPI-linked proteins often partition into so-called lipid rafts (detergent-resistant membrane microdomains with high cholesterol content, or DRMs), areas of the PM where retroviral budding is thought to be enriched,[62] tetherin is perfectly suited to being recruited into the membrane of nascent virus particles. Tetherin can indeed be found in DRMs isolated from transfected cells.[63] Evidence that membrane cholesterol might be important to tetherin and Vpu function is encouraged by the observation that a cholesterol-binding antifungal agent, amphotericin-B-methyl ester (AME), that has potent antiviral action against HIV-1, inhibits Vpu-mediated tetherin antagonism.[63]

Since tetherin localizes to viral budding domains,[49,64] it was highly likely that tetherin was at least a constituent of the tethers. Several recent microscopy studies have demonstrated that tetherin is incorporated into virion membranes.[61,65–67] A striking observation in one of these studies[61] showed that protease-released virions from tetherin-expressing cells bore vestigial dimeric tetherin fragments, suggesting that tetherin dimers directly crosslinked cellular and virion membranes. Furthermore, it was possible to design an "artificial" tetherin molecule from the dimeric N-terminus of the transferrin receptor, a coiled-coil motif from DMPK, and the GPI-anchor from uPAR.[61] Such a chimeric protein mediates tethering of virus particles and strongly implies that tetherin itself, simply by virtue of its topological structure, can crosslink virions to cells in the absence of a cellular cofactor and suggests that parallel tetherin dimers might be key to this [Fig. 3(c)]. Another recent study identified BCA/rabring7 as a tetherin-interacting factor, and implicated it in the internalization of restricted virus particles, rather than tethering virions *per se*.[68]

MECHANISM OF Vpu's ANTAGONISM OF TETHERIN

The mechanism by which Vpu antagonizes tetherin is not clear at present. Vpu and tetherin readily co-localize in over-expressing cells.[49] Vpu causes a downregulation of cell-surface tetherin levels,[50] and in infected HeLa cells and macrophages tetherin seems to be degraded[69] in a manner sensitive to inhibition of the proteasome.[70–72] However, whether these events are absolutely required for Vpu antagonism of tetherin-mediated restriction is less apparent. The CKII phosphorylation mutant of Vpu, Vpu2/6, is incapable of mediating cell surface downregulation of tetherin but retains activity to promote virus release.[36,50,73] Additionally, a recent study demonstrated that in infected T-cell lines, tetherin surface expression was only mildly downregulated.[69]

Vpu's anti-tetherin function occurs in a post-ER compartment.[36] Vpu localizes to the TGN as well as some endosomal structures. Furthermore, disruption of the recycling endosome using dominantly negative mutants of Rab11a or the tail fragment of Myosin Vb inhibits Vpu-mediated virion release.[17] This would suggest that Vpu trafficking through the endosomal network is required for antagonizing tetherin. In support of this, a recent study suggests that Vpu trafficking between the TGN and endosomal compartments is essential for Vpu to counteract tetherin. Mutations in Vpu that impair its TGN localization (either R30, K31, or truncations of the second cytoplasmic alpha-helix) have intermediate effects on tetherin antagonism, and a concomitant stablization of Vpu from endsomal degradation.[74] A third phosphorylation site in the cytoplasmic tail of Vpu has also been implicated in Vpu stability and modulating virion release.[75] How these disparate observations relate to the underlying mechanism of Vpu-mediated tetherin antagonism and whether they are required for Vpu function in primary T-cells and macrophages are an outstanding questions at present. Given that Vpu is a known adaptor of an E3-ubiquitin ligase (SCF/Cullin 1) for CD4 degradation,[27] it is an open question whether there is a role for the ubiquitin-proteasome system in the antagonism of tetherin,[70] although treatment of cells with proteasomal inhibitors does not restore tetherin surface expression in cells expressing Vpu.[50] Finally, in contrast to data suggesting that tetherin is degraded via the proteasome, other recent studies[73,76,77] show that tetherin degradation in response to Vpu requires β-TRCP2 and takes place in lysozomes. However, lysozomal inhibition or β-TRCP depletion only define the difference in activity of Vpu compared to Vpu2/6, which retains a degree tetherin antagonism. Thus, the mechanism of Vpu's anti-tetherin activity and whether the subsequent degradation observed explains this antagonism of restriction still require clarification.

Another open question is whether Vpu's ion channel activity is required for antagonism of tetherin. Scrambling the TM domain of Vpu prevents it promoting virion release.[78] Interestingly, evidence from the Stephens laboratory has shown that replacement of the TM of Vpu with that of influenza M2 yields a chimeric protein capable of mediating CD4 degradation.[79] Treatment of cells infected with a virus encoding this VpuM2 protein with the influenza antiviral drug rimantidine, which blocks the M2 proton channel, inhibits virus replication and leads to accumulation of virions on the cell surfaces. Further studies showed that a single mutation, replacing A18 of the Vpu TM with a histidine residue to mimic the proton sensitive "gate" of M2, was sufficient to render virus growth sensitive to rimantidine.[80] These data suggest that Vpu may indeed form an ion-permeable channel in infected cell that is essential for its function, but how ion flux itself relates to tetherin antagonism is unknown at present. However, it is intriguing that amiloride-based compounds related to rimantidine also inhibit Vpu function,[15,16] and this raises the possibility that pharmcological targeting of the putative TM ion channel has potential for the development of HIV antivirals.

SPECIES-SPECIFIC VARIANTS OF TETHERIN

Tetherin orthologues are present in the equivalent chromosomal locus of all the mammalian genomes so far sequenced (*www.ensembl.org*). Unlike the TRIM5 or APOBEC3 loci that encode other lentiviral restriction factors,[81] there is as yet no evidence of gene duplication in most species. The broad activity of human tetherin against retroviruses[64] suggested that it might be a significant barrier to cross-species transmission. An interesting aspect of retroviral restriction factors is that they are generally unable to restrict the wild-type virus that replicates in that species but are often highly potent against retroviruses from another species[81] — for example, the rhesus macaque APOBEC3G protein does not restrict SIVmac replication but does restrict HIV-1 because the Vif protein of the human virus cannot target it for proteasomal destruction. This difference in sensitivity to viral countermeasures stems from the high degree of positive selection that has acted on these genes during recent primate evolution, as evidenced by the high proportion of nucleotide differences between species orthologues that results in a concommitant amino acid change (ratio of non-synononomous to synonomous changes, or *dN/dS*).[82,83] TRIM5 and APOBEC3G genes have amongst the highest *dN/dS* of any primate genes, implying that they have been evolving to adapt to constantly changing selective pressures. These pressures are presumed to be infectious agents which themselves are under pressure to avoid host inhibitory responses that target essential stages of their replication.

Alignment of the tetherin genes of primates reveals some striking differences in the distribution of amino acid variations[84] (Fig. 4). While the ectodomains of the proteins are reasonably well conserved, the cytoplasmic tails and transmembrane domains are considerably different. By comparison to the human protein, those tetherin molecules of other non-human primates bear an insertion of five amino acids in the cytoplasmic tail but lack a glycine-isoluecine (GI) motif in the N terminus of the TM domain.[84] In addition to this, many single amino acid positions differ in both the cytoplasmic tail and TM domain. Analysis of the *dN/dS*

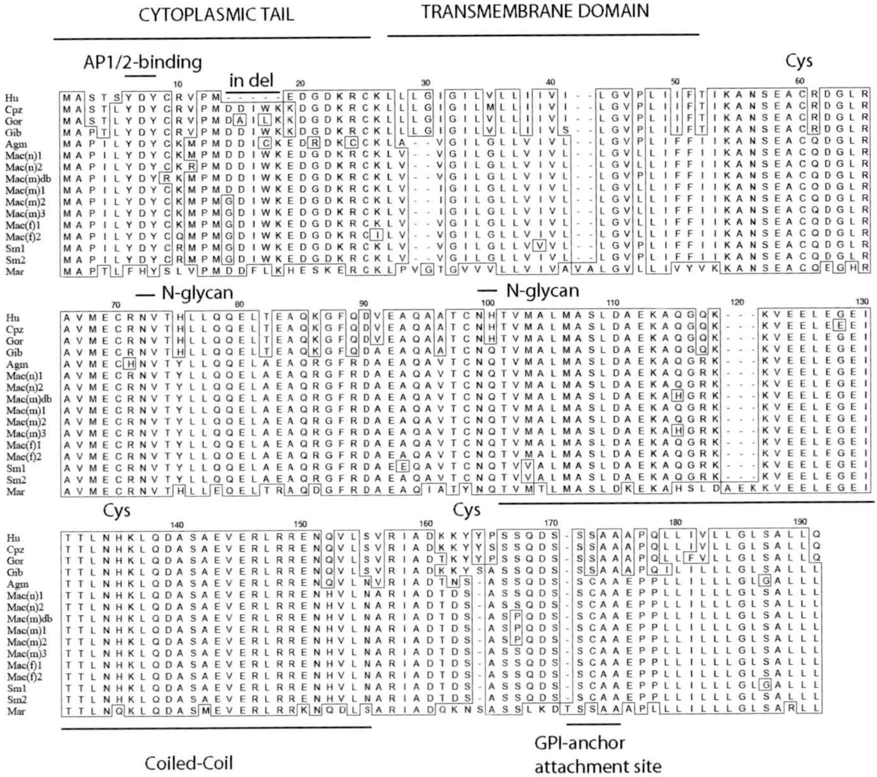

Fig. 4. Alignment of tetherins from humans and primates (adapted from Ref. 84). Known features are annotated. Note that the variability of tetherin orthologues is disproportionately concentrated within the cytoplasmic tail and transmembrane domain. This variability accounts for the insensitivity of most primate tetherins to HIV-1 Vpu and potentially results from selection pressures imposed by enveloped viruses during primate evolution.

ratios for each position relative to other primate species reveals that many of these variable positions result from a high degree of positive selection that appears to be mainly focused on these areas of the protein.[84] This is intriguing because these areas of tetherin are likely to be the most accessible to targeting by viral proteins. And the sensitivity of these primate tetherins to HIV-1 Vpu provides an interesting illustration of this.

All mammalian tetherin molecules tested so far can restrict the release of Vpu(−) HIV-1 particles.[70,84] However, only human and chimpanzee tetherins are sensitive to HIV-1 Vpu.[84] This is consistent with the previous observation that type I IFN treatment of COS7 (AGM) cells gave rise to a tethering phenotype that equally restricted wild-type HIV-1 and the Vpu(−) mutant.[49] Two studies used these observations to identify the differences in monkey tetherins that rendered them insensitive to Vpu.[71,84] In the first, swapping the TM domains of human, AGM and rhesus tetherin molecules swapped their abilities to be able to discriminate between Vpu(+) and Vpu(−) HIV-1.[84] Then, by sequentially mutating/deleting the human tetherin TM domain to that of rhesus tetherin, it was shown that a combination of changes at two positions was sufficient to produce a Vpu-resistant human tetherin.[84] Changing the T residue at position 45 (I in rhesus tetherin), the final predicted residue of the TM domain, in combination with the in-frame deletion of a glycine–isoleucine gave a similar Vpu-resistant phenotype.[84] Several other positions when mutated similarly could affect the sensitivity of tetherin to Vpu. Interestingly, all these residues were shown to be under high positive selection. A second group arrived at a similar conclusion by using positive selection analyses as a guide to predicting residues associated with Vpu sensitivity,[71] and the tetherin TM domain as a target of Vpu has been confirmed by others.[85,86] The results of both these studies suggest that tetherin has been under high positive selection throughout mammalian evolution. The localization of human tetherin's Vpu sensitivity to positive selected residues in the TM domain suggests that Vpu may directly associate with tetherin via its own TM domain (known to be critical for enhancing virus release) and suggests that Vpu-like proteins that target the TM-domain of tetherin may have driven this variation. Furthermore, the lack of contributions to Vpu sensitivity by those positively selected residues in the cytoplasmic tail implies that they may have been selected by other viral countermeasures against tetherin. And, as described below, Vpu is not the only protein from mammalian viruses that has the ability to antagonize tetherin-mediated restriction.

RESTRICTION OF OTHER ENVELOPED VIRUSES BY TETHERIN AND THEIR EVASION MECHANISMS

The seeming lack of tetherin specificity for any HIV protein implies that it could be a general inhibitor of enveloped virus assembly and release. In fact, this "non-specificity" is a very powerful attribute for an innate immune effector: it requires that any virus encountering tetherin will be under pressure to evolve a direct targeting mechanism because it cannot mutate its structural proteins to avoid it. It is well known that Vpu stimulates the release of many diverse retroviral particles,[35] and in fact all retroviral-like particles (so far without exception) are sensitive to tetherin.[49,64] This includes viruses with different assembly pathways: betaretroviruses, which preassemble particles in the cytosol prior to budding, and spumaviruses, which bud through internal membranes through an obligate interaction with their envelope proteins.[64] This can be extended to the VLPs of several RNA viruses. Filamentous Ebolavirus (EBOV) and Marburg virus (MV) VLPs can be generated by expressing the VP40 matrix protein. EBOV VLP release was shown to be poor in HeLa cells and IFN treated 293T unless Vpu was co-expressed. Several groups directly demonstrated tetherin-mediated restriction of EBOV, MV and arenavirus particles recently.[64,87,88] Thus, it appears that tetherin has the capacity to restrict diverse enveloped viruses, supporting a mechanism through which tetherin is passively incorporated into budding virions without a direct interaction with viral proteins. Doubtless, this list of sensitive viruses will grow, as it would appear that tetherin only needs to be localized to the membrane through which the virus buds.

The above studies were conducted with VLPs. Since EBOV, MV and some arenaviruses cause lethal pathologies with concomitant production of large quantities of infectious virus, for tetherin to be relevant to their pathogenesis, these viruses should encode a tetherin antagonist. P. Bates's group has recently identified the surface glycoprotein (GP) of EBOV as a tetherin antagonist.[87] EBOV-GP, in addition to mediating virus entry, has a general ability to remove host cell-surface proteins, including those for immune recognition and cell adhesion,[89] and induce vascular permeability,[90] a hallmark of hemorrhagic fever. EBOV-GP expression is sufficient to rescue Vpu(−) HIV-1 particle release from tetherin-expressing cells. Co-immunprecipitation studies revealed an association between EBOV-GP and tetherin in the cell membrane.[87] While GP did not require its proteolytic processing from GP0 to its mature GP1 and GP2 form to target tetherin, it required membrane anchoring and was independent of its mucin-like domains that are required for the downregulation of other cell surface molecules. Whether this attribute of the EBOV-GP is shared by the MV-GP or the envelope protein of arenaviruses is yet to be determined.

Tetherin was identified in a proteomic screen for the targets of K5, a membrane-bound RING-CH (MARCH) domain E3 ubiquitin ligase encoded by human herpesvirus 8, the aetiological agent of Kaposi's sarcoma (HHV8, also known as Kaposi's sarcoma-associated herpesvirus, or KSHV).[51] K5, along with its close relative K3, are again viral proteins whose role appears to be immunomodulatory. K5 again targets a variety of cell-surface molecules involved in the immune recognition of virally infected cells — class I MHC proteins, adhesion molecules and NK cell-receptor ligands.[91,92] K5 reduces the cellular content of tetherin and is a functional homologue of the cellular protein MARCH-VIII.[51] Herpesviruses have a complex envelopment strategy involving sequential budding through the nuclear-ER membrane.[93,94] That HHV8 might be sensitive to tetherin is intriguing, but completely consistent with its cellular tropism. HHV8 infects B lymphocytes and establishes a latent infection. When the B cell matures into an antibody-producing plasma cell, the unfolded-protein response, induced by the sudden upregulation of secreted protein synthesis, triggers HHV8 reactivation and lytic replication.[94,95] Mature B cells and plasma cells express high levels of tetherin, and thus could be a significant impediment to the release of new HHV8 virions during this process. MARCH E3 ligases are encoded by several γ2-herpesviruses and also various poxviruses[92] (viruses whose complex envelopment processes could be severely compromised by a tetherin-like molecule[96]). K5 expression mediates ESCRT-dependent tetherin degradation through the direct ubiquitination of its cytoplasmic tail lysine at position 18 and rescues both HHV8 and Vpu-defective HIV-1 from tetherin-mediated restriction.[97,98]

The Vpu gene is peculiar to HIV-1 and its evolutionary precursors, SIVcpz, SIVgsn and SIVmon. However, the majority of primate immunodeficiency viruses do not encode a Vpu protein, including SIVsmm, the zoonotic spread of which to humans gave rise to HIV-2.[99] Interestingly, the HIV-2 envelope surface glycoprotein (and at least one HIV-1 isolate) has long been known to harbor a Vpu-like function in promoting virion release from cells by countering a host restriction.[100–103] The HIV-2 ROD envelope (Env), but not that of SIVmac, antagonizes human tetherin in a cell-type specific way.[104] Like Vpu, it induces tetherin downregulation from the cell surface. However, there is no apparent degradation; rather tetherin is sequestered in intracellular compartments, particularly the TGN.[104] Similarly, the Env of SIVtan can antagonize a range of primate tetherins leading to intracellular sequestration.[105] The molecular determinants of this are complex: the HIV-2 Env Vpu-like function requires proper processing of the Env precursor[104,106] and a GYxxθ motif within its cytoplasmic tail that binds to the clathrin-adaptor AP-2.[104,107] However, this motif is common to most immunodeficiency virus Envs including those that do not have a Vpu-like activity. Moreover, chimeric protein analysis indicates multiple determinants in the ectodomain of the HIV-2-Env contribute to this activity.[103,104] Consistent with this, HIV-2 Env and tetherin interact in co-immunoprecipitation experiments[104] and a single point mutation in the extracellular domain of tetherin modulates its sensitivity to SIVtan Env antagonism.[105] At present, we do not know how widespread Env-mediated tetherin antagonism is in primate immunodeficiency viruses, but as described below, Vpu is not the only accessory protein that can target tetherin.

The accessory gene Nef is encoded by all HIV-1, HIV-2 and SIV strains. Nef is another immunodulatory adaptor protein, targeted via myristoylation to the cytosolic face of cell membranes, which acts to remove cell-surface proteins involved in immune recognition, including CD4, MHCI and MHCII.[21] Recent evidence demonstrates that the Nef proteins of many SIV isolates have a broader repertoire of surface targets than HIV-1 Nef.[108] Moreover, these Nef proteins come from viruses that do not encode a Vpu protein.[108] Given that all Nef and Vpu alleles share a common function in the targeting of CD4, these observation perhaps suggest that Vpu and Nef fulfil overlapping functions in SIV replication and pathogenesis.[3] Two recent studies indicate that this overlapping of Vpu and Nef functions extends to tetherin antagonism.[85,109] The Nef proteins of Vpu-negative viruses SIVmac,[85,109] SIVagm, SIVsmm and SIVblu[109] all target primate orthologues of tetherin, most efficiently the protein from the virus' host species. Notably, none of these Nef proteins can rescue Vpu-defective HIV-1 from human tetherin. All primate tetherins up to and including that from chimpanzees have a five amino acid patch (G/DDIWK) that has been deleted in the human protein. The specificity of SIV Nefs for primate tetherins maps to this motif, and re-insertion into human tetherin renders it sensitive to SIVmac Nef.[109] It is tempting to speculate that the deletion of this motif in human tetherin has arisen through selective pressures exerted by Nef-like proteins during ancestral viral epidemics. A further provocative study shows that of the three known HIV-1 Groups (M, N and O) that arose from separate SIVcpz zoonoses, only viruses from the pandemic Group M appear to counteract human tetherin efficiently.[110] SIVcpz Vpu proteins are not tetherin antagonists in their host species — that role is fulfilled by Nef. SIVcpz is thought to be a recombinant between the SIVgsn/mon lineage and that of SIVrcm, with the former providing a Vpu protein with tetherin-antagonizing potential that was subsequently lost, and the latter contributing a Nef protein with a similar activity. The authors speculate that upon zoonotic transfer to humans, SIVcpz was forced to adapt to a tetherin that lacked the Nef targeting sequence, and re-acquired both tetherin antagonism and CD4 downregulation in Vpu. That this occurred in Group M and not Groups N and O, which have thus far caused minor localized epidemics in Africa, has led the authors to speculate that this adaption in Vpu was essential for the establishment of the human pandemic. Whether this was really the case requires far more study, but nonetheless the evolutionary pressures that gave rise to analogous functions in three immunodeficiency virus proteins (Vpu, Nef and the HIV-2/SIVtan envelope protein — summarized in Table 1) underscore the potential importance of tetherin evasion in the pathogenesis of these viruses.

ROLES OF Vpu AND TETHERIN IN THE PATHOGENESIS OF HIV/AIDS

So far, all the data concerning tetherin and its antagonism by Vpu comes from assays in transformed cell types. The role of Vpu in HIV-1 pathogenesis is thus an important question for evaluating its function as a potential target for novel antiretroviral therapeutics for the treatment of HIV/AIDS. In principle, interfering with the ability of the virus to target innate restriction factors may potently block the production and spread of HIV-1 *in vivo*. A clear example of this for Vpu is the fact that Vpu(−) viruses are hypersensitive to type I IFN in primary $CD4^+$ T cells.[6] Targeting HIV-1 accessory genes has several attractions. Firstly, we know that other

Table 1. Summary of Lentiviral and Non-retroviral Tetherin Antagonists with Known Mechanistic Information (see text for details)

Tetherin Countermeasure	Species Specificity	Cell surface Downregulation	Intracellular Sequestration	Degradation	Target Sequence of Tetherin
HIV-1 Vpu	Human	Yes	Yes (TGN)	Yes (βTRCP2-dependent). Proteasomal or lysozomal?	Transmembrane domain
HIV-2 Env	Human	Yes	Yes (TGN)	No	Extracellular?
SIVtan Env	Human and primate	Yes	Yes (TGN)	No	Extracellular (A100)
SIV Nef (various)	Primate (not human)	Yes	?	?	DDIWK in cytoplasmic tail
HHV8 K5	Human	Yes	No	Yes (ESCRT-dependent)	Ubiquitination of K18 in cytoplasmic tail
EBOV G	Human and mouse	?	?	?	?

retroviral restriction factors provide a very powerful barrier to zoonotic infection — HIV-1 cannot replicate in rhesus macaque T cells without being engineered to avoid both TRIM5 and the APOBECs 3G, 3F and 3H.[111] Thus, inhibiting Vif or Vpu *in vivo* may achieve the same potent effects in patients. Secondly, since targeting APOBECs and tetherin is the role of specialized accessory proteins,[112] there is less scope for cross-resistance by mutations in the major viral targets of antiretroviral therapy, reverse transcriptase, protease and integrase. Thirdly, and more speculatively, restriction factors are part of the innate antiviral immune response. Their importance in HIV-1 pathogenesis is likely to be resolved by placing these factors in the context of the wider immune response. The interdependence of these arms of the immune system in the innate sensing of pathogen components (pattern recognition) and, through the induction of inflammatory responses, the mobilization of adaptive immunity has formed an exciting part of immunology research in recent years. It is therefore of interest in the case of HIV-1 to suggest that inhibiting the action of the viral accessory proteins that target innate immune factors might have knock-on effects in enhancing the antiviral immune response in general, in addition to blocking viral replication on a cellular level. In the case of Vpu and tetherin, understanding their roles in HIV-1 pathogenesis requires the answering of fundamental questions regarding their biology in primary target cells and tissues. Tetherin has recently been shown to act as a ligand for the leukocyte inhibitory receptor ILT7 on plasmacytoid dendritic cells, thereby inhibiting the activation of Toll-like receptors 7 and 9.[57] This raises the intriguing possibility that tetherin–HIV interactions might have a role in modulating the recognition of viral products by pattern recognition receptors.

HIV-1 primarily targets CD4$^+$ T cells and macrophages. The ability of tetherin to restrict HIV-1 particle release highlights our current understanding of how the virus spreads from one infected cell to another and is disseminated from the initial site of infection in the genital/anal mucosa to the systemic lymphoid tissue. In CD4$^+$ T cells, while they can obviously produce cell-free virions, a major route of spread between infected and uninfected cells is through direct cell-to-cell contact. Infected lymphocytes form polarized contacts, similar to immunological synapses, in which adhesion molecules and viral components are directed to the contact point, a closed membrane bound gap known as the virological synapse.[113] Viral assembly and secretion occurs across these synapses, and disruption of these contacts reduces replicative spread of HIV-1 in CD4$^+$ T-cell cultures. HIV-1 transfer between T cells has also been suggested to occur via membrane nanotubes,[114] similar to cytonemes induced in adherent cells by murine retroviruses.[115] How, therefore, might tetherin and Vpu affect cell-to-cell transmission of virus across virological synapses and/or nanotubes? Vpu-defective HIV-1 retains the capacity to spread in some T-cell cultures despite the reduction in cell-free virus production.[32,34] One interesting study performed almost 10 years ago is highly suggestive that Vpu can modulate cell-free versus cell-to-cell transmission. *In vitro* selection for HIV-1 mutants that require cell contact for efficient replication in CD4$^+$ T-cell lines led not only to the acquisition of changes to the envelope protein gp120, but also induced what would seem to be inactivating mutations in Vpu.[116] This might suggest that Vpu provides a balance between cell-free and cell-to-cell transfer. However, how tetherin factors into such a model is unclear, particularly because of the sensitivity of Vpu(−) viruses to type I IFN. Therefore, how tetherin affects the formation of virological synapses and whether tetherin-restricted viruses on T-cell surfaces are still amenable for cell-to-cell transfer are key outstanding questions.

In macrophages, HIV-1 replication is far less cytopathic than in CD4$^+$ T cells, with infected cells able to persist in culture for weeks.[117] Virus particles accumulate in deep invaginations of the plasma membrane where they might be sequestered and released on contact with T cells across synapse structures.[118] Vpu-defective HIV-1 replicates poorly in macrophages, which express high levels of tetherin as they mature.[69] HIV-1 Gag-VLPs clearly bud at the PM in macrophages, but over time accumulate in phagosomes.[44] Thus, it is likely that there is a difference in the site of intracellular accumulation of Vpu(−) HIV-1 and wild-type virus. Since macrophages are antigen presenting cells, it will be interesting to see whether Vpu(−) viruses are targeted for proteolytic destruction in macrophage phagosomes and viral components processed for class II MHC antigen presentation. Such a process would intimately link tetherin-mediated HIV-1 restriction to the augmentation of an antiviral adaptive immune response.

How might these cell biological observations be relevant to Vpu/tetherin interactions *in vivo*? HIV-1 is usually transmitted by sexual contact, with the viral inoculum penetrating the epithelium of the mucosal barrier to infect resident $CD4^+$ target cells. For the virus to establish a persistent systemic infection, new virions from this localized replication must spread to the lymphoid tissue, predominantly that of the gut, where the majority of CD4+CCR5+ T cells reside. It is unclear how this spread is achieved, but it is likely to require virus transport by both infected cell migration and the production of cell-free virions that can be spread by the circulation or by interaction with, and carriage by, dendritic cells. HIV+ individuals often have high levels of plasma type I IFN, and recent data in acutely infected individuals demonstrates a systemic induction of IFN levels as early as a week following exposure.[119] Given this mobilization of the innate immune response and the strength of the restriction observed in culture, tetherin may impart a significant block to the establishment of a systemic infection. Indeed, studies in pig-tailed macaques using SIVmac/HIV-1 chimeric viruses (SHIVs that encode HIV-1 Env and Vpu proteins) show that disruption of Vpu leads to a virus with reduced pathogenicity.[120]

Vpu sequence is surprisingly variable amongst primary strains of HIV-1, especially between clades. While CD4 downregulation is common to most Vpu alleles, whether the same is true of Vpu-mediated tetherin antagonism is unknown. Conservation of tetherin evasion amongst Vpu alleles will be important to determine. Cytotoxic T-cell responses in HIV+ patients are often targeted to Vpu,[121] raising the possibility that sequence variation in Vpu might be driven in part by immune escape. It will be interesting to determine whether variability of Vpu function might be associated with disease progression in HIV+ individuals. Temporal conservation of tetherin evasion between virus isolated early during infection and later after systemic establishment, and whether Vpu-defective viruses are transmitted between individuals, will all provide evidence of the importance of Vpu and tetherin *in vivo* and provide a rationale for the design of novel antiviral agents to target Vpu. Such investigations are clearly important: HIV-1 is a moving drug target, constantly developing resistance to current therapies, so there is still much scope for the exploration of new therapeutic compounds. Inhibiting the action of HIV-1 accessory genes such as Vpu or Vif will lay the virus open to natural mediators of innate antiviral immunity, a potentially useful augmentation to currently used drugs that target the HIV-1 replicative enzymes. We already know that amiloride-based compounds[16] and AME[63] effect Vpu-mediated virus release and have antiviral activity. Perhaps derivatives of these compounds will provide avenues to explore Vpu-targeting as a new arm in antiretroviral treatment. Finally, given the potential for tetherin to be a general antiviral inhibitor of the release of enveloped viruses, understanding how other human viral pathogens deal with this factor could well establish this part of viral replication as a common target for new antiviral treatments.

ACKNOWLEDGEMENTS

I would like to thank Juan Martin-Serrano and members of my laboratory for critical reading of the manuscript and helpful comments, and Paul Bieniasz, for sharing unpublished data. Research in my laboratory is funded by a Wellcome Research Career Development Award RCDF WT082274MA.

REFERENCES

1. Cohen EA, Terwilliger EF, Sodroski JG, Haseltine WA. (1988) Identification of a protein encoded by the vpu gene of HIV-1. *Nature* **334**: 532–534
2. Strebel K, Klimkait T, Martin MA. (1988) A novel gene of HIV-1, Vpu, and its 16-kilodalton product. *Science* **241**: 1221–1223.
3. Kirchhoff F. (2009) Is the high virulence of HIV-1 an unfortunate coincidence of primate lentiviral evolution? *Nat Rev Microbiol.*
4. Schwartz S, Felber BK, Fenyo EM, Pavlakis GN. (1990) Env and Vpu proteins of human immunodeficiency virus type 1 are produced from multiple bicistronic mRNAs. *J Virol* **64**: 5448–5456.
5. Neil SJ, Eastman SW, Jouvenet N, Bieniasz PD. (2006) HIV-1 Vpu promotes release and prevents endocytosis of nascent retrovirus particles from the plasma membrane. *PLoS Pathog* **2**: e39.
6. Neil SJ, Sandrin V, Sundquist WI, Bieniasz PD. (2007) An interferon-alpha-induced tethering mechanism inhibits HIV-1 and Ebola virus particle release but is counteracted by the HIV-1 Vpu protein. *Cell Host Microbe* **2**: 193–203.

7. Varthakavi V, Smith RM, Bour SP, *et al.* (2003) Viral protein U counteracts a human host cell restriction that inhibits HIV-1 particle production. *Proc Natl Acad Sci USA* **100**: 15154–15159.
8. Schubert U, Clouse KA, Strebel K. (1995) Augmentation of virus secretion by the human immunodeficiency virus type 1 Vpu protein is cell type independent and occurs in cultured human primary macrophages and lymphocytes. *J Virol* **69**: 7699–7711.
9. Schubert U, Henklein P, Boldyreff B, *et al.* (1994) The human immunodeficiency virus type 1 encoded Vpu protein is phosphorylated by casein kinase-2 (CK-2) at positions Ser52 and Ser56 within a predicted alpha-helix-turn-alpha-helix-motif. *J Mol Biol* **236**: 16–25.
10. Schubert U, Schneider T, Henklein P, *et al.* (1992) Human-immunodeficiency-virus-type-1-encoded Vpu protein is phosphorylated by casein kinase II. *Eur J Biochem* **204**: 875–883.
11. Ewart GD, Sutherland T, Gage PW, Cox GB. (1996) The Vpu protein of human immunodeficiency virus type 1 forms cation-selective ion channels. *J Virol* **70**: 7108–7115.
12. Schubert U, Ferrer-Montiel AV, Oblatt-Montal M, *et al.* (1996) Identification of an ion channel activity of the Vpu transmembrane domain and its involvement in the regulation of virus release from HIV-1-infected cells. *FEBS Lett* **398**: 12–18.
13. Gonzalez ME, Carrasco L. (2003) Viroporins. *FEBS Lett* **552**: 28–34.
14. Marassi FM, Ma C, Gratkowski H, *et al.* (1999) Correlation of the structural and functional domains in the membrane protein Vpu from HIV-1. *Proc Natl Acad Sci USA* **96**: 14336–14341.
15. Ewart GD, Mills K, Cox GB, Gage PW. (2002) Amiloride derivatives block ion channel activity and enhancement of virus-like particle budding caused by HIV-1 protein Vpu. *Eur Biophys J* **31**: 26–35.
16. Ewart GD, Nasr N, Naif H, *et al.* (2004) Potential new anti-human immunodeficiency virus type 1 compounds depress virus replication in cultured human macrophages. *Antimicrob Agents Chemother* **48**: 2325–2330.
17. Varthakavi V, Smith RM, Martin KL, *et al.* (2006) The pericentriolar recycling endosome plays a key role in Vpu-mediated enhancement of HIV-1 particle release. *Traffic* **7**: 298–307.
18. Pacyniak E, Gomez ML, Gomez LM, *et al.* (2005) Identification of a region within the cytoplasmic domain of the subtype B Vpu protein of human

immunodeficiency virus type I (HIV-1) that is responsible for retention in the golgi complex and its absence in the Vpu protein from a subtype C HIV-1. *AIDS Res Hum Retroviruses* **21**: 379–394.

19. Ruiz A, Hill MS, Schmitt K, *et al.* (2008) Requirements of the membrane proximal tyrosine and dileucine-based sorting signals for efficient transport of the subtype C Vpu protein to the plasma membrane and in virus release. *Virology* **378**: 58–68.
20. Willey RL, Maldarelli F, Martin MA, Strebel K. (1992) Human immunodeficiency virus type 1 Vpu protein induces rapid degradation of CD4. *J Virol* **66**: 7193–7200.
21. Foster JL, Garcia JV. (2008) HIV-1 Nef: at the crossroads. *Retrovirology* **5**: 84.
22. Willey RL, Maldarelli F, Martin MA, Strebel K. (1992) Human immunodeficiency virus type 1 Vpu protein regulates the formation of intracellular gp160-CD4 complexes. *J Virol* **66**: 226–234.
23. Chen MY, Maldarelli F, Karczewski MK, *et al.* (1993) Human immunodeficiency virus type 1 Vpu protein induces degradation of CD4 in vitro: the cytoplasmic domain of CD4 contributes to Vpu sensitivity. *J Virol* **67**: 3877–3884.
24. Bour S, Schubert U, Strebel K. (1995) The human immunodeficiency virus type 1 Vpu protein specifically binds to the cytoplasmic domain of CD4: implications for the mechanism of degradation. *J Virol* **69**: 1510–1520.
25. Yao XJ, Friborg J, Checroune F, *et al.* (1995) Degradation of CD4 induced by human immunodeficiency virus type 1 Vpu protein: a predicted alpha-helix structure in the proximal cytoplasmic region of CD4 contributes to Vpu sensitivity. *Virology* **209**: 615–623.
26. Tiganos E, Yao XJ, Friborg J, *et al.* (1997) Putative alpha-helical structures in the human immunodeficiency virus type 1 Vpu protein and CD4 are involved in binding and degradation of the CD4 molecule. *J Virol* **71**: 4452–4460.
27. Margottin F, Bour SP, Durand H, *et al.* (1998) A novel human WD protein, h-beta TrCp, that interacts with HIV-1 Vpu connects CD4 to the ER degradation pathway through an F-box motif. *Mol Cell* **1**: 565–574.
28. Meusser B, Sommer T. (2004) Vpu-mediated degradation of CD4 reconstituted in yeast reveals mechanistic differences to cellular ER-associated protein degradation. *Mol Cell* **14**: 247–258.

29. Binette J, Dube M, Mercier J, *et al.* (2007) Requirements for the selective degradation of CD4 receptor molecules by the human immunodeficiency virus type 1 Vpu protein in the endoplasmic reticulum. *Retrovirology* **4**: 75.
30. Schubert U, Anton LC, Bacik I, *et al.* (1998) CD4 glycoprotein degradation induced by human immunodeficiency virus type 1 Vpu protein requires the function of proteasomes and the ubiquitin-conjugating pathway. *J Virol* **72**: 2280–2288.
31. Binette J, Dube M, Mercier J, *et al.* (2007) Requirements for the selective degradation of CD4 receptor molecules by the human immunodeficiency virus type 1 Vpu protein in the endoplasmic reticulum. *Retrovirology* **4**: 75.
32. Terwilliger EF, Cohen EA, Lu YC, *et al.* (1989) Functional role of human immunodeficiency virus type 1 vpu. *Proc Natl Acad Sci USA* **86**: 5163–5167.
33. Strebel K, Klimkait T, Maldarelli F, Martin MA. (1989) Molecular and biochemical analyses of human immunodeficiency virus type 1 vpu protein. *J Virol* **63**: 3784–3791.
34. Klimkait T, Strebel K, Hoggan MD, *et al.* (1990) The human immunodeficiency virus type 1-specific protein Vpu is required for efficient virus maturation and release. *J Virol* **64**: 621–629.
35. Gottlinger HG, Dorfman T, Cohen EA, Haseltine WA. (1993) Vpu protein of human immunodeficiency virus type 1 enhances the release of capsids produced by gag gene constructs of widely divergent retroviruses. *Proc Natl Acad Sci USA* **90**: 7381–7385.
36. Schubert U, Strebel K. (1994) Differential activities of the human immunodeficiency virus type 1-encoded Vpu protein are regulated by phosphorylation and occur in different cellular compartments. *J Virol* **68**: 2260–2271.
37. Tiganos E, Friborg J, Allain B, *et al.* (1998) Structural and functional analysis of the membrane-spanning domain of the human immunodeficiency virus type 1 Vpu protein. *Virology* **251**: 96–107.
38. Gottlinger HG, Dorfman T, Sodroski JG, Haseltine WA. (1991) Effect of mutations affecting the p6 gag protein on human immunodeficiency virus particle release. *Proc Natl Acad Sci USA* **88**: 3195–3199.
39. Callahan MA, Handley MA, Lee YH, *et al.* (1998) Funct ional interaction of human immunodeficiency virus type 1 Vpu and Gag with a novel member of the tetratricopeptide repeat protein family. *J Virol* **72**: 5189–5197.
40. Hsu K, Seharaseyon J, Dong P, *et al.* (2004) Mutual functional destruction of HIV-1 Vpu and host TASK-1 channel. *Mol Cell* **14**: 259–67.

41. Sheehy AM, Gaddis NC, Choi JD, Malim MH. (2002) Isolation of a human gene that inhibits HIV-1 infection and is suppressed by the viral Vif protein. *Nature* **418**: 646–650.
42. Simon JH, Gaddis NC, Fouchier RA, Malim MH. (1998) Evidence for a newly discovered cellular anti-HIV-1 phenotype. *Nat Med* **4**: 1397–400.
43. Harila K, Prior I, Sjoberg M, *et al.* (2006) Vpu and Tsg101 regulate intracellular targeting of the human immunodeficiency virus type 1 core protein precursor Pr55gag. *J Virol* **80**: 3765–3772.
44. Jouvenet N, Neil SJ, Bess C, *et al.* (2006) Plasma membrane is the site of productive HIV-1 particle assembly. *PLoS Biol* **4**: e435.
45. Jouvenet N, Bieniasz PD, Simon SM. (2008) Imaging the biogenesis of individual HIV-1 virions in live cells. *Nature* **454**: 236–240.
46. Kornbluth RS, Oh PS, Munis JR, *et al.* (1989) Interferons and bacterial lipopolysaccharide protect macrophages from productive infection by human immunodeficiency virus *in vitro. J Exp Med* **169**: 1131–1151
47. Poli G, Orenstein JM, Kinter A, *et al.* (1989) Interferon-alpha but not AZT suppresses HIV expression in chronically infected cell lines. *Science* **244**: 575–577.
48. Smith MS, Thresher RJ, Pagano JS. (1991) Inhibition of human immunodeficiency virus type 1 morphogensis in T cells by alpha interferon. *Antimicrob Agents Chemother* **35**: 62–67.
49. Neil SJ, Zang T, Bieniasz PD. (2008) Tetherin inhibits retrovirus release and is antagonized by HIV-1 Vpu. *Nature* **451**: 425–430.
50. Van Damme N, Goff D, Katsura C, *et al.* (2008) The interferon-induced protein BST-2 restricts HIV-1 release and is downregulated from the cell surface by the viral Vpu protein. *Cell Host Microbe* **3**: 245–252.
51. Bartee E, McCormack A, Fruh K. (2006) Quantitative membrane proteomics reveals new cellular targets of viral immune modulators. *PLoS Pathog* **2**: e107.
52. Kupzig S, Korolchuk V, Rollason R, *et al.* (2003) Bst-2/HM1.24 is a raft-associated apical membrane protein with an unusual topology. *Traffic* **4**: 694–709.
53. Ohtomo T, Sugamata Y, Ozaki Y, *et al.* (1999) Molecular cloning and characterization of a surface antigen preferentially overexpressed on multiple myeloma cells. *Biochem Biophys Res Commun* **258**: 583–591.
54. Blasius AL, Giurisato E, Cella M, *et al.* (2006) Bone marrow stromal cell antigen 2 is a specific marker of type I IFN-producing cells in the naive

mouse, but a promiscuous cell surface antigen following IFN stimulation. *J Immunol* **177**: 3260–3265.

55. Ozaki S, Kosaka M, Wakatsuki S, *et al.* (1997) Immunotherapy of multiple myeloma with a monoclonal antibody directed against a plasma cell-specific antigen, HM1.24. *Blood* **90**: 3179–3186.
56. Ono K, Ohtomo T, Yoshida K, *et al.* (1999) The humanized anti-HM1.24 antibody effectively kills multiple myeloma cells by human effector cell-mediated cytotoxicity. *Mol Immunol* **36**: 387–395.
57. Cao W, Bover L, Cho M, *et al.* (2009) Regulation of TLR7/9 responses in plasmacytoid dendritic cells by BST2 and ILT7 receptor interaction. *J Exp Med* **206**: 1603–1614.
58. Rollason R, Korolchuk V, Hamilton C, *et al.* (2009) A CD317/tetherin-RICH2 complex plays a critical role in the organization of the subapical actin cytoskeleton in polarized epithelial cells. *J Cell Biol* **184**: 721–736.
59. Rollason R, Korolchuk V, Hamilton C, *et al.* (2007) Clathrin-mediated endocytosis of a lipid-raft-associated protein is mediated through a dual tyrosine motif. *J Cell Sci* **120**: 3850–3858.
60. Andrew AJ, Miyagi E, Kao S, Strebel K. (2009) The formation of cysteine-linked dimers of BST-2/tetherin is important for inhibition of HIV-1 virus release but not for sensitivity to Vpu. *Retrovirology* **6**: 80.
61. Perez-Caballero D, Zang T, Ebrahimi A, *et al.* (2009) Tetherin inhibits HIV-1 release by directly tethering virions to cells. *Cell* **139**: 499–511.
62. Ono A, Freed EO. (2001) Plasma membrane rafts play a critical role in HIV-1 assembly and release. *Proc Natl Acad Sci USA* **98**: 13925–13930.
63. Waheed AA, Ablan SD, Soheilian F, *et al.* (2008) Inhibition of human immunodeficiency virus type 1 assembly and release by the cholesterol-binding compound amphotericin B methyl ester: Evidence for Vpu dependence. *J Virol* **82**: 9776–9781.
64. Jouvenet N, Neil SJ, Zhadina M, *et al.* (2009) Broad-spectrum inhibition of retroviral and filoviral particle release by tetherin. *J Virol* **83**: 1837–1844.
65. Hammonds J, Wang JJ, Yi H, Spearman P. (2010) Immunoelectron microscopic evidence for Tetherin/BST2 as the physical bridge between HIV-1 virions and the plasma membrane. *PLoS Pathog* **6**: e1000749.
66. Fitzpatrick K, Skasko M, Deerinck TJ, *et al.* (2010) Direct restriction of virus release and incorporation of the interferon-induced protein BST-2 into HIV-1 particles. *PLoS Pathog* **6**: e1000701.

67. Habermann A, Krijnse Locker J, Oberwinkler H, *et al.* (2010) CD317/Tetherin is Enriched in the HIV-1 Envelope and Downregulated from the Plasma Membrane upon Virus Infection. *J Virol.*
68. Miyakawa K, Ryo A, Murakami T, *et al.* (2009) BCA2/Rabring7 promotes tetherin-dependent HIV-1 restriction. *PLoS Pathog* **5**: e1000700.
69. Miyagi E, Andrew AJ, Kao S, Strebel K. (2009) Vpu enhances HIV-1 virus release in the absence of Bst-2 cell surface down-modulation and intracellular depletion. *Proc Natl Acad Sci USA* **106**: 2868–2873.
70. Goffinet C, Allespach I, Homann S, *et al.* (2009) HIV-1 antagonism of CD317 is species specific and involves Vpu-mediated proteasomal degradation of the restriction factor. *Cell Host Microbe* **5**: 285–297.
71. Gupta RK, Hue S, Schaller T, *et al.* (2009) Mutation of a single residue renders human tetherin resistant to HIV-1 Vpu-mediated depletion. *PLoS Pathog* **5**: e1000443.
72. Mangeat B, Gers-Huber G, Lehmann M, *et al.* (2009) HIV-1 Vpu neutralizes the antiviral factor Tetherin/BST-2 by binding it and directing its beta-TrCP2-dependent degradation. *PLoS Pathog* **5**: e1000574.
73. Mitchell RS, Katsura C, Skasko MA, *et al.* (2009) Vpu antagonizes BST-2-mediated restriction of HIV-1 release via beta-TrCP and endo-lysosomal trafficking. *PLoS Pathog* **5**: e1000450.
74. Dube M, Roy BB, Guiot-Guillain P, *et al.* (2009) Suppression of tetherin-restricting activity on HIV-1 particle release correlates with localization of Vpu in the trans-Golgi network. *J Virol* **83**(9) 4574–4590.
75. Estrabaud E, Le Rouzic E, Lopez-Verges S, *et al.* (2007) Regulated degradation of the HIV-1 Vpu protein through a betaTrCP-independent pathway limits the release of viral particles. *PLoS Pathog* **3**: e104.
76. Douglas JL, Viswanathan K, McCarroll MN, *et al.* (2009) Vpu directs the degradation of the human immunodeficiency virus restriction factor BST-2/Tetherin via a {beta}TrCP-dependent mechanism. *J Virol* **83**: 7931–7947.
77. Iwabu Y, Fujita H, Kinomoto M, *et al.* (2009) HIV-1 accessory protein Vpu internalizes cell-surface BST-2/tetherin through transmembrane interactions leading to lysosomes. *J Biol Chem* **284**: 35060–35072.
78. Schubert U, Bour S, Ferrer-Montiel AV, *et al.* (1996) The two biological activities of human immunodeficiency virus type 1 Vpu protein involve two separable structural domains. *J Virol* **70**: 809–819.

79. Hout DR, Gomez ML, Pacyniak E, *et al.* (2006) Substitution of the transmembrane domain of Vpu in simian-human immunodeficiency virus (SHIVKU1bMC33) with that of M2 of influenza A results in a virus that is sensitive to inhibitors of the M2 ion channel and is pathogenic for pig-tailed macaques. *Virology* **344**: 541–559.
80. Hout DR, Gomez LM, Pacyniak E, *et al.* (2006) A single amino acid substitution within the transmembrane domain of the human immunodeficiency virus type 1 Vpu protein renders simian-human immunodeficiency virus (SHIV(KU-1bMC33)) susceptible to rimantadine. *Virology* **348**: 449–461.
81. Wolf D, Goff SP. (2008) Host restriction factors blocking retroviral replication. *Annu Rev Genet* **42**: 143–163.
82. Sawyer SL, Emerman M, Malik HS. (2004) Ancient adaptive evolution of the primate antiviral DNA-editing enzyme APOBEC3G. *PLoS Biol* **2**: E275.
83. Sawyer SL, Wu LI, Emerman M, Malik HS. (2005) Positive selection of primate TRIM5alpha identifies a critical species-specific retroviral restriction domain. *Proc Natl Acad Sci USA* **102**: 2832–2837.
84. McNatt MW, Zang T, Hatziioannou T, *et al.* (2009) Species-specific activity of HIV-1 Vpu and positive selection of tetherin transmembrane domain variants. *PLoS Pathog* **5**: e1000300.
85. Jia B, Serra-Moreno R, Neidermyer W, *et al.* (2009) Species-specific activity of SIV Nef and HIV-1 Vpu in overcoming restriction by tetherin/BST2. *PLoS Pathog* **5**: e1000429.
86. Rong L, Zhang J, Lu J, *et al.* (2009) The transmembrane domain of BST-2 determines its sensitivity to down-modulation by human immunodeficiency virus type I Vpu. *J Virol* **83**: 7536–7546.
87. Kaletsky RL, Francica JR, Agrawal-Gamse C, Bates P. (2009) Tetherin-mediated restriction of filovirus budding is antagonized by the Ebola glycoprotein. *Proc Natl Acad Sci USA* **106**: 2886–2891.
88. Sakuma T, Noda T, Urata S, *et al.* (2009) Inhibition of Lassa and Marburg virus production by tetherin. *J Virol* **83**: 2382–2385.
89. Simmons G, Wool-Lewis RJ, Baribaud F, *et al.* (2002) Ebola virus glycoproteins induce global surface protein down-modulation and loss of cell adherence. *J Virol* **76**: 2518–2528.
90. Yang ZY, Duckers HJ, Sullivan NJ, *et al.* (2000) Identification of the Ebola virus glycoprotein as the main viral determinant of vascular cell cytotoxicity and injury. *Nat Med* **6**: 886–889.

91. Nathan JA, Lehner PJ. (2008) The trafficking and regulation of membrane receptors by the RING-CH ubiquitin E3 ligases. *Exp Cell Res* **315**: 1593–1600.
92. Lehner PJ, Hoer S, Dodd R, Duncan LM. (2005) Downregulation of cell surface receptors by the K3 family of viral and cellular ubiquitin E3 ligases. *Immunol Rev* **207**: 112–125.
93. Mettenleiter TC. (2002) Herpesvirus assembly and egress. *J Virol* **76**: 1537–1547.
94. Jenner RG, Maillard K, Cattini N, *et al.* (2003) Kaposi's sarcoma-associated herpesvirus-infected primary effusion lymphoma has a plasma cell gene expression profile. *Proc Natl Acad Sci USA* **100**: 10399–10404.
95. Wilson SJ, Tsao EH, Webb BL, *et al.* (2007) X box binding protein XBP-1s transactivates the Kaposi's sarcoma-associated herpesvirus (KSHV) ORF50 promoter, linking plasma cell differentiation to KSHV reactivation from latency. *J Virol* **81**: 13578–13586.
96. Smith GL, Law M. (2004) The exit of vaccinia virus from infected cells. *Virus Res* **106**: 189–197.
97. Mansouri M, Viswanathan K, Douglas JL, *et al.* (2009) Molecular mechanism of BST2/tetherin downregulation by K5/MIR2 of Kaposi's sarcoma-associated herpesvirus. *J Virol* **83**: 9672–9681.
98. Pardieu C, Vigan R, Wilson SJ, *et al.* (2010) The RING-CH ligase K5 antagonizes restriction of KSHV and HIV-1 particle release by mediating ubiquitin-dependent endosomal degradation of tetherin. *PLoS Pathogens* In Press.
99 de Silva TI, Cotten M, Rowland-Jones Sl. (2008) HIV-2: The forgetton AIDS virus. *Trends Microbiol* **16**: 588–595.
100. Bour S, Schubert U, Peden K, Strebel K. (1996) The envelope glycoprotein of human immunodeficiency virus type 2 enhances viral particle release: A Vpu-like factor? *J Virol* **70**: 820–829.
101. Bour S, Strebel K. (1996) The human immunodeficiency virus (HIV) type 2 envelope protein is a functional complement to HIV type 1 Vpu that enhances particle release of heterologous retroviruses. *J Virol* **70**: 8285–8300.
102. Schubert U, Bour S, Willey RL, Strebel K. (1999) Regulation of virus release by the macrophage-tropic human immunodeficiency virus type 1 AD8 isolate is redundant and can be controlled by either Vpu or Env. *J Virol* **73**: 887–896.

103. Abada P, Noble B, Cannon PM. (2005) Functional domains within the human immunodeficiency virus type 2 envelope protein required to enhance virus production. *J Virol* **79**: 3627–3638.
104. Le Tortorec A, Neil SJ. (2009) Antagonism to and intracellular sequestration of human tetherin by the human immunodeficiency virus type 2 envelope glycoprotein. *J Virol* **83**: 11966–11978.
105. Gupta RK, Mlcochova P, Pelchen-Matthews A, *et al.* (2009) Simian immunodeficiency virus envelope glycoprotein counteracts tetherin/BST-2/CD317 by intracellular sequestration. *Proc Natl Acad Sci USA* Epub 2009/10/30.
106. Bour S, Akari H, Miyagi E, Strebel K. (2003) Naturally occurring amino acid substitutions in the HIV-2 ROD envelope glycoprotein regulate its ability to augment viral particle release. *Virology* **309**: 85–98.
107. Noble B, Abada P, Nunez-Iglesias J, Cannon PM. (2006) Recruitment of the adaptor protein 2 complex by the human immunodeficiency virus type 2 envelope protein is necessary for high levels of virus release. *J Virol* **80**: 2924–2932.
108. Schindler M, Munch J, Kutsch O, *et al.* (2006) Nef-mediated suppression of T cell activation was lost in a lentivirasl lineage that gave rise to HIV-1. *Cell* **125**: 1055–1067.
109. Zhang F, Wilson SJ, Landford WC, *et al.* (2009) Nef proteins from Simian immunodeficiency viruses are tetherin antagonists. *Cell Host Microbe* **6**: 54–67.
110. Sauter D, Schindler M, Specht A, *et al.* (2009) Tetherin-driven adaptation of Vpu and Nef function and the evolution of pandemic and nonpandemic HIV-1 strains. *Cell Host Microbe* **6**: 409–421.
111. Hatziioannou T, Princiotta M, Piatak M, Jr., *et al.* (2006) Generation of simian-tropic HIV-1 by restriction factor evasion. *Science* **314**: 95.
112. Malim MH, Emerman M. (2008) HIV-1 accessory proteins — ensuring viral survival in a hostile environment. *Cell Host Microbe* **3**: 388–398.
113. Piguet V, Sattentau Q. (2004) Dangerous liaisons at the virological synapse. *J Clin Invest* **114**: 605–610.
114. Sowinski S, Jolly C, Berninghausen O, *et al.* (2008) Membrane nanotubes physically connect T cells over long distances presenting a novel route for HIV-1 transmission. *Nat Cell Biol* **10**: 211–219.
115. Sherer NM, Lehmann MJ, Jimenez-Soto LF, *et al.* (2007) Retroviruses can establish filopodial bridges for efficient cell-to-cell transmission. *Nat Cell Biol* **9**: 310–315.

116. Gummuluru S, Kinsey CM, Emerman M. (2000) An *in vitro* rapid-turnover assay for human immunodeficiency virus type 1 replication selects for cell-to-cell spread of virus. *J Virol* **74**: 10882–10891.
117. Marsh M, Theusner K, Pelchen-Matthews A. (2009) HIV assembly and budding in macrophages. *Biochem Soc Trans* **37**: 185–189.
118. Deneka M, Pelchen-Matthews A, Byland R, *et al.* (2007) In macrophages, HIV-1 assembles into an intracellular plasma membrane domain containing the tetraspanins CD81, CD9, and CD53. *J Cell Biol* **177**: 329–341.
119. Stacey AR, Norris PJ, Qin L, *et al.* (2009) Induction of a striking systemic cytokine cascade prior to peak viremia in acute human immunodeficiency virus type 1 infection, in contrast to more modest and delayed responses in acute hepatitis B and C virus infections. *J Virol* **83**: 3719–3733.
120. Hout DR, Gomez ML, Pacyniak E, *et al.* (2005) Scrambling of the amino acids within the transmembrane domain of Vpu results in a simian-human immunodeficiency virus (SHIVTM) that is less pathogenic for pig-tailed macaques. *Virology* **339**: 56–69.
121. Addo MM, Yu XG, Rosenberg ES, *et al.* (2002) Cytotoxic T-lymphocyte (CTL) responses directed against regulatory and accessory proteins in HIV-1 infection. *DNA Cell Biol* **21**: 671–678.

CHAPTER 10

Retrovirus Replication: New Perspectives on Enzyme and Substrate Dynamics

Jason W. Rausch*, Elio Abbondanzieri†, Shixin Liu†, Xiaowei Zhuang†,‡,§ and Stuart F. J. Le Grice*,¶

SUMMARY

Although individual steps of the retroviral reverse transcription process can be accurately reproduced *in vitro,* gaps still remain in our understanding of how certain of these are accomplished at the molecular level. Since intimate cross-talk between the viral enzyme and its nucleic acid substrates underlies such events, our recent research focuses on applying chemical biology to manipulate and characterize nucleoprotein complexes representing different steps in reverse transcription. Examples of this include targeted insertion of nucleoside analogs to examine the contribution of nucleic acid geometry to enzyme recognition, site-specific insertion of unnatural

* RT Biochemistry Section, HIV Drug Resistance Program, National Cancer Institute, Frederick, MD 21702, USA.

† Department of Chemistry and Chemical Biology, Harvard University, Cambridge, MA 02138, USA.

‡ Department of Physics, Harvard University, Cambridge, MA 02138, USA.

§ Howard Hughes Medical Institute, Harvard University, Cambridge, MA 02138, USA.

¶ Corresponding author, e-mail: slegrice@ncifcrf.gov.

amino acids that introduce novel functionality into reverse transcriptase (RT), and developing mass spectroscopic techniques to dissect protein: ligand complexes. More recently, these methods have been augmented by examining conformational dynamics of reverse transcription via single molecule spectroscopy. While this chapter concentrates on RT from retroviruses and long-terminal repeat (LTR)-containing retrotransposons, the approaches described herein have broad applicability and should encourage molecular biologists and biochemists to implement new tools that examine protein/nucleic acid interactions with greater precision than previously possible.

INTRODUCTION

Synthesis of integration-competent, duplex DNA from the single-stranded RNA genome of retroviruses and long terminal repeat (LTR)-containing retrotransposons reflects a highly orchestrated program of events catalyzed by reverse transcriptase (RT).[1] The dynamics whereby (i) newly-synthesized DNA is transferred within or between templates, (ii) the plus strand, polypurine tract (PPT) primers are accurately recognized by catalytic domains located at either RT terminus, and (iii) DNA synthesis terminates at phased A-tracts in the center of the duplex DNA product (Fig. 1) might best be described as "retroviral gymnastics," requiring intimate communication between RT and its single-stranded, hybrid and duplex nucleic acid substrates. While crystallographic studies have provided incisive insights into the structures of the human immunodeficiency virus type 1 (HIV-1) RT apoenzyme[2] and various complexes of the enzyme bound to nucleic acid substrates[3–5] and DNA polymerase inhibitors,[6] our recent efforts have centered on developing and exploiting new experimental tools from which high resolution structural data can be gleaned in the absence of a crystal structure, and revealing interesting dynamics of the enzyme that are critical for its function.

As examples of our strategy, increased availability of nucleoside analogs, coupled with the ease with which they can be incorporated into chemically synthesized DNA or RNA, has allowed us to examine contributions of the nucleobase, sugar or phosphodiester backbone to nucleic acid geometry and recognition of important regulatory signals, e.g., the central termination sequence (CTS) or PPT-containing RNA/DNA hybrid.

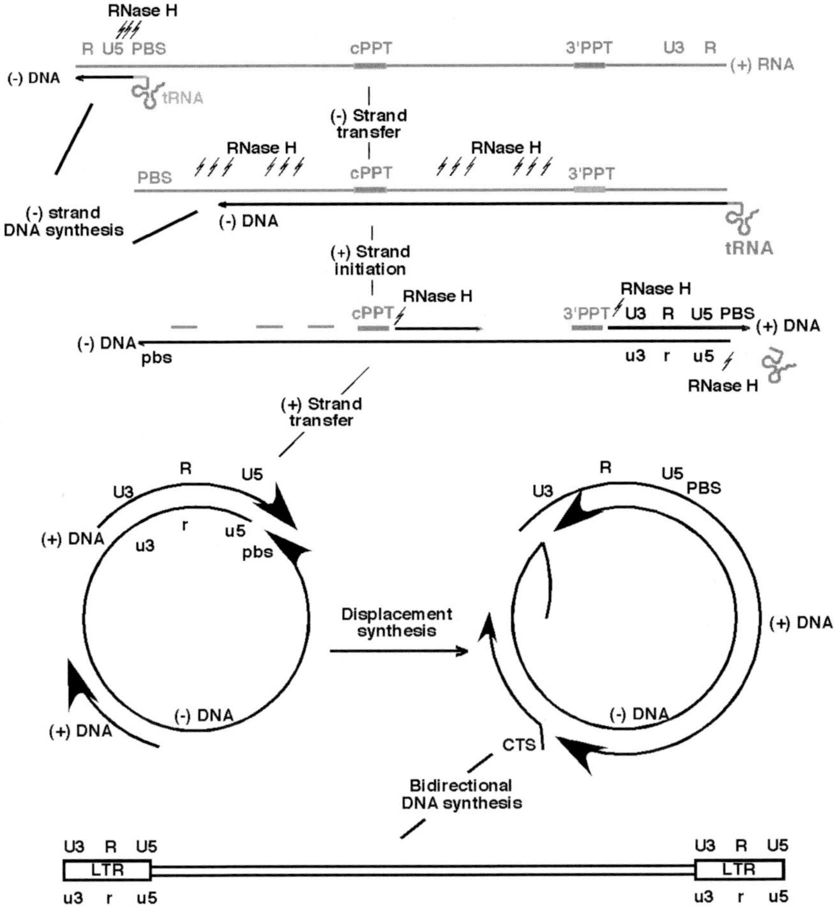

Fig. 1. RT-catalyzed synthesis of double-stranded, integration-competent DNA from the single-stranded HIV-1 viral RNA genome. Minus (−) strand DNA synthesis, initiated from a host-coded tRNA bound to the viral primer binding site (PBS), proceeds to the RNA 5′ terminus, copying repeat (R) and unique 5′ (U5) sequences. Concomitantly, RNase H activity hydrolyzes the ensuing RNA/DNA hybrid. Complementary R sequences of the genome termini promote transfer of nascent DNA to the 3′ terminus of the genome and continued DNA synthesis. Hydrolysis of the RNA/DNA replication intermediate continues, with the exception of the 3′ and central PPTs, which prime plus (+) strand, DNA-dependent DNA synthesis up to and including 18 nucleotides of the tRNA primer, creating a complementary (+) strand PBS sequence. The PPT and tRNA primers are excised, and PBS complementarity supports a second strand transfer event, after which bidirectional DNA synthesis creates double-stranded proviral DNA flanked by the hallmark long terminal repeat (LTR) sequences. The presence of a central PPT in HIV creates a "central flap" which is processed by host-coded enzymes.

Conversely, the ability to site-specifically introduce unnatural amino acids with novel functionalities into proteins by translational suppression[7] affords the opportunity of examining protein structure with a precision unattainable by classical site-directed mutagenesis. In addition, nucleoprotein complexes mimicking important steps in reverse transcription, while too large to be examined by X-ray crystallography (e.g., the RT/tRNA/viral RNA initiation complex), can be examined by mass spectrometric footprinting with respect to both the protein and nucleic acid components. Finally, as we have advanced these techniques, they have now facilitated a detailed examination of the interaction between HIV-1 RT and a variety of nucleic acid duplexes by single molecule spectroscopy, revealing enzyme "sliding" and "flipping" as an integral feature of the polymerization process. In reviewing our current endeavors, it is important to emphasize that the technologies we describe have general applicability to the area of protein–nucleic acid interactions, hopefully providing a new repertoire of molecular techniques with which biological systems can be investigated.

NUCLEOSIDE ANALOGS AS STRUCTURAL PROBES OF RT–SUBSTRATE INTERACTIONS

Incorporating base, sugar and phosphate mimetics into RNA and DNA, a variety of which are available for enzymatic incorporation or chemical synthesis (Fig. 2), is a powerful experimental tool for exploring interactions between HIV-1 RT and its various substrates. Nucleoside analogs can be introduced to locally alter the structure or chemical bonding characteristics of nucleic acid in a defined and predictable manner. Furthermore, using high-resolution RT crystal structures as a guide, critical nucleoprotein contacts can be identified and characterized to a degree not previously possible. Alternatively, modified nucleosides embedded within an RT substrate may possess an inherent functionality that can be exploited for biochemical detection or conjugation, exemplifying another tool for probing RT–RNA/DNA interactions at the sub-molecular level.[8] We have exploited a number of such analogs to probe the HIV-1 RT–substrate interaction in a variety of contexts, with particular emphasis on understanding the structural basis for plus-strand initiation from the plus-strand PPT RNA primer.

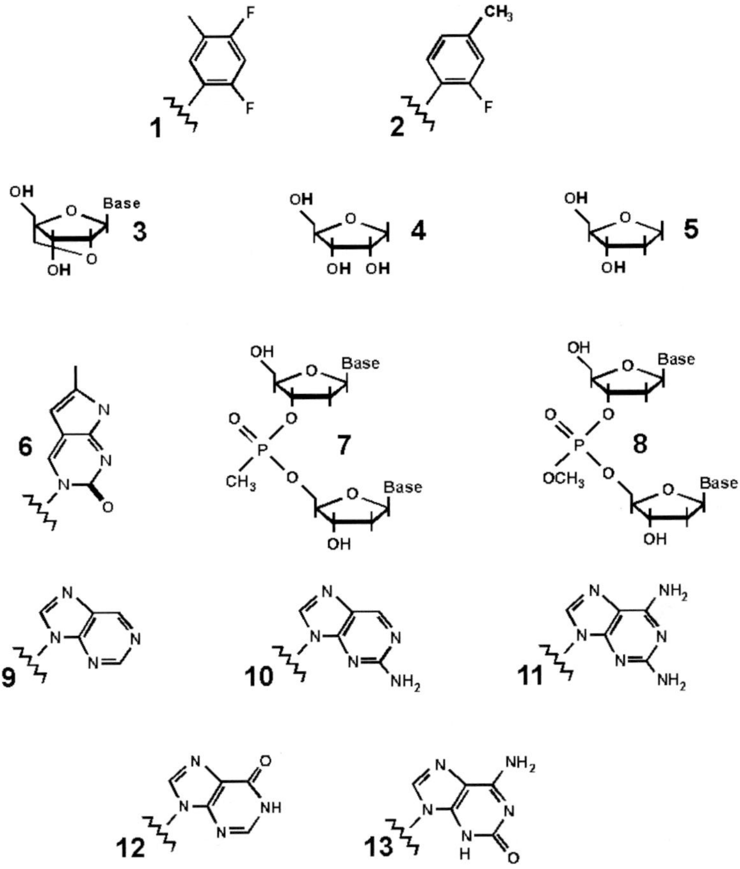

Fig. 2. Examples of base, sugar and phosphate mimics. (**1**) 2,4-difluoro-5-methylbenzene (F); (**2**) 2-fluoro-4-methyl benzene (D); (**3**) locked nucleic acid (LNA); (**4**) abasic riboside (rAb); (**5**) abasic deoxyriboside (dAb); (**6**) pyrrolo-C (pC); (**7**) methylphosphonate linkage; (**8**) methylphosphotriester linkage; (**9**) purine; (**10**) 2-aminopurine (2-AP); (**11**) 2,6-diaminopurine (2,6-DAP); (**12**) inosine (I); (**13**) isoguanine (i-G).

Non-Hydrogen-Bonding Pyrimidine Isosteres

2,4-difluoro-5-methylbenzene (dF, Fig. 2, No. 1) and 2-fluoro-4-methylbenzene (dD, Fig. 2, No. 2) are isosteric (shape) mimics of thymine and cytosine, respectively.[8,9] While sterically equivalent to their naturally occurring counterparts, dF and dD do not participate in hydrogen bonding in the context of duplex nucleic acid. To examine how eliminating hydrogen bonding affects plus-strand primer processing, we introduced single or tandem dF and dD substitutions into the PPT minus-strand DNA template [Fig. 3(a)]. Remarkably, in apparent competition with normal PPT binding determinants, the RNase H catalytic center was consistently relocated to cleave 3 nt downstream from the site of analog substitution [Fig. 3(b)]. This effect was observed on multiple substrates, although there was a direct relationship between the extent of RNA hydrolysis and the proximity of the analogs to the PPT-U3 junction. In many cases, anomalous cleavage occurred at the expense of normal PPT processing, while RT failed to cleave substrates containing dF opposite the PPT 5′ end.

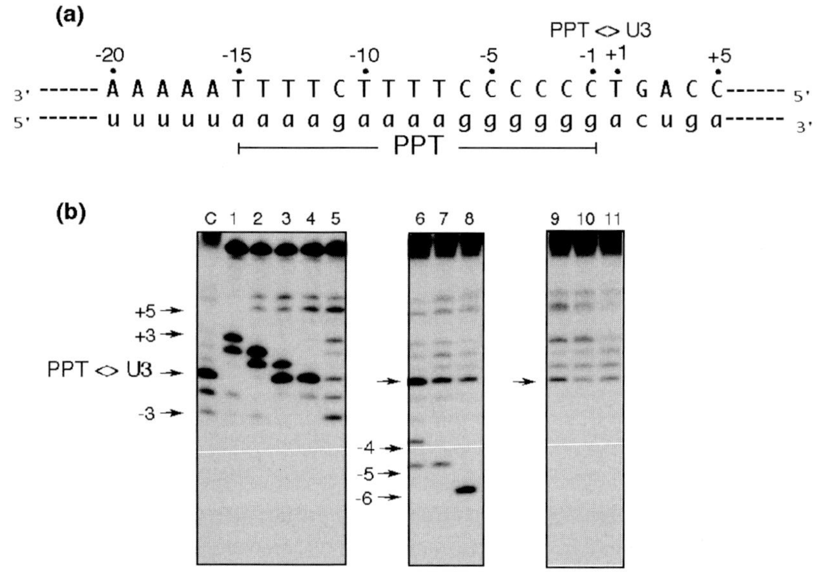

Fig. 3. Modification of PPT cleavage specificity by targeted, pairwise insertion of non-hydrogen-bonding pyrimidine isosteres into the (–) DNA template. (**a**) Sequence of the HIV-1 PPT-containing RNA/DNA hybrid. T and C isosteres were added pairwise throughout the T_4 and C_6 tracts of the DNA template. PPT<>U3 denotes the scissile phosphodiester bond. (**b**) Hydrolysis profiles of modified PPT hybrids. Lane 1, –1D/–2D; Lane 2, –2D/–3D; Lane 3, –3D/–4D; Lane 4, –4D/–5D; Lane 5, –5D/–6D; Lane 6, –7F/–8F; Lane 7, –8F/–9F; Lane 8, –9F/–10F; Lane 9, –12/–13F; Lane 10, –13F/–14F; Lane 11, –14F/–15F. Lane C, unmodified DNA.

Clearly, tandem insertion of pyrimidine isosteres into the DNA strand of a PPT/DNA hybrid is sufficient to relocate the RNase H catalytic center to an alternative site(s) of hydrolysis. Although the precise mechanism by which this unusual cleavage occurs is uncertain, it may reflect increased affinity between RT and nucleic acid at the site(s) of analog insertion. An RT motif that most likely participates in such an interaction is the RNase H primer grip, which X-ray crystallography indicates contacts the DNA strand of an RNA/DNA hybrid 3–8 bp from the RNase H active site.[5] As potential binding sites within the PPT-containing hybrid are "sampled" by RT, one in which a more flexible region can be accommodated by the RNase H primer grip (due to hydrogen bond disruption) may be favored, leading to cleavage of the RNA strand 3 bp downstream from the site of analog substitution.

Locked Nucleic Acid (LNA)

LNAs are nucleoside analogs whose ribose 2′ OH group is replaced by a 2′O–4′C methylene linkage, locking the sugar ring in the C3′-*endo* configuration typical of RNA (Fig. 2, No. 3). When embedded in DNA oligonucleotides, LNAs enhance the stability of DNA/RNA hybrids by as much as 9°C per analog insertion.[10] Introducing LNA monomers into duplex nucleic acid also forces adjacent nucleotides to assume the C3′-*endo* sugar configuration. This conformational change also extends to the opposite strand, such that a stable, localized B-form to A-form helical transition occurs one base pair to either side of the substitution. Consequently, LNA substitution at every third nucleotide in either strand will impart A-form geometry into duplex DNA.[11]

LNAs have been utilized to explore how localized structural changes within a PPT-containing RNA/DNA hybrid affect RNase H cleavage specificity by introducing a localized region of A-form geometry into a hybrid that normally assumes an overall non-A/non-B helical configuration. The methylene bridge(s) would also decrease local helical flexibility and introduce a steric impediment to enzyme contacts with the DNA sugar-phosphate backbone.

HIV-1 RT cleavage profiles of PPT variants revealed that when LNAs were introduced into the DNA between positions −12 and −16, or −1 and

−4 (defining −1 as the first ribonucleotide 5′ of the PPT/U3 junction), RNA cleavage was either displaced downstream from the PPT/U3 junction or impaired. In contrast, introducing LNAs at positions −8 to −11, a region of the duplex previously shown to have unusual base pairing properties,[5] had no effect. Interestingly, the regions affected by LNA substitution are predicted sites of contact between nucleic acid and the thumb and C-terminal RNase H primer grip of an enzyme whose p66 RNase H catalytic center is positioned for hydrolysis at the PPT/U3 junction. Such contacts are most likely disrupted by local changes in sugar pucker, reduced helical flexibility, or steric conflict, relocating the enzyme to sites that are unfavorable for hydrolysis.

Abasic Tetrahydrofuran Linkages

Abasic tetrahydrofuran linkages possess a sugar moiety that chemically resembles the 2′-deoxyribose or ribose sugar in RNA (Fig. 2, No. 4) or DNA (Fig. 2, No. 5), but lack a nucleobase, thus disrupting intra-strand base stacking and inter-strand hydrogen bonding. To determine the role of individual nucleobases in PPT recognition, RNA/DNA hybrids were constructed with abasic substitutions in the DNA and RNA strands around, and upstream of, the PPT/U3 junction. Our analysis[12] demonstrated that abasic substitutions between positions −11 and −15 locally altered base-pairing, decreasing cleavage efficiency at the PPT-U3 junction, and increasing cleavage further downstream. Within the DNA strand, these effects were most pronounced when the abasic lesion was introduced at position −15.

Nucleobase removal from the DNA strand opposite the PPT-U3 junction had little effect on cleavage efficiency or specificity, supporting structural studies that indicated minimal contact between RT and DNA opposite the scissile phosphate.[5] In contrast, abasic lesions in the RNA strand at positions −1 and +1 completely inhibited PPT/U3 cleavage. Nucleobase removal at these positions most likely interferes with catalytically important contacts between RNA and the conserved primer grip residues Gln475 and Arg448. In the RT-PPT/DNA co-crystal structure, these residues interact with RNA nucleobases flanking the scissile phosphate at positions −2 and +1, respectively.

Pyrrolo dC

Pyrrolo dC (pdC) is an environmentally sensitive, fluorescent dC analog which, despite its bicyclic structure (Fig. 2, No. 6), pairs normally with dG or rG and does not disturb helix geometry. While pdC is maximally fluorescent as a free nucleoside, it is partially quenched by adjacent bases in an oligonucleotide, and even more so by participation in Watson–Crick base pairing, a property that can be exploited to probe altered PPT base pairing. Using this strategy,[13] we demonstrated altered base pairing when pdC was introduced at position −11 of the HIV-1 PPT, supporting the notion of unpaired bases in this region.[5] Within the PPT rG:dC tract, base pairing strength was variable, with the most significant deviation at position −2. The latter observation supported previous studies showing that altering −2rG significantly affects PPT selection and plus-strand DNA synthesis both *in vitro* and *in vivo.*[14,15]

Phosphodiester Backbone Modification

The structure of *Bacillus halodurans* RNase H in a catalytically competent complex with hybrid substrate[16] suggested a structural motif, which was designated the phosphate binding pocket, was critical for substrate recognition and binding. Although many findings of this work were applicable to retroviral RNases H, structural divergence between these and bacterial enzymes outside the catalytic domain prevented definitive conclusions regarding the equivalent motif of viral enzymes. The *B. halodurans* phosphate binding pocket corresponds to residues Thr473 and Gln475 in the HIV-1 RNase H primer grip, which were shown by crystallographic analysis to contact phosphates in the DNA strand 3–4 bp from the scissile bond.[5] To determine how specific phosphate contacts affected HIV RNase H efficiency and specificity, RNA/DNA hybrids containing methylphosphonate-substituted DNA were examined. The methylphosphonate linkage (Fig. 2, No. 7) replaces a phosphodiester bond non-bridging oxygen with a methyl group, neutralizing charge at that position, thereby assessing if specific protein–phosphate contact sites are charge dependent, and whether disrupting these contacts affects tracking of the RNA strand into the RNase H catalytic center.

Methylphosphonate scanning[17] indicated RNase H cleavage is sensitive to charge neutralization of the phosphate backbone, particularly when introduced immediately opposite the scissile phosphodiester bond. In addition, a triple methylphosphonate substitution, which creates a "neutral patch", was found to significantly impair DNA synthesis. Taken together, these results suggest that contact between RT and the primer strand in the context of either RNase H or DNA synthesis activity is highly sensitive to charge neutralization at specific positions in the phosphate backbone. The finding that these sites of interaction consistently mapped to the region of nucleic acid contacted by the p66 RNase H primer grip supports a functional role for the phosphate binding pocket of HIV-1 RT.

Examining HIV-1 RT Contacts with the Single-Stranded Template

The manner in which the single-stranded template accesses the DNA polymerase active site and is subsequently positioned for catalysis can also be evaluated by nucleoside analog interference. Dash *et al.*[18] examined the outcome of the polymerizing enzyme encountering nucleotide analogs in the template overhang designed to reduce its flexibility (LNA), stacking interactions (abasic deoxyriboside) or charge (methyphosphonate linkage). These studies are summarized in Fig. 4. They demonstrated that the position and extent of pausing around the sites of substitution were dependent on both the identity of the analog and multiplicity of substitution. For example, a single abasic lesion at template position +8 induced almost complete cessation of DNA synthesis after addition of eight nucleotides to the primer. However, dual +8/+9 and triple +8/+9/+10 abasic lesions cause RT to pause one nucleotide *earlier* [Fig. 4(b)]. At the same time, general accumulation of +6 and +7 pause products suggested the N-terminal fingers subdomain of HIV-1 RT "senses" the template anomaly before it reaches the active site. This notion was supported with single or tandem LNA insertions of the template [Fig. 4(c)]. Although these were less stringent in blocking polymerization, DNA synthesis transiently paused opposite template nucleotide +6, i.e., two nucleotides *before* the analog was encountered. Lastly, neutralizing the phosphate backbone progressively stalled the polymerizing machinery, although enzyme pausing as it approached the template anomaly was less severe [Fig. 4(d)].

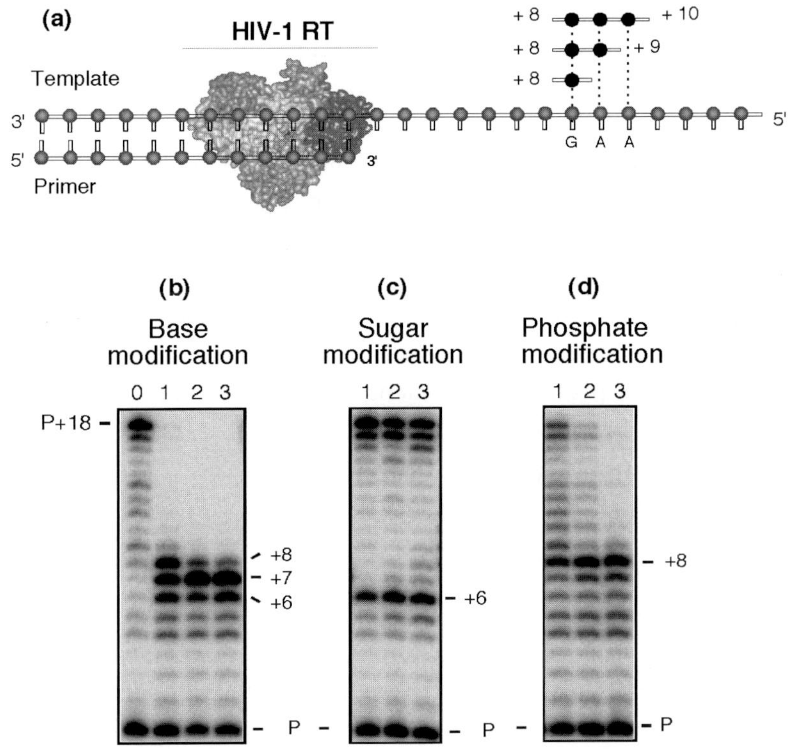

Fig. 4. Examining the interaction of HIV-1 RT with single-stranded template ahead of the DNA polymerase catalytic center by nucleoside analog interference. The experimental strategy is summarized in (**a**), involving single, dual and triple analog insertions between template positions +8 and +10. Analogs were chosen to address the contribution of base (abasic deoxyriboside, (**b**)), sugar (locked nucleic acid, (**c**)) or phosphodiester backbone moieties (methylphosphonate linkage, (**d**)). Data of (**b**)–(**d**) illustrate the unique DNA synthesis signature in response to each class of nucleoside analog.

"Saturation" Nucleoside Analog Interference

Finally, substrates containing one or more purine analog substitution were examined in a comprehensive study of plus-strand primer selection specificity. Ribonucleosides containing purine, 2-amino-purine, diaminopurine, inosine and isoguanosine (Fig. 2, Nos. 9–13, respectively), along with their appropriate DNA complements, were substituted for one or more rG:dC pairs of the PPT $(rG)_6$:$(dC)_6$ tract to determine which bases were critical for maintaining PPT-U3 cleavage specificity, in addition to their chemical features. Introducing analogs at positions +1, −2 and −4 (relative to the PPT-U3 junction) adversely affected both the specificity and efficiency of RNase H mediated cleavage; i.e., cleavage at the junction was reduced, while abnormal cleavage in the immediate vicinity was observed.[15] For PPT positions +1rA and −2rG, the identity of the exocyclic group at position 2 of the purine ring was found to be critical. Specifically, at position −2rG, a 2-NH_2 group in the purine ring was essential, while the same chemical moiety at position +1A was not tolerated. In contrast, substrates modified at PPT position −4rG required a carbonyl group at the purine ring 6-position for precise and efficient cleavage at the PPT-U3 junction.

STRUCTURAL STUDIES OF PROTEIN:LIGAND COMPLEXES BY MASS SPECTROMETRY

Although several chemical probing techniques are available for studying nucleic acid structure,[19,20] the most recent of which can be used to rapidly examine entire retroviral RNA genomes,[21] developing equivalent strategies to understand ligand-induced conformational changes in proteins presents a significantly greater challenge. In view of this, our attention has focused on combining site-specific chemical modification of proteins with proteolysis and mass spectrometry to analyze complexes of HIV enzymes with either nucleic acid or small molecule antagonists.

Protein/Nucleic Acid Complexes

HIV-1 RT poses an additional challenge for protein footprinting, since its 66 and 51 kDa polypeptides derive from the same gene, differing in that p51 results from proteolytic cleavage of p66 between Phe440 and Tyr441.[22] Thus, with the exception of the C-terminal p66 RNase H domain, footprinting approaches must unambiguously determine the subunit from which peptides are derived. The solution to this required the stepwise strategy illustrated in Fig. 5(a), comprising (i) chemical modification of p66/p51 RT in the absence and presence of ligand, (ii) separating the modified p66 and p51 subunits by SDS/polyacrylamide

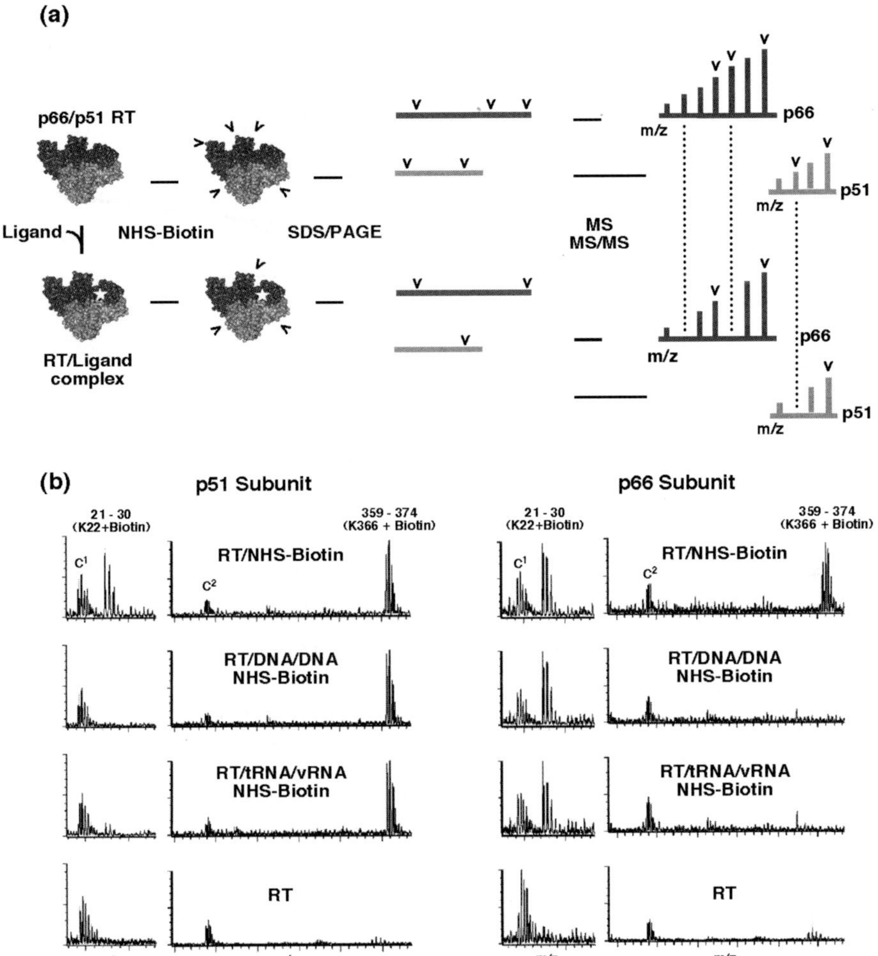

Fig. 5. Protein footprinting of the HIV RT-tRNA/viral RNA complex. The method is presented in (**a**), and involves subjecting RT and the RT/ligand complex to NHS-biotin, which modifies surface-exposed lysine residues. Subsequent SDS/PAGE separates the modified p66 and p51 RT subunits. Gel slices containing each subunit are homogenized and subjected to in-gel proteolysis by trypsin, with residues protected by NHS-biotin rendered insensitive to proteolysis. Peptides extracted from the gel homogenate are further examined by mass spectrometry to identify sites protected from chemical modification. Panel (**b**) indicates regions of p51 (left) and p66 (right) RT that were protected from modification in the presence of the tRNA/viral RNA duplex, as well as internal control peaks (C^1 and C^2) common to all reactions.

gel electrophoresis, (iii) in-gel proteolysis, and (iv) tandem mass spectrometry to determine the sequence of peptides recovered from the gel. A particular advantage of this approach was the significantly smaller amounts of starting material required compared with standard biophysical techniques. This strategy allowed assignment of peptides to the p66 and p51 subunits following biotinylation of surface accessible lysine residues with N-hydroxysuccinimidobiotin (NHS-biotin) and trypsin treatment (since biotinylated lysines are insensitive to trypsin), and was applied to a complex of HIV-1 RT containing the tRNA/viral RNA duplex[23] [Fig. 5(b)]. The results indicate that Lys22 of a p51-derived peptide covering amino acids 21–30 is protected from chemical modification, while its p66 counterpart retains NHS-biotin sensitivity. Conversely, Lys366 of a p66-derived peptide covering amino acids 359–374 is protected from NHS-biotin modification in the presence of the tRNA/viral RNA duplex, while its p51 counterpart remains sensitive. The power of this approach was further demonstrated by our finding that p66 residues Lys30 and Lys64 were NHS-biotin-accessible in the presence of duplex DNA but protected from modification when this was replaced with the tRNA/viral RNA duplex.[23]

RT/RNase H Inhibitor Complexes

The absolute requirement for RT-associated RNase H activity for virus replication[24] offers another therapeutic strategy to augment the current armament of anti-RT drugs. Indeed, our high throughput screening efforts have identified a variety of natural products[25–28] and chemical entities[29,30] that inhibit RNase H activity *in vitro* in the sub-micromolar range. Protein footprinting was used to determine the binding site for the vinylogous urea 2-amino-5,6,7,8-tetrahydro-4*H*-cyclohepta[*b*]thiophene-3-carboxamide (NSC727447, Fig. 6), and implicated residues 278–284 of the p51 thumb subdomain in inhibitor binding, with no evidence for occupancy of the RNase H active site. While this result was unexpected, crystal structures of HIV-1 RT[3,6] show that the p51 thumb provides structural support for the RNase H domain of p66. Mass spectrometric footprinting thus provided evidence of "off-site" inhibition of RNase H activity, presumably by altering

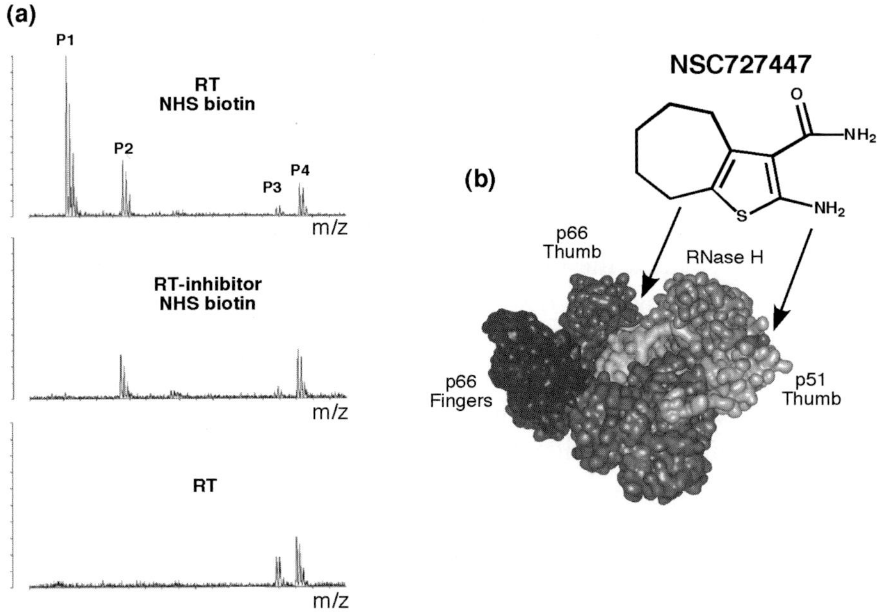

Fig. 6. Determining the vinylogous urea NSC727447 binding site on HIV-1 RT by protein footprinting. (**a**) Representative segments of MALDI-TOF mass spectra of RT treated with NHS-biotin (upper), the RT-NSC727447 complex treated with NHS-biotin (middle) and unmodified RT (lower). Peaks indicated represent the following peptides: P1, aa 278–284 containing modified Cys280 and Lys281; P2, aa 66–72 containing modified Lys66 and Lys70; P3, unmodified aa 264–275; P4, unmodified aa 144–154. Comparing these panels indicates that peptide P1 ($_{268}$Gln-Leu-Cys-Lys-Leu-Leu-Arg$_{274}$) was specifically shielded from chemical modification in the RT-inhibitor complex. Data of Ref. 30 indicate that NSC727447 interacts with the equivalent regions of the p66 and p51 RT thumbs. (**b**) Location of two NSC727447 binding sites (arrows) on the thumb subdomains of HIV-1 RT. The p66 fingers subdomain and RNaseH domain are also indicated.

the coordination geometry of the active site divalent metal ions. Data of Fig. 6 also indicate that the equivalent region of the p66 subunit was also affected by inhibitor binding. Since the overall geometry of the p51 and p66 thumb subdomains is similar, the idea that NSC727447 occupies two sites on HIV-1 RT is not unreasonable, raising the intriguing possibility that two sites of RT outside the DNA polymerase and RNase H catalytic centers can be simultaneously targeted with a single inhibitor.

HIV Integrase/Inhibitor Complexes

Another HIV drug target receiving attention is integrase (IN), with the potent and specific IN inhibitor raltegravir having recently received FDA approval.[31] However, in contrast to the vast body of crystallographic data on HIV protease and RT, a high resolution structure of the intact viral IN protein is currently unavailable. In the absence of such data, we have successfully investigated the binding of two small molecule antagonists by protein footprinting. The first of these studies[32] took advantage of the observation that molecules bearing aryl di-*O*-acetyl groups can acetylate Cys, Lys or Tyr residues. Proteolytic digestion and mass spectrometry was therefore used to identify IN peptides that had been acetylated in the presence of inhibitor, after which the position of acetylation was determined by tandem MS sequencing. Figure 7(a) indicates that IN residue Lys173 was the target of the acetylated inhibitor while Fig. 7(b) is a proposed structural model for inhibitor binding.

A second study[33] extended early findings that pyridoxal 5′-phosphate (PLP) is a moderately potent inhibitor of HIV-1 IN, although the exact site of action was unknown.[34] Our studies took advantage of the property of PLP to form a Schiff base with a primary amine, which again provided a convenient means of marking the enzyme. As illustrated in Fig. 7(c), a PLP-containing IN peptide comprised of residues 241–258 was identified, within which MS/MS sequencing indicated Lys244 as the site of chemical modification. Figure 7(d) shows a proposed model of the IN/PLP interaction.

Mass Spectrometric Footprinting of Nucleic Acids

In an earlier section, we reviewed our efforts to examine structural features of the PPT-containing RNA/DNA hybrid that dictate specific recognition by the RNase H activity of RT to create the plus-strand RNA primer. These studies made extensive use of nucleoside analog substitutions of either the (−) strand DNA template or (+) strand RNA primer, and raised the possibility that an unusual hybrid geometry flanking both extremities of the PPT might provide a "signature" for recognition by, and orientation of, HIV-1 RT. In order to examine PPT architecture in greater detail and correlate this with enzyme binding, a unique mass spectroscopic approach

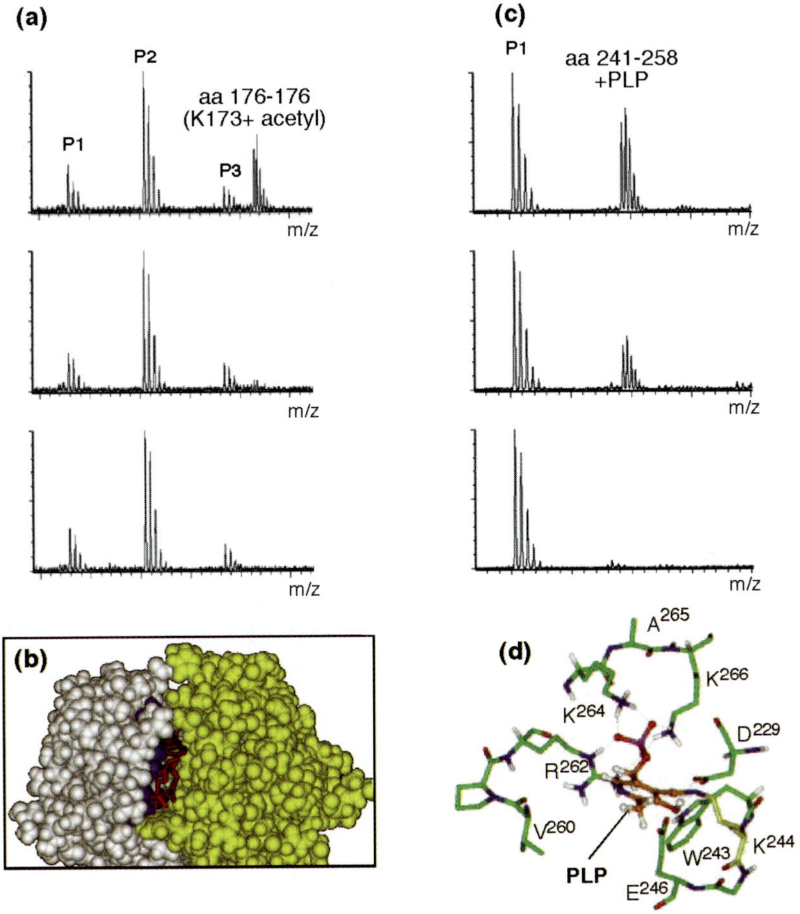

Fig. 7. (**a**) MS data showing specific inactivation of HIV-1 IN by acetylation with the aryl di-*O*-acetyl inhibitor methyl N,O-bis(3,4-diacetyloxycinnamoyl)-L-serinate. Upper: IN + 3 μM inhibitor; middle: IN + 3 μM methyl 3,4-diacetyloxycinnamate as a control; lower: Free IN. Treating IN with 3 μM methyl N,O-bis(3,4-diacetyloxycinnamoyl)-L-serinate produced a single new peak, P4, representing aa 167–185, within which MS/MS fragmentation indicated acetylation of Lys173. P1–P3 are unmodified tryptic peptides. (**b**) Proposed IN:inhibitor complex. The model depicts the inhibitor (red) bound at the dimer interface, contacting residues Lys173, Thr174 and Met178 (blue). IN monomers are colored in yellow and white. (**c**) MS analysis of PLP binding to HIV-1 IN. Upper: IN + 200 μM PLP; middle: IN + 15 μM PLP; lower: free IN. A single new peak of mass 2215 Da was observed among tryptic fragments of the IN/PLP complex, corresponding to residues 241–258 (1,984 Da) plus one PLP adduct (231 Da). P1, unmodified IN tryptic peptide. Only a portion of the Q-TOF spectrum is presented. (**d**) Proposed model of the PLP binding site on HIV-1 IN. See Ref. 33 for additional details.

was devised that allowed for examination of small molecule binding to the PPT with respect to both stoichiometry and position.

One advantage of this approach[35] was the capacity to multiplex with respect to the duplexes under examination, based on the fact that each exhibited a unique mass signature. Specifically, this strategy allowed a simultaneous assessment of ligand binding to several PPT variants (i.e., wild type, duplex DNA, duplex RNA, and inter-conversion of the RNA and DNA strands) in a single reaction [Fig. 8(a)]. Examination of several archetypical nucleic acid ligands[35] indicated that the aminoglycoside

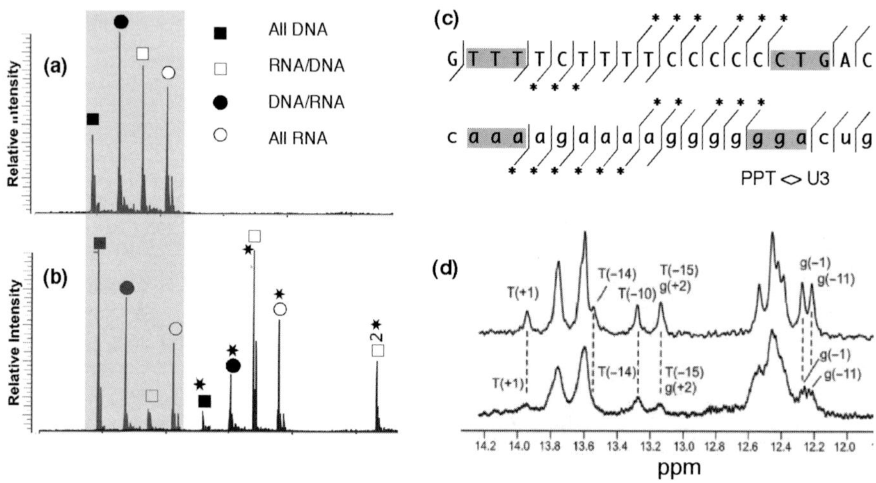

Fig. 8. Examining the interaction of neomycin B with the HIV-1 PPT by mass spectrometry. (**a**) Nanospray FT-ICR mass spectra of the initial mixture of four PPT-containing duplexes, color-coded as indicated, where RNA/DNA represents the native PPT-RNA/DNA hybrid. (**b**) Mass spectra of the same substrates in the presence of a 10-fold excess of neomycin B. Modified masses indicate that the wild-type PPT exclusively binds two neomycin B equivalents. (**c**) Tandem mass spectrum of the non-covalent complex of HIV-1 PPT containing two equivalents of neomycin B. Asterisks identify fragmentation products containing ligand, while lines indicate the phosphodiester bond cleaved by gas-phase dissociation, pointing toward the respective end included in each fragment. Two distinct gaps are evident in the fragmentation pattern of the PPT DNA and RNA components (shaded gray), indicating two distinct binding sites. (**d**) One-dimensional 1H NMR spectra of the imino region of the wild-type PPT in the absence (upper trace) or presence (lower trace) of HIV-1 RT. Imino protons were assigned using 2D NOESY experiments; assignments are listed above each peak. Additional details can be found in Ref. 35.

neomycin B occupied two sites on the wild-type PPT RNA/DNA hybrid with greater affinity than control duplexes [Fig. 8(b)]. Subsequent tandem mass spectrometry of the PPT/neomycin B complex allowed fragmentation of both the DNA and RNA strands and indicated two positions where the cleavage pattern was interrupted [Fig. 8(c)]. These regions corresponded to the PPT/U3 junction and the distal $(rA)_4$:$(dT)_4$ tract between positions −13 and −15. Furthermore, by extrapolating crystallographic data for the HIV-1 RT/PPT complex[22] onto our model system, these structural anomalies would be predicted to contact the p66 thumb subdomain and the RNase H catalytic center of an enzyme positioned for cleavage at the PPT/U3 junction. To test this experimentally, 1H-NMR spectra of the HIV-1 PPT in the absence [Fig. 8(d), upper], and presence of HIV-1 RT [Fig. 8(d), lower] were obtained. Significant broadening was observed for a select subset of imino proton resonances, which was consistent with direct contacts between RT and the 5′ $(rA)_4$:$(dT)_4$ tract and the PPT-U3 junction. While the resolution provided by the NMR analysis of the RT-PPT complex was limited, the observation that adding RT selectively perturbed a subset of imino resonances at positions similar to those interacting with neomycin B binding suggests that they contact common regions of the PPT.

INCORPORATING UNNATURAL AMINO ACIDS INTO HIV-1 RT

Site-specific introduction of unnatural amino acids with novel biophysical properties into proteins by translational suppression, either through coupled transcription/translation systems[36] or genetically-engineered *E. coli*,[7] permits analysis of protein structure and function with a level of resolution previously unattainable. Examples of our application of these strategies to HIV-1 RT are summarized below.

Active Site Mutagenesis

Tyr^{183} of the p66 DNA polymerase active site $–Y^{183}–M^{184}–D^{185}–D^{186}–$ motif contributes to catalysis through interactions with both the sugar and nucleobase of the primer terminus.[3] Based on reports that, for evolutionary purposes, the active site of the Klenow fragment of DNA polymerase

I has more steric room than necessary,[37] we substituted Tyr183 with a variety of tyrosine analogs (e.g., *m*-Tyr, *m*-O-Me-Tyr, *nor*-Tyr). Surprisingly, although differences in polymerization efficiency were evident, we observed remarkable tolerance at this position of the active site, in particular the ability to accept 1- and 2-naphthol-Tyr substitutions which significantly increase steric bulk at the active site (G. Klarmann and S. Le Grice, unpublished observations). However, the properties of HIV-1 RT containing a Tyr^{183}–*nor*-Tyr^{183} mutation were especially interesting, in that the mutant enzyme retained DNA-dependent DNA polymerase activity but was severely impaired in RNA-dependent DNA polymerase function.[38] At the same time, enzyme harboring a Tyr^{183}–*m*-F-Tyr^{183} retained both activities. Thus, by restricting side chain flexibility by removing a methylene group, template specificity was affected.

Altering the "Steric Gate" of HIV-1 RT Induces 3TC Resistance

Tyr^{115} of p66 HIV-1 RT has been designated the "steric gate", inasmuch as it serves to discriminate between deoxy- and ribonucleoside triphosphates.[39] We investigated whether introducing tyrosine analogs at this position affected such discrimination and found that p66/p51 aminomethyl-Phe^{115} RT incorporated dCTP more efficiently than the wild-type enzyme.[39] In addition, when examined in a steady-state fidelity assay, Fig. 9(a) shows that the mutant enzyme was resistant to the chain terminating nucleoside analog (–)-β-2′, 3′ dideoxy-3′-thiacytidine triphosphate (3TCTP), a central component of current antiretroviral therapy. Molecular modeling of RT containing an aminomethyl-Phe at position 115 [Figs. 9(b)–(d)] suggested that introducing the aminomethyl group provided new hydrogen bonds through the minor groove to the incoming dNTP and the template residue of the terminal base pair. These hydrogen bonds likely contribute to the increased efficiency of dCTP incorporation. A second RT mutant containing a 2-naphthyl-Tyr^{115} substitution inefficiently incorporated dCTP at low concentrations and was kinetically slower with a variety of dCTP analogs. In contrast, models of HIV-1 RT containing 2-naphthyl-Tyr^{115} revealed a steric clash with Pro157 of the p66 palm subdomain, necessitating rearrangement of the DNA polymerase active site.

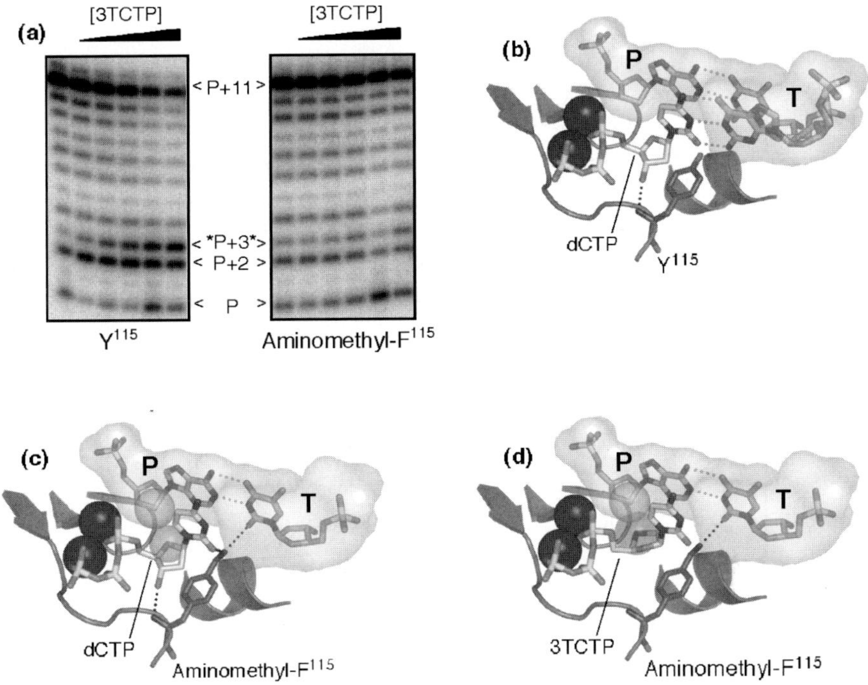

Fig. 9. Unnatural amino acid substitution at the steric gate of HIV-1 RT induces NRTI resistance. (**a**) DNA-dependent DNA polymerase activity of wt HIV-1 RT (left) and a containing Tyr115-aminomethyl-Phe115 variant (right). The assay examines pausing opposite a "target" G (defined as extension product P+3) in the DNA template at increasing concentrations of the chain terminator 3TCTP. Modification of the steric gate allows the mutant enzyme to override the NRTI-induced pause. (**b**)–(**d**) Models of tyrosine analogs at position 115 of p66 HIV-1 RT. (**b**) WT active site. (**c**) Active site containing aminomethyl-Phe115 and dCTP. (**d**) Active site containing aminomethyl-Phe115 RT and 3TCTP. See Ref. 39 for additional details.

Unnatural Amino Acid Mutagenesis of the RNase H Primer Grip

As indicated earlier, crystallographic studies[5] suggest the p66 RNase H primer grip interacts with the DNA primer to confer a trajectory on the RNA template strand appropriate for positioning of the scissile phosphodiester bond at the RNase H active site. Within this motif, Tyr[501] is highly conserved, replacement of which has been correlated with defects in HIV-1 RT *in vitro*[40] and viral replication kinetics *in vivo*.[41] The hydroxylated

(a)

NSC18806 NSC727447

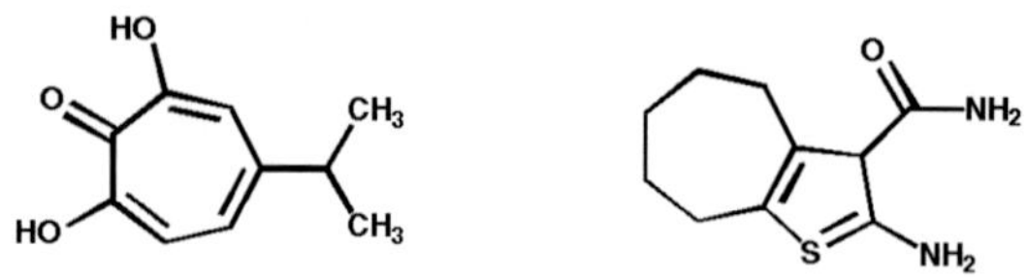

(b)

aa	Structure	IC_{50} NSC18806	IC_{50} NSC727447
Y	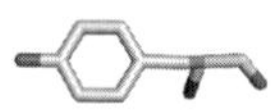	308 nM	6.6 μM
F		356 nM	20.0 μM
W	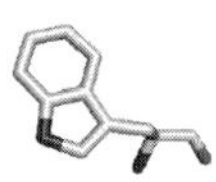	470 mM	70.6 μM
Az-F	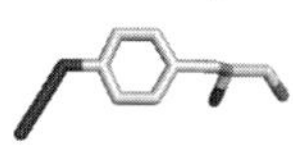	386 nM	89.6 μM
Bp-F	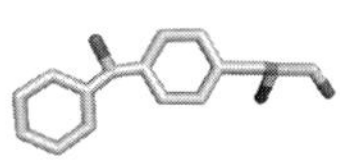	>300 mM	197 μM

Fig. 10. Unnatural amino acid substitution at the RNase H primer grip of p66 HIV-1 RT induces resistance to RNase H inhibitors NSC18806 and NSC727447. (a) Inhibitor structures. (b) IC_{50} values for the two inhibitor classes. Tyr501 (Y) of wt HIV-1 RT was replaced with Phe (F), Trp (W), azido-Phe (Az-F) and benzoyl-Phe (Bp-F). Additional details can be found in Ref. 30.

tropolone *β*-thujaplicinol (NSC18806) and the vinylogous urea NSC727447 [Fig. 10(a)] were reported as modestly potent and selective inhibitors of HIV-1 RNase H activity, the former presumably by chelating divalent metals at the active site[26] and the latter through an "off-site" allosteric mechanism.[30] In order to study the effect of these inhibitors on the interaction of HIV-1 RT with nucleic acid, the unnatural amino acids azido-Phe (Az-F) and benzoyl-Phe (Bp-F) were substituted for Y^{501}. Data of Fig. 10(b) indicate that RNase H activity of HIV-1 RT is equally sensitive to NSC188056, when Tyr^{501} was replaced with Phe, Trp or Az-Phe. This was particularly interesting for the $Az\text{-}Phe^{501}$ mutant, considering it contained a rigid azido moiety. However, introducing a benzophenone at this position resulted in a 1000-fold decrease in inhibitor sensitivity, suggesting it restricted access of the hydroxytropolone to the active site.

Inhibitor binding to wild-type RT may be mediated by coordination of a catalytic Mg^{2+} ion by carbonyl and hydroxyl oxygens of NSC18806, and stabilized by π-stacking between the central ring of the inhibitor and the side chain of Tyr^{501}. The latter interaction would require rotation of the Tyr^{501} side chain around the $C\alpha$-$C\beta$ and $C\beta$-$C\gamma$ bonds from the position observed in RT apoenzyme, which, together with the inhibitor, would create a steric impediment to substrate binding. Conversely, resistance of $p66Bp\text{-}F^{501}/p51$ RT to NSC18806 suggests the inhibitor cannot direct the benzophenone moiety to a position amenable for stacking and, as a consequence, fails to stably bind. It is unlikely that benzophenone substitution at position 501 would itself preclude substrate binding, given the predicted trajectory of the unnatural side chain.

Substituting Tyr501 with the same series of natural and unnatural amino acids resulted in a stepwise drop in NSC727447 sensitivity, with maximally a ~30-fold decrease for mutant $Bp\text{-}Phe^{501}$ (Fig. 10). Although indirect, the significantly different IC_{50} values for the two inhibitors with RT mutant $p66Bp\text{-}F^{501}/p51$ also suggest they are not targeted to the same region.

Introducing Biophysical Probes into HIV-1 RT via Unnatural Amino Acids

While the preceding studies examined the effects of unnatural amino acids on HIV RT function, our current research focuses on introducing biophysical probes, underscoring the potential of this exciting technology. Introducing fluorophores into proteins via site-specific alkylation of a uniquely positioned cysteine has promoted a variety of fluorescence resonance energy transfer studies. Although we have successfully applied this to HIV RT which has only two non-essential cysteines per subunit,[42,43] the strategy becomes impractical as the number of cysteines in the recombinant proteins increases. To circumvent this, we are substituting uniquely positioned phenylalanine residues with Az-Phe, after which Click chemistry[44] can be used to covalently attach alkyne-containing fluorophores (e.g., alkynyl-By3). Secondly, to complement our ongoing structural studies with RNase H inhibitors, phenylalanine and tyrosine residues in the vicinity of the active site are being replaced with trifluoromethoxy-Phe, allowing analysis of inhibitor-induced conformational changes by ^{19}F-NMR.[45] Efficient production of proteins in recombinant *E coli,* coupled with the versatility of orthogonal tRNA-aminoacyl tRNA synthetase pairs for unnatural amino acid incorporation, has widespread utility and adds a new and exciting dimension to understanding protein structure and function.

MEASURING INTERACTIONS OF HIV-1 RT WITH NUCLEIC ACID BY SINGLE-MOLECULE SPECTROSCOPY

Advances in fluorescence spectroscopy have made it possible to utilize fluorescence resonance energy transfer (FRET[46,47]) to continuously measure the distance between an individual pair of dye molecules in real time.[48] In combination with the ability to site-specifically attach dyes to both proteins and nucleic acids, single-molecule FRET provides a new and sensitive probe of enzyme-substrate interactions. By monitoring the reaction of individual molecules or molecular complexes, features are revealed that are hard to detect by traditional ensemble approaches such as parallel reactions pathways, short-lived intermediate states, and heterogeneity among molecules. In recent years, single-molecule FRET has been applied to a variety of biological systems and has revealed interesting dynamic details of the biological processes.[49,50] An early single-molecule study of RT:DNA complexes demonstrated that RT bound duplex DNA substrate in multiple conformations, including a conformation that has not yet been crystallized.[51] Here we discuss two more recent applications of single-molecule FRET to study specific steps performed by HIV-1 RT during the replication process, which revealed intriguing dynamics of the enzyme–substrate complexes that are critical for the function of HIV-1 RT.[42,43]

An Assay for Real-Time Observation of Dynamic RT:Substrate Interactions

In order to monitor the action of individual HIV-1 RT molecules on nucleic acid, the protein was site-specifically labeled with the FRET donor dye cyanine-3 (Cy3) using a cysteine-maleimide reaction. Nucleic acid substrates containing one amine-modified base were labeled with the FRET acceptor dye cyanine-5 (Cy5) using an N-hydroxysuccinimide ester. Biotinylated nucleic acids were immobilized on a streptavidin-coated quartz surface and immersed in a solution containing Cy3-labeled RT. Fluorescence from individual RT-substrate complexes was monitored by total-internal-reflection fluorescence (TIRF) microscopy [Fig. 11(a)]. The observed FRET value during each binding event

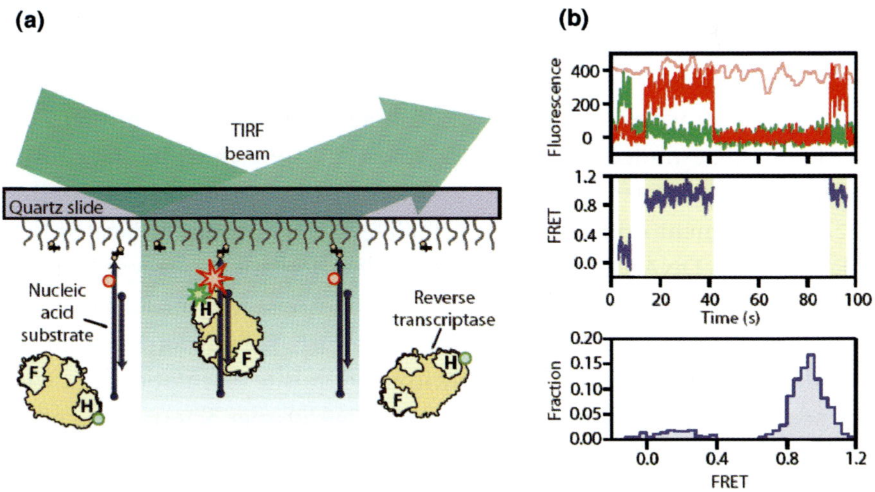

Fig. 11. Single-molecule FRET assay for probing the RT-nucleic acid interactions. (**a**) Single-molecule detection of Cy3-labeled RT (green) binding to surface-immobilized, Cy5-labeled nucleic-acid substrates (red). F and H on RT represent the p66 fingers subdomain and RNase H domain, respectively. (**b**) FRET analysis for RT binding to an individual substrate molecule. Upper: fluorescence time traces from Cy3 (green) and Cy5 (red) under excitation at 532 nm and that from Cy5 (pink) under excitation at 635 nm. Middle: FRET value calculated over the duration of the binding events (yellow shaded regions). Lower: FRET distribution histogram created for the binding events.

allowed the configuration/orientation of the enzyme relative to the nucleic acid duplex to be determined [Fig. 11(b)]. Moreover, since the interacting complexes were immobilized, we could also track changes in conformation as a function of time, which allowed us to extract kinetic information.[42,43]

Orientational Dynamics of HIV-1 RT on DNA and RNA Substrates

During the reverse transcription process, RT must obey the following primer-selection rules to avoid creating non-functional DNA strands: (i) DNA primers should readily engage the DNA polymerase active center, (ii) generic RNA primers must not be extended by RT but should readily engage the RNase H activity of RT in the context of the RNA/DNA replication intermediate, and (iii) the 3′ and central PPT RNA primers, from which plus-strand DNA synthesis is specifically initiated, should direct both DNA polymerase activity and, subsequently, site-specific RNase H activity to promote their removal from nascent DNA. To examine the mechanism by which HIV-1 RT discriminates among these primers, we used our single-molecule FRET assay to examine enzyme binding configurations on substrates comprising a variety of 19-nt primers hybridized to their complement on 50-nt DNA templates.[42] Our studies indicated that RT binds to a DNA/DNA duplex in a single orientation corresponding to the polymerization-competent binding mode observed in RT:DNA co-crystal structures [Fig. 12(a)].[3,4,42] In contrast, the enzyme bound an RNA/DNA hybrid almost exclusively in the opposite orientation, i.e., with the p66 RNase H domain located over the RNA primer 3′ terminus and its fingers subdomain in the vicinity of the 5′ terminus. This orientation clearly cannot support primer extension activity but can support RNase H cleavage activity [Fig. 12(b)].[42,52]

Our observation that RT bound to the DNA and RNA primers with opposite orientations suggests that primer extension activity is determined by the binding orientation of the enzyme. To test this hypothesis, we probed the DNA polymerase activity of RT on all-DNA (19D) and all-RNA (19R) primers as well as chimeric primers 5′-9D:10R-3′ (9D:10R) and 5′-10R:9D-3′ (10R:9D). RT rapidly extended both the 19D and

9D:10R primers, even though the latter contained a ribose sugar backbone at its 3′ terminus. In contrast, polymerase activity was strongly inhibited for both 19R and 10R:9D. Interestingly, the 9D:10R and 10R:9D primers supported binding orientations of RT that closely resembled the orientational distributions of RT bound to pure DNA and RNA primers, respectively [Fig. 12(c)]. Furthermore, the rate of primer extension is quantitatively correlated with the fraction of time for which HIV-1 RT bound in the polymerase-competent orientation [Fig. 12(d)]. These results indicate that binding orientation is a primary determinant of primer-extension activity.[42]

We also used our single-molecule FRET assay to study interactions between RT and the PPT sequence. Unlike generic RNA primers, PPT RNA primers were found to support both of the RT binding orientations described above. This observation helps explain why RT can perform both DNA polymerase and RNase H activities on the PPT.[42,53] Surprisingly, single RT molecules exhibited spontaneous transitions between the two orientations while continuously associated with a PPT substrate, indicating that RT can "flip" on PPT primers [Fig. 12 (e)].[42] Flipping behavior was also observed on other substrates such as DNA annealed to chimeric primers.

The observation that HIV-1 RT spontaneously alternated its orientation during a single binding event was unexpected in view of the extensive network of contacts between the RT and its substrates observed in crystal structures.[3,5,54] To explore what structural rearrangements within the enzyme may be required to promote such events, we tested the effects of small molecule ligands on flipping kinetics. In the presence of the incoming dNTP and a chain-terminated DNA primer, the polymerization-competent binding orientation of RT was significantly stabilized, with a concomitant reduction in the rate of flipping. In contrast, binding of the non-nucleoside RT inhibitor (NNRTI) nevirapine destabilized the polymerization-competent orientation [Fig. 12(f)] and increased the rate of flipping. Structurally, NNRTI and cognate dNTP have opposite effects on the conformation of RT near the polymerase active site: While nucleotide binding tightens the clamp of the fingers and thumb subdomains around substrate to form a ternary complex,[54] binding of NNRTI at the base of the

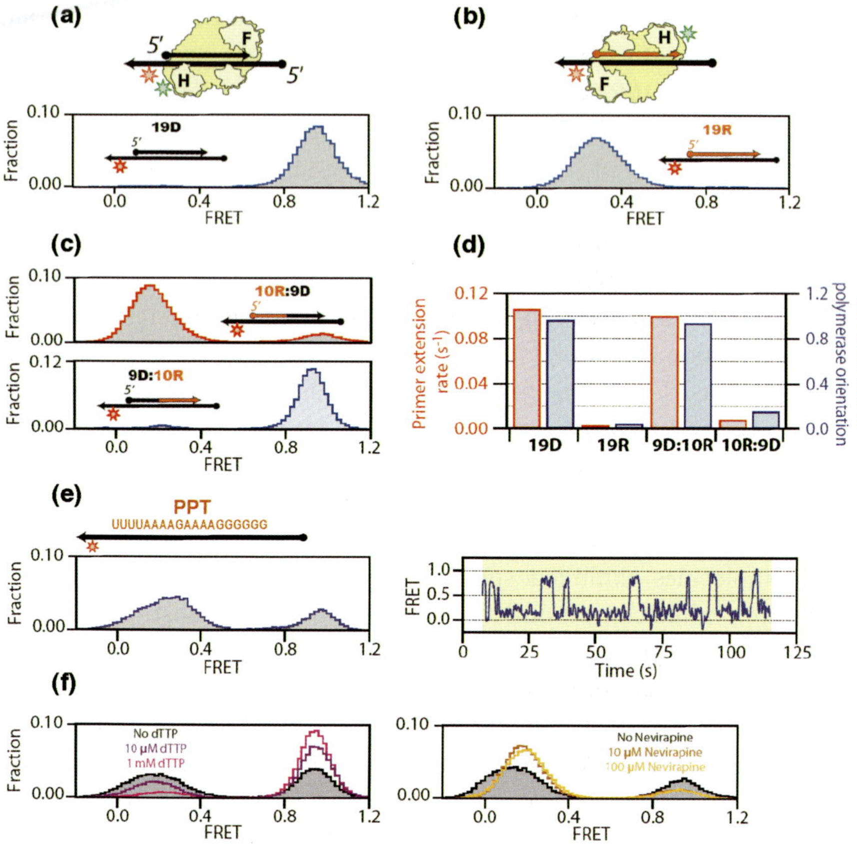

Fig. 12. Dynamic RT binding orientations. (**a**) RT with Cy3 (green star) conjugated to the RNase H domain was bound to duplex DNA (black arrows) with Cy5 (red star) attached near the primer 5′ terminus. Binding in a polymerization-competent orientation positions the RNase H domain near the primer 5′ terminus, yielding high FRET values (~0.94). (**b**) The DNA primer was replaced with a RNA primer (orange). Enzyme binding in the opposite orientation positions the RNase H domain close to the primer 3′ terminus, yielding low FRET values (~0.27). (**c**) Binding to chimeric RNA:DNA primers. Duplexes containing the 5′-10R:9D-3′ and 5′-9D:10R-3′ primers support binding orientations resembling those of RT bound to pure RNA and DNA primers, respectively. (**d**) Primer extension rate constants (red) of four primers (19D, 19R, 9D:10R, and 10R:9D) correlate with the fraction of time that RT bound in the high-FRET orientation conducive to polymerization (blue). (**e**) RT bound to the PPT primer. The PPT sequence is shown in orange. The PPT primer supports both binding orientations (left). The FRET time trace (right, taken in the presence of nevirapine) shows spontaneous flipping transitions between these orientations.

p66 thumb has the consequence of loosening the grip on DNA.[6,55] Our data[42] therefore suggest a potential pathway for the flipping transition that requires relaxation of the fingers-thumb grip.

Sliding Dynamics of RT on Longer Substrates

In order to better approximate the extended duplex substrates encountered during the majority of the reverse transcription cycle, we broadened our investigation to include nucleic acid duplexes longer than 19 bp.[43] On extended DNA/RNA substrates, HIV-1 RT was observed to stably bind either at the front end of the hybrid (poised for DNA primer extension) or at the back end (placing the RNase H domain close to the 3′ end of the RNA template). FRET time traces showed repeated transitions between the front- and back-end bound states during a single binding event [Fig. 13 (a)]. These transitions are consistent with a sliding motion, and many transitions featured short-lived pauses at intermediate positions along the hybrid. RT was also observed to slide on DNA/DNA duplexes. Similar to the flipping dynamics, sliding kinetics were strongly regulated in the presence of either the incoming dNTP or NNRTIs.[43]

HIV-1 RT has been previously observed to cleave RNA at multiple sites within a DNA/RNA hybrid.[52,56] Sliding may provide a mechanism by which these cleavage sites can be rapidly accessed. Sliding may also provide a mechanism for RT to rapidly bind a long duplex and perform a one-dimensional target search for the primer terminus in order to continue DNA synthesis. Additional single-molecule FRET experiments support this hypothesis. On longer duplex DNA substrates (550 bp), the enzyme first bound at a distance from the polymerization site and subsequently moved to the primer terminus [Fig. 13(b)].[43] If the enzyme first binds to the duplex internally, lack of a directional bias will frequently cause RT to approach the primer terminus in an orientation incompatible with

---

Fig. 12. (*Continued*) (f) Left: FRET histograms of RT bound to PPT:dd2 substrates in the presence of 0 (black), 10 μM (purple) and 1 mM (pink) cognate nucleotide dTTP; right: FRET histograms of RT bound to PPT:dd2 substrates or in the presence of 0 (black), 10 (orange) and 100 μM (yellow) nevirapine.

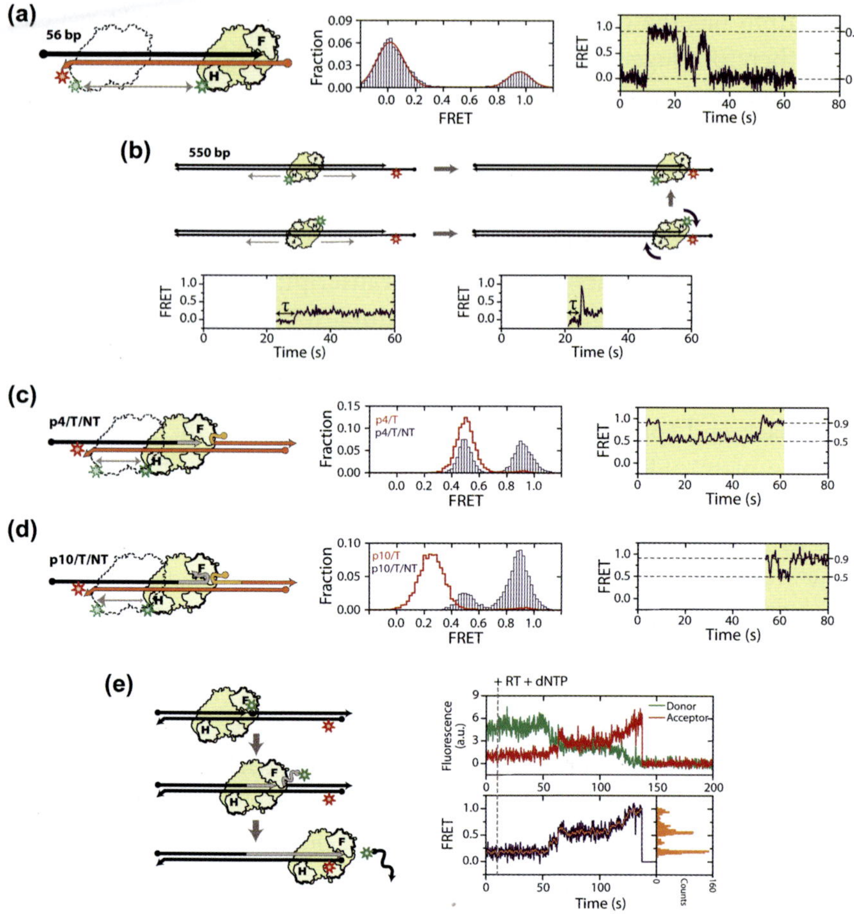

Fig. 13. Dynamic shuttling of RT on nucleic acid duplexes. (**a**) Left: RT with Cy3 (green star) conjugated to the RNase H domain was bound to a 56-bp DNA primer (black)/RNA template (orange) hybrid labeled with Cy5 at the RNA 3′ terminus. The FRET histogram (middle) was fit with two Gaussian peaks (red line). RT can stably bind either to the front end or the back end of the hybrid, yield FRET values of 0 or 0.95, respectively. The FRET time traces (right) show gradual transitions between the 0 or 0.95 FRET states and the preferred intermediates near FRET ~0.3 to 0.5. (**b**) Upper: RT was added to a ~550-bp DNA duplex with Cy5 attached near the front end. Binding to the polymerization site is expected in give a FRET value of 0.3. The majority of the binding events (lower, left and right) initiated with a FRET value of 0, reaching 0.3 only after a time delay (t), suggesting RT first bound to the DNA outside the polymerization site and subsequently slid to the primer terminus. Among these traces, nearly 50 percent (lower, right) shows a transient 0.9 FRET state

continued DNA synthesis. This notion was borne out experimentally, where we noted that RT arrived at the primer 3′ terminus in the opposite orientation in about half of all events. Remarkably, the enzyme had an inherent capacity to spontaneously flip into the polymerization-competent orientation [Fig. 13(b)]. Our single-molecule FRET analysis therefore indicates that flipping and sliding can be combined in an intricate series of movements to enhance the efficacy of the enzyme. Such a binding procedure can increase the target searching efficiency on long duplexes where the primer terminus constitutes only a tiny fraction of the duplex.

Polymerization site targeting may be even more challenging during strand displacement synthesis, as template secondary structures could displace the nascent primer terminus, thereby occluding it. This could be especially problematic during intrastrand displacement synthesis on the RNA template (e.g., over the TAR loop) as duplex RNA is more stable than a DNA/RNA hybrid.[56] To investigate these issues, we designed a series of experiments using triple-stranded substrates mimicking the substrates during RNA strand displacement synthesis, comprising a DNA primer (p),

---

Fig. 13. (***Continued***) between and 0 and 0.3 FRET states, suggesting RT arrived at the polymerization target site in the opposite orientation and subsequently flipped to the polymerization-competent orientation. (**c**) RT bound to RNA displacement substrate consisting of a DNA primer extended by 4 nt (p4), an RNA template (T), and an RNA nontemplate strand (NT). The corresponding FRET histogram (middle, blue bars) and time trace (right) show dynamic transitions between two FRET states (0.5 and 0.9). Overlaid in red line (middle) is the FRET distribution of RT bound to the p4/T substrate lacking the NT strand. RT was able to slide between the back end and front end of the p4/T duplex, but always bind to the back end first before sliding to the front end. (**d**) Left: the substrate is replaced with p10/T/NT, with a primer extended by 10 nt. In this case, RT was unable to slide all the way to the front end of the primer, as indicated by the significant difference between its lower FRET peak (blue bar) and the FRET peak (red line) observed for the p10/T substrate (middle). This inhibits further extension of the primer. (**e**) Left: single-molecule detection of DNA displacement synthesis. Cy3 and Cy5 were attached to the non-template and template strand, respectively. RT and dNTPs were added to initiate DNA synthesis. Right: FRET gradually increased as reaction progressed, reflecting displacement of the non-template strand. Completion of the reaction was marked by the abrupt loss of fluorescence reflecting dissociation of the donor-labeled non-template strand. Plateaus were observed in the fluorescence and FRET traces, indicating pausing during DNA synthesis.

an RNA template (T), and an RNA nontemplate strand (NT) complementary to T downstream of the primer. In the absence of enzyme, a stable T/NT RNA duplex was formed, separating the nascent extended primer from the template. When RT was added, we observed that the enzyme first bound to the back end of the p/T duplex and subsequently slid forward to the front end of the duplex, poised to engage in DNA synthesis [Fig. 13(c)]. In this case, the sliding activity is required for RT to access the disrupted polymerization site, thereby facilitating RNA displacement synthesis.

Interestingly, when the strand displacement synthesis exceeded a certain number of nucleotides, HIV-1 RT failed to slide as far as the primer 3′ terminus [Fig. 13 (d)]. This result predicted that the efficiency of strand displacement synthesis would drop accordingly, which was directly confirmed by the ensemble primer extension assay.[43,57] This barrier was caused by the energetic difference between the DNA/RNA and RNA/RNA duplexes. Such an energetic difference does not exist in DNA strand displacement synthesis. RT was thus able to access the primer terminus during DNA strand displacement synthesis regardless of the length of the primer extension, consistent with its ability to catalyze displacement synthesis through long DNA duplexes.[57] Finally, a variation of the single-molecule FRET assay made it possible to monitor DNA displacement synthesis in real time, allowing us to directly observe many detailed features of the reaction, such as the kinetic pausing events [Fig. 13 (e)].[43]

FUTURE PERSPECTIVES

While data reviewed here focuses on two clinically significant HIV enzymes, it is important to stress the widespread applicability of techniques we have developed and extended. The relative ease with which DNA and RNA are now chemically synthesized, coupled with a constantly expanding repertoire of structurally-diverse nucleoside analogs, can be regarded as a means of "fine tuning" nucleic acid to decipher cross-talk within nucleoprotein complexes. Similarly, site-specific incorporation of unnatural amino acids represents expansion of the genetic code to precisely examine the contribution of individual functional groups to substrate recognition and enzyme activity. Extending these strategies into areas such as protein footprinting and single molecule spectroscopy clearly provides a valuable array of technologies with which molecular biologists can explore protein/nucleic interactions with a level of precision previously unattainable.

ACKNOWLEDGMENTS

S. F. J. Le Grice is funded by the Intramural Research Program of the Center for Cancer Research, National Cancer Institute, National Institutes of Health. This work is supported in part by NIH Grant GM 068518 (to X. Zhuang). X. Zhuang is a Howard Hughes Medical Institute investigator. E. Abbondanzieri is a Jane Coffin Childs postdoctoral investigator.

REFERENCES

1. Telesnitsky A, Goff SP. (1997) Reverse transcriptase and the generation of retroviral DNA. In: Coffin JM, Hughes SH, Varmus HE (eds), *Retroviruses*, pp. 121–160. Cold Spring Harbor Laboratory Press, Plainview, NY.
2. Rodgers DW, Gamblin SJ, Harris BA, *et al.* (1995) The structure of unliganded reverse transcriptase from the human immunodeficiency virus type 1. *Proc Natl Acad Sci USA* **92**(4): 1222–1226.
3. Jacobo-Molina A, Ding J, Nanni RG, *et al.* (1993) Crystal structure of human immunodeficiency virus type 1 reverse transcriptase complexed with double-stranded DNA at 3.0 A resolution shows bent DNA. *Proc Natl Acad Sci USA* **90**(13): 6320–6324.
4. Huang H, Harrison SC, Verdine GL. (2000) Trapping of a catalytic HIV reverse transcriptase*template:primer complex through a disulfide bond. *Chem Biol* **7**(5): 355–364.
5. Sarafianos SG, Das K, Tantillo C, *et al.* (2001) Crystal structure of HIV-1 reverse transcriptase in complex with a polypurine tract RNA:DNA. *Embo J* **20**(6): 1449–1461.
6. Kohlstaedt LA, Wang J, Friedman JM, *et al.* (1992) Crystal structure at 3.5 Å resolution of HIV-1 reverse transcriptase complexed with an inhibitor. *Science* **256**(5065): 1783–1790.
7. Ryu Y, Schultz PG. (2006) Efficient incorporation of unnatural amino acids into proteins in *Escherichia coli. Nat Methods* **3**(4): 263–265.

8. Moran S, Ren RX, Kool ET. (1997) A thymidine triphosphate shape analog lacking Watson-Crick pairing ability is replicated with high sequence selectivity. *Proc Natl Acad Sci USA* **94**(20): 10506–10511.
9. Guckian KM, Krugh TR, Kool ET. (1998) Solution structure of a DNA duplex containing a replicable difluorotoluene-adenine pair. *Nat Struct Biol* **5**(11): 954–959.
10. Bondensgaard K, Petersen M, Singh SK, *et al.* (2000) Structural studies of LNA:RNA duplexes by NMR: Conformations and implications for RNase H activity. *Chemistry* **6**(15): 2687–2695.
11. Petersen M, Nielsen CB, Nielsen KE, *et al.* (2000) The conformations of locked nucleic acids (LNA). *J Mol Recognit* **13**(1): 44–53.
12. Yi-Brunozzi HY, Le Grice SF. (2005) Investigating HIV-1 polypurine tract geometry via targeted insertion of abasic lesions in the (-)-DNA template and (+)-RNA primer. *J Biol Chem* **280**(20): 20154–20162.
13. Dash C, Rausch JW, Le Grice SF. (2004) Using pyrrolo-deoxycytosine to probe RNA/DNA hybrids containing the human immunodeficiency virus type-1 3′ polypurine tract. *Nucleic Acids Res* **32**(4): 1539–1547.
14. Julias JG, McWilliams MJ, Sarafianos SG, *et al.* (2004) Effects of mutations in the G tract of the human immunodeficiency virus type 1 polypurine tract on virus replication and RNase H cleavage. *J Virol* **78**(23): 13315–13324.
15. Rausch JW, Le Grice SF. (2007) Purine analog substitution of the HIV-1 polypurine tract primer defines regions controlling initiation of plus-strand DNA synthesis. *Nucleic Acids Res* **35**(1): 256–268.
16. Nowotny M, Gaidamakov SA, Ghirlando R, *et al.* (2007) Structure of human RNase H1 complexed with an RNA/DNA hybrid: Insight into HIV reverse transcription. *Mol Cell* **28**(2): 264–276.
17. Dash C, Scarth BJ, Badorrek C, *et al.* (2008) Examining the ribonuclease H primer grip of HIV-1 reverse transcriptase by charge neutralization of RNA/DNA hybrids. *Nucleic Acids Res* **36**(20): 6363–6371.
18. Dash C, Fisher TS, Prasad VR, Le Grice SF. (2006) Examining interactions of HIV-1 reverse transcriptase with single-stranded template nucleotides by nucleoside analog interference. *J Biol Chem* **281**(38): 27873–27881.
19. Ricci MS, El-Deiry WS. (2003) DNA footprinting. *Methods Mol Biol* **223**: 117–127.
20. Tijerina P, Mohr S, Russell R. (2007) DMS footprinting of structured RNAs and RNA-protein complexes. *Nat Protoc* **2**(10): 2608–2623.

21. Wilkinson KA, Gorelick RJ, Vasa SM, *et al.* (2008) High-throughput SHAPE analysis reveals structures in HIV-1 genomic RNA strongly conserved across distinct biological states. *PLoS Biol* **6**(4): e96.
22. Le Grice SF, Ette R, Mills J, Mous J. (1989) Comparison of the human immunodeficiency virus types 1 and 2 proteases by hybrid gene construction and trans-complementation. *J Biol Chem* **264**(25): 14902–14908.
23. Kvaratskhelia M, Miller JT, Budihas SR, *et al.* (2002) Identification of specific HIV-1 reverse transcriptase contacts to the viral RNA:tRNA complex by mass spectrometry and a primary amine selective reagent. *Proc Natl Acad Sci USA* **99**(25): 15988–15993.
24. Tisdale M, Schulze T, Larder BA, Moelling K. (1991) Mutations within the RNase H domain of human immunodeficiency virus type 1 reverse transcriptase abolish virus infectivity. *J Gen Virol* **72**(Pt 1): 59–66.
25. Bokesch HR, Wamiru A, Le Grice SF, *et al.* (2008) HIV-1 ribonuclease H inhibitory phenolic glycosides from Eugenia hyemalis. *J Nat Prod* **71**(9): 1634–1636.
26. Budihas SR, Gorshkova I, Gaidamakov S, *et al.* (2005) Selective inhibition of HIV-1 reverse transcriptase-associated ribonuclease H activity by hydroxylated tropolones. *Nucleic Acids Res* **33**(4): 1249–1256.
27. Dat NT, Bae K, Wamiru A, *et al.* (2007) A dimeric lactone from Ardisia japonica with inhibitory activity for HIV-1 and HIV-2 ribonuclease H. *J Nat Prod* **70**(5): 839–841.
28. Takada K, Bermingham A, O'Keefe BR, *et al.* (2007) An HIV RNase H inhibitory 1,3,4,5–tetragalloylapiitol from the African plant Hylodendron gabunensis. *J Nat Prod* **70**(10): 1647–1649.
29. Marchand C, Beutler JA, Wamiru A, *et al.* (2008) Madurahydroxylactone derivatives as dual inhibitors of human immunodeficiency virus type 1 integrase and RNase H. *Antimicrob Agents Chemother* **52**(1): 361–364.
30. Wendeler M, Lee HF, Bermingham A, *et al.* (2008) Vinylogous ureas as a novel class of inhibitors of reverse transcriptase-associated ribonuclease H activity. *ACS Chem Biol* **3**(10): 635–644.
31. Summa V, Petrocchi A, Bonelli F, *et al.* (2008) Discovery of raltegravir, a potent, selective orally bioavailable HIV-integrase inhibitor for the treatment of HIV-AIDS infection. *J Med Chem* **51**(18): 5843–5855.
32. Shkriabai N, Patil SS, Hess S, *et al.* (2004) Identification of an inhibitor-binding site to HIV-1 integrase with affinity acetylation and mass spectrometry. *Proc Natl Acad Sci USA* **101**(18): 6894–6899.

33. Williams KL, Zhang Y, Shkriabai N, *et al.* (2005) Mass spectrometric analysis of the HIV-1 integrase-pyridoxal 5′-phosphate complex reveals a new binding site for a nucleotide inhibitor. *J Biol Chem* **280**(9): 7949–7955.
34. Mazumder A, Neamati N, Sommadossi JP, *et al.* (1996) Effects of nucleotide analogues on human immunodeficiency virus type 1 integrase. *Mol Pharmacol* **49**(4): 621–628.
35. Turner KB, Brinson RG, Yi-Brunozzi HY, *et al.* (2008) Structural probing of the HIV-1 polypurine tract RNA:DNA hybrid using classic nucleic acid ligands. *Nucleic Acids Res* **36**(8): 2799–2810.
36. Anderson RD, 3rd, Zhou J, Hecht SM. (2002) Fluorescence resonance energy transfer between unnatural amino acids in a structurally modified dihydrofolate reductase. *J Am Chem Soc* **124**(33): 9674–9675.
37. Kim TW, Delaney JC, Essigmann JM, Kool ET. (2005) Probing the active site tightness of DNA polymerase in subangstrom increments. *Proc Natl Acad Sci USA* **102**(44): 15803–15808.
38. Klarmann GJ, Eisenhauer BM, Zhang Y, *et al.* (2004) Site- and subunit-specific incorporation of unnatural amino acids into HIV-1 reverse transcriptase. *Protein Expr Purif* **38**(1): 37–44.
39. Klarmann GJ, Eisenhauer BM, Zhang Y, *et al.* (2007) Investigating the "steric gate" of human immunodeficiency virus type 1 (HIV-1) reverse transcriptase by targeted insertion of unnatural amino acids. *Biochemistry* **46**(8): 2118–2126.
40. Rausch JW, Lener D, Miller JT, *et al.* (2002) Altering the RNase H primer grip of human immunodeficiency virus reverse transcriptase modifies cleavage specificity. *Biochemistry* **41**(15): 4856–4865.
41. Julias JG, McWilliams MJ, Sarafianos SG, *et al.* (2003) Mutation of amino acids in the connection domain of human immunodeficiency virus type 1 reverse transcriptase that contact the template-primer affects RNase H activity. *J Virol* **77**(15): 8548–8554.
42. Abbondanzieri EA, Bokinsky G, Rausch JW, *et al.* (2008) Dynamic binding orientations direct activity of HIV reverse transcriptase. *Nature* **453**(7192): 184–189.
43. Liu S, Abbondanzieri EA, Rausch JW, *et al.* (2008) Slide into action: Dynamic shuttling of HIV reverse transcriptase on nucleic acid substrates. *Science* **322**(5904): 1092–1097.
44. Kolb HC, Finn MG, Sharpless KB. (2001) Click chemistry: Diverse chemical function from a few good reactions. *Angew Chem Int Ed Engl* **40**(11): 2004–2021.

45. Cellitti SE, Jones DH, Lagpacan L, *et al.* (2008) *In vivo* incorporation of unnatural amino acids to probe structure, dynamics, and ligand binding in a large protein by nuclear magnetic resonance spectroscopy. *J Am Chem Soc* **130**(29): 9268–9281.
46. Stryer L, Haugland RP. (1967) Energy transfer: A spectroscopic ruler. *Proc Natl Acad Sci USA* **58**(2): 719–726.
47. Selvin PR. (2000) The renaissance of fluorescence resonance energy transfer. *Nat Struct Biol* **7**(9): 730–734.
48. Weiss S. (2000) Measuring conformational dynamics of biomolecules by single molecule fluorescence spectroscopy. *Nat Struct Biol* **7**(9): 724–729.
49. Zhuang X. (2005) Single-molecule RNA science. *Annu Rev Biophys Biomol Struct* **34**: 399–414.
50. Roy R, Hohng S, Ha T. (2008) A practical guide to single-molecule FRET. *Nat Methods* **5**(6): 507–516.
51. Rothwell PJ, Berger S, Kensch O, *et al.* (2003) Multiparameter single-molecule fluorescence spectroscopy reveals heterogeneity of HIV-1 reverse transcriptase: primer/template complexes. *Proc Natl Acad Sci USA* **100**(4): 1655–1660.
52. Wisniewski M, Balakrishnan M, Palaniappan C, *et al.* (2000) Unique progressive cleavage mechanism of HIV reverse transcriptase RNase H. *Proc Natl Acad Sci USA* **97**(22): 11978–11983.
53. Rausch JW, Le Grice SF. (2004) 'Binding, bending and bonding': Polypurine tract-primed initiation of plus-strand DNA synthesis in human immunodeficiency virus. *Int J Biochem Cell Biol* **36**(9): 1752–1766.
54. Ren J, Esnouf R, Garman E, *et al.* (1995) High resolution structures of HIV-1 RT from four RT-inhibitor complexes. *Nat Struct Biol* **2**(4): 293–302.
55. DeStefano JJ, Mallaber LM, Fay PJ, Bambara RA. (1993) Determinants of the RNase H cleavage specificity of human immunodeficiency virus reverse transcriptase. *Nucleic Acids Res* **21**(18): 4330–4338.
56. Hall KB, McLaughlin LW. (1991) Thermodynamic and structural properties of pentamer DNA.DNA, RNA.RNA, and DNA.RNA duplexes of identical sequence. *Biochemistry* **30**(44): 10606–10613.
57. Whiting SH, Champoux JJ. (1994) Strand displacement synthesis capability of Moloney murine leukemia virus reverse transcriptase. *J Virol* **68**(8): 4747–4758.

CHAPTER 11

Non-Human Primates in HIV-1 Research

Willy M. J. M. Bogers*

SUMMARY

The primary cause of AIDS, the human immunodeficiency virus type 1 (HIV-1), is a pathogen that is highly specific for humans and generally does not infect or cause disease in other species. The search for an animal model for AIDS started soon after the discovery of HIV-1 as the etiological agent of AIDS. Since the discovery of simian immunodeficiency virus (SIV) causing AIDS-like diseases in Asian macaques, non-human primates (NHP) have played an important role, not only providing key insights into HIV pathogenesis, but also in developing and evaluating new antiretroviral therapies and vaccines. Although these models are useful for particular applications, the fact that SIV is a distinct virus compared with HIV-1 represents limitations to their use.

INTRODUCTION

AIDS is among the most devastating diseases of our time, with approximately 35 million infected cases and more than 20 million deaths

---

*Department of Virology, Biomedical Primate Research Centre, Lange Kleiweg 139, 2288 GJ Rijswijk, The Netherlands. E-mail: bogers@bprc.nl.

recorded so far (UNAIDS/WHO 2008 report on global AIDS epidemic). Estimations of 2.5 million newly infected individuals and over 2 millions dying from AIDS every year prompted scientists to find a cure for this life-threatening disease. More than a quarter of a century later, the correlates of immune protection against HIV are still not known, while an effective vaccine is not yet available. In fact, recent results from clinical vaccine trials were disappointing[1–4] and it is clear therefore that a better understanding of pathogenesis is required to devise new strategies for HIV control. The use of animal models in the study of HIV pathogenesis has obvious advantages: the disease can be studied more invasively and thoroughly, while drug and vaccine toxicity and efficacy in animal models can provide strategies for ensuing trials in human individuals, limiting the risk, time and cost of trials in humans.

The HIV-1 virus belongs to the lentivirus genus of the Retroviridae family and has spread globally. It is now generally accepted that HIV evolved from multiple cross-species transmission events from African primates to humans.[5–8] Phylogeny studies have shown that HIV-1 evolved from a strain of SIV named SIVcpz, found in the chimpanzee species *Pan troglodytes troglodytes*,[9–11] while HIV-2 (a closely related lentivirus that shares many characteristic with HIV-1) originated from SIVsm of sooty mangabey.[12,13] HIV-1 does not replicate in most animal species tested, including favoured model species such as rodents[14–17] and small non-human primates (NHPs). Chimpanzees can be experimentally infected with HIV-1, but infection by HIV-1 is typically non-pathogenic.[18–20] Chimpanzees rarely develop HIV-associated immunodeficiency despite the fact that various strains of HIV-1 have been evaluated. Only sporadic cases of development of AIDS-like symptoms have been reported in chimpanzees experimentally infected with a combination of different HIV-1 strains.[21,22] Therefore, the difficulties in achieving a productive infection, the cost of this animal model and the fact that chimpanzees are endangered makes them an impractical animal model in AIDS research and vaccine development.

HIV infection in other NHPs was already attempted in the early 1990s, but failed in most models proposed or when animals became infected, virus replication *in vivo* was at low levels and did not cause disease[23,24] with the exception of HIV-2 in baboons.[25,26] Recently, several gene products have been identified that inhibit retroviral replication in non-human primate cells. These host restriction factors for HIV-1 replication could explain the poor replication of HIV-1 in NHPs. Macaque TRIM-5α, a component of cytoplasmic bodies, blocks HIV-1 replication at a step after viral entry, but prior to reverse transcription.[27,28] Also APOBEC3G, a single-strand DNA-editing enzyme inducing hypermutation and DNA degradation, acts as a potent antiviral that can be counteracted by the viral *vif* gene product.[21,29] Further understanding of species-specific host restriction factors and their interactions with viral protein targets may point to new approaches to adapt HIV-1 for more efficient replication in macaques.

Even though most non-human primates are resistant to HIV-1 infection, they do harbor relatives of the human virus: SIV. SIV was isolated in the early 1980s from monkeys with AIDS-like diseases or lymphoma.[30–32] The majority of African monkey species are hosts for various SIVs that have been named from the species they were first isolated from: SIVmac from rhesus macaques, SIVsmm from sooty mangabeys, or SIVmne from pig-tailed macaques. These isolates share a common ancestor, SIVsmm, a virus that is endemic and generally non-pathogenic in its natural host, the sooty mangabey.[33] However, in the mid-80s it was accidentally discovered that infection of rhesus macaques with SIVsmm led to a disease that is remarkably similar to human AIDS.[31,32,34] Molecular clones of this virus, e.g., SIVmac251, were generated and shown to cause disease in inoculated animals.[35–37] Later, it became clear that as for HIV-1, SIVmac infection can be transmitted horizontally during pregnancy[38,39] and from mother to child via breastfeeding.[40,41] Like the most common infection route for HIV-1, SIV infection can also be established via mucosal surfaces, such as oral,[42] rectal and vaginal tissues.[43,44] Altogether, experimental inoculation of SIV into a number of Asian macaque species (rhesus, pig-tailed and cynomolgous), results in a spectrum of pathological responses similar to AIDS in humans, characterized by massive and acute depletion of memory CD4 T cells from the gut,[45–47] from which the immune system usually cannot recover.[48–50] Because of its ability to cause AIDS-like diseases in relatively accessible primate species, SIV infection of macaques has made them the animal model of choice for studying pathogenesis and AIDS vaccine research.[29,51–54]

BIOLOGY OF HIV AND SIV

Several findings have established the similarities between SIV infection in macaques and HIV-1 infection in humans. Clinical disease progression in HIV-infected humans and SIV-infected macaques is slow despite generally robust viral replication. In HIV-1 infected individuals, disease progression occurs within an average of 8–10 years, and for macaques infected with pathogenic SIV, within 0.5–3 years. As already mentioned, the most common infection route for both HIV-1 and SIV infection is established via mucosal surfaces, such as oral,[42] rectal and vaginal tissues.[43,44] To infect CD4 T cells and macrophages, the major target populations, two cellular receptors are required in combination: the CD4 molecule and a chemokine receptor, the two most common being CCR5 and CXCR4.[55–57] Like most HIV-1 isolates, the majority of SIV isolates utilize the CCR5 co-receptor for viral entry.[58–61] Infection by SIV is characterized by massive, rapid, and selective depletion of memory T cells in gut associated lymphoid tissues, a finding which was confirmed in HIV infection.[45–47,62–65] Both SIV and HIV replicate in activated and proliferating T cells as well as in resting T cells.[66] While the naive CD4 T-cells and the entire CD8 compartment remain largely unaffected by the virus in the course of acute HIV/SIV infection, these cell types undergo excessive activation (likely due to microbial translocation[67]) and turnover, followed by eventual exhaustion. Recent work suggests that this exhaustion occurs by progressive conversion of central memory CD4 T cells into effector memory, rather than by activation of naive CD4 T cells, which become infected by the virus and are subsequently killed.[68] Furthermore, acute infection in HIV-1 and SIV models will lead to antigen-specific immune responses,[69–72] and both viruses utilize similar tactics to escape from host immune responses, which includes mutations in neutralization and CTL determinants[73–75]as well as modification of glycosylation patterns of the viral envelope protein.[76–79] Also characteristic of both HIV and SIV infections is the rapid establishment of a peak viral load within 2–3 weeks post infection, followed by a decrease in plasma viral loads. This viral load after the acute phase, the "viral set point," in combination with peripheral CD4 T-cell depletion often predicts the rapidity of disease progression and the onset of AIDS-defining events (e.g., opportunistic infections, neoplastic diseases, hematological or neurological disorders).[80,81]

Despite the fact that both HIV and SIV are closely related lentiviruses that cause AIDS in humans and macaques, respectively, the SIV model has shortcomings. SIVmac shares approximately 80 percent genomic sequences with HIV-2 and only 45 percent with HIV-1.[82] However, the genomic sequence, gene regulation and biology of HIV-1 and SIVmac are sufficiently related to use the SIV-rhesus macaque model to determine the importance and function of at least some viral genes *in vivo*. Some viral properties of HIV-1 like the *vpu* and *vpr* functions are not present in SIVmac, while this virus presents a *vpx* function that is absent in HIV-1.[83] These gene differences restrict the ability to study their relevance to pathogenicity or their utility as a vaccine component *in vivo*. Serological cross-reactivity between SIV and HIV-1 is limited.[84] The V3 loop of the HIV-1 Envelope (Env) protein, the main target for neutralizing antibody (Nab) development,[85–87] is much more variable than the homologous region in SIVmac,[88,89] suggesting that Nabs could be quantitatively different.[90] Finally, immunodominant CTL epitopes within HIV-1 are often not present in SIVmac. The efficacy of HIV-1 based vaccines, therefore, cannot be directly evaluated in the SIV model.

Another point of concern is the use of SIV isolates that have been passaged multiple times in macaques in order to increase virulence and disease progression.[74,91,92] Viruses with enhanced virulence may allow for a more rapid and uniform challenge outcome, but the relevance of these viruses to natural HIV infection is not clear, and the reliance on these models for challenge studies may underestimate vaccine efficacy. This is also true for differences between experimental inoculation of animals and the natural transmission in humans. Most efficacy studies use standardized cell-free virus inocula that might not mimic the natural transmission, which likely involves both cell-free and cell-associated virus. In earlier studies, the virus inoculae were given intravenously, which gave highly reproducible viral loads. This could mimic blood-borne HIV transmission but not the natural route of HIV transmission that occurs at mucosal sites. In order to obtain uniform infection via the mucosa, relatively high doses of virus inoculum were given. The relevance of this approach can be debated since natural sexual transmission through mucosa will take place as a much lower probability event.[93] This aspect has now been recognized by the introduction of the so-called repeated low-dose mucosal exposure model.[60,94–97]

Finally, the choice of macaque species and their genotype needs to be considered. The use of animals with certain defined genotypes, such as rhesus macaques with major histocompatibility complex (MHC) I alleles Mamu-A*01 and Manu-B*17, may bias the result of challenge with SIVmac infection because these alleles have been linked to a better disease outcome.[98–104] Also, differences between macaque species derived from different areas will have an influence on the infection outcome. Infection of rhesus macaques from Chinese origin is characterized by lower viral loads and reduced CD4 T-cell depletion compared to macaques of Indian origin.[105–108]

NATURAL SIV INFECTION

As mentioned, several findings have established the similarities between SIV infection in macaques and HIV-1 infection in humans. But what can we learn from non-progressive SIV infections in their natural hosts? All studies performed thus far reported high levels of SIV replication in natural African primate hosts during both acute and chronic infection (nicely reviewed in Ref. 54). The levels of viral replication during primary infection are similar to those observed in pathogenic HIV-1 and SIVmac infections in humans and macaques, respectively.[109–113] During chronic infection, the levels of SIV replication in natural hosts are slightly higher than those observed in HIV-1 infected individuals or SIV-infected macaques that eventually progress to AIDS.[54,111] Furthermore, during the chronic phase, viral replication is quite constant in natural hosts,[114,115] whereas it shows significant variation among individuals and increases with disease progression in pathogenic infections.[54,116] This result indicates that the ability of natural hosts to resist disease progression does not reside in low levels of SIV replication. With a relatively high viral load, one would expect that this would correlate with low peripheral $CD4^+$ T cells. However, during chronic SIV infection in natural hosts, the $CD4^+$ T-cell population is in general well preserved.[109,110,117,118] Therefore, the general preservation of $CD4^+$ T cells in naturally infected African hosts is one of the major differences with pathogenic HIV-1 and SIVmac infections, where the level of peripheral $CD4^+$ T cells is one of the parameters defining disease progression. It has been observed that African species that are natural hosts for SIVs have very low levels of co-receptor CCR5 expression on $CD4^+$ T cells, especially in the intestine, in contrast to progressive hosts showing high percentages of $CD4^+$ T cells expressing CCR5 at mucosal sites.[119] In addition, some African species harbor low levels of $CD4^+$ T cells, especially in the intestine, even in the absence of SIV infection.[111] Thus, the low level of $CCR5^+$ $CD4^+$ T cells in natural infections may be one reason for the sustained low level of the target cell population and may therefore also be involved in preventing virus transmission.[115]

As already mentioned, infection by SIV and HIV is characterized by massive, rapid, and selective depletion of memory T cells in gut associated lymphoid tissues.[45–47,62–65] Studies have reported that in natural hosts a massive mucosal $CD4^+$ T-cell depletion also occurs during acute SIV infection, and of the same magnitude as reported for pathogenic infections.[120,121] The observation that an acute mucosal $CD4^+$ T-cell depletion is a common feature for all types of SIV infection (progressive, nonprogressive) suggests that acute mucosal $CD4^+$ T-cell depletion is not predictive for disease progression in NHPs.

One might consider that enhanced immune responses induced in natural SIV infections could prevent against a pathogenic disease course. However, it has been observed that the levels of SIV-specific T-cell responses in sooty mangabeys and African green monkeys are not stronger or broader that those observed in pathogenic SIV or HIV infections.[122,123] Also, humoral responses are similar in the natural host compared with pathogenic infections.[124,125] Therefore, the lack of disease progression in natural SIV hosts does not rely on strong or different immune responses. So what could be the reason that natural hosts of SIVs are resistant to disease progression? More and more evidence has been shown that the establishment of a state of chronic, generalized immune activation is a major determinant of disease progression in pathogenic HIV and SIVmac infections. Some of these parameters in pathogenic infections are: increased LPS levels, increased levels of inflammatory cytokines like IL-10 and Fox-P3 (T-regulatory cell maker) and proinflammatory cytokines such as IFNγ, TNFα and IL12, increased levels of HLA-DR expression on both $CD4^+$ and $CD8^+$ T cells, increased levels of T-cell proliferation (Ki-67^+ $CD8^+$ T cells) and increased apoptosis of $CD4^+$ T cells in the intestine.[54,126–129] All these different activation markers suggest that natural hosts of SIV infection routinely maintain low levels of immune activation that prevent chronic immune exhaustion and eventual progression to AIDS. Therefore, understanding the mechanisms contributing to the lack of immune activation during natural SIV infection has become a major objective of current studies.

SHIV ANIMAL MODELS

To overcome some of the limitations of the SIVmac model, chimeric SIV-HIV viruses (SHIV) have been constructed that have features important for the development of an AIDS vaccine. SHIVs include genes from HIV-1 and, in particular, they encode the HIV-1 Env glycoprotein, allowing for more detailed studies on the role of Nabs in protection following challenge. The first SHIV was a chimera containing the HIV-1 *tat*, *rev*, *vpu* and *env* genes inserted into the genome of a pathogenic molecular clone of SIVmac.[130] The first generation of SHIVs were able to infect rhesus macaques but were unable to cause AIDS-like disease.[130–132] Subsequently, researchers created more pathogenic SHIVs that cause disease in animals after serial *in vivo* passages.[133–135] Many versions of this type of SHIV chimera have been made since and have been evaluated in various studies.[51,136] Although there is a concern that some SHIVs have unnatural properties, they have helped considerably in the design and *in vivo* testing of vaccine candidates that specifically target HIV-1 proteins. In particular, characterization of HIV-1 Env-Nabs in animals during SHIV infection[137,138] and the effect of passive immunity on SHIV infection and pathogenesis have benefited most from this model.[139–141] Furthermore, the SHIV model provided evidence that novel vaccine approaches could attenuate the infection without providing sterilizing immunity.[142,143]

There are also differences between the commonly used SIV and the SHIV strains. The early SHIVs use CXCR4 (X4) as co-receptor or are dual tropic (X4R5),[134,144] in contrast to SIV and HIV-1 early isolates, which utilize CCR5 (R5).[144] Differences in co-receptor usage is reflected in target cell populations after infection and during the disease course.[145] The kinetics of CD4 depletion after SHIV infection differs from that observed in the primary infection with HIV-1 or SIV. The SHIVs can cause a rapid and irreversible loss of CD4 T cells within three weeks after infection,[146,147] while death from immunodeficiency generally occurred within several months.[148,149] In order to mimic the natural infection of HIV-1, pathogenic SHIVs were constructed more recently that incorporate HIV-1 Env variants that use the CCR5 (R5) co-receptor.[150–153] As most primary HIV-1 isolates are R5-tropic, these SHIVs could be more relevant to the study of early HIV-1 infection and pathogenesis, in addition to mucosal transmission. Like HIV-1 and SIV, R5-SHIVs cause an initial loss of intestinal CD4 T cells, followed by a gradual depletion in peripheral CD4 T cells. This model has already proven to be very useful, not only in vaccine evaluation, but also in the development of microbicides to prevent HIV-1 transmission. Promising preclinical testing of these components has been performed in the context of mucosal challenges with R5-tropic SHIVs.[154]

Generally, standard antiretroviral therapy (ART) developed for humans cannot be evaluated in macaques infected with SIVmac because these anti-HIV-1 drugs do not inhibit this virus. Therefore, SIV containing HIV-1 RT in place of the SIV RT was designed for *in vivo* drug studies.[155,156] These RT-SHIVs are now being used in triple therapy regimens *in vivo* but can also be used to evaluate novel compounds, alone or in combination with other inhibitors, and to evaluate RT inhibitors as chemoprophylactic or microbicidal treatment.

The SHIV animal model has greatly advanced HIV-1 research and forms a useful addition to the SIVmac model. However, these viruses still consist primarily of SIV sequences and the use of SHIVs in HIV-1 vaccine efficacy studies is limited to Env-directed responses. The RT-SHIVs can only be used to evaluate HIV-1 RT-specific drugs, but not drugs that target other HIV-1 proteins, such as protease inhibitors. A new SHIV model in which most or even all of the viral genome is derived from HIV-1 would overcome many of these limitations. More recently, recombinant SHIVs have been constructed that meet these criteria. By substituting the genomic area coding for HIV-1 Gag and Vif by the SIVmac counterpart Vif, researchers were able to generate infectious virus for which approximately 90 percent of the genome was derived from HIV-1.[157–159] By making these changes in the HIV-1 genome, the so-called simian tropic

HIV-1 (stHIV) becomes resistant to the TRIM5α and APOBEC3 proteins in rhesus macaque lymphocytes, the two major restriction factors that block HIV-1, but not SIVmac infection in rhesus macaques.[21,27–29] These stHIV viruses replicate in primary rhesus lymphocytes *in vitro*[158,159] and in pig-tailed macaques,[157,160] where the acute viremia reaches levels approaching those observed in HIV-1 infected humans. Furthermore, the potential utility of this HIV-1 based animal model was evaluated with an HIV-1 therapeutic regimen that was used as chemoprophylaxis, protecting pig-tailed macaques from stHIV-1 infection after a rigorous high-dose challenge.[157]

Although the currently available animal models of human AIDS have served investigators very well, it is important that animal models of HIV infection should be developed and refined further to more accurately reflect the biological properties of the virus infection in humans.

NON-HUMAN PRIMATES IN AIDS VACCINE DEVELOPMENT

Despite all the limitations mentioned, multiple SIV and SHIV isolates have been developed for AIDS studies in macaques, and a broad spectrum of vaccines and vaccination strategies have been evaluated, including live-attenuated viruses, subunit vaccines, DNA vaccines, prime-boost strategies with different viral and bacterial vectors and novel adjuvants.[29,51,52,136] However, the results of these studies are sometimes difficult to compare directly, due to the lack of standardization, differences in macaque species, route of challenge and the choice of challenge virus.[51] For example, while immunization with adenovirus serotype 5 (Ad5)-expressing HIV-1 proteins fully protected macaques from challenge with SHIV89.6p,[161,162] a similar protocol using SIVmac239 as challenge virus failed to provide protection.[99,100]

The predictive value of any animal model depends on validating data from human trials. One of the first approaches to developing an HIV vaccine focused on purified recombinant forms of gp120, the outer Env protein of HIV, as vaccine candidate. A gp120 subunit protein vaccine has been shown to elicit neutralizing antibodies and to protect chimpanzees against HIV-1 IIIB challenge,[163,164] but the failure of this vaccine in two Phase 3 efficacy trials indicates that these type-specific antibody responses were insufficient to protect against HIV-1 infection in humans,[165–167] probably because the vaccine failed to induce antibodies that neutralize a broad array of primary HIV isolates. These results have been viewed as evidence against the usefulness of NHP models. However, lots of key findings from NHP models are remarkably consistent with results from clinical trials. Although gp120-elicited antibodies neutralized T-cell line adapted viruses and other highly sensitive HIV isolates, sera from immunized chimpanzees and humans failed to neutralize typical primary HIV-1 isolates.[168,169] In this regard, chimpanzees are suitable for immunogenicity assessment but not for challenge studies. A similar SIV Env protein vaccine failed to elicit neutralizing antibodies and to protect macaques against SIVmac251 infection,[170] indicating that available data from NHP models are consistent with the outcome of human efficacy trial. More recently, we evaluated the vaccine candidate NYVAC-C, a pox virus vector expressing various HIV clade C antigens in NHP,[171,172] and found that the induced T-cell responses were comparable with the responses in human volunteers,[173–175] indicating again the predictive value of NHP studies.

Despite the urgent need for an HIV-1 vaccine, only two vaccine concepts have completed clinical efficacy studies so far. The first vaccine concept, which was already mentioned, was the monomeric HIV-1 Env gp120 protein, and the aim of this strategy was to induce Env-specific humoral immune responses.[165–169,176] To date, no HIV-1 vaccine has induced broadly reactive neutralizing antibodies in clinical trials. The development of immunogens that induce broadly reactive neutralizing antibodies is perhaps the most important priority for the HIV-1 vaccine field.[177] Proof-of-concept passive transfer studies in non-human primates have shown that administration of high doses of broadly reactive monoclonal antibodies can afford sterilizing protection from infection, thus demonstrating the potential of virus-specific humoral immunity.[139,140] However, it has not been possible to induce such broadly reactive neutralizing antibodies by vaccination so far. Although there has been substantial progress in our understanding of the Env protein structure and function, there are currently no vaccine candidates that are aimed at eliciting broadly reactive Env-specific neutralizing antibodies in clinical trials. It is likely that the next-generation Env immunogens will be engineered antigens. Strategies that are being pursued include generating biochemically stabilized Env trimers, constraining Env immunogens in structurally defined conformations, scaffolding conserved neutralization epitopes onto foreign proteins, developing methods to circumvent immunoregulation, and designing immunogens to target specific regions such as the CD4-binding site, the membrane proximal ectodomain region (MPER), and structurally conserved elements of the V3 loop. The relevance of non-neutralizing antibodies that mediate other effector functions such as antibody-dependent cell-mediated virus inhibition, complement activation and phagocytosis is also being investigated.

A second vaccine concept that has completed clinical efficacy studies involved replication-incompetent recombinant adenovirus serotype 5 (rAd5) vectors expressing HIV-1 Gag, Pol and Nef. The aim of this strategy was to elicit HIV-1-specific cellular immune responses. Early-phase safety and immunogenicity clinical trials demonstrated that rAd5 vector-based vaccines elicited cell-mediated immune responses against the HIV-1 Gag, Pol and Nef proteins in most subjects, although these responses were partially suppressed in individuals with pre-existing Ad5-specific neutralizing antibodies.[178] Phase 2b test-of-concept efficacy trials known as STEP, sponsored by Merck and the HIV Vaccine Trials Network (HVTN), were unexpectedly terminated when the first planned interim analysis showed that this vaccine failed to protect against infection or to reduce viral loads after infection. Even a completely unexpected observation emerged from the STEP trail: vaccines with pre-existing Ad5-specific neutralizing antibodies exhibited an enhanced rate of HIV-1 acquisition (Fauci AS. (2007) The release of new data from the HVTN 502 (STEP) HIV vaccine study. *NIH News* (http://www3.niaid.nih.gov/news/newsrelease/topics/director/). This recent failure of the Merck vaccine STEP trial could be interpreted in light of the results obtained in macaque vaccine trials using SHIV89.6p or SIVmac239.[179,180] While immunization with Ad5-expressing HIV-1 proteins fully protects macaques from challenge with SHIV89.6p,[161,162] a similar protocol performed using SIVmac239 as the challenge virus failed to provide protection.[99,100] These observations stress the importance of the choice of a challenge virus in a given simian model. Some combinations can predict the outcome of human trials, while other combinations cannot. The results of the STEP trial were thus consistent with, and predicted by, studies of analogous SIV vaccines in rhesus macaques. This suggests that SIV challenge of monkeys can be a valuable filter for choosing which vaccine candidates should advance to clinical trials. Altogether, these results have highlighted new scientific challenges and have led to substantial debate regarding the optimal path forward for the HIV-1 vaccine field.[180–182] Various experts in the field have come to a consensus to develop a strategy for the effective use of NHPs in preclinical testing of candidate HIV-1 T cell-based vaccines.[182] This strategy contains three steps. Step 1: A NHP challenge trial to evaluate the potential efficacy of each candidate HIV-1 vaccine would be required. The NHP challenge study must include three elements: an SIV analogue of the candidate HIV-1 vaccine, an appropriate pathogenic SIV challenge isolate, and use of NHPs with diverse MHC class I haplotypes. Pathogenic SHIVs using CCR5 as co-receptors that recapitulate the key features of HIV-1 infection might be appropriate challenge viruses to estimate the ability of a neutralizing antibody-based HIV-1 vaccine candidate to prevent infection in NHPs. NHPs with diverse MHC class I haplotypes must be used to minimize selection of animals with strong natural capacity to control SIV replication even without vaccination. Step 2: NHP immunogenicity testing of a candidate HIV-1 vaccine should only be undertaken if statistically significant protection from uncontrolled SIV replication is observed for vaccinated compared to unvaccinated NHPs in Step 1. Importantly, the immunogenicity of the HIV-1 vaccine product from a seed stock intended for clinical trials must be tested in NHPs. If strong and consistent immunogenicity of the candidate HIV-1 vaccine is observed in NHPs in Step 2, preclinical testing of toxicity and stability could proceed. Step 3: If a candidate HIV-1 vaccine is found to be immunogenic in NHPs, and nontoxic and stable, then the vaccine product could be made using good manufacturing practice (GMP) procedures, vialed, and released for Phase I trials in humans.

In the past, there has been a reluctance to give NHP models the "gatekeeper" status for advancing vaccine candidates into clinical efficacy studies because of concerns that potentially effective HIV-1 vaccine candidates might thereby be missed and because the capacity of this model to predict the results of clinical efficacy studies remains unclear. However, it would seem reasonable to give relative priority to the development of vaccine candidates that lead to durable control of set point viral loads after

SIVmac239 or SIVmac251 challenge. The last few years have seen the development of vaccines that are more consistently immunogenic in NHPs. Hence, screening for the strongest vaccine candidates, while at the same time selecting against those vaccine candidates that have low efficacy, has now become an important element of the preclinical HIV-1 vaccine evaluation process.

The number of HIV-1 vaccine candidates now available for Phase I testing has increased, resulting in the need to develop cost-effective strategies to place filters in the vaccine pipeline. As such, the role of NHP models in the process of selecting HIV-1 vaccines for clinical trials is currently being reconsidered.[180–182]

ACKNOWLEDGMENTS

I am very grateful to Thea de Koning for help and support in putting together this manuscript. I would also like to thank the staff of the Virology department of the BPRC for their contribution to part of the work described in this chapter, including Jonathan Heeney, Gerrit Koopman, Petra Mooij, Ernst Verschoor, David Davis, all technical personnel and the excellent veterinary and animal care staff. Part of the work described in this chapter has been made possible by grants BMH4-CT95-0206, BMH4-CT97-2067, the EU Grant EUROVAC QLRT-PL1999-01321, and NIH grants 1 P01 AI48225-01A2 and 5 P01 AI066287-02.

REFERENCES

1. Sekaly RP. (2008) The failed HIV Merck vaccine study: A step back or a launching point for future vaccine development? *J Exp Med* **205**: 7–12.
2. Ledford H. (2007) HIV vaccine may raise risk. *Nature* **450**: 325.
3. Cohen J. (2007) AIDS research. Did Merck's failed HIV vaccine cause harm? *Science* **318**: 1048–1049.
4. Cohen J. (2007) AIDS research. Promising AIDS vaccine's failure leaves field reeling. *Science* **318**: 28–29.
5. Gao F, Bailes E, Robertson DL, *et al.* (1999) Origin of HIV-1 in the chimpanzee *Pan troglodytes troglodytes. Nature* **397**: 436–441.

6. Gao F, Yue L, White AT, *et al.* (1992) Human infection by genetically diverse SIVSM-related HIV-2 in west Africa. *Nature* **358**: 495–499.
7. Marx PA, Li Y, Lerche NW, *et al.* (1991) Isolation of a simian immunodeficiency virus related to human immunodeficiency virus type 2 from a west African pet sooty mangabey. *J Virol* **65**: 4480–4485.
8. Santiago ML, Lukasik M, Kamenya S, *et al.* (2003) Foci of endemic simian immunodeficiency virus infection in wild-living eastern chimpanzees (*Pan troglodytes schweinfurthii*). *J Virol* **77**: 7545–7562.
9. Heeney JL, Dalgleish AG, Weiss RA. (2006) Origins of HIV and the evolution of resistance to AIDS. *Science* **313**: 462–466.
10. Keele BF, Van Heuverswyn F, Li Y, *et al.* (2006) Chimpanzee reservoirs of pandemic and nonpandemic HIV-1. *Science* **313**: 523–526.
11. Sharp PM, Shaw GM, Hahn BH. (2005) Simian immunodeficiency virus infection of chimpanzees. *J Virol* **79**: 3891–3902.
12. Damond F, Worobey M, Campa P, *et al.* (2004) Identification of a highly divergent HIV type 2 and proposal for a change in HIV type 2 classification. *AIDS Res Hum Retrovir* **20**: 666–672.
13. Santiago ML, Range F, Keele BF, *et al.* (2005) Simian immunodeficiency virus infection in free-ranging sooty mangabeys (*Cercocebus atys atys*) from the Tai Forest, Cote d'Ivoire: Implications for the origin of epidemic human immunodeficiency virus type 2. *J Virol* **79**: 12515–12527.
14. Baenziger S, Tussiwand R, Schlaepfer E, *et al.* (2006) Disseminated and sustained HIV infection in CD34+ cord blood cell-transplanted Rag2-/-gamma c-/- mice. *Proc Natl Acad Sci USA* **103**: 15951–15956.
15. Jamieson BD, Zack JA. (1999) Murine models for HIV disease. *Aids* **13** (Suppl A): S5–11.
16. McCune JM, Namikawa R, Shih CC, *et al.* (1990) Pseudotypes in HIV-infected mice. *Science* **250**: 1152–1154.
17. Mosier DE, Gulizia RJ, Baird SM, *et al.* (1991) Human immunodeficiency virus infection of human-PBL-SCID mice. *Science* **251**: 791–794.
18. Fultz PN, McClure HM, Swenson RB, *et al.* (1986) Persistent infection of chimpanzees with human T-lymphotropic virus type III/lymphadenopathy-associated virus: A potential model for acquired immunodeficiency syndrome. *J Virol* **58**: 116–124.
19. Gardner MB, Luciw PA. (1989) Animal models of AIDS. *Faseb J* **3**: 2593–2606.

20. Rutjens E, Balla-Jhagjhoorsingh S, Verschoor E, *et al.* (2003) Lentivirus infections and mechanisms of disease resistance in chimpanzees. *Front Biosci* **8**: d1134–1145.
21. Novembre FJ, Saucier M, Anderson DC, *et al.* (1997) Development of AIDS in a chimpanzee infected with human immunodeficiency virus type 1. *J Virol* **71**: 4086–4091.
22. O'Neil SP, Novembre FJ, Hill AB, *et al.* (2000) Progressive infection in a subset of HIV-1-positive chimpanzees. *J Infect Dis* **182**: 1051–1062.
23. Agy MB, Frumkin LR, Corey L, *et al.* (1992) Infection of *Macaca nemestrina* by human immunodeficiency virus type-1. *Science* **257**: 103–106.
24. Otten RA, Brown BG, Simon M, *et al.* (1994) Differential replication and pathogenic effects of HIV-1 and HIV-2 in *Macaca nemestrina. Aids* **8**: 297–306.
25. Barnett SW, Murthy KK, Herndier BG, Levy JA. (1994) An AIDS-like condition induced in baboons by HIV-2. *Science* **266**: 642–646.
26. Locher CP, Witt SA, Herndier BG, *et al.* (2003) Increased virus replication and virulence after serial passage of human immunodeficiency virus type 2 in baboons. *J Virol* **77**: 77–83.
27. Stremlau M. (2007) GE Prize-winning essay. Why Old World monkeys are resistant to HIV-1. *Science* **318**: 1565–1566.
28. Stremlau M, Owens CM, Perron MJ, *et al.* (2004) The cytoplasmic body component TRIM5alpha restricts HIV-1 infection in Old World monkeys. *Nature* **427**: 848–853.
29. Ambrose Z, KewalRamani VN, Bieniasz PD, Hatziioannou T. (2007) HIV/AIDS: In search of an animal model. *Trends Biotechnol* **25**: 333–337.
30. Benveniste RE, Arthur LO, Tsai CC, *et al.* (1986) Isolation of a lentivirus from a macaque with lymphoma: Comparison with HTLV-III/LAV and other lentiviruses. *J Virol* **60**: 483–490.
31. Daniel MD, Letvin NL, King NW, *et al.* (1985) Isolation of T-cell tropic HTLV-III-like retrovirus from macaques. *Science* **228**: 1201–1204.
32. Murphey-Corb M, Martin LN, Rangan SR, *et al.* (1986) Isolation of an HTLV-III-related retrovirus from macaques with simian AIDS and its possible origin in asymptomatic mangabeys. *Nature* **321**: 435–437.
33. Hirsch VM, Olmsted RA, Murphey-Corb M, *et al.* (1989) An African primate lentivirus (SIVsm) closely related to HIV-2. *Nature* **339**: 389–392.

34. Letvin NL, Daniel MD, Sehgal PK, *et al.* (1985) Induction of AIDS-like disease in macaque monkeys with T-cell tropic retrovirus STLV-III. *Science* **230**: 71–73.
35. Hirsch VM, Dapolito G, Johnson PR, *et al.* (1995) Induction of AIDS by simian immunodeficiency virus from an African green monkey: Species-specific variation in pathogenicity correlates with the extent of *in vivo* replication. *J Virol* **69**: 955–967.
36. Kestler H, Kodama T, Ringler D, *et al.* (1990) Induction of AIDS in rhesus monkeys by molecularly cloned simian immunodeficiency virus. *Science* **248**: 1109–1112.
37. Naidu YM, Kestler HW, 3rd, Li Y, *et al.* (1988) Characterization of infectious molecular clones of simian immunodeficiency virus (SIVmac) and human immunodeficiency virus type 2: Persistent infection of rhesus monkeys with molecularly cloned SIVmac. *J Virol* **62**: 4691–4696.
38. Davison-Fairburn B, Blanchard J, Hu FS, *et al.* (1990) Experimental infection of timed-pregnant rhesus monkeys with simian immunodeficiency virus (SIV) during early, middle, and late gestation. *J Med Primatol* **19**: 381–393.
39. Fazely F, Sharma PL, Fratazzi C, *et al.* (1993) Simian immunodeficiency virus infection via amniotic fluid: A model to study fetal immunopathogenesis and prophylaxis. *J Acquir Immune Defic Syndr* **6**: 107–114.
40. Amedee AM, Lacour N, Ratterree M. (2003) Mother-to-infant transmission of SIV via breast-feeding in rhesus macaques. *J Med Primatol* **32**: 187–193.
41. Rychert J, Lacour N, Amedee AM. (2006) Genetic analysis of simian immunodeficiency virus expressed in milk and selectively transmitted through breastfeeding. *J Virol* **80**: 3721–3731.
42. Baba TW, Trichel AM, An L, *et al.* (1996) Infection and AIDS in adult macaques after nontraumatic oral exposure to cell-free SIV. *Science* **272**: 1486–1489.
43. Miller CJ, Alexander NJ, Sutjipto S, *et al.* (1989) Genital mucosal transmission of simian immunodeficiency virus: Animal model for heterosexual transmission of human immunodeficiency virus. *J Virol* **63**: 4277–4284.
44. Miller CJ, Marthas M, Torten J, *et al.* (1994) Intravaginal inoculation of rhesus macaques with cell-free simian immunodeficiency virus results in persistent or transient viremia. *J Virol* **68**: 6391–6400.

45. Brenchley JM, Schacker TW, Ruff LE, *et al.* (2004) CD4+ T cell depletion during all stages of HIV disease occurs predominantly in the gastrointestinal tract. *J Exp Med* **200**: 749–759.
46. Mehandru S, Poles MA, Tenner-Racz K, *et al.* (2004) Primary HIV-1 infection is associated with preferential depletion of CD4+ T lymphocytes from effector sites in the gastrointestinal tract. *J Exp Med* **200**: 761–770.
47. Veazey RS, DeMaria M, Chalifoux LV, *et al.* (1998) Gastrointestinal tract as a major site of CD4+ T cell depletion and viral replication in SIV infection. *Science* **280**: 427–431.
48. Brenchley JM, Price DA, Douek DC. (2006) HIV disease: Fallout from a mucosal catastrophe? *Nat Immunol* **7**: 235–239.
49. Picker LJ, Watkins DI. (2005) HIV pathogenesis: The first cut is the deepest. *Nat Immunol* **6**: 430–432.
50. Veazey RS, Lackner AA. (2005) HIV swiftly guts the immune system. *Nat Med* **11**: 469–470.
51. Baroncelli S, Negri DR, Michelini Z, Cara A. (2008) *Macaca mulatta, fascicularis* and *nemestrina* in AIDS vaccine development. *Expert Rev Vaccines* **7**: 1419–1434.
52. Hu SL. (2005) Non-human primate models for AIDS vaccine research. *Curr Drug Target Infect Disord* **5**: 193–201.
53. Pandrea I, Silvestri G, Apetrei C. (2009) AIDS in African nonhuman primate hosts of SIVs: A new paradigm of SIV infection. *Curr HIV Res* **7**: 57–72.
54. Pandrea I, Sodora DL, Silvestri G, Apetrei C. (2008) Into the wild: Simian immunodeficiency virus (SIV) infection in natural hosts. *Trends Immunol* **29**: 419–428.
55. Moore JP, Kitchen SG, Pugach P, Zack JA. (2004) The CCR5 and CXCR4 coreceptors — central to understanding the transmission and pathogenesis of human immunodeficiency virus type 1 infection. *AIDS Res Hum Retrovir* **20**: 111–126.
56. Scarlatti G, Tresoldi E, Bjorndal A, *et al.* (1997) *In vivo* evolution of HIV-1 co-receptor usage and sensitivity to chemokine-mediated suppression. *Nat Med* **3**: 1259–1265.
57. van't Wout AB, Kootstra NA, Mulder-Kampinga GA, *et al.* (1994) Macrophage-tropic variants initiate human immunodeficiency virus type 1 infection after sexual, parenteral, and vertical transmission. *J Clin Invest* **94**: 2060–2067.

58. Zhang Y, Chirmule N, Gao G, Wilson J. (2000) CD40 ligand-dependent activation of cytotoxic T lymphocytes by adeno-associated virus vectors *in vivo*: Role of immature dendritic cells. *J Virol* **74**: 8003–8010.
59. Chen Z, Gettie A, Ho DD, Marx PA. (1998) Primary SIVsm isolates use the CCR5 coreceptor from sooty mangabeys naturally infected in west Africa: A comparison of coreceptor usage of primary SIVsm, HIV-2, and SIVmac. *Virology* **246**: 113–124.
60. Kim CN, Adams DR, Bashirian S, *et al.* (2006) Repetitive exposures with simian/human immunodeficiency viruses: Strategy to study HIV pre-clinical interventions in non-human primates. *J Med Primatol* **35**: 210–216.
61. Zhang L, He T, Huang Y, *et al.* (1998) Chemokine coreceptor usage by diverse primary isolates of human immunodeficiency virus type 1. *J Virol* **72**: 9307–9312.
62. Veazey RS, Tham IC, Mansfield KG, *et al.* (2000) Identifying the target cell in primary simian immunodeficiency virus (SIV) infection: Highly activated memory CD4(+) T cells are rapidly eliminated in early SIV infection *in vivo*. *J Virol* **74**: 57–64.
63. Mattapallil JJ, Douek DC, Hill B, *et al.* (2005) Massive infection and loss of memory CD4+ T cells in multiple tissues during acute SIV infection. *Nature* **434**: 1093–1097.
64. Mattapallil JJ, Smit-McBride Z, McChesney M, Dandekar S. (1998) Intestinal intraepithelial lymphocytes are primed for gamma interferon and MIP-1beta expression and display antiviral cytotoxic activity despite severe CD4(+) T-cell depletion in primary simian immunodeficiency virus infection. *J Virol* **72**: 6421–6429.
65. Picker LJ, Hagen SI, Lum R, *et al.* (2004) Insufficient production and tissue delivery of CD4+ memory T cells in rapidly progressive simian immunodeficiency virus infection. *J Exp Med* **200**: 1299–1314.
66. Zhang Z, Schuler T, Zupancic M, *et al.* (1999) Sexual transmission and propagation of SIV and HIV in resting and activated CD4+ T cells. *Science* **286**: 1353–1357.
67. Brenchley JM, Price DA, Schacker TW, *et al.* (2006) Microbial translocation is a cause of systemic immune activation in chronic HIV infection. *Nat Med* **12**: 1365–1371.
68. Okoye A, Meier-Schellersheim M, Brenchley JM, *et al.* (2007) Progressive CD4+ central memory T cell decline results in CD4+ effector memory

insufficiency and overt disease in chronic SIV infection. *J Exp Med* **204**: 2171–2185.

69. Jin X, Bauer DE, Tuttleton SE, *et al.* (1999) Dramatic rise in plasma viremia after CD8(+) T cell depletion in simian immunodeficiency virus-infected macaques. *J Exp Med* **189**: 991–998.
70. Kuroda MJ, Schmitz JE, Charini WA. (1999) Emergence of CTL coincides with clearance of virus during primary simian immunodeficiency virus infection in rhesus monkeys. *J Immunol* **162**: 5127–5133.
71. Schmitz JE, Kuroda MJ, Santra S, *et al.* (1999) Control of viremia in simian immunodeficiency virus infection by CD8+ lymphocytes. *Science* **283**: 857–860.
72. Schmitz JE, Kuroda MJ, Santra S, *et al.* (2003) Effect of humoral immune responses on controlling viremia during primary infection of rhesus monkeys with simian immunodeficiency virus. *J Virol* **77**: 2165–2173.
73. Allen TM, Altfeld M, Geer SC, *et al.* (2005) Selective escape from CD8+ T-cell responses represents a major driving force of human immunodeficiency virus type 1 (HIV-1) sequence diversity and reveals constraints on HIV-1 evolution. *J Virol* **79**: 13239–13249.
74. Goulder PJ, Watkins DI. (2004) HIV and SIV CTL escape: Implications for vaccine design. *Nat Rev Immunol* **4**: 630–640.
75. O'Connor DH, McDermott AB, Krebs KC, *et al.* (2004) A dominant role for CD8+-T-lymphocyte selection in simian immunodeficiency virus sequence variation. *J Virol* **78**: 14012–14022.
76. Chackerian B, Rudensey LM, Overbaugh J. (1997) Specific N-linked and O-linked glycosylation modifications in the envelope V1 domain of simian immunodeficiency virus variants that evolve in the host alter recognition by neutralizing antibodies. *J Virol* **71**: 7719–7727.
77. Johnson WE, Desrosiers RC. (2002) Viral persistance: HIV's strategies of immune system evasion. *Annu Rev Med* **53**: 499–518.
78. Richman DD, Wrin T, Little SJ, Petropoulos CJ. (2003) Rapid evolution of the neutralizing antibody response to HIV type 1 infection. *Proc Natl Acad Sci USA* **100**: 4144–4149.
79. Wei X, Decker JM, Wang S, *et al.* (2003) Antibody neutralization and escape by HIV-1. *Nature* **422**: 307–312.

80. Lifson JD, Nowak MA, Goldstein S, *et al.* (1997) The extent of early viral replication is a critical determinant of the natural history of simian immunodeficiency virus infection. *J Virol* **71**: 9508–9514.
81. Watson A, Ranchalis J, Travis B, *et al.* (1997) Plasma viremia in macaques infected with simian immunodeficiency virus: Plasma viral load early in infection predicts survival. *J Virol* **71**: 284–290.
82. Hirsch VM, Dapolito G, Goeken R, Campbell BJ. (1995) Phylogeny and natural history of the primate lentiviruses, SIV and HIV. *Curr Opin Genet Dev* **5**: 798–806.
83. Miller RJ, Cairns JS, Bridges S, Sarver N. (2000) Human immunodeficiency virus and AIDS: Insights from animal lentiviruses. *J Virol* **74**: 7187–7195.
84. Kanki PJ, McLane MF, King NW, Jr., *et al.* (1985) Serologic identification and characterization of a macaque T-lymphotropic retrovirus closely related to HTLV-III. *Science* **228**: 1199–1201.
85. Emini EA, Graham DJ, Gotlib L, *et al.* (1993) HIV and multidrug resistance. *Nature* **364**: 679.
86. Letvin NL, Robinson S, Rohne D, *et al.* (2001) Vaccine-elicited V3 loop-specific antibodies in rhesus monkeys and control of a simian-human immunodeficiency virus expressing a primary patient human immunodeficiency virus type 1 isolate envelope. *J Virol* **75**: 4165–4175.
87. Pantophlet R, Burton DR. (2006) GP120: Target for neutralizing HIV-1 antibodies. *Annu Rev Immunol* **24**: 739–769.
88. Almond N, Jenkins A, Heath AB, *et al.* (1992) The genetic evolution of the envelope gene of simian immunodeficiency virus in cynomolgus macaques infected with a complex virus pool. *Virology* **191**: 996–1002.
89. Javaherian K, Langlois AJ, Schmidt S, *et al.* (1992) The principal neutralization determinant of simian immunodeficiency virus differs from that of human immunodeficiency virus type 1. *Proc Natl Acad Sci USA* **89**: 1418–1422.
90. Yuste E, Sanford HB, Carmody J, *et al.* (2006) Simian immunodeficiency virus engrafted with human immunodeficiency virus type 1 (HIV-1)-specific epitopes: Replication, neutralization, and survey of HIV-1-positive plasma. *J Virol* **80**: 3030–3041.
91. Schultz AM, Hu SL. (1993) Primate models for HIV vaccines. *Aids* **7** (Suppl 1): S161–170.

92. Staprans SI, Feinberg MB. (2004) The roles of nonhuman primates in the preclinical evaluation of candidate AIDS vaccines. *Expert Rev Vaccines* **3**: S5–32.
93. Gray RH, Wawer MJ, Brookmeyer R, *et al.* (2001) Probability of HIV-1 transmission per coital act in monogamous, heterosexual, HIV-1-discordant couples in Rakai, Uganda. *Lancet* **357**: 1149–1153.
94. Aidoo M, Otten RA, Rodriguez V, *et al.* (2007) Absence of SHIV infection in gut and lymph node tissues in rhesus monkeys after repeated rectal challenges following HIV-1 DNA/MVA immunizations. *Vaccine* **25**: 6474–6481.
95. McDermott AB, Mitchen J, Piaskowski S, *et al.* (2004) Repeated low-dose mucosal simian immunodeficiency virus SIVmac239 challenge results in the same viral and immunological kinetics as high-dose challenge: A model for the evaluation of vaccine efficacy in nonhuman primates. *J Virol* **78**: 3140–3144.
96. Otten RA, Adams DR, Kim CN, *et al.* (2005) Multiple vaginal exposures to low doses of R5 simian-human immunodeficiency virus: Strategy to study HIV preclinical interventions in nonhuman primates. *J Infect Dis* **191**: 164–173.
97. Wilson NA, Reed J, Napoe GS, *et al.* (2006) Vaccine-induced cellular immune responses reduce plasma viral concentrations after repeated low-dose challenge with pathogenic simian immunodeficiency virus SIVmac239. *J Virol* **80**: 5875–5885.
98. Bontrop RE, Watkins DI. (2005) MHC polymorphism: AIDS susceptibility in non-human primates. *Trends Immunol* **26**: 227–233.
99. Casimiro DR, Wang F, Schleif WA, *et al.* (2005) Attenuation of simian immunodeficiency virus SIVmac239 infection by prophylactic immunization with DNA and recombinant adenoviral vaccine vectors expressing Gag. *J Virol* **79**: 15547–15555.
100. McDermott AB, O'Connor DH, Fuenger S, *et al.* (2005) Cytotoxic T-lymphocyte escape does not always explain the transient control of simian immunodeficiency virus SIVmac239 viremia in adenovirus-boosted and DNA-primed Mamu-A*01-positive rhesus macaques. *J Virol* **79**: 15556–15566.
101. Mothe BR, Weinfurter J, Wang C, *et al.* (2003) Expression of the major histocompatibility complex class I molecule Mamu-A*01 is associated with

control of simian immunodeficiency virus SIVmac239 replication. *J Virol* 77: 2736–2740.

102. O'Connor DH, Mothe BR, Weinfurter JT, *et al.* (2003) Major histocompatibility complex class I alleles associated with slow simian immunodeficiency virus disease progression bind epitopes recognized by dominant acute-phase cytotoxic-T-lymphocyte responses. *J Virol* 77: 9029–9040.
103. Pal R, Venzon D, Letvin NL, *et al.* (2002) ALVAC-SIV-*gag-pol-env*-based vaccination and macaque major histocompatibility complex class I (A*01) delay simian immunodeficiency virus SIVmac-induced immunodeficiency. *J Virol* **76**: 292–302.
104. Yant LJ, Friedrich TC, Johnson RC, *et al.* (2006) The high-frequency major histocompatibility complex class I allele Mamu-B*17 is associated with control of simian immunodeficiency virus SIVmac239 replication. *J Virol* **80**: 5074–5077.
105. Joag SV, Stephens EB, Adams RJ, *et al.* (1994) Pathogenesis of SIVmac infection in Chinese and Indian rhesus macaques: Effects of splenectomy on virus burden. *Virology* **200**: 436–446.
106. Ling B, Veazey RS, Luckay A, *et al.* (2002) SIV(mac) pathogenesis in rhesus macaques of Chinese and Indian origin compared with primary HIV infections in humans. *Aids* **16**: 1489–1496.
107. Marthas ML, Lu D, Penedo MC, *et al.* (2001) Titration of an SIVmac251 stock by vaginal inoculation of Indian and Chinese origin rhesus macaques: Transmission efficiency, viral loads, and antibody responses. *AIDS Res Hum Retrovir* **17**: 1455–1466.
108. Reimann KA, Parker RA, Seaman MS, *et al.* (2005) Pathogenicity of simian-human immunodeficiency virus SHIV-89.6P and SIVmac is attenuated in cynomolgus macaques and associated with early T-lymphocyte responses. *J Virol* **79**: 8878–8885.
109. Diop OM, Gueye A, Dias-Tavares M, *et al.* (2000) High levels of viral replication during primary simian immunodeficiency virus SIVagm infection are rapidly and strongly controlled in African green monkeys. *J Virol* **74**: 7538–7547.
110. Onanga R, Kornfeld C, Pandrea I, *et al.* (2002) High levels of viral replication contrast with only transient changes in CD4(+) and CD8(+) cell numbers during the early phase of experimental infection with simian immunodeficiency virus SIVmnd-1 in *Mandrillus sphinx*. *J Virol* **76**: 10,256–10,263.

111. Pandrea I, Apetrei C, Dufour J, *et al.* (2006) Simian immunodeficiency virus SIVagm.sab infection of Caribbean African green monkeys: A new model for the study of SIV pathogenesis in natural hosts. *J Virol* **80**: 4858–4867.
112. Silvestri G, Fedanov A, Germon S, *et al.* (2005) Divergent host responses during primary simian immunodeficiency virus SIVsm infection of natural sooty mangabey and nonnatural rhesus macaque hosts. *J Virol* **79**: 4043–4054.
113. Souquiere S, Onanga R, Makuwa M, *et al.* (2009) Simian immunodeficiency virus types 1 and 2 (SIV mnd 1 and 2) have different pathogenic potentials in rhesus macaques upon experimental cross-species transmission. *J Gen Virol* **90**: 488–499.
114. Apetrei C, Gautam R, Sumpter B, *et al.* (2007) Virus subtype-specific features of natural simian immunodeficiency virus SIVsmm infection in sooty mangabeys. *J Virol* **81**: 7913–7923.
115. Pandrea I, Ribeiro RM, Gautam R, *et al.* (2008) Simian immunodeficiency virus SIVagm dynamics in African green monkeys. *J Virol* **82**: 3713–3724.
116. VandeWoude S, Apetrei C. (2006) Going wild: Lessons from naturally occurring T-lymphotropic lentiviruses. *Clin Microbiol Rev* **19**: 728–762.
117. Broussard SR, Staprans SI, White R, *et al.* (2001) Simian immunodeficiency virus replicates to high levels in naturally infected African green monkeys without inducing immunologic or neurologic disease. *J Virol* **75**: 2262–2275.
118. Kornfeld C, Ploquin MJ, Pandrea I, *et al.* (2005) Antiinflammatory profiles during primary SIV infection in African green monkeys are associated with protection against AIDS. *J Clin Invest* **115**: 1082–1091.
119. Pandrea I, Apetrei C, Gordon S, *et al.* (2007) Paucity of CD4+CCR5+ T cells is a typical feature of natural SIV hosts. *Blood* **109**: 1069–1076.
120. Gordon SN, Klatt NR, Bosinger SE, *et al.* (2007) Severe depletion of mucosal CD4+ T cells in AIDS-free simian immunodeficiency virus-infected sooty mangabeys. *J Immunol* **179**: 3026–3034.
121. Pandrea IV, Gautam R, Ribeiro RM, *et al.* (2007) Acute loss of intestinal CD4+ T cells is not predictive of simian immunodeficiency virus virulence. *J Immunol* **179**: 3035–3046.
122. Dunham R, Pagliardini P, Gordon S, *et al.* (2006) The AIDS resistance of naturally SIV-infected sooty mangabeys is independent of cellular immunity to the virus. *Blood* **108**: 209–217.

123. Wang Z, Metcalf B, Ribeiro RM, *et al.* (2006) Th-1-type cytotoxic CD8+ T-lymphocyte responses to simian immunodeficiency virus (SIV) are a consistent feature of natural SIV infection in sooty mangabeys. *J Virol* **80**: 2771–2783.
124. Hirsch VM. (2004) What can natural infection of African monkeys with simian immunodeficiency virus tell us about the pathogenesis of AIDS? *AIDS Rev* **6**: 40–53.
125. Onanga R, Souquiere S, Makuwa M, *et al.* (2006) Primary simian immunodeficiency virus SIVmnd-2 infection in mandrills (*Mandrillus sphinx*). *J Virol* **80**: 3301–3309.
126. Deeks SG, Kitchen CM, Liu L, *et al.* (2004) Immune activation set point during early HIV infection predicts subsequent CD4+ T-cell changes independent of viral load. *Blood* **104**: 942–947.
127. Giorgi JV, Hultin LE, McKeating JA, *et al.* (1999) Shorter survival in advanced human immunodeficiency virus type 1 infection is more closely associated with T lymphocyte activation than with plasma virus burden or virus chemokine coreceptor usage. *J Infect Dis* **179**: 859–870.
128. Hazenberg MD, Otto SA, Hamann D, *et al.* (2003) Depletion of naive CD4 T cells by CXCR4-using HIV-1 variants occurs mainly through increased T-cell death and activation. *Aids* **17**: 1419–1424.
129. Sodora DL, Silvestri G. (2008) Immune activation and AIDS pathogenesis. *Aids* **22**: 439–446.
130. Shibata R, Kawamura M, Sakai H, *et al.* (1991) Generation of a chimeric human and simian immunodeficiency virus infectious to monkey peripheral blood mononuclear cells. *J Virol* **65**: 3514–3520.
131. Sakuragi S, Shibata R, Mukai R, *et al.* (1992) Infection of macaque monkeys with a chimeric human and simian immunodeficiency virus. *J Gen Virol* **73** **(Pt 11)**: 2983–2987.
132. Li J, Lord CI, Haseltine W, *et al.* (1992) Infection of cynomolgus monkeys with a chimeric HIV-1/SIVmac virus that expresses the HIV-1 envelope glycoproteins. *J Acquir Immune Defic Syndr* **5**: 639–646.
133. Karlsson GB, Halloran M, Li J, *et al.* (1997) Characterization of molecularly cloned simian-human immunodeficiency viruses causing rapid CD4+ lymphocyte depletion in rhesus monkeys. *J Virol* **71**: 4218–4225.
134. Reimann KA, Li JT, Veazey R, *et al.* (1996) A chimeric simian/human immunodeficiency virus expressing a primary patient human immunodeficiency

virus type 1 isolate *env* causes an AIDS-like disease after *in vivo* passage in rhesus monkeys. *J Virol* **70**: 6922–6928.

135. Luciw PA, Pratt-Lowe E, Shaw KE, *et al.* (1995) Persistent infection of rhesus macaques with T-cell-line-tropic and macrophage-tropic clones of simian/human immunodeficiency viruses (SHIV). *Proc Natl Acad Sci USA* **92**: 7490–7494.
136. Bogers WM, Cheng-Mayer C, Montelaro RC. (2000) Developments in preclinical AIDS vaccine efficacy models. *Aids* **14** (Suppl 3): S141–151.
137. Etemad-Moghadam B, Karlsson GB, Halloran M, *et al.* (1998) Characterization of simian-human immunodeficiency virus envelope glycoprotein epitopes recognized by neutralizing antibodies from infected monkeys. *J Virol* **72**: 8437–8445.
138. Montefiori DC, Reimann KA, Wyand MS, *et al.* (1998) Neutralizing antibodies in sera from macaques infected with chimeric simian-human immunodeficiency virus containing the envelope glycoproteins of either a laboratory-adapted variant or a primary isolate of human immunodeficiency virus type 1. *J Virol* **72**: 3427–3431.
139. Baba TW, Liska V, Hofmann-Lehmann R, *et al.* (2000) Human neutralizing monoclonal antibodies of the IgG1 subtype protect against mucosal simian-human immunodeficiency virus infection. *Nat Med* **6**: 200–206.
140. Mascola JR, Stiegler G, VanCott TC, *et al.* (2000) Protection of macaques against vaginal transmission of a pathogenic HIV-1/SIV chimeric virus by passive infusion of neutralizing antibodies. *Nat Med* **6**: 207–210.
141. Shibata R, Igarashi T, Haigwood N, *et al.* (1999) Neutralizing antibody directed against the HIV-1 envelope glycoprotein can completely block HIV-1/SIV chimeric virus infections of macaque monkeys. *Nat Med* **5**: 204–210.
142. Amara RR, Villinger F, Altman JD, *et al.* (2001) Control of a mucosal challenge and prevention of AIDS by a multiprotein DNA/MVA vaccine. *Science* **292**: 69–74.
143. Belyakov IM, Hel Z, Kelsall B, *et al.* (2001) Mucosal AIDS vaccine reduces disease and viral load in gut reservoir and blood after mucosal infection of macaques. *Nat Med* **7**: 1320–1326.
144. Zhang Y, Hatse S, De Clercq E, Schols D. (2000) CXC-chemokine receptor 4 is not a coreceptor for human herpesvirus 7 entry into CD4(+) T cells. *J Virol* **74**: 2011–2016.

145. Harouse JM, Gettie A, Tan RC, *et al.* (1999) Distinct pathogenic sequela in rhesus macaques infected with CCR5 or CXCR4 utilizing SHIVs. *Science* **284**: 816–819.
146. Endo Y, Igarashi T, Nishimura Y, *et al.* (2000) Short- and long-term clinical outcomes in rhesus monkeys inoculated with a highly pathogenic chimeric simian/human immunodeficiency virus. *J Virol* **74**: 6935–6945.
147. Igarashi T, Brown CR, Byrum RA, *et al.* (2002) Rapid and irreversible CD4+ T-cell depletion induced by the highly pathogenic simian/human immunodeficiency virus SHIV(DH12R) is systemic and synchronous. *J Virol* **76**: 379–391.
148. Borsetti A, Baroncelli S, Maggiorella MT, *et al.* (2008) Viral outcome of simian-human immunodeficiency virus SHIV-89.6P adapted to cynomolgus monkeys. *Arch Virol* **153**: 463–472.
149. Joag SV, Li Z, Foresman L, *et al.* (1996) Chimeric simian/human immunodeficiency virus that causes progressive loss of CD4+ T cells and AIDS in pig-tailed macaques. *J Virol* **70**: 3189–3197.
150. Harouse JM, Gettie A, Eshetu T, *et al.* (2001) Mucosal transmission and induction of simian AIDS by CCR5-specific simian/human immunodeficiency virus SHIV(SF162P3). *J Virol* **75**: 1990–1995.
151. Hsu M, Buckner C, Harouse J, *et al.* (2003) Antigenic variations in the CD4 induced sites of the CCR5-tropic, pathogenic SHIVsf162p3 gp120 variants. *J Med Primatol* **32**: 211–217.
152. Hsu M, Ho SH, Balfe P, *et al.* (2005) A CCR5-tropic simian-HIV molecular clone capable of inducing AIDS in rhesus macaques. *J Acquir Immune Defic Syndr* **40**: 383–387.
153. Song RJ, Chenine AL, Rasmussen RA, *et al.* (2006) Molecularly cloned SHIV-1157ipd3N4: A highly replication-competent, mucosally transmissible R5 simian-human immunodeficiency virus encoding HIV clade C Env. *J Virol* **80**: 8729–8738.
154. Lederman MM, Veazey RS, Offord R, *et al.* (2004) Prevention of vaginal SHIV transmission in rhesus macaques through inhibition of CCR5. *Science* **306**: 485–487.
155. Ambrose Z, Boltz V, Palmer S, *et al.* (2004) *In vitro* characterization of a simian immunodeficiency virus-human immunodeficiency virus (HIV) chimera expressing HIV type 1 reverse transcriptase to study antiviral resistance in pigtail macaques. *J Virol* **78**: 13,553–13,561.

156. Uberla K, Stahl-Hennig C, Bottiger D, *et al.* (1995) Animal model for the therapy of acquired immunodeficiency syndrome with reverse transcriptase inhibitors. *Proc Natl Acad Sci USA* **92**: 8210–8214.
157. Hatziioannou T, Ambrose Z, Chung NP, *et al.* (2009) A macaque model of HIV-1 infection. *Proc Natl Acad Sci USA* **106**: 4425–4429.
158. Hatziioannou T, Princiotta M, Piatak M, Jr., *et al.* (2006) Generation of simian-tropic HIV-1 by restriction factor evasion. *Science* **314**: 95.
159. Kamada K, Igarashi T, Martin MA, *et al.* (2006) Generation of HIV-1 derivatives that productively infect macaque monkey lymphoid cells. *Proc Natl Acad Sci USA* **103**: 16,959–16,964.
160. Igarashi T, Donau OK, Imamichi H, *et al.* (2007) Although macrophage-tropic simian/human immunodeficiency viruses can exhibit a range of pathogenic phenotypes, a majority of isolates induce no clinical disease in immunocompetent macaques. *J Virol* **81**: 10,669–10,679.
161. Barouch DH, Santra S, Schmitz JE, *et al.* (2000) Control of viremia and prevention of clinical AIDS in rhesus monkeys by cytokine-augmented DNA vaccination. *Science* **290**: 486–492.
162. Shiver JW, Fu TM, Chen L, *et al.* (2002) Replication-incompetent adenoviral vaccine vector elicits effective anti-immunodeficiency-virus immunity. *Nature* **415**: 331–335.
163. Berman PW, Gregory TJ, Riddle L, *et al.* (1990) Protection of chimpanzees from infection by HIV-1 after vaccination with recombinant glycoprotein gp120 but not gp160. *Nature* **345**: 622–625.
164. Bruck C, Thiriart C, Fabry L, *et al.* (1994) HIV-1 envelope-elicited neutralizing antibody titres correlate with protection and virus load in chimpanzees. *Vaccine* **12**: 1141–1148.
165. Flynn NM, Forthal DN, Harro CD, *et al.* (2005) Placebo-controlled phase 3 trial of a recombinant glycoprotein 120 vaccine to prevent HIV-1 infection. *J Infect Dis* **191**: 654–665.
166. Gilbert PB, Peterson ML, Follmann D, *et al.* (2005) Correlation between immunologic responses to a recombinant glycoprotein 120 vaccine and incidence of HIV-1 infection in a phase 3 HIV-1 preventive vaccine trial. *J Infect Dis* **191**: 666–677.
167. Pitisuttithum P, Gilbert P, Gurwith M, *et al.* (2006) Randomized, double-blind, placebo-controlled efficacy trial of a bivalent recombinant

glycoprotein 120 HIV-1 vaccine among injection drug users in Bangkok, Thailand. *J Infect Dis* **194**: 1661–1671.

168. Hanson CV. (1994) Measuring vaccine-induced HIV neutralization: Report of a workshop. *AIDS Res Hum Retrovir* **10**: 645–648.
169. Mascola JR, Snyder SW, Weislow OS, *et al.* (1996) Immunization with envelope subunit vaccine products elicits neutralizing antibodies against laboratory-adapted but not primary isolates of human immunodeficiency virus type 1. The National Institute of Allergy and Infectious Diseases AIDS Vaccine Evaluation Group. *J Infect Dis* **173**: 340–348.
170. Giavedoni LD, Planelles V, Haigwood NL, *et al.* (1993) Immune response of rhesus macaques to recombinant simian immunodeficiency virus gp130 does not protect from challenge infection. *J Virol* **67**: 577–583.
171. Corbett M, Bogers WM, Heeney JL, *et al.* (2008) Aerosol immunization with NYVAC and MVA vectored vaccines is safe, simple, and immunogenic. *Proc Natl Acad Sci USA* **105**: 2046–2051.
172. Mooij P, Balla-Jhagjhoorsingh S, Beenhakker N, *et al.* (2009) Comparison of human and rhesus macaque T-cell responses elicited by boosting with NYVAC encoding HIV-1 clade C immunogens. *J Virol* **83**: 5881–5889.
173. Bart PA, Goodall R, Barber T, *et al.* (2008) EV01: A phase I trial in healthy HIV negative volunteers to evaluate a clade C HIV vaccine, NYVAC-C undertaken by the EuroVacc Consortium. *Vaccine* **26**: 3153–3161.
174. Harari A, Bart PA, Stohr W, *et al.* (2008) An HIV-1 clade C DNA prime, NYVAC boost vaccine regimen induces reliable, polyfunctional, and long-lasting T cell responses. *J Exp Med* **205**: 63–77.
175. McCormack S, Stohr W, Barber T, *et al.* (2008) EV02: A Phase I trial to compare the safety and immunogenicity of HIV DNA-C prime-NYVAC-C boost to NYVAC-C alone. *Vaccine* **26**: 3162–3174.
176. Moore JP, Cao Y, Qing L, *et al.* (1995) Primary isolates of human immunodeficiency virus type 1 are relatively resistant to neutralization by monoclonal antibodies to gp120, and their neutralization is not predicted by studies with monomeric gp120. *J Virol* **69**: 101–109.
177. Montefiori D, Sattentau Q, Flores J, *et al.* (2007) Antibody-based HIV-1 vaccines: Recent developments and future directions. *PLoS Med* **4**: e348.

178. Priddy FH, Brown D, Kublin J, *et al.* (2008) Safety and immunogenicity of a replication-incompetent adenovirus type 5 HIV-1 clade B *gag/pol/nef* vaccine in healthy adults. *Clin Infect Dis* **46**: 1769–1781.
179. Koff WC, Johnson PR, Watkins DI, *et al.* (2006) HIV vaccine design: Insights from live attenuated SIV vaccines. *Nat Immunol* 7: 19–23.
180. Watkins DI, Burton DR, Kallas EG, *et al.* (2008) Nonhuman primate models and the failure of the Merck HIV-1 vaccine in humans. *Nat Med* **14**: 617–621.
181. Fauci AS, Johnston MI, Dieffenbach CW, *et al.* (2008) HIV vaccine research: The way forward. *Science* **321**: 530–532.
182. Morgan C, Marthas M, Miller C, *et al.* (2008) The use of nonhuman primate models in HIV vaccine development. *PLoS Med* 5: e173.

CHAPTER 12

Lentiviral Integration and the Role of the Cellular Cofactors LEDGF/p75 and Transportin-SR2

Bart van Heertum*,†, Jan De Rijck† and Zeger Debyser†

INTRODUCTION

Infection with HIV-1 continues to be a major health problem worldwide. Current antiretroviral therapy can control but not eradicate HIV infection. Resistance development is urging a search for new targets for antiretrovirals. Drugs in the clinic target the viral enzymes reverse transcriptase, protease and integrase, as well as viral entry. Co-receptor inhibitors are the first examples of drugs targeting host proteins to block HIV infection. Because of its limited genetic make-up, HIV-1 uses the host-cell machinery to complete its life cycle. From entry to budding, cellular proteins assist viral replication. Identification and study of these virus–host interactions will reveal new targets for antiretroviral therapy. In this chapter, we take a closer look into the integration step of HIV-1 and explain the roles of two of the best studied integrase cofactors: LEDGF/p75 and Transportin-SR2. LEDGF/p75 targets the viral pre-integration complex to the host

*Corresponding author.
†Molecular Medicine, K.U. Leuven, Flanders, Belgium.

chromatin, and Transportin-SR2 is involved in nuclear import of the virus. We will discuss in detail our current knowledge about these cofactors and the potential to use their interactions with integrase as new targets for antiretroviral therapy.

ON COFACTORS AND RESTRICTION FACTORS

Much like other viruses, HIV-1 has only a limited genetic make-up (~10 kb), though it has to carry out numerous and diverse functions during its replication cycle. Consequently, the virus takes advantage of cellular proteins and cellular pathways to complete the different steps in its life cycle. For viral entry in the target cell, HIV-1 is dependent on the CD4 cellular receptor and a suitable co-receptor, mainly CXCR4 or CCR5. Several cellular proteins have been proposed to intervene during the different steps of the viral life cycle. Besides the cellular proteins that aid virus replication, referred to as cofactors, the cell also harbors proteins counteracting HIV-1 replication, the so-called restriction factors. One of the best studied cellular restriction factors is the cellular cytidine deaminase, APOBEC3G. The presence of APOBEC3G results in hypermutation of the viral genome. This activity is counteracted by the HIV-1 virion infectivity factor (Vif).[1] Besides this restriction at the level of reverse transcription, additional cellular defense mechanisms appear to target the incoming retroviral particle. The alpha spliced variant of tripartite interaction motif 5 (TRIM5α), a factor that confers resistance to HIV-1 infection in rhesus monkeys, is likely to be an important natural barrier for cross-species retroviral transmission.[2]

Integration, catalyzed by the viral integrase (IN) protein, is a key step in the life cycle of all retroviruses, including HIV. After completion of reverse transcription, the linear viral DNA is assembled together with IN and other viral and cellular proteins into the pre-integration complex (PIC).[3,4] Although purified HIV-1 IN is sufficient to catalyze the two spatially and temporally distinct reactions of 3'-processing and strand transfer of the integration reaction in the absence of any other cofactor, a number of studies indicate that cellular proteins participate in the integration of the reverse transcribed viral DNA in the host-cell genome.[5–8] The particular host-cell proteins that assist IN in establishing the provirus in a host chromosome are termed "integration cofactors." Several studies indicate that certain IN mutants show pleiotropic defects, implying an effect on other stages in the replication cycle than the integration step.[9,10] Therefore, integration cofactors may also play a role in steps other than the integration process.

Cellular cofactors of integration have been identified by various means: by *in vitro* reconstitution of the enzymatic activity of salt-stripped PICs,[11,12] by yeast two-hybrid screening,[13,14] and by co-immunoprecipitation (co-IP).[15,16] The discovery of candidate cofactors of integration was originally based on the *in vitro* reconstitution of enzymatic activity of salt-stripped PICs. PICs are defined as high molecular weight nucleoprotein complexes isolated from infected cells that can catalyze the integration of endogenous viral DNA into an exogenous recombinant target DNA *in vitro*.[17–21] Purified PICs from infected cells manifest two properties that are not reproduced in assays using recombinant integrase and synthetic DNA substrates: PICs efficiently insert both viral DNA ends into a target DNA in a concerted way,[21,22] whereas reactions with recombinant integrase mainly result in the insertion of a single viral DNA end in a single strand of the duplex target DNA[23,24] and PICs preferentially integrate the viral DNA intermolecularly into a target DNA, thereby avoiding suicidal intramolecular autointegration.[21] *In vitro* reactions using purified PICs therefore better resemble the *in vivo* situation. High salt treatment of purified PICs gives rise to integration defective complexes, the activity of which can be restored upon addition of host-cell cytoplasmic extracts, suggesting that cellular factors assisting retroviral integration are removed from the PICs by the relatively high salt concentrations. By protein fractionation and reconstitution of the salt stripped PICs, the cofactors BAF and HMGA1 were originally identified.[11,12] Alternative methods for identifying interacting partners of IN are the yeast two-hybrid assay[14] and co-IP.[15,16] Since both methods are based on different principles, they often yield different hits. In fact, each method has its specific advantages and disadvantages. Although yeast two-hybrid (Y2H) is very sensitive, it measures protein–protein interactions in the nucleus of yeast. Y2H is also notorious for a relative high rate of false positives. With co-IP, the natural HIV-1 target cells can be used instead of yeast, but the extraction procedures may interfere with protein–protein interactions. Co-IP is often used to validate initial hits from Y2H.

Many integrase cofactors have been proposed but for the majority of putative cofactors no clear role during *in vivo* integration has been demonstrated. An exception is the lens epithelium-derived growth factor/p75 (LEDGF/p75). This cofactor was identified by co-IP and Y2H,[13,15,25] has been intensively validated in cell culture experiments, and was shown to tether HIV-1 integrase to the chromosomes. Recently, a second cofactor, Transportin-SR2, identified by Y2H and confirmed by co-IP, was validated as a cofactor regulating PICs transport into the nucleus. For other cofactors, however, a direct role during integration remains to be shown. Barrier-to-autointegration factor (BAF) prevents suicidal autointegration *in vitro*,[26,27] the minimal integrase-interaction domain of integrase interactor 1 (INI1) can act as a specific transdominant inhibitor of the late steps of HIV-1 replication,[28,29] but the role of the endogenous protein for viral replication remains controversial. A last cofactor, the high mobility group chromosomal protein A1 (HMGA1), may affect transcription rather than integration.[30,31]

After a short overview of the integrase protein and the integration reaction, we will discuss the well-studied integrase cofactor LEDGF/p75 and the recently discovered cofactor Transportin-SR2 in more detail.

THE HIV-1 INTEGRASE PROTEIN AND INTEGRATION

HIV-1 IN is composed of three functional domains: the N-terminal domain (NTD), the catalytic core domain (CCD) and the C-terminal domain (CTD)[32–34] (Fig. 1). The CCD spans the most conserved region of IN and shows close structural homology to prokaryotic transposases.[35] The active site of the enzyme is comprised of three invariant acidic residues (the $DDX_{35}E$ motif) that coordinate a pair of Mg^{2+} cations during catalysis.[36] The NTD forms a three-helical bundle that folds around a zinc atom coordinated by His and Cys residues of an HHCC motif.[37] The CTD features an SH3-like fold, is rich in basic residues, and is likely involved in DNA binding.[38,39] The structure of the full length IN has not yet been elucidated, despite considerable efforts. The active form of IN is at least a tetramer,[40] and a plausible tetramer model for the enzyme was proposed based on a crystal structure of a two-domain fragment of HIV-1 IN containing the NTD and CCD.[41]

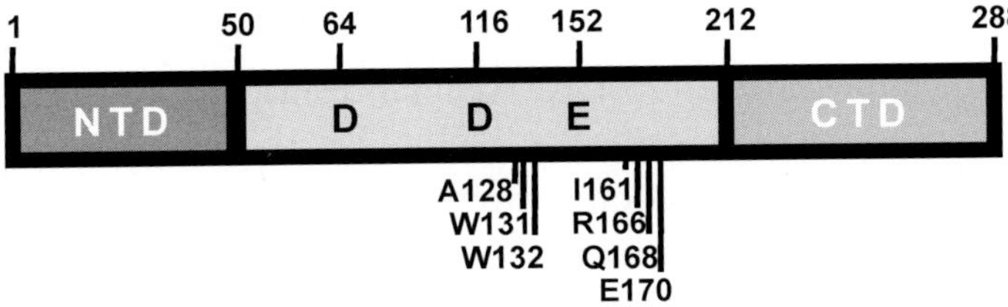

Fig. 1. Domain organization of HIV-1 IN. IN consists of three distinct domains[138]: the N-terminal zinc binding domain, the catalytic core containing the catalytically active DDE motif, and the positively charged C-terminal domain.[33,139] The latter displays DNA binding activity and (like the other two domains) is involved in multimerization.[140] Amino acids in the catalytic core that are important for interaction with LEDGF/p75 are indicated.[73,86,88,103,140]

HIV integration is a multistep process taking place in different compartments of the cell. It can be divided into the following steps: (i) formation of the integrase–viral DNA complex, (ii) 3′-processing of viral DNA ends, (iii) nuclear import of the PIC, (iv) target site selection, (v) strand transfer into the host chromosome, and (vi) repair of the single-stranded DNA gaps remaining at the viral–chromosomal DNA junctions after integration.

In the cytoplasm, the viral integrase removes a pGT dinucleotide from each 3' end of the viral long terminal repeats (LTRs), adjacent to a highly conserved CA dinucleotide. This reaction is called 3'-processing. Next, the PIC has to be transported into the nucleus. For most retroviruses, the PIC is unable to access the nucleus unless the host cell is in mitosis, possibly because nuclear envelope breakdown allows unrestricted passage of the PIC to the nucleus.[42] Since HIV can infect non-dividing cells like macrophages, dendritic cells and helper T cells, transport through the nuclear pore complex (NPC) is required to reach the host-cell chromosomes. The nuclear pore complex regulates cargo transport between the cytoplasm and the nucleus. Each vertebrate NPC is a ~120 MDa macromolecular assembly comprising ~30 different proteins, known as nucleoporins, that surround a central pore measuring ~40 nm in diameter in the nuclear envelope.[43] The passive diffusion size limit of the NPC is about 9 nm in diameter. Small molecules such as ions, metabolites and intermediate-sized macromolecules can pass the NPC by diffusion.[10] For the PIC, which has an estimated diameter of at least 50 nm,[4] passive diffusion is impossible and active cellular transport mechanisms are required.

The strand transfer reaction occurs in the nucleus following nuclear import of the PIC. The viral IN mediates a concerted nucleophilic attack by the 3′-hydroxyl residues of the viral DNA on phosphodiester bridges located on either side of the major groove in the target DNA. Next, the processed CA-3′-OH viral DNA ends are ligated to the 5′-O-phosphate ends of the target DNA. The 3′-ends of the target DNA remain gapped after the strand transfer reaction, forming an intermediate product in which the 5′-phosphate ends of the viral DNA are not attached to the 3′-OH ends of the host DNA.[44] The integration process is completed by cleavage of the unpaired dinucleotides from the 5′-ends of the viral DNA and repair of the single-stranded gaps created between the viral and target DNA. This repair is most likely mediated by cellular DNA repair enzymes.[45]

LENS EPITHELIUM DERIVED GROWTH FACTOR (LEDGF)/p75 TETHERS THE HIV-1 PRE-INTEGRATION COMPLEX TO THE CHROMOSOMES

Cherepanov *et al.* first reported LEDGF/p75 as a binding partner of HIV-1 IN following a study of the HIV-1 IN complexes present in nuclei of human cells that stably overexpress the viral integrase from a synthetic gene.[15] In a series of biochemical experiments, the authors demonstrated that the complexes isolated were associated with a cellular protein with an apparent molecular mass of 76 kDa. This novel IN interactor proved to be identical to the previously identified protein lens epithelium-derived growth factor (LEDGF/p75).

Prior to the elucidation of the link with HIV integration, both LEDGF/p75 and the highly homologous but shorter p52 had been independently identified by at least three groups. Both proteins were first identified as 75 kDa and 52 kDa proteins, respectively, that copurified with the transcriptional co-activator PC4, suggesting a role in transcriptional regulation.[46] Independently, a cDNA clone coding for a protein identical to p75 had been isolated from a lens epithelium cell (LEC) library using antibodies from a cataract patient.[47] This protein was named "lens epithelium-derived growth factor," and it was proposed that addition of the protein to the culture medium of LECs, cos7 cells, skin fibroblasts and keratinocytes stimulated their growth and prolonged cell survival. Although GFP-LEDGF was reported to be secreted in culture medium and subsequently internalized into cells, these observations could not be independently reproduced.[48] Based on sequence similarity, LEDGF/p75 is a member of the hepatoma-derived growth factor (HDGF) related protein (HRP) family (reviewed in Ref. 49). HRPs are characterized by a conserved N-terminal PWWP domain, a 90- to 135-amino acid module found in a variety of nuclear proteins.[50,51] Seven human HRP family members have been described: HDGF, HRP1, HRP2, HRP3, HRP4, LEDGF/p75, and LEDGF/p52,[52–54] of which two, LEDGF/p75 and HRP2, possess affinity for HIV-1 IN. LEDGF/p75 is also known as a common nuclear autoantigen in a variety of inflammatory conditions and autoimmune disorders, including atopic dermatitis, asthma, and interstitial cystitis.[55] Originally, the nuclear autoantigen was designated dense fine speckles 70 antigen (DFS70), based on the staining of dense fine speckles in the nucleus by the autoantibodies and the detection of a ~70 kDa protein by these antibodies.[56] Analysis of protein sequence databases revealed that DFS70 and the transcriptional co-activator p75 and LEDGF/p75 are identical. The pro-survival role of LEDGF/p75 is exerted via transcriptional activation of stress-related/anti-apoptotic proteins, such as heat-shock protein 27 (Hsp27), αB-crystallin, Hsp90 and antioxidant protein 2 (AOP2).[57] The cultivation of lens epithelial cells under heat or oxidative stress resulted in elevated levels of LEDGF mRNA and protein, associated with a higher level of stress-related proteins.[58] LEDGF/p75 can be cleaved by caspases-3 and -7, the main effector caspases in apoptosis, resulting in two cleavage fragments of 65 and 58 kDa.[59] Cellular over-expression of these cleaved LEDGF/p75 fragments clearly abrogated the pro-survival role of LEDGF/p75.[60] By preventing cells from undergoing apoptotic cell death, the protein seems to play a role in oncogenesis as well.[61–65] Through a link with the mixed-lineage leukemia (MLL) histone methyltransferase, LEDGF/p75 was recently shown to be essential for MLL-dependent transcription and leukemic transformation.[66] Therefore, LEDGF/p75 probably plays a key role in the balance between cell survival and cell death. The cellular protein is now generally referred to as LEDGF/p75, although the protein is probably not a growth factor and neither is it specific to lens epithelium. We therefore propose to use the acronym LEDGF/p75 without any longer referring to its original ill-chosen meaning.

LEDGF/p75 is predominantly localized in the nucleus, where it is intimately associated with the chromosomes.[67] It was proposed to have affinity for heat-shock and stress-related DNA elements.[68] The LEDGF gene (*PSIP1*) also encodes a smaller splice variant, p52, which shares a region of 325 residues with LEDGF/p75 at the N-terminus but contains eight additional C-terminal amino acids.[46] In contrast to LEDGF/p75, p52 has a stronger and more general transcriptional co-activator activity[46] and, interestingly, fails to interact with HIV-1 IN *in vitro* as well as in living cells.[69]

LEDGF/p75 contains 530 amino acids and the following functional domains (Fig. 2): (i) a PWWP domain of approximately 90 residues in the N-terminal part of LEDGF/p75 that probably functions as a chromatin interaction domain[51]; (ii) a functional nuclear localization signal (NLS), GRKRKAEKQ (amino acids: 148–156)[70]; (iii) two AT-hook motifs; and (iv) in accord with its ability to interact with HIV-1 IN, an evolutionary conserved integrase-binding domain (IBD) of approximately 80 amino acids (amino acids 347–429) mapped to the C-terminus.[52] An NMR solution structure of the IBD has been resolved.[71] The IBD is formed by a compact right-handed bundle composed of five α-helices topologically similar to a pair of HEAT repeats. Next, the crystal structure of an IN CCD dimer interacting with two IBD monomers was resolved.[72] The most critical interacting residues of the IBD, Ile-365, Asp-366 and Phe-406, are located in the interhelical loops of the IBD. In the integrase CCD dimer, two regions were identified to be involved in the interaction with LEDGF/p75. The first region centers around residues W131 and W132 while the second extends from I161 up to E170.[71,73] Recently, a crystal structure was resolved of the NTD and CCD domains of HIV-2 IN in complex with the IBD of LEDGF. This model elucidated the charge–charge interactions between the NTD and the IBD.[74] Together, these findings provide a structural basis for the IN–LEDGF interaction.

Interestingly, JPO2 and pogZ, two cellular protein-interaction partners of LEDGF/p75 have recently been identified by Y2H[75] and

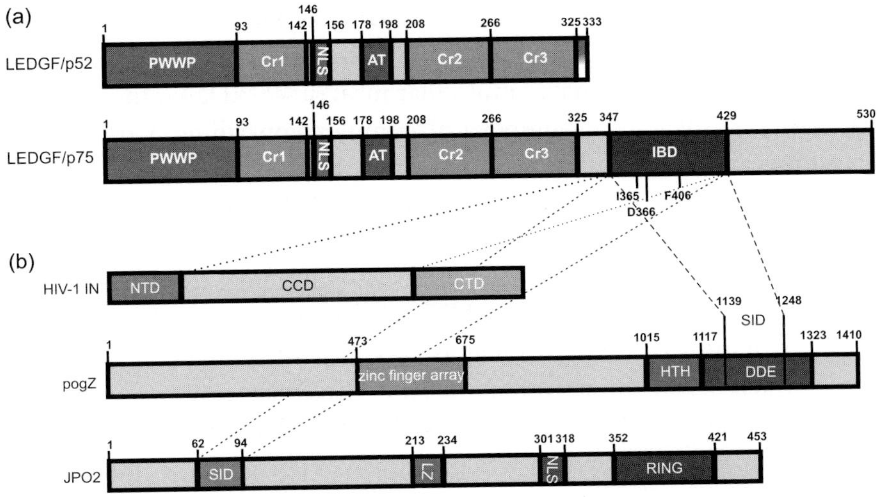

Fig. 2. Domain organization of LEDGF and cellular and viral interacting proteins. (a) The N-terminal part of LEDGF proteins, including the PWWP domain, two AT-hook motifs and charged regions (Cr) contribute to chromatin binding.[25,141,142] The PWWP domain most probably functions as the main chromatin tether. LEDGF proteins have a functional nuclear localization signal.[70,143] In the C-terminal part of the protein, an evolutionary conserved integrase binding domain (IBD) is present.[52] Amino acids important for interaction with HIV-1 IN are indicated. (b) Domain organization of interactors of LEDGF/p75. HIV-1 IN interacts with LEDGF/p75 through its CCD. PogZ contains a six-zinc finger array and in its C-terminal domain a DDE motif is present. The SID, part of the DDE motif, is responsible for the interaction with IBD.[77] JPO2 contains a leucine zipper (LZ), a NLS and a RING-finger domain. Interaction with IBD is mediated by a N-terminal SID.[76]

subsequently confirmed through co-IP.[76] They both interact with the IBD-domain of LEDGF/p75 (Fig. 2). It is worth noting that the interaction of JPO2 is not inhibited significantly by mutations of the Ile-365, Asp-366 and Phe-406 residues which abrogate the interaction with IN.[75,76] In contrast, PogZ shares more interacting IBD amino acids with IN; mutation of Ile-365 or Phe-406 abrogate its interaction with IBD. The Asp-366 mutation in the IBD is exclusively linked to a loss of interaction with lentiviral integrases.[77]

As mentioned above, LEDGF/p75 was originally identified as a cellular partner in HIV-1 IN complexes isolated from the nuclei of 293T cells stably expressing the viral integrase.[15] To date the precise stoichiometry of the nuclear IN–LEDGF complex has not been elucidated, but the simplest model, based on an estimated molecular mass of 400 kDa for the complex in isolation, suggests a symmetrical complex containing a pair of IN tetramers and two subunits of LEDGF/p75. HIV-1 IN is known to localize predominantly in the nucleus[78] and the distribution of nuclear IN perfectly matches with that of LEDGF/p75.[15,69] The IN–LEDGF/p75 direct interaction was confirmed by *in vitro* pull-down using the recombinant proteins. Moreover, the addition of recombinant LEDGF/p75 to an *in vitro* mini-HIV–based integrase assay clearly enhanced the strand transfer activity of the recombinant HIV-1 IN,[15] suggesting a contribution of LEDGF/p75 during the HIV-1 integration process. It is noteworthy that LEDGF/p75 was previously reported to be upregulated in HIV-infected cells.[79] The IN–LEDGF/p75 interaction was shown to be specific for lentiviridae and not a general characteristic of all retroviral integrases.[72,80,81]

To investigate in more detail the precise role of the interaction between LEDGF/p75 and HIV-1 IN, RNAi technology was employed to study the effect of a partial depletion of endogenous LEDGF/p75 on the cellular localization of HIV-1 IN and on the outcome of HIV-1 replication. Knock-down of endogenous LEDGF/p75 abolished the nuclear localization of HIV-1 IN as well as its association with chromosomes in cells transiently transfected with IN fused to eGFP.[69] Co-localization studies of different HIV-1 eGFP-IN deletion mutants in the absence or presence of over-expressed LEDGF/p75 revealed that both the N-terminal zinc binding domain and the core domain of HIV-1 IN are involved in the interaction with LEDGF/p75. In fact, the core domain proved to be the main determinant for the interaction since over-expression of LEDGF/p75 was able to restore nuclear/chromosomal localization of the IN core domain but not that of the N-terminus. Llano *et al.*[81] confirmed the requirement of LEDGF/p75 for the nuclear localization of integrases derived from HIV-1 and feline immunodeficiency virus (FIV). The knock-down of endogenous LEDGF/p75 by small interfering RNA (siRNA) did not affect the cytoplasmic distribution of the non-karyophilic integrase of MoMLV.

A second consequence of LEDGF/p75 knock-down was an enhanced proteosomal degradation of integrase. In fact, as earlier observations had suggested,[69] the presence of LEDGF/p75 increases the stability of HIV-1 integrase in the cell. Knock-down of LEDGF/p75 resulted in an increase in ubiquitinated HIV-1 IN that could be rescued by restoration of LEDGF/p75 levels.[82] This protection of HIV-1 integrase from the proteasome by LEDGF/p75 did not require chromatin tethering or nuclear localization of the IN–LEDGF complex. The only prerequisite for protection from proteasomal degradation was the interaction between LEDGF/p75 and HIV-1 IN. Finally, addition of a proteasome inhibitor to cells defective for LEDGF/p75 restored IN accumulation in the nucleus; nuclear LEDGF/p75 does apparently protect IN from proteolytic degradation in the nucleus.

Although LEDGF/p75 knock-down was associated with a clear phenotype of integrase re-localization, the effects on HIV-1 replication and/or lentiviral transduction initially led to conflicting results. In single-round infections with HIV-1 and FIV-derived vectors in stable cell lines depleted for LEDGF/p75, no difference in reporter gene expression was detected.[81] These findings were at odds with data by Emiliani *et al.*[13] and later Vandekerckhove *et al.*[83] Mutations in IN that abolish the interaction with p75 were identified using yeast two-hybrid analysis. Although the Q168A recombinant IN displayed normal integrase activity *in vitro*, viruses containing IN Q168A were defective for replication due to a specific block at the integration step,[13] while nuclear import was not hampered. In a strictly controlled RNAi knock-down study using stable shRNA LEDGF/p75 knock-down cell lines, Vandekerckhove *et al.* showed a reproducible two- to four-fold reduction in HIV-1 replication which was alleviated upon back-complementation of LEDGF/p75.[83] It was later shown that only a small amount of residual LEDGF/p75 in the cell is sufficient for HIV-1 replication.[84] Western blot analysis is usually the preferred technique to analyze the potency of an siRNA for knock-down of a gene. However, at least in the case of LEDGF/p75, a protein level that is undetectable by western blotting corresponds with enough protein to aid replication of HIV-1. With hindsight, the initial controversy surrounding the potential role of LEDGF/p75 in HIV replication was caused by unknown limitations and misinterpretation of RNAi technology. We should keep in mind that the same may hold true for other cellular integration cofactors whose roles have been invalidated using RNAi technology. For any abundant cellular protein, high level knock-down (preferably measured by western blotting and Q-RT-PCR) is required prior to making conclusions about the importance of the protein during integration.

Recently, Shun *et al.* reported on the generation of a knockout (KO) mouse for murine LEDGF/p75 and the isolation of KO mouse embryonic fibroblasts (LEDGF/p75 KO MEFs).[85] Although HIV-1 is not a murine virus, the domains in LEDGF/p75 important for its function in HIV-1 replication are 100 percent conserved in the murine orthologue. Infection of a single-round Luc-reporter HIV-1 virus was reduced to 1.5 percent of wild-type (WT) levels in the LEDGF/p75 KO MEFs, whereas MLV replication was not hampered at all. Reintroduction of human LEDGF/p75 in these cells alleviated the replication block whereas the D336N mutation, defective for interaction with integrase,[86] did not. Deletion mutants of LEDGF/p75 were analyzed for reconstitution of HIV-1 replication. While a PWWP-deletion mutant showed an 83 percent defect in replication, AT-hook-deletions or mutations showed no defect in replication. When deletions in both domains were introduced, no rescue of the block in HIV-1 replication was detected in the LEDGF/p75 KO MEFs. These data corroborate the previous findings of Llano *et al.* in their knock-down cell lines, where deletion of both PWWP and ATH was necessary for depletion of chromatin-association of LEDGF/p75.[84] QPCR analysis of HIV-1 derived DNA forms revealed equal amounts of total DNA and 2-LTR circles but provirus numbers were reduced to 8.5 percent of WT. These findings again argue against a block in nuclear import and for a role of LEDGF/p75 in the integration process.

Together, these RNAi-, knockout- and mutant-based experiments thus point to an important role of LEDGF/p75 in HIV replication. A separate approach was undertaken by stably over-expressing the IBD-domain of LEDGF/p75 fused to eGFP in HeLaP4 and MT4 cells. De Rijck *et al.* showed that by competing with endogenous LEDGF/p75 for IN binding a potent block of HIV-1 replication at the integration step was established.[87] In a follow-up virological study, Hombrouck *et al.* described and characterized virus strains selected to be resistant against this block in replication.[88]

Interestingly, the integrase mutations A128T and E170G at key positions in the LEDGF/p75–integrase interface resulted in a virus with lower affinity for the over-expressed C-terminal fragment of LEDGF/p75 while still recognizing the endogenous LEDGF/p75 but with a lower affinity. Importantly, the selected virus in the IBD cell lines remained dependent on LEDGF/p75 for its replication, at a cost of reduced replication fitness. Apart from corroborating the importance of the interaction between integrase and LEDGF/p75 for HIV-1 replication, these experiments also provide a proof-of-principle that the LEDGF/p75–integrase interaction constitutes a genuine anti-HIV-1 target. The study showed again the power of molecular evolution. We recommend the use of this approach in validating other cofactors of HIV replication.

LEDGF/p75 has been well characterized as a chromatin binding protein that tethers integrase to the chromatin and that stimulates integrase catalysis *in vitro*. Together with the data provided from the knockout, knock-down and transdominant inhibition experiments, the most important role for LEDGF/p75 at this point seems to be the tethering to, and selectively positioning of, the incoming (lentiviral) PIC on the chromosome. Data support a direct link between both activities; by increasing the affinity of integrase for DNA 40-fold, LEDGF/p75 may stimulate integration activity.[72] Moreover, the tendency of LEDGF/p75 for preferential interaction with transcriptionally active regions of the chromatin may indirectly lead to a statistically significant targeting to these regions. At present, we believe that tethering, targeting and increased integration are coupled activities of LEDGF/p75. Although a direct role in nuclear import is at odds with experimental data, a role of LEDGF/p75 in the stability of the PIC in human cells cannot be excluded.[88]

Fusion of the IBD to phage lambda repressor protein (lambdaR) could substantially redirect integration of HIV-1 integrase to sites near lambdaR-binding sites *in vitro*[89] consistent with the notion that the LEDGF/p75–integrase interaction is important for targeting HIV-1 integration to specific chromatin regions. This study also provided a proof-of-principle for possible redirection of HIV-1 integration. In time, this knowledge could lead to the construction of lentiviral vectors that can be selectively targeted to regions of choice in the chromosome, providing safe lentiviral vectors for gene therapy.

Early efforts to determine retroviral integration site preferences were done by analysis of 524 integration sites of wild-type or HIV-1 vector based integrations in SupT1 cells.[79] The analysis revealed that 69 percent of the HIV-1 integration sites reside in transcription units. Given that approximately 30 percent of the genome is transcribed, this highlights a strong preference for HIV-1 integration in actively transcribed regions. In a similar study, integration sites of HIV-1 and MLV were determined in HeLa cells.[90] The results for HIV-1 integration sites paralleled the ones identified by Schroder *et al.*[79] MLV integration sites showed a less pronounced preference towards transcriptional active regions, but preferred transcription start sites and CpG islands.[91,92] Interestingly, integration sites of other lentiviruses, such as EIAV and SIV, correspond well with those observed for HIV-1[93,94] in contrast to those of MLV and spumaretroviral foamy virus.[95,96] A recent publication revealed a positive association between HIV-1 integration sites and markers of transcriptionally active chromatin.[97]

Finally, a clear role for LEDGF/p75 in chromosomal positioning was provided by sequencing of integration sites in the LEDGF/p75 KO MEFs and the knock-down cell lines. Depletion of LEDGF/p75 showed a redistribution of HIV-1 integration sites from its characteristic distribution in transcription units (TUs) outside the promotor regions to a more aspecific distribution resembling integration patterns of MLV and ASLV.[85,98,99]

As it becomes more and more clear that LEDGF/p75 is a genuine cellular cofactor of HIV-1 replication, exerting its function through interaction with HIV-1 integrase, efforts have recently started to understand the exact mechanism through which LEDGF/p75 acts. In *in vitro* experiments using recombinant proteins, a stimulatory role of LEDGF/p75 on strand transfer as well as 3′-processing was shown. Yu *et al.* showed that the stimulatory effect by LEDGF/p75 was only possible when LEDGF/p75 was added to integrase before an integrase/donor DNA complex was formed[100] suggesting an engagement of integrase by LEDGF/p75 before it becomes actively bound to the reverse transcribed lentiviral genome in the cell. In another study, inhibition of pre-assembled complexes was not seen and a role for LEDGF/p75 in the viral life cycle after the formation of the pre-integration synaptic complexes was put forward.[101] Pandey *et al.* showed that the activation of the integration reaction by LEDGF/p75 was optimal when low molar ratios of LEDGF/p75 and IN were used.[102] Surprisingly, the IBD on its own was able to stimulate the integration reaction albeit at lower efficiency.

The protein–protein interaction between LEDGF/p75 and HIV-1 IN has a well-defined and compact area, as evidenced by crystal structures and mutagenesis studies.[13,52,71,73,74,86,88,103] This interaction hot spot defines a potential target for development of small molecule protein–protein interaction inhibitors (SMPPIIs). SMPPIIs blocking interactions between host and HIV proteins can become a valuable addition to the current regime of antiretrovirals. Prior to the identification of LEDGF/p75 as a binding partner of HIV-1 IN, a pocket in integrase was identified that could accommodate a small molecule inhibitor.[104] Using X-ray crystallography, compounds were identified that bind at the dimer interface of the HIV-1 IN catalytic domain, in a small cleft about 5 Å deep. Two compounds were found and defined as integrase inhibitors, but only one weakly inhibited *in vitro* enzymatic activity. The interaction of the IBD with HIV-1 IN overlaps with this small binding cleft, making it a favorable target for SMPPIIs. Using photoaffinity and mass spectrometry the same binding site in integrase was identified; coumarin derivatives were shown to bind as well to this region of IN.[105] Hou *et al.* described the further elaboration of an AlphaScreen interaction test, originally described by De Rijck *et al.*,[87] into a high throughput screening assay. 700,000 compounds were analyzed and several compounds showing *in vitro* activity in the low micromolar range were identified.[106] It is unclear whether these compounds demonstrate antiviral activity. Du *et al.* have recently reported a small molecular compound called D77 that potently inhibits the IN–LEDGF/p75 reaction in both a yeast and mammalian two-hybrid assay and that affects nuclear distribution of eYFP-integrase, possibly by blocking the interaction between EYFP-IN and LEDGF/p75.[107] D77, a benzoic acid derivative, also exhibited antiretroviral activity in MT-4 cells in the micromolar range and interacts with the IN core domain in a surface plasmon resonance assay, supporting the feasibility of a small molecule inhibiting HIV replication by blocking the IN–LEDGF/p75 interaction. However, whether this activity can indeed be attributed to the disruption of the IN–LEDGF/p75 interaction awaits further confirmation. These studies provide a proof-of-concept for the identification of genuine integrase–LEDGF/p75 inhibitors.

TRANSPORTIN-SR2 IMPORTS HIV-1 INTO THE NUCLEUS

Lentiviruses such as HIV-1 have the capacity to infect non-dividing cells.[108–110] The viral DNA, packed in the PIC, is recognized by the cellular nuclear transport machinery and traverses the nuclear envelope in an active and energy-dependent manner to integrate into human chromatin. Although nuclear import is a critical step in the lentiviral replication cycle and the search for the import factor guiding the PIC into the nucleus has been a long-standing focus of interest, this research field has remained highly controversial for a long time. In addition to different known cellular import factors, three HIV-1 PIC-associated proteins [matrix (MA), viral protein R (Vpr), and integrase (IN)] and the DNA-flap have been put forward to play a role in this process (for reviews, see Refs. 111 and 112).

IN was originally claimed to contain a bipartite NLS in the C-terminal domain.[113] Later studies, however, identified a non-classical NLS sequence in the catalytic core domain.[114] Classical *in vitro* import assays with digitonin-permeabilized cells resulted in conflicting hypotheses. Whereas some described an ATP-dependent import mechanism not involving any known karyophylins, others accredited the nuclear import of IN to an importin α- or importin α/β-dependent mechanism.[115] However, all different theories have remained unproven so far and lack final confirmation in cellular HIV replication experiments. Instead of studying the nuclear import of isolated viral proteins, Fassati *et al.* used reverse-transcription complexes (RTCs) in digitonin-permeabilized cells. The inhibition of HIV replication but not that of MoMLV by siRNA-mediated knock-down of importin 7 (Imp7) suggested a role in mediating the RTC nuclear import.[116] Although these results could not be confirmed in macrophages,[117] Imp7 was identified as a cellular binding partner of HIV-1 integrase by co-IP. Even though strong siRNA-mediated knock-down of Imp7 was achieved, only modest effects on HIV-1 replication were observed, leaving space for speculation as to whether Imp7 is indeed a physiological import factor of HIV-1.[118] Christ *et al.* performed a Y2H screen in order to identify novel interaction partners of HIV IN, with special interest toward members of the nuclear-import factor family. In this study, Transportin-SR2 (TRN-SR2, TNPO3) was identified as a karyophilic interaction partner of IN.[119] Three full genome siRNA screens were recently performed to identify cellular proteins influencing viral replication.[120–122] In two screens, TRN-SR2 was also identified as an important cofactor of HIV replication.[120,121]

The 975-amino acid Transportin-SR (TRN-SR1) (Fig. 3) was first identified in a Y2H screen for proteins binding to the arginine/serine (RS)-rich domain of SR (serine/arginine rich) proteins by Kataoka *et al.* The presence of this domain, containing numerous arginine/serine dipeptide repeats, characterizes the protein as an essential mRNA splicing factor.[123] TRN-SR1 proteins were shown to be capable of binding several SR proteins and to be responsible for the nuclear import of RS-domain containing fusion proteins, suggesting that TRN-SR1 is part of a common import pathway used by many of the SR proteins. In a different study, Lai *et al.* used a Y2H screen to find interaction partners for the RS-rich hinge region of the E2 human papillomavirus type 5 protein. In this study TRN-SR2, almost identical to the earlier described TRN-SR1, was identified as a binding partner of the E2 protein.[124] TRN-SR2 contains only 923 amino acids (Fig. 3), lacking two regions of 30 amino acids, compared to the 975 amino acids of TRN-SR1. TRN-SR2 is also capable of binding several SR proteins through an interaction with phosphorylated RS domains in these proteins. The interaction is disrupted by the binding of Ran-GTP.[124,125] TRN-SR1 and TRN-SR2 are alternatively spliced

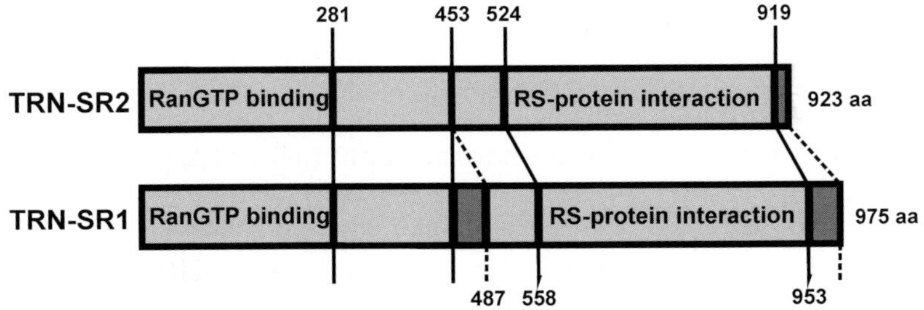

Fig. 3. Two splice variants of the *TNPO3* gene. TRN-SR2 is 52 amino acids shorter than the TRN-SR1 protein, missing 34 amino acids at position 453–487 and 18 amino acids at the C-terminus. A RanGTP binding motif is present in the N-terminus and an RS-protein interaction motif in the C-terminus of both proteins.[123,124] No domain organization is known for TRN-SR proteins.

products from the same gene, *tnpo3*, which could give rise to a different distribution pattern in mammalian cells. TRN-SR2 is predominantly expressed in most cells, with a higher abundance in testis, while the TRN-SR1 splice variant remained undetected.[126] Surprisingly, Lai *et al.* identified multiple non-SR proteins in another Y2H screen as binding partners of TRN-SR2; the most prominent was RNA binding motif protein 4 (RBM4), a zinc finger protein involved in the regulation of splicing.[127]

The effect on HIV replication was evaluated by siRNA and shRNA mediated knock-down of TRN-SR2; HIV-1 and HIV-2 replication were severely impaired in both dividing and non-dividing cells including primary macrophages. A small effect was seen with MLV replication, pointing towards a possible lentivirus-specific role for TRN-SR2.[119] Effects on RNA splicing and viral production were excluded. Instead, the effect of TRN-SR2 knock-down on HIV replication was pinpointed to the nuclear import step by Q-PCR analysis of viral DNA species in knock-down cells; a reduction in 2-LTR circles was observed leading to a reduction in viral integration. In addition, by tracking fluorescently labeled virus during HIV infection,[128] a direct role for TRN-SR2 in nuclear import of the HIV PIC was corroborated[119] (Fig. 4).

Various import factors have been implicated in the nuclear import of HIV.[111,112] Transportin-SR2 is the only import factor to date that has been identified in several screens[120,121] and of which the relevance for import has been corroborated in a cellular import assay.[119] The effect of other potential import factors on HIV replication has been tested in digitonin-permeabilized cells; these studies only provide limited information because these import assays use recombinant IN. Therefore, it is not surprising that several karyopherins have been put forward as putative import factors of HIV-1 PICs.[113,115,129] Only RNAi mediated knock-down of importin 7 achieved inhibition of HIV-1 replication.[129] The strong effect on HIV-1 replication seen upon knock-down of TRN-SR2 can be explained by the observation that nuclear import is the rate limiting step during HIV-1 replication. This strong replication defect provides a rationale for the identification of SMPPIIs interfering with loading of the PIC onto this nuclear import factor.

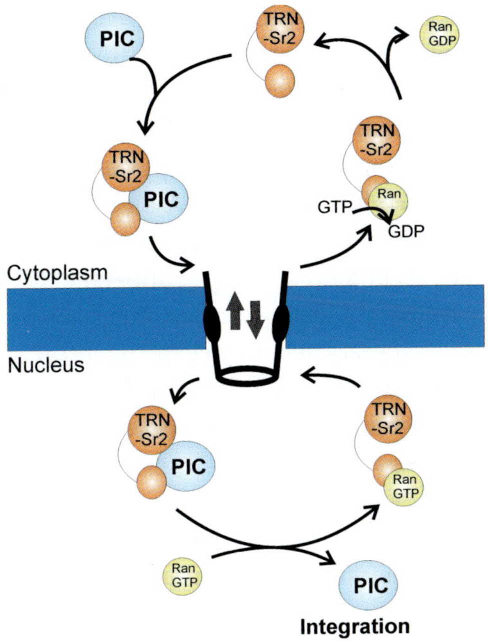

Fig. 4. TRN-SR2 mediated import model for HIV. TRN-SR2 binds to HIV-1 IN present in the PIC and shuttles the protein DNA complex through the nuclear pore into the nucleus. In the nucleus, TRN-SR2 releases its cargo upon RanGTP binding and the TRN-SR2/RanGTP is exported to the cytoplasm where RanGTP is hydrolyzed to RanGDP, returning TRN-SR2 to the state in which it can bind to its cargo again.

CONCLUSIONS

The increasing knowledge about the role of cellular proteins during HIV replication has opened up a new area of research in virology. Several new methods to identify and validate cofactors have been introduced and with these techniques an increasing number of cellular co- and restriction factors have been identified.[7,120–122,130] Co-IP and mass spectrometry can identify cellular proteins physically interacting with viral proteins. Genome wide Y2H screening can be used to identify direct interaction partners of viral proteins. Recently, functional genomics screens based on RNAi have emerged as a valuable tool for identification of proteins involved in a specific pathway. With all these screening techniques in place, researchers have powerful tools at hand to identify new cofactors of HIV-1 IN. Still, each method has some disadvantages, shown for instance by the absence of the well-established cofactor LEDGF/p75 in three RNAi screens.[120–122] The new cofactor of HIV-1 replication, TRN-SR2, was identified in two out of three functional siRNA screens. Moreover, care must be taken because further analysis is definitely required to validate each cofactor and evaluate its potential as future anti-HIV drug target. Whereas LEDGF/p75 was originally identified as an integrase binding partner in 2003,[69,131] it took several years to convince the scientific community about the role of LEDGF/p75 in HIV replication and to convince pharmaceutical companies about the validity of the IN–LEDGF/p75 interaction as a genuine antiviral target.[7]

Proof-of-principle for targeting protein–protein interactions between HIV and cellular proteins in antiviral treatment comes from a new class of HIV inhibitors on the market, namely enfuvirtide and maraviroc. These compounds are classified as entry inhibitors. Although enfuvirtide and maraviroc are not protein–protein interaction inhibitors *pur sang*, since they interfere with ligand–receptor interactions, these therapeutics illustrate the great potential of blocking the interaction between viral and cellular proteins for antiviral therapy. Enfuvirtide (T-20, Fuzeon) binds to the viral gp41 protein and inhibits the conformational changes needed for fusion of the virus and cell membrane. Therefore, it is also referred to as a fusion inhibitor.[132] In September 2007, maraviroc (Celsentri) was approved by the FDA as an entry inhibitor and chemokine receptor antagonist. This small molecule inhibitor specifically blocks the chemokine receptor CCR5, preventing its interaction with gp120.[133] Other HIV entry inhibitors like monoclonal antibodies and small molecules are being developed and analyzed in clinical trials (for a review, see Ref. 134). Several other interactions between host and viral proteins in different stages of HIV-1 replication are currently being investigated for their therapeutic potential. These include the interaction of capsid with cyclophillin A[135] and the interaction between Vif and APOBEC3G.[136]

What makes a particular protein–protein interaction an attractive target for drug discovery? Several points need to be addressed: (i) the host protein has to be validated as an important cofactor of HIV-1 replication,

(ii) inhibition of the interaction between the viral and cellular protein should not be associated with toxicity, (iii) structural information about the interaction surface should be available, and (iv) the identification of a genuine and selective inhibitor with antiviral activity in cell culture finally proves the concept.

When targeting cellular cofactors, cellular toxicity should be investigated. So far, knock-down of LEDGF/p75 has not been associated with apparent toxicity in human cell lines. LEDGF/p75 appears to be important during embryogenesis, since the majority of LEDGF/p75-null mice die prenatally; the survivors display a range of abnormal phenotypes, compatible with defects in homeodomain proteins.[137] However, in LEDGF/p75-null mice, both LEDGF/p75 and LEDGF/p52 are depleted, whereas an HIV therapeutic strategy would only target IN or LEDGF/p75. Indeed, HIV drug discovery aims at inhibiting protein–protein interaction with HIV-1 IN without affecting the cellular function of the cofactor. Based on our experience gained by studying the interaction between HIV-1 IN and LEDGF/p75 and the early discovery of specific inhibitors targeting the interaction between viral and host protein, we believe that the new paradigm of cofactor inhibitors (cofins) awaits a bright future in antiviral therapy.

REFERENCES

1. Sheehy AM, Gaddis NC, Choi JD, Malim MH. (2002) Isolation of a human gene that inhibits HIV-1 infection and is suppressed by the viral Vif protein. *Nature* **418**: 646–650.
2. Stremlau M, Owens CM, Perron MJ, *et al.* (2004) The cytoplasmic body component TRIM5alpha restricts HIV-1 infection in Old World monkeys. *Nature* **427**: 848–853.
3. Bowerman B, Brown PO, Bishop JM, Varmus HE. (1989) A nucleoprotein complex mediates the integration of retroviral DNA. *Genes Dev* **3**: 469–478.
4. Miller MD, Farnet CM, Bushman FD. (1997) Human immunodeficiency virus type 1 preintegration complexes: Studies of organization and composition. *J Virol* **71**: 5382–5390.
5. Ciuffi A, Bushman FD. (2006) Retroviral DNA integration: HIV and the role of LEDGF/p75. *Trends Genet* **22**: 388–395.

6. Turlure F, Devroe E, Silver PA, Engelman A. (2004) Human cell proteins and human immunodeficiency virus DNA integration. *Front Biosci* **9**: 3187–3208.
7. Van Maele B, Busschots K, Vandekerckhove L, *et al.* (2006) Cellular co-factors of HIV-1 integration. *Trends Biochem Sci* **31**: 98–105.
8. Vandegraaff N, Engelman A. (2007) Molecular mechanisms of HIV integration and therapeutic intervention. *Expert Rev Mol Med* **9**: 1–19.
9. Wilkinson TA, Januszyk K, Phillips ML, *et al.* (2009) Identifying and characterizing a functional HIV-1 reverse transcriptase-binding site on integrase. *J Biol Chem* **284**: 7931–7939. doi 10.1074/jbc.M806241200.
10. Zhu K, Dobard C, Chow SA. (2004) Requirement for integrase during reverse transcription of human immunodeficiency virus type 1 and the effect of cysteine mutations of integrase on its interactions with reverse transcriptase. *J Virol* **78**: 5045–5055.
11. Lee MS, Craigie R. (1994) Protection of retroviral DNA from autointegration: Involvement of a cellular factor. *Proc Natl Acad Sci USA* **91**: 9823–9827.
12. Farnet CM, Bushman FD. (1997) HIV-1 cDNA integration: Requirement of HMG I(Y) protein for function of preintegration complexes *in vitro*. *Cell* **88**: 483–492.
13. Emiliani S, Mousnier A, Busschots K, *et al.* (2005) Integrase mutants defective for interaction with LEDGF/p75 are impaired in chromosome tethering and HIV-1 replication. *J Biol Chem* **280**: 25517–25523.
14. Kalpana GV, Marmon S, Wang W, *et al.* (1994) Binding and stimulation of HIV-1 integrase by a human homolog of yeast transcription factor SNF5. *Science* **266**: 2002–2006.
15. Cherepanov P, Maertens G, Proost P, *et al.* (2003) HIV-1 integrase forms stable tetramers and associates with LEDGF/p75 protein in human cells. *J Biol Chem* 2003, **278**: 372–381.
16. Lin CW, Engelman A. (2003) The barrier-to-autointegration factor is a component of functional human immunodeficiency virus type 1 preintegration complexes. *J Virol* **77**: 5030–5036.
17. Fujiwara T, Mizuuchi K. (1988) Retroviral DNA integration: Structure of an integration intermediate. *Cell* **54**: 497–504.
18. Brown PO, Bowerman B, Varmus HE, Bishop JM. (1987) Correct integration of retroviral DNA *in vitro*. *Cell* **49**: 347–356.

19. Brown PO, Bowerman B, Varmus HE, Bishop JM. (1989) Retroviral integration: Structure of the initial covalent product and its precursor, and a role for the viral IN protein. *Proc Natl Acad Sci USA* **86**: 2525–2529.
20. Ellison V, Abrams H, Roe T, *et al.* (1990) Human immunodeficiency virus integration in a cell-free system. *J Virol* **64**: 2711–2715.
21. Farnet CM, Haseltine WA. (1990) Integration of human immunodeficiency virus type 1 DNA *in vitro*. *Proc Natl Acad Sci USA* **87**: 4164–4168.
22. Lee YM, Coffin JM. (1991) Relationship of avian retrovirus DNA synthesis to integration *in vitro*. *Mol Cell Biol* **11**: 1419–1430.
23. Bushman FD, Craigie R. (1991) Activities of human immunodeficiency virus (HIV) integration protein *in vitro*: Specific cleavage and integration of HIV DNA. *Proc Natl Acad Sci USA* **88**: 1339–1343.
24. Craigie R, Fujiwara T, Bushman F. (1990) The IN protein of Moloney murine leukemia virus processes the viral DNA ends and accomplishes their integration *in vitro*. *Cell* **62**: 829–837.
25. Turlure F, Maertens G, Rahman S, *et al.* (2006) A tripartite DNA-binding element, comprised of the nuclear localization signal and two AT-hook motifs, mediates the association of LEDGF/p75 with chromatin *in vivo*. *Nucl Acid Res* **34**: 1663–1675.
26. Lee MS, Craigie R. (1998) A previously unidentified host protein protects retroviral DNA from autointegration. *Proc Natl Acad Sci USA* **95**: 1528–1533.
27. Chen H, Engelman A. (1998) The barrier-to-autointegration protein is a host factor for HIV type 1 integration. *Proc Natl Acad Sci USA* **95**: 15270–15274.
28. Yung E, Sorin M, Pal A, *et al.* (2001) Inhibition of HIV-1 virion production by a transdominant mutant of integrase interactor 1. *Nat Med* **7**: 920–926.
29. Yung E, Sorin M, Wang EJ, *et al.* (2004) Specificity of interaction of INI1/hSNF5 with retroviral integrases and its functional significance. *J Virol* **78**: 2222–2231.
30. Henderson A, Bunce M, Siddon N, *et al.* (2000) High-mobility-group protein I can modulate binding of transcription factors to the U5 region of the human immunodeficiency virus type 1 proviral promoter. *J Virol* **74**: 10523–10534.
31. Henderson A, Holloway A, Reeves R, Tremethick DJ. (2004) Recruitment of SWI/SNF to the human immunodeficiency virus type 1 promoter. *Mol Cell Biol* **24**: 389–397.

32. Bushman FD, Engelman A, Palmer I, *et al.* (1993) Domains of the integrase protein of human immunodeficiency virus type 1 responsible for polynucleotidyl transfer and zinc binding. *Proc Natl Acad Sci USA* **90**: 3428–3432.
33. Engelman A, Bushman FD, Craigie R. (1993) Identification of discrete functional domains of HIV-1 integrase and their organization within an active multimeric complex. *EMBO J* **12**: 3269–3275.
34. Engelman A, Hickman AB, Craigie R. (1994) The core and carboxyl-terminal domains of the integrase protein of human immunodeficiency virus type 1 each contribute to nonspecific DNA binding. *J Virol* **68**: 5911–5917.
35. Dyda F, Hickman AB, Jenkins TM, *et al.* (1994) Crystal structure of the catalytic domain of HIV-1 integrase: Similarity to other polynucleotidyl transferases. *Science* **266**: 1981–1986.
36. Goldgur Y, Dyda F, Hickman AB, *et al.* (1998) Three new structures of the core domain of HIV-1 integrase: An active site that binds magnesium. *Proc Natl Acad Sci USA* **95**: 9150–9154.
37. Cai M, Zheng R, Caffrey M, *et al.* (1997) Solution structure of the N-terminal zinc binding domain of HIV-1 integrase. *Nat Struct Biol* **4**: 567–577.
38. Eijkelenboom AP, Lutzke RA, Boelens R, *et al.* (1995) The DNA-binding domain of HIV-1 integrase has an SH3-like fold. *Nat Struct Biol* **2**: 807–810.
39. Lodi PJ, Ernst JA, Kuszewski J, *et al.* (1995) Solution structure of the DNA binding domain of HIV-1 integrase. *Biochemistry* **34**: 9826–9833.
40. Faure A, Calmels C, Desjobert C, *et al.* (2005) HIV-1 integrase crosslinked oligomers are active *in vitro*. *Nucleic Acids Res* **33**: 977–986.
41. Wang JY, Ling H, Yang W, Craigie R. (2001) Structure of a two-domain fragment of HIV-1 integrase: Implications for domain organization in the intact protein. *EMBO J* **20**: 7333–7343.
42. Roe T, Reynolds TC, Yu G, Brown PO. (1993) Integration of murine leukemia virus DNA depends on mitosis. *EMBO J* **12**: 2099–2108.
43. Beck M, Forster F, Ecke M, *et al.* (2004) Nuclear pore complex structure and dynamics revealed by cryoelectron tomography. *Science* **306**: 1387–1390.
44. Engelman A, Mizuuchi K, Craigie R. (1991) HIV-1 DNA integration: Mechanism of viral DNA cleavage and DNA strand transfer. *Cell* **67**: 1211–1221.
45. Yoder KE, Bushman FD. (2000) Repair of gaps in retroviral DNA integration intermediates. *J Virol* **74**: 11191–11200.

46. Ge H, Si Y, Roeder RG. (1998) Isolation of cDNAs encoding novel transcription coactivators p52 and p75 reveals an alternate regulatory mechanism of transcriptional activation. *EMBO J* **17**: 6723–6729.
47. Singh DP, Ohguro N, Kikuchi T, *et al.* (2000) Lens epithelium-derived growth factor: Effects on growth and survival of lens epithelial cells, keratinocytes, and fibroblasts. *Biochem Biophys Res Commun* **267**: 373–381.
48. Singh DP, Ohguro N, Chylack LT, Jr, Shinohara T. (1999) Lens epithelium-derived growth factor: Increased resistance to thermal and oxidative stresses. *Invest Ophthalmol Vis Sci* **40**: 1444–1451.
49. Dietz F, Franken S, Yoshida K, *et al.* (2002) The family of hepatoma-derived growth factor proteins: Characterization of a new member HRP-4 and classification of its subfamilies. *Biochem J* **366**: 491–500.
50. Qiu C, Sawada K, Zhang X, Cheng X. (2002) The PWWP domain of mammalian DNA methyltransferase Dnmt3b defines a new family of DNA-binding folds. *Nat Struct Biol* **9**: 217–224.
51. Stec I, Nagl SB, van Ommen GJ, den Dunnen JT. (2000) The PWWP domain: A potential protein–protein interaction domain in nuclear proteins influencing differentiation? *FEBS Lett* **473**: 1–5.
52. Cherepanov P, Devroe E, Silver PA, Engelman A. (2004) Identification of an evolutionarily conserved domain in human lens epithelium-derived growth factor/transcriptional co-activator p75 (LEDGF/p75) that binds HIV-1 integrase. *J Biol Chem* **279**: 48883–48892.
53. Ikegame K, Yamamoto M, Kishima Y, *et al.* (1999) A new member of a hepatoma-derived growth factor gene family can translocate to the nucleus. *Biochem Biophys Res Commun* **266**: 81–87.
54. Izumoto Y, Kuroda T, Harada H, *et al.* (1997) Hepatoma-derived growth factor belongs to a gene family in mice showing significant homology in the amino terminus. *Biochem Biophys Res Commun* **238**: 26–32.
55. Ganapathy V, Casiano CA. (2004) Autoimmunity to the nuclear autoantigen DFS70 (LEDGF): What exactly are the autoantibodies trying to tell us? *Arthritis Rheum* **50**: 684–688.
56. Ochs RL, Muro Y, Si Y, *et al.* (2000) Autoantibodies to DFS 70 kd/transcription coactivator p75 in atopic dermatitis and other conditions. *J Allergy Clin Immunol* **105**: 1211–1220.
57. Shinohara T, Singh DP, Fatma N. (2002) LEDGF, a survival factor, activates stress-related genes. *Prog Retin Eye Res* **21**: 341–358.

58. Sharma P, Singh DP, Fatma N, *et al.* (2000) Activation of LEDGF gene by thermal-and oxidative-stresses. *Biochem Biophys Res Commun* **276**: 1320–1324.
59. Brown-Bryan TA, Leoh LS, Ganapathy V, *et al.* (2008) Alternative splicing and caspase-mediated cleavage generate antagonistic variants of the stress oncoprotein LEDGF/p75. *Mol Cancer Res* **6**: 1293–1307.
60. Wu X, Daniels T, Molinaro C, *et al.* (2002) Caspase cleavage of the nuclear autoantigen LEDGF/p75 abrogates its pro-survival function: Implications for autoimmunity in atopic disorders. *Cell Death Differ* **9**: 915–925.
61. Daniels T, Zhang J, Gutierrez I, *et al.* (2005) Antinuclear autoantibodies in prostate cancer: Immunity to LEDGF/p75, a survival protein highly expressed in prostate tumors and cleaved during apoptosis. *Prostate* **62**: 14–26.
62. Daugaard M, Kirkegaard-Sorensen T, Ostenfeld MS, *et al.* (2007) Lens epithelium-derived growth factor is an Hsp70-2 regulated guardian of lysosomal stability in human cancer. *Cancer Res* **67**: 2559–2567.
63. Grand FH, Koduru P, Cross NC, Allen SL. (2005) NUP98-LEDGF fusion and t(9;11) in transformed chronic myeloid leukemia. *Leuk Res* **29**: 1469–1472.
64. Hussey DJ, Moore S, Nicola M, Dobrovic A. (2001) Fusion of the NUP98 gene with the LEDGF/p52 gene defines a recurrent acute myeloid leukemia translocation. *BMC Genet* **2**: 20.
65. Morerio C, Acquila M, Rosanda C, *et al.* (2005) t(9;11)(p22;p15) with NUP98-LEDGF fusion gene in pediatric acute myeloid leukemia. *Leuk Res* **29**: 467–470.
66. Yokoyama A, Cleary ML. (2008) Menin critically links MLL proteins with LEDGF on cancer-associated target genes. *Cancer Cell* **14**: 36–46.
67. Nishizawa Y, Usukura J, Singh DP, *et al.* (2001) Spatial and temporal dynamics of two alternatively spliced regulatory factors, lens epithelium-derived growth factor (LEDGF/p75) and p52, in the nucleus. *Cell Tissue Res* **305**: 107–114.
68. Singh DP, Fatma N, Kimura A, *et al.* (2001) LEDGF binds to heat shock and stress-related element to activate the expression of stress-related genes. *Biochem Biophys Res Commun* **283**: 943–955.
69. Maertens G, Cherepanov P, Pluymers W, *et al.* (2003) LEDGF/p75 is essential for nuclear and chromosomal targeting of HIV-1 integrase in human cells. *J Biol Chem* **278**: 33528–33539.

70. Maertens G, Cherepanov P, Debyser Z, *et al.* (2004) Identification and characterization of a functional nuclear localization signal in the HIV-1 integrase interactor LEDGF/p75. *J Biol Chem* **279**: 33421–33429.
71. Cherepanov P, Sun ZY, Rahman S, *et al.* (2005) Solution structure of the HIV-1 integrase-binding domain in LEDGF/p75. *Nat Struct Mol Biol* **12**: 526–532.
72. Busschots K, Vercammen J, Emiliani S, *et al.* (2005) The interaction of LEDGF/p75 with integrase is lentivirus-specific and promotes DNA binding. *J Biol Chem* **280**: 17841–17847.
73. Busschots K, Voet A, De Maeyer M, *et al.* (2007) Identification of the LEDGF/p75 binding site in HIV-1 integrase. *J Mol Biol* **365**: 1480–1492.
74. Hare S, Shun MC, Gupta SS, *et al.* (2009) A novel co-crystal structure affords the design of gain-of-function lentiviral integrase mutants in the presence of modified PSIP1/LEDGF/p75. *PLoS Pathog* **5**: e1000259.
75. Maertens GN, Cherepanov P, Engelman A. (2006) Transcriptional co-activator p75 binds and tethers the Myc-interacting protein JPO2 to chromatin. *J Cell Sci* **119**: 2563–2571.
76. Bartholomeeusen K, De Rijck J, Busschots K, *et al.* (2007) Differential interaction of HIV-1 integrase and JPO2 with the C terminus of LEDGF/p75. *J Mol Biol* **372**: 407–421.
77. Bartholomeeusen K, Gijsbers R, Christ F, *et al.* (2009) Lens epithelium derived growth factor/p75 interacts with the transposase derived DDE domain of pogZ. *J Biol Chem* **284**: 11467–11477. doi: 10.1074/jbc.M807781200.
78. Pluymers W, Cherepanov P, Schols D, *et al.* (1999) Nuclear localization of human immunodeficiency virus type 1 integrase expressed as a fusion protein with green fluorescent protein. *Virology* **258**: 327–332.
79. Schroder AR, Shinn P, Chen H, *et al.* (2002) HIV-1 integration in the human genome favors active genes and local hotspots. *Cell* **110**: 521–529.
80. Cherepanov P. (2007) LEDGF/p75 interacts with divergent lentiviral integrases and modulates their enzymatic activity *in vitro. Nucl Acid Res* **35**: 113–124.
81. Llano M, Vanegas M, Fregoso O, *et al.* (2004) LEDGF/p75 determines cellular trafficking of diverse lentiviral but not murine oncoretroviral integrase proteins and is a component of functional lentiviral preintegration complexes. *J Virol* **78**: 9524–9537.

82. Llano M, Delgado S, Vanegas M, Poeschla EM. (2004) LEDGF/p75 prevents proteasomal degradation of HIV-1 integrase. *J Biol Chem* **8**: 8.
83. Vandekerckhove L, Christ F, Van Maele B, *et al.* (2006) Transient and stable knockdown of the integrase cofactor LEDGF/p75 reveals its role in the replication cycle of human immunodeficiency virus. *J Virol* **80**: 1886–1896.
84. Llano M, Saenz DT, Meehan A, *et al.* (2006) An essential role for LEDGF/p75 in HIV integration. *Science* **314**: 461–464.
85. Shun MC, Raghavendra NK, Vandegraaff N, *et al.* (2007) LEDGF/p75 functions downstream from preintegration complex formation to effect gene-specific HIV-1 integration. *Genes Dev* **21**: 1767–1778.
86. Cherepanov P, Ambrosio AL, Rahman S, *et al.* (2005) Structural basis for the recognition between HIV-1 integrase and transcriptional coactivator p75. *Proc Natl Acad Sci USA* **102**: 17308–17313.
87. De Rijck J, Vandekerckhove L, Gijsbers R, *et al.* (2006) Over-expression of the lens epithelium-derived growth factor/p75 integrase binding domain inhibits human immunodeficiency virus replication. *J Virol* **80**: 11498–11509.
88. Hombrouck A, De Rijck J, Hendrix J, *et al.* (2007) Virus evolution reveals an exclusive role for LEDGF/p75 in chromosomal tethering of HIV. *PLoS Pathog* **3**: e47.
89. Ciuffi A, Diamond TL, Hwang Y, *et al.* (2006) Modulating target site selection during human immunodeficiency virus DNA integration *in vitro* with an engineered tethering factor. *Hum Gene Ther* **17**: 960–967.
90. Wu X, Li Y, Crise B, Burgess SM. (2003) Transcription start regions in the human genome are favored targets for MLV integration. *Science* **300**: 1749–1751.
91. Bushman FD. (2002) Integration site selection by lentiviruses: Biology and possible control. *Curr Top Microbiol Immunol* **261**: 165–177.
92. Wu X, Burgess SM. (2004) Integration target site selection for retroviruses and transposable elements. *Cell Mol Life Sci* **61**: 2588–2596.
93. Crise B, Li Y, Yuan C, *et al.* (2005) Simian immunodeficiency virus integration preference is similar to that of human immunodeficiency virus type 1. *J Virol* **79**: 12199–12204.
94. Hacker CV, Vink CA, Wardell TW, *et al.* (2006) The integration profile of EIAV-based vectors. *Mol Ther* **14**: 536–545.

95. Nowrouzi A, Dittrich M, Klanke C, *et al.* (2006) Genome-wide mapping of foamy virus vector integrations into a human cell line. *J Gen Virol* **87**: 1339–1347.
96. Trobridge GD, Miller DG, Jacobs MA, *et al.* (2006) Foamy virus vector integration sites in normal human cells. *Proc Natl Acad Sci USA* **103**: 1498–1503.
97. Wang GP, Ciuffi A, Leipzig J, *et al.* (2007) HIV integration site selection: Analysis by massively parallel pyrosequencing reveals association with epigenetic modifications. *Genome Res* **17**: 1186–1194.
98. Ciuffi A, Llano M, Poeschla E, *et al.* (2005) A role for LEDGF/p75 in targeting HIV DNA integration. *Nat Med* **11**: 1287–1289.
99. Marshall HM, Ronen K, Berry C, *et al.* (2007) Role of PSIP1/LEDGF/p75 in lentiviral infectivity and integration targeting. *PLoS ONE* **2**: e1340.
100. Yu F, Jones GS, Hung M, *et al.* (2007) HIV-1 integrase preassembled on donor DNA is refractory to activity stimulation by LEDGF/p75. *Biochemistry (Mosc)* **46**: 2899–2908.
101. Raghavendra NK, Engelman A. (2007) LEDGF/p75 interferes with the formation of synaptic nucleoprotein complexes that catalyze full-site HIV-1 DNA integration *in vitro*: Implications for the mechanism of viral cDNA integration. *Virology* **360**: 1–5.
102. Pandey KK, Sinha S, Grandgenett DP. (2007) Transcriptional coactivator LEDGF/p75 modulates human immunodeficiency virus type 1 integrase-mediated concerted integration. *J Virol* **81**: 3969–3979.
103. Rahman S, Lu R, Vandegraaff N, *et al.* (2007) Structure-based mutagenesis of the integrase-LEDGF/p75 interface uncouples a strict correlation between *in vitro* protein binding and HIV-1 fitness. *Virology* **357**: 79–90.
104. Molteni V, Greenwald J, Rhodes D, *et al.* (2001) Identification of a small-molecule binding site at the dimer interface of the HIV integrase catalytic domain. *Acta Crystallogr D Biol Crystallogr* **57**: 536–544.
105. Al-Mawsawi LQ, Fikkert V, Dayam R, *et al.* (2006) Discovery of a small-molecule HIV-1 integrase inhibitor-binding site. *Proc Natl Acad Sci USA* **103**: 10080–10085.
106. Hou Y, McGuinness DE, Prongay AJ, *et al.* (2008) Screening for antiviral inhibitors of the HIV integrase-LEDGF/p75 interaction using the AlphaScreen luminescent proximity assay. *J Biomol Screen* **13**: 406–414.
107. Du L, Zhao Y, Chen J, *et al.* (2008) D77, one benzoic acid derivative, functions as a novel anti-HIV-1 inhibitor targeting the interaction between

integrase and cellular LEDGF/p75. *Biochem Biophys Res Commun* **375**: 139–144.

108. Lewis P, Hensel M, Emerman M. (1992) Human immunodeficiency virus infection of cells arrested in the cell cycle. *EMBO Journal* **11**: 3053–3058.
109. Vodicka MA, Koepp DM, Silver PA, Emerman M. (1998) HIV-1 Vpr interacts with the nuclear transport pathway to promote macrophage infection. *Genes Dev* **12**: 175–185.
110. Weinberg JB, Matthews TJ, Cullen BR, Malim MH. (1991) Productive human immunodeficiency virus type 1 (HIV-1) infection of nonproliferating human monocytes. *J Exp Med* **174**: 1477–1482.
111. De Rijck J, Vandekerckhove L, Christ F, Debyser Z. (2007) Lentiviral nuclear import: A complex interplay between virus and host. *Bioessays* **29**: 441–451.
112. Suzuki Y, Craigie R. (2007) The road to chromatin — nuclear entry of retroviruses. *Nat Rev Microbiol* **5**: 187–196.
113. Gallay P, Hope T, Chin D, Trono D. (1997) HIV-1 infection of nondividing cells through the recognition of integrase by the importin/karyopherin pathway. *Proc Natl Acad Sci USA* **94**: 9825–9830.
114. Armon-Omer A, Graessmann A, Loyter A. (2004) A synthetic peptide bearing the HIV-1 integrase 161–173 amino acid residues mediates active nuclear import and binding to importin alpha: Characterization of a functional nuclear localization signal. *J Mol Biol* **336**: 1117–1128.
115. Hearps AC, Jans DA. (2006) HIV-1 integrase is capable of targeting DNA to the nucleus via an importin alpha/beta-dependent mechanism. *Biochem J* **398**: 475–484.
116. Fassati A, Gorlich D, Harrison I, *et al.* (2003) Nuclear import of HIV-1 intracellular reverse transcription complexes is mediated by importin 7. *EMBO J* **22**: 3675–3685.
117. Zielske SP, Stevenson M. (2005) Importin 7 may be dispensable for human immunodeficiency virus type 1 and simian immunodeficiency virus infection of primary macrophages. *J Virol* **79**: 11541–11546.
118. Ao Z, Huang G, Yao H, *et al.* (2007) Interaction of human immunodeficiency virus type 1 integrase with cellular nuclear import receptor importin 7 and its impact on viral replication. *J Biol Chem* **282**: 13456–13467. doi: 10.1074/jbc.M610546200.
119. Christ F, Thys W, De Rijck J, *et al.* (2008) Transportin-SR2 imports HIV into the nucleus. *Curr Biol* **18**: 1192–1202.

120. Brass AL, Dykxhoorn DM, Benita Y, *et al.* (2008) Identification of host proteins required for HIV infection through a functional genomic screen. *Science* **319**: 921–926.

121. Konig R, Zhou Y, Elleder D, *et al.* (2008) Global analysis of host-pathogen interactions that regulate early-stage HIV-1 replication. *Cell* **135**: 49–60.

122. Zhou H, Xu M, Huang Q, *et al.* (2008) Genome-scale RNAi screen for host factors required for HIV replication. *Cell Host Microbe* **4**: 495–504.

123. Kataoka N, Bachorik JL, Dreyfuss G. (1999) Transportin-SR, a nuclear import receptor for SR proteins. *J Cell Biol* **145**: 1145–1152.

124. Lai MC, Lin RI, Huang SY, *et al.* (2000) A human importin-beta family protein, transportin-SR2, interacts with the phosphorylated RS domain of SR proteins. *J Biol Chem* **275**: 7950–7957.

125. Lai MC, Lin RI, Tarn WY. (2001) Transportin-SR2 mediates nuclear import of phosphorylated SR proteins. *Proc Natl Acad Sci USA* **98**: 10154–10159.

126. Yun CY, Velazquez-Dones AL, Lyman SK, Fu XD. (2003) Phosphorylation-dependent and -independent nuclear import of RS domain-containing splicing factors and regulators. *J Biol Chem* **278**: 18050–18055.

127. Lai MC, Kuo HW, Chang WC, Tarn WY. (2003) A novel splicing regulator shares a nuclear import pathway with SR proteins. *EMBO J* **22**: 1359–1369.

128. Albanese A, Arosio D, Terreni M, Cereseto A. (2008) HIV-1 pre-integration complexes selectively target decondensed chromatin in the nuclear periphery. *PLoS ONE* **3**: e2413.

129. Ao Z, Huang G, Yao H, *et al.* (2007) Interaction of human immunodeficiency virus type 1 integrase with cellular nuclear import receptor importin 7 and its impact on viral replication. *J Biol Chem* **282**: 13456–13467. doi: 10.1074/jbc.M610546200.

130. Goff SP. (2008) Knockdown screens to knockout HIV-1. *Cell* **135**: 417–420.

131. Emiliani S, Rain JC, Maroun M, *et al.* (2003) Identification of LEDGF/p75 as an essential factor for HIV replication that binds integrase: A novel target for anti-HIV drug discovery. Oral communication at: *The 10th Conference on Retroviruses and Opportunistic Infections, Boston.* February 10, 2003.

132. Kilby JM, Hopkins S, Venetta TM, *et al.* (1998) Potent suppression of HIV-1 replication in humans by T-20, a peptide inhibitor of gp41-mediated virus entry. *Nat Med* **4**: 1302–1307.

133. Dorr P, Westby M, Dobbs S, *et al.* (2005) Maraviroc (UK-427,857), a potent, orally bioavailable, and selective small-molecule inhibitor of chemokine

receptor CCR5 with broad-spectrum anti-human immunodeficiency virus type 1 activity. *Antimicrob Agents Chemother* **49**: 4721–4732.

134. Agrawal L, Lu X, Jin Q, Alkhatib G. (2006) Anti-HIV therapy: Current and future directions. *Curr Pharm Des* **12**: 2031–2055.
135. Li J, Tang S, Hewlett I, Yang M. (2007) HIV-1 capsid protein and cyclophilin A as new targets for anti-AIDS therapeutic agents. *Infect Disord Drug Target* **7**: 238–244.
136. Nathans R, Cao H, Sharova N, *et al.* (2008) Small-molecule inhibition of HIV-1 Vif. *Nat Biotechnol* **26**: 1187–1192.
137. Sutherland HG, Newton K, Brownstein DG, *et al.* (2006) Disruption of Ledgf/Psip1 results in perinatal mortality and homeotic skeletal transformations. *Mol Cell Biol* **26**: 7201–7210.
138. Asante-Appiah E, Skalka AM. (1999) HIV-1 integrase: Structural organization, conformational changes, and catalysis. *Adv Virus Res* **52**: 351–369.
139. Engelman A, Craigie R. (1992) Identification of conserved amino acid residues critical for human immunodeficiency virus type 1 integrase function *in vitro*. *J Virol* **66**: 6361–6369.
140. Jenkins TM, Engelman A, Ghirlando R, Craigie R. (1996) A soluble active mutant of HIV-1 integrase: Involvement of both the core and carboxyl-terminal domains in multimerization. *J Biol Chem* **271**: 7712–7718.
141. Botbol Y, Raghavendra NK, Rahman S, *et al.* (2008) Chromatinized templates reveal the requirement for the LEDGF/p75 PWWP domain during HIV-1 integration *in vitro*. *Nucl Acid Res* **36**: 1237–1246.
142. Llano M, Vanegas M, Hutchins N, *et al.* (2006) Identification and characterization of the chromatin-binding domains of the HIV-1 integrase interactor LEDGF/p75. *J Mol Biol* **360**: 760–773.
143. Vanegas M, Llano M, Delgado S, *et al.* (2005) Identification of the LEDGF/p75 HIV-1 integrase-interaction domain and NLS reveals NLS-independent chromatin tethering. *J Cell Sci* 2005, **118**: 1733–1743.

CHAPTER 13

Retrovirus Restriction Factors

Reuben S. Harris*,‡, Rebecca S. LaRue*,§ and Thomas J. Hope†

SUMMARY

One of the major advances in the last decade of retrovirus research has been the discovery of cellular proteins that function to block viral replication. Retrovirus restriction factors such as APOBEC3G, TRIM5α and BST2 are dominant, variable, and interferon-inducible, and they have forced susceptible viruses to evolve intricate escape measures. This chapter will focus on these prototypical restriction factors and highlight several common themes.

INTRODUCTION

Retrovirus restriction factors can be generally regarded as specific mediators of the broader innate immune response to viral infections. Retrovirus

* Department of Biochemistry, Molecular Biology and Biophysics, University of Minnesota, Minneapolis, Minnesota 55455, USA. E-mail: ‡rsh@umn.edu.

† Department of Cell and Molecular Biology, Northwestern University, Chicago, Illinois 60611, USA. E-mail: §larue005@umn.edu; †thope@northwestern.edu.

restriction factors are distinguished from other cellular proteins by four major criteria. First, they dominantly interfere with viral replication at a specific point in the retroviral life cycle. Second, they show especially high levels of genetic diversity (positive selection) indicative of direct participation in ancient and likely ongoing conflicts with viruses. Third, the expression of these proteins is induced by viral infection and/or interferon. Fourth, potentially susceptible viruses will possess a means of evading or neutralizing the restriction mechanisms of the host to which they are adapted.

Three proteins in particular — APOBEC3G, TRIM5α and BST2/TETHERIN — have led the way as prototypes defining the new field of restriction factors. This chapter will concentrate on these three important proteins and their mechanisms of action. However, we emphasize that these are likely to be only the first of many additional related and as-yet-undiscovered factors with roles in retrovirus restriction.

APOBEC3G

In 2002, three groups independently discovered APOBEC3G. The first reported its genomic location and membership to a larger family of proteins.[1] The second cloned APOBEC3G and demonstrated that it triggered a specific type of DNA mutation, C/G-to-T/A transitions, by catalyzing the conversion of cytidine to uridine within single-strand DNA.[2] The third showed that APOBEC3G could inhibit the replication of HIV-1 lacking its accessory protein Vif.[3]

An important key to the latter discovery was the early realization that Vif-deficient HIV-1 can only replicate on a fraction of T-cell lines that are permissive for the wild-type virus.[4] Permissive and non-permissive cell fusion experiments revealed that the latter cells express a dominant factor that is neutralized by Vif.[5] Subtractive hybridization experiments between the permissive line CEM-SS and its parental non-permissive line CEM identified APOBEC3G.[3] Stable introduction of APOBEC3G into CEM-SS cells rendered them non-permissive for Vif-deficient HIV-1, thus establishing APOBEC3G as a *bona fide* retrovirus restriction factor.[3]

Retrovirus Restriction by DNA Cytidine Deamination

Human APOBEC3G is a 384-amino-acid protein with two conserved Zn^{2+}-coordinating domains [Fig. 1(a)]. Three observations indicated that antiviral activity of APOBEC3G might be due to a viral cDNA cytidine-to-uridine (C-to-U) deamination mechanism.[6–9] First, APOBEC3G was able to catalyze a specific type of base substitution mutation C/G-to-T/A transition.[2] This occurs when APOBEC3G catalyzes the conversion of DNA C-to-U, which irreversibly alters the base pairing capacity such that the U's now template the insertion of A's during DNA replication (C/G → U/G → U/A → T/A). Second, APOBEC3G was similar in amino acid sequence to AID, an enzyme that deaminates DNA C-to-U to trigger antibody gene diversification by somatic hypermutation and class switch recombination.[10] Finally, a viral cDNA deamination mechanism was consistent with the fact that patient-derived HIV-1 DNA sequences frequently contain so-called G-to-A hypermutations (see e.g., Ref. 11).

In the next few years, three major lines of evidence arose to strongly support a retroviral cDNA cytidine deamination mechanism [Fig. 1(b)]. First, APOBEC3G expression in virus-producing cells was sufficient to render the resulting particles up to 100- to 1000-fold less infectious.[6,8,9] Second, the viral cDNA sequences from such an experiment were found to contain extraordinary numbers of G-to-A transition mutations.[6,8,9] Third, based on previously solved structures of related deaminases such as *E. coli* cytidine deaminase, which converts the free nucleoside C-to-U, the putative catalytic residue of APOBEC3G Glu 259 was shown to be required for restriction.[8] The following paragraphs will highlight key steps in the mechanism of Vif-deficient HIV-1 restriction by APOBEC3G.

APOBEC3G is expressed predominantly in the cytoplasm of cells. At some point during retrovirus assembly, cytoplasmic APOBEC3G binds assembling Gag-ribonucleoprotein complexes and literally "hitch-hikes" into viral particles (reviewed in Ref. 12). The APOBEC3G amino-terminal pseudo-catalytic domain and the nucleocapsid region of Gag (NC) are required for this activity. However, the interaction between these proteins is probably indirect because the APOBEC3G–NC interaction is sensitive to RNase digestion.[13–15] Thus, APOBEC3G requires RNA for packaging, but

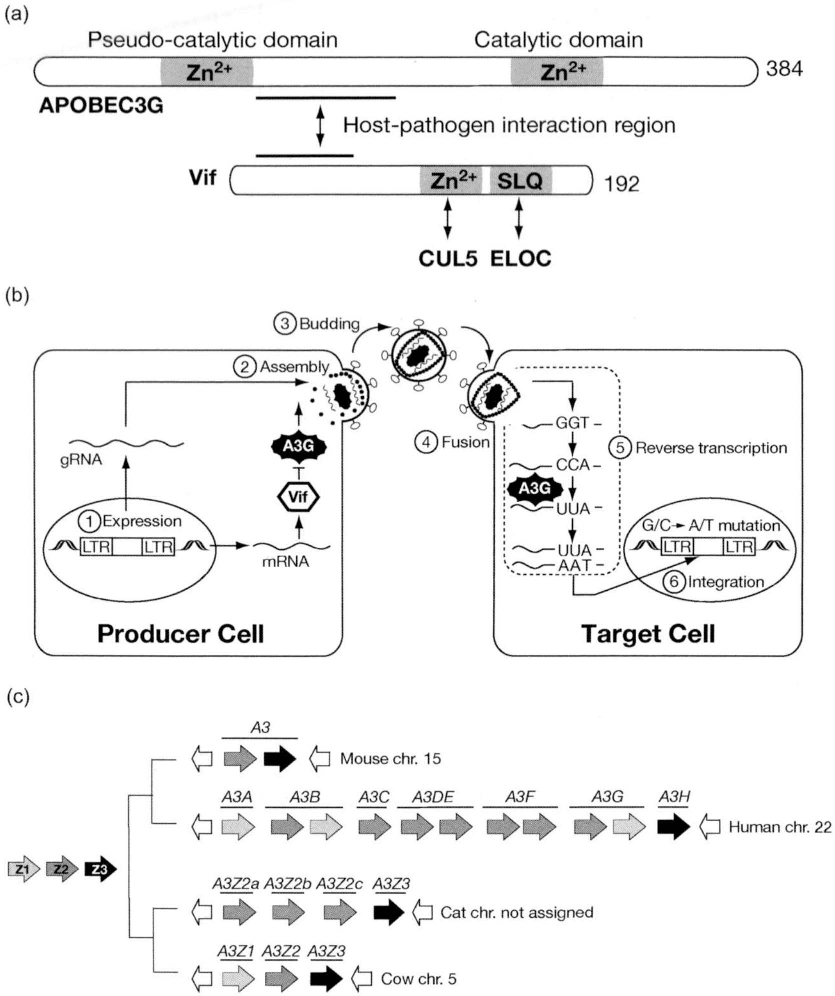

Fig. 1. APOBEC3G protein domains, function and phylogenetic distribution. (a) Schematic of human APOBEC3G (NP_068594.1) and HIV-1 Vif (NP_057851) indicating some relevant motifs. (b) A current model for HIV-1 restriction by human APOBEC3G depicted within the context of a simplified six-step viral life cycle. Producer cell APOBEC3G encapsidates through a Gag ribonucleoprotein interaction (not shown), travels with nascent viral particles until a new target cell is infected, and deaminates viral cDNA during reverse transcription (see the text for additional details). The dotted line surrounding the reverse transcription and DNA deamination steps indicates that these are generally thought to occur within the confines of a semi-porous capsid shell. (c) A schematic of the APOBEC3 locus in mice, humans, cats and cattle (adapted from Refs. 43 and 44). The dog APOBEC3

the identity of the critical RNA molecule(s) is still contested. The answer to this question may relate to the ability of some abundant cellular RNAs to be packaged preferentially, such as 7SL, Alu (which is derived from 7SL) and/or Y elements (see e.g., Refs. 16 and 17; reviewed in Ref. 12).

At least one mechanistic step exists between APOBEC3G packaging and cDNA deamination: core penetration. To pose a threat to the retroviral genome, APOBEC3G must gain access to the site of reverse transcription, the nucleic acid-containing, capsid-enclosed core of the retroviral particle. Chimeras between APOBEC3G and the related restriction factor APOBEC3F provided the first evidence that this critical step is distinct from packaging.[18] An APOBEC3G–APOBEC3F chimera restricted Vif-deficient HIV-1, whereas the reciprocal APOBEC3F–APOBEC3G chimera was defective in restriction despite the fact that it gained access to particles and showed wild-type levels of DNA deaminase activity. Complementary studies indicated that non-encapsidated DNA deaminases such as APOBEC3A could be fused to the amino-terminal half of APOBEC3G (the pseudo-catalytic domain) or to Vpr (an accessory protein that encapsidates), and that the resulting chimeric proteins were now capable of encapsidation, viral cDNA deamination and HIV-1 restriction.[19,20]

The next major step in HIV-1 restriction occurs during reverse transcription, when the nascent viral cDNA is transiently single-stranded [Fig. 1(b)]. It is at this stage that encapsidated APOBEC3G deaminates cDNA C-to-U, which subsequently template the insertion of adenosines during plus-strand DNA synthesis. The potency of APOBEC3G is remarkable, as it can deaminate up to 10 percent of the retroviral cDNA cytidines in a single round of reverse transcription (see e.g., Ref. 21). There is no doubt that the DNA deaminase activity of APOBEC3G is a major part of the HIV-1 restriction mechanism. For instance, mutants of APOBEC3G that lack the catalytic Glu 259 are competent for every measurable activity except catalysis, and they are grossly defective in HIV-1 restriction (see

---

Fig. 1. (*Continued*) locus is not yet fully annotated, but it appears similar to that of the cow with at least one of each Z-type (LaRue and Harris, unpublished observations). The Z1-, Z2- and Z3-type APOBEC3 domains are shaded light gray, dark gray and black, respectively, and the gene names are indicated above. Open arrows depict conserved flanking genes CBX6 and CBX7.

e.g., Refs. 8 and 22–25). Besides hypermutation, APOBEC3G-dependent DNA deamination also triggers the degradation of the retroviral cDNA.[26] More work needs to be done to define the degradation pathway, but it is unlikely to involve uracil excision by the uracil DNA glycosylase UNG2 (see e.g., Refs. 24, 26 and 27).

In addition, there is mounting evidence that APOBEC3G may have some DNA deamination-independent activities (reviewed in Refs. 28–30). Such activities include inhibiting reverse transcriptase progression, preventing tRNA annealing, blocking first and second strand transfer reactions, and altering the cDNA end-trimming reactions. All of these inevitably lead to lower levels of cDNA accumulation and integration. However, it is doubtful that all of these potential mechanisms are physiologically relevant, as they may be due at least in part to APOBEC3G over-expression artifacts. It is easy to envisage that an abundance of any encapsidated single-strand nucleic acid binding protein (RNA or DNA) will negatively impact reverse transcription. Nevertheless, more work may ultimately provide mechanistic ballast for one or more of these deamination-independent mechanisms.

While a lot is known about APOBEC3G function and mechanism, several observations have yet to be explained. For instance, a prominent study indicated that target cell APOBEC3G is capable of providing a block to incoming HIV-1.[31] However, others have been unable to confirm or define this alternative restriction mechanism.[32] Additional studies are also needed to resolve this issue.

Vif Neutralizes APOBEC3G Activity

The virion infectivity factor (Vif) of HIV-1 is required to counteract the potent restriction activity of APOBEC3G. Shortly after the discovery of APOBEC3G, several groups found that Vif triggers its degradation through a proteasome-dependent pathway (reviewed in Refs. 28–30) [Fig. 1(b)]. These studies were followed by a key mechanistic breakthrough from the Yu laboratory, which used affinity purification and mass spectrometry to reveal that Vif associates with a ubiquitin ligation complex consisting of CUL5, ELOB, ELOC and RBX.[33] A current model posits that Vif itself acts as a molecular bridge between APOBEC3G and the ubiquitin ligation complex that mediates APOBEC3G degradation. However, the precise target lysine for ubiquitination has not been identified, raising some questions as to whether APOBEC3G, Vif, and/or another component of the complex may actually be poly-ubiquitinated.

Other studies have indicated that Vif may also suppress the antiretroviral activity of APOBEC3G by proteasome-independent mechanisms. For instance, Vif has been shown to interfere with APOBEC3G mRNA translation[34] and with APOBEC3G activity by direct binding.[35–37] The details of these alternative mechanisms are currently being resolved by further studies. Nevertheless, the important broad conclusion is that Vif clearly functions to minimize APOBEC3G encapsidation and it thereby renders the cores of viral particles "safe" for reverse transcription.

APOBEC3G Expression Profile

Consistent with its ability to engage assembling HIV-1 ribonucleoprotein complexes, the APOBEC3G protein is predominantly cytoplasmic.[8] This original localization report has been confirmed by a number of laboratories using many distinct epitope tags and cell lines. Residues within the amino-terminal pseudo-catalytic domain of APOBEC3G are required for its localization (see e.g., Ref. 38). The distribution endogenous APOBEC3G in cells or tissues has not yet been reported due to lack of specific antibodies.

RNA blots and quantitative RT-PCR experiments have indicated that APOBEC3G is expressed in T cells, B cells and many other cell types (see e.g., Refs. 2 and 3). APOBEC3G mRNA is modestly induced in these cell types by type-1 interferons (see e.g., Refs. 39–41). Together, the broad and potentially ubiquitous expression profile and the interferon-inducibility of APOBEC3G are consistent with a role for this protein in innate immunity against not just HIV-1, but also many other parasitic elements.

Evolutionary Perspective of the Mammalian APOBEC3 Family

APOBEC3 diversity in mammals was first suggested with the realization that humans have seven proteins APOBEC3A through APOBEC3H (DE was later found to be one protein), whereas mice have only one.[42]

APOBEC3 proteins possess either one or two zinc-coordinating domains, which can be identified by a conserved His-x_1-Glu-$x_{23\text{-}28}$-Cys-$x_{2\text{-}4}$-Cys motif [x represents nearly any amino acid; Fig. 1(a)]. The histidine and the two cysteines coordinate zinc, and the glutamate shuttles protons during catalysis. With the growing number of available mammalian genomic sequences, it is clear that the APOBEC3 locus is one of the most dynamic loci in mammals [Fig. 1(c)]. First, the zinc (Z)-coordinating domains of the APOBEC3s form three distinct phylogenetic clusters: Z1, Z2 and Z3.[43] These Z-domains may be mixed and matched to form the two domain proteins (e.g., cow A3Z2 and A3Z3 can be expressed as two single domain proteins or one double-domain protein). Second, the APOBEC3s show extreme copy number variation, ranging from two to eleven Z-domains in mice and humans, respectively (one to seven genes, respectively). Other mammals such as cattle and cats show intermediate copy numbers. Third, like many other genes, several of the APOBEC3s have transcriptional variants due to internal promoter activity, alternative splicing and/or termination read-through (see e.g., Refs. 44 and 45). Fourth, APOBEC3G and other human and mammalian APOBEC3s show extreme levels of amino acid variability (i.e., positive selection[44–46]). Finally, evidence is starting to emerge that the APOBEC3s have distinct basal and inducible gene expression profiles. For instance, human APOBEC3G is broadly expressed and modestly interferon inducible while human APOBEC3A is strictly confined to myeloid lineage cells and massively interferon inducible (see e.g., Refs. 47 and Stenglein & Harris, unpublished observations).

Why such tremendous variation? All of the aforementioned data are consistent with the mammalian APOBEC3 proteins forming a discrete arm of the innate immune response that functions to limit the spread of mobile nucleic acids. Tremendous variation is needed to cope with the fact that parasitic elements are themselves continuously changing, with HIV-1 being one of the most extreme examples. It is easy to envisage how a failure to defend against a particularly pathogenic element could lead to a speciation bottleneck or even to species extinction.

Although APOBEC3G and HIV-1 are the main focus of this section, it is clear that APOBEC3G and the other APOBEC3 proteins are capable of restricting a huge number of parasitic elements, including many retroviruses and retrotransposons (reviewed in Refs. 28–30). Demonstrated substrates include: (i) lentiviruses such as HIV-1, HIV-2, SIV, MVV, FIV and EIAV; (ii) alpha-, beta-, gamma- and delta-retroviruses such as RSV, MPMV/MMTV, MuLV/FeLV and HTLV/BLV, respectively; (iii) spumaviruses such as PFV and FFV; (iv) hepadnaviruses such as HBV; (v) endogenous retroviruses and LTR-retrotransposons such as HERV-K, IAP, MusD and PERV; (vi) non-LTR retrotransposons such as L1 and Alu; and (vii) DNA viruses such as AAV and HPV. Thus, it is reasonable to postulate that no single parasitic element will be the physiological substrate of a given APOBEC3 protein. Rather, each APOBEC3 protein will have a broad number of physiological substrates, and only experimentation in an appropriate model system (or ideally *in vivo*) will determine which APOBEC3 protein or combination of APOBEC3 proteins is relevant to each specific parasitic element.

Therapeutic Prospects

The discovery of APOBEC3G and its role in HIV-1 restriction has helped establish a new paradigm for virus–host interactions — some host proteins function to dominantly and potently interfere with retrovirus replication. The fact that HIV-1 depends on APOBEC3G neutralization via a direct interaction with Vif and the subsequent recruitment of a cellular E3 ubiquitin ligation complex strongly suggests multiple points for therapeutic exploitation. Indeed, the first of hopefully many small molecule antivirals that work through the APOBEC3G-Vif axis was recently reported.[48] This compound called RN-18 appeared to work by causing the Vif-E3 ubiquitin ligation complex to "back-fire", triggering the degradation of itself rather than APOBEC3G. Although much future work will be necessary to develop RN-18 and additional novel small molecules, the prospects are very good for the development of novel HIV/AIDS and other anti-viral therapeutics that work by exploiting this elegant retrovirus restriction pathway.

TRIM5

It has been known since 1991 that rhesus macaque cells are resistant to HIV-1 infection.[49] Early experiments revealed that this infection block is not due to defective entry; rather that infection arrests at an early stage of viral replication in the cytoplasm.[50,51] Through the work of a number of laboratories, it was determined that this HIV-1 infection block manifests as an inhibition of the accumulation of late reverse transcription products and that it was due to the presence of a dominant inhibitory factor present in target cells.[52,53]

A number of similar interspecies blocks have also been identified. For example, HIV-1 cannot infect cells from several different monkey species, while some types of SIV are capable of infecting human cells.[53] Similarly, N-tropic murine leukemia virus (MLV) cannot infect human cells, while B-tropic MLV is unrestricted.[54] Like the restriction of HIV-1 by macaque cells, all of these restrictions appeared to manifest as a block to the formation of late reverse transcription products, and they were deduced to involve an interaction with the capsid component of the incoming retrovirus. Moreover, all of these systems are saturable, with low-titer infections being efficiently blocked and high-titer infections capable of overwhelming the restriction mechanism.[53,54]

These post-entry blocks to retroviral infection are similar to another restriction system in murine cells known as Fv1.[55–57] Fv1 blocks infection of certain strains of MLV but not others. The Fv1 block differs slightly in that it does not inhibit reverse transcription but rather the nuclear localization of the incoming MLV particle.[58] However, it does have similarities in terms of capsid recognition and saturability. Interestingly, there is also a host genetic component as different mouse strains show differential resistance to the different types of MLV.[55–57] Positional cloning led to the identification of the Fv1 and, surprisingly, it is derived from the Gag gene of an endogenous retrovirus.[59]

A major breakthrough in understanding the molecular mechanism of these post-entry restriction blocks occurred in 2004 when the Sodroski laboratory identified TRIM5α as the factor in rhesus macaque cells that potently inhibits HIV-1 infection.[60] Rhesus TRIM5α is a 497-amino-acid protein with four prominent motifs, an amino-terminal RING finger ubiquitin ligase motif, a B-box motif, a coiled-coil motif (these three constituting the "<u>tri</u>partite <u>m</u>otif" namesake) and a carboxyl-terminal B30.2/SPRY domain [Fig. 2(a)]. Humans express a 493-amino-acid form of TRIM5α that is incapable of blocking infection by HIV-1, consistent with the species specificity described above. Through the quick work of a

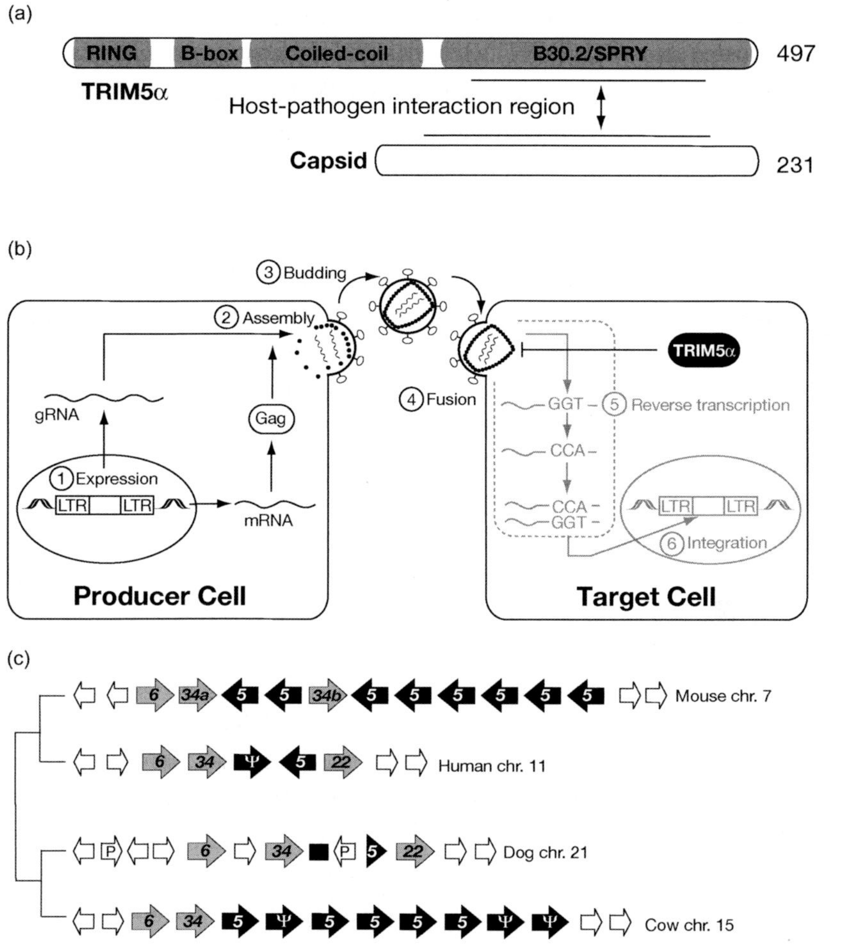

Fig. 2. TRIM5 protein domains, function and phylogenetic distribution. (a) A schematic of rhesus TRIM5α (NP_001028082) and HIV-1 Capsid (NP_579880) indicating some relevant motifs. (b) A current model for HIV-1 restriction by rhesus TRIM5α depicted within the context of a simplified six-step viral life cycle. Rhesus TRIM5α is thought to bind the incoming viral capsid and prevent the completion of reverse transcription (see the text for details). The dotted line surrounding the reverse transcription steps indicates that these are generally thought to occur within the confines of a semi-porous capsid shell. It is noteworthy that human TRIM5α is 92 percent similar to rhesus TRIM5α, but it cannot inhibit HIV-1 replication.[60] (c) A schematic of the TRIM5 genomic region in mice, humans, dogs and cattle (adapted from Refs. 108 and 109). TRIM5 genes or pseudogenes are shaded black, and TRIM6, TRIM34 and TRIM22 are shaded gray. Open arrows depict flanking (and sometimes interrupting) olfactory receptor (OR) genes and, in the dog locus only, the PNRC1 gene (labeled "P").

number of groups, it became clear that TRIM5α was the critical factor mediating a variety of previously defined interspecies restrictions. Stable introduction of TRIM5α into permissive target cells protected them from infection by TRIM5α-susceptible viruses, thus establishing TRIM5α as a *bona fide* retrovirus restriction factor.[60–64]

Further confirmation of the role of TRIM5α in retrovirus restriction came from the Luban laboratory through the identification of a unique cellular factor that prevents HIV-1 from infecting owl monkey cells.[65] Early studies had demonstrated that restriction of HIV-1 in owl monkey cells was unique in that it could be prevented by the addition of the immunosuppressive drug cyclosporine A (CsA).[66] The target of CsA is the prolyl-isomerase cyclophilin A (CypA). Previous studies had demonstrated that CypA binds to the outer surface of the incoming HIV-1 capsid and that this interaction can inhibits viral infectivity.[67] The mysterious nature of this restriction phenomenon was resolved by the identification of a modified form of TRIM5α expressed in owl monkeys in which a portion of TRIM5α is fused to CypA.[65] A retroelement had reverse-transcribed the CypA mRNA and inserted a copy into a terminal region of the TRIM5 gene. This insertion results in a natural fusion protein with the binding specificity of CypA and the retrovirus restriction capacity of TRIM5α. In addition to being a demonstration of the wonder of molecular evolution, the discovery of TRIM-Cyp also emphasized the important function of TRIM5α in retrovirus restriction.

Mechanism of Retrovirus Restriction by TRIM5α

Prior to the discovery of TRIM5α, genetic studies had suggested that its function required the retroviral capsid protein. The best demonstration of this was the observation that it is possible to saturate restriction by pre-infection of target cells with defective particles carrying restriction-susceptible capsid.[68,69] The saturation was specific to the incoming capsid because pre-infection with non-restrictable virus failed to saturate the restriction mechanism. The discovery of TRIM5α enabled this interaction to be dissected. Genetic experiments showed that the carboxyl-terminal B30.2/SPRY domain of TRIM5α specified the ability to recognize capsid. Changing a single amino acid in this region of human TRIM5α renders it capable of restricting HIV-1 infection.[70] TRIM5α is thought to recognize a higher-order capsid assembly because restriction is not saturated by over-expressing capsid alone and an interaction between TRIM5α and monomeric capsid has not been demonstrated. In contrast, the species-specific interaction of TRIM5α and the multimeric retroviral capsid assembly is apparent *in vitro*. Collectively, these data indicate that TRIM5α interacts directly with the intact structure of the retroviral capsid protein.

The ability of CsA to regulate the interaction between TRIM-Cyp and capsid facilitated kinetic studies of restriction by the Bieniasz laboratory.[71] Time of addition and washout experiments revealed that TRIM-Cyp acts quickly to mediate restriction consistent with it binding an intact capsid structure. Removal of the TRIM-Cyp inhibitor, CsA, at later time points failed to recover restriction, indicating that the target of restriction is transitory. Since the incoming capsid only remains intact for a limited time following fusion, it was inferred that the TRIM-Cyp binds the intact capsid structure. Thus, all models for the mechanism of restriction mediated by TRIM5α begin with the interaction of TRIM5α with the retroviral capsid [Fig. 2(b)].

It is generally hypothesized that the inhibition of reverse transcription is a consequence of TRIM5α disrupting the incoming capsid, but precisely how this occurs remains controversial. At present there are two schools of thought. The first model proposes that TRIM5α engages the retroviral capsid and causes its premature disassembly, a process that has been called "accelerated uncoating".[72] Evidence for this model comes from a "fate-of-capsid" assay, which measures the quantity of capsid-containing structures that penetrate a 40 percent sucrose cushion and accumulate as pellets at the bottom of an ultracentrifuge tube. In contrast to the sizable pellet derived from a normal 24-hour continuous infection, capsid-containing complexes fail to accumulate in the presence of TRIM5α. The (dis)appearance of these complexes has the expected species specificity correlating with known TRIM5α restriction profiles.[72] The pelleted complexes are therefore postulated to be intact cytoplasmic retroviral capsid structures, and the loss of these complexes has been interpreted as TRIM5α inducing accelerated uncoating of the retroviral capsid structure.

An alternative model posits that the disruption of the HIV-1 capsid is mediated by the action of the cellular proteasome. The E3 ubiquitin ligase (RING) domain located at the amino-terminus of TRIM5 proteins originally implicated the proteasome. Support for such a model comes from the observation that proteasomal inhibitors enable reverse transcription to be completed in the presence of TRIM5α, although the overall infection is still blocked.[73,74] This retroviral replication complex does not appear to reach the nucleus because there is a clear inhibition of the accumulation of 2-LTR circles, a commonly utilized surrogate to evaluate the nuclear localization of the HIV-1 DNA genome. Nevertheless, fully functional pre-integration complexes (PICs) appear to form and live-cell imaging studies reveal the accumulation of cytoplasmic complexes containing TRIM5α and HIV-1 capsid.[75] These studies support a model in which proteasome inhibition causes an accumulation of conical capsids sequestered in TRIM5α shells. In this shielded state, the PIC is unlikely to interact with the cellular factors required for nuclear import, thus inhibiting viral integration.

The ability of the proteasome to influence the formation of late reverse transcription products supports a two-phase model for HIV-1 restriction. In the first phase, the occurrence of proteasome-independent intermediates suggests that TRIM5α binds and stabilizes the retroviral capsid, a process that may actually inhibit uncoating (rather than accelerating it). Capsid sequestration traps the reverse transcribing genome and prevents its nuclear localization and integration. In the second, proteasome-dependent phase, TRIM5α-bound capsid structures are destroyed by the action of the proteasome. Consistent with this two-phase model, multiple permutations of susceptible capsid and restrictive TRIM5α are influenced by the proteasome.[76] However, to date there is no evidence that capsid itself is ubiquitinated or degraded by the proteasome. Thus, the proteasome may indirectly force capsid disassembly by ripping bound TRIM5α from the surface of viral replication complexes and thereby expose core proteins to non-specific degradation pathways.

In the end, there may not be much difference between the two models. Both call for a destruction of the capsid structure in the cytoplasmic compartment of the cell with the only major difference being a role for the proteasome. Consistent with a commonality of the two models, it has recently been shown that the accelerated MLV uncoating detected by the fate-of-capsid assay is also sensitive to proteasome inhibition.[77] However, the results of this experiment are difficult to interpret because proteasomal inhibition causes a general increase in pelletable capsid. Future studies will undoubtedly resolve these differences and clarify the mechanism of retrovirus restriction by TRIM5α.

TRIM5 Expression Profile

TRIM5α is expressed constitutively in all tissues analyzed to date and, like other restriction factors, its expression is induced by interferon.[78] Although TRIM5α is the predominantly expressed isoform, alternative splicing can generate several others.[60] The potential functional differences of the various isoforms are not clear but some evidence suggests that the delta isoform, which lacks the B30.2/SPRY region, can function as a dominant negative inhibitor of TRIM5α restriction.[61]

TRIM5α and related TRIM family proteins are frequently observed in cellular structures called "bodies"[79] or "cytoplasmic bodies."[60] These cellular substructures and their potential importance in the capability of TRIM5α to restrict retroviral replication have been controversial.[71,80] Debate has been fueled partially by the lack of antibodies that can accurately determine the localization of endogenous TRIM5α and also by a lack of a clear definition of what constitutes a cytoplasmic body.

Work with exogenously expressed TRIM5α has indicated that cytoplasmic body formation may be a function of expression levels, with higher levels causing a proportional increase in the number and size of the cytoplasmic bodies.[71] However, additional studies have since shown that cellular TRIM5α is highly dynamic, capable of shuttling between cytoplasmic pools and body structures.[81] Live cell characterization of the interaction of incoming HIV-1 particles and rhesus TRIM5α showed that TRIM5α can form body-like structures on viral complexes *de novo*.[75] Moreover, restriction was impaired in recent studies of mutant forms of TRIM5α that are unable to form higher-order oligomers or cytoplasmic bodies.[82] Together, these observations suggest that the propensity of this protein to form higher-order structures is a key part of its function.

Dynamics of TRIM5 Evolution

Following the discovery of TRIM5α, phylogenetic analyses showed that the TRIM5 gene has undergone extensive positive selection in primates.[83] Positive selection can be identified by aligning the same gene from related species and identifying differences in coding capacity. TRIM5 alignments have identified a 13-amino-acid region in the carboxyl-terminal B30.2/SPRY domain that is highly divergent among primates.[83] Consistent with its ability to directly bind capsid, the TRIM5α B30.2/SPRY domain is a key determinant of the species specificity of this interaction.[70,84] It is also notable that TRIM5 gene copy numbers vary dramatically among different mammals [e.g., Fig. 2(c)]. Both positive selection at the amino acid level and gene copy number changes are consistent with TRIM5 being involved in a continual conflict with retroviral pathogens.

As discussed above, an extraordinary example of molecular evolution is the owl monkey TRIM-Cyp fusion protein. The existence of this chimeric restriction factor leads to questions about functional significance, as owl monkey TRIM-Cyp appears to only inhibit HIV-1, despite evidence that HIV-like retroviruses do not exist in this animal's natural habitat in South America. Amazingly, an independent TRIM-Cyp fusion event was identified recently in three distinct Asian macaque species.[85–88] Distinct TRIM-Cyp fusion proteins in two separate primate lineages strongly suggests that this endows a significant fitness advantage, but the natural substrate(s) of these fusion proteins have not yet been identified. It is also worth noting that TRIM5α is a member of a very large gene family, with over 60 mostly uncharacterized members. Initial studies have indicated that at least a handful of these other TRIM family members may also be potent retrovirus restriction factors (e.g., Refs. 89 and 90).

Therapeutic Prospects for TRIM5

Human TRIM5α has the ability to restrict some retroviruses, but unfortunately not HIV-1. However, it has been suggested that human TRIM5α may influence HIV-2 replication.[91] Therefore, it is conceptually feasible that the potent restriction activity of TRIM5α could be modulated to inhibit replication of HIV-1. Since replacing a charged arginine in the capsid recognition domain of human TRIM5α renders it competent to restrict HIV-1, a compound that specifically masks the positive charge of this residue could enable it to interact more efficiently with the HIV-1 capsid. Alternatively, a HIV-1 restricting allele of human TRIM5α could be delivered to pluripotent bone marrow cells by gene therapy, thereby generating a population of immune cells resistant to HIV-1 infection.

BST2/TETHERIN

HIV-1 Vpu is an 81-amino-acid integral membrane protein that was identified originally by the Martin and Haseltine groups.[92] In the ensuing two decades, Vpu had been ascribed many functions, but a general consensus favors roles in down-regulating CD4 and MHC class II and in enhancing virus release (reviewed in Ref. 93). A major advance in understanding Vpu function occurred in 2008 when the Bieniasz and Guatelli laboratories showed that it antagonizes BST2 and thereby facilitates the release of viral particles from cells.[94,95] BST2 is also called Tetherin, CD317 or HM1.24, but the official GenBank designation BST2 will be used hereon. In the absence of Vpu, BST2 literally "tethers" viral particles to the cell surface. This section will compare the general features of the BST2 HIV-1 restriction mechanism to those of APOBEC3G and TRIM5α. A much more detailed review of BST2 can be found in Chapter 9.

Retrovirus Restriction by Particle Tethering

Like the APOBEC3G and TRIM5α stories, permissive and non-permissive cell lines figured prominently in identifying BST2.[94,95] Certain cell lines such as COS-7 permitted the efficient production of Vpu-deficient HIV-1, whereas others such as HeLa restricted its replication. Chimeric lines created by the fusion of COS-7 and HeLa were restrictive, suggesting the presence of a dominant restriction factor.[96] Together with observations indicating that the critical cellular factor is interferon-inducible, candidates were systematically whittled down until BST2 was identified.[94,95]

Human BST2 is a 180-amino-acid membrane protein with an amino-terminal cytoplasmic tail and a hydrophobic membrane-spanning domain, a central extra-cellular domain, which may be modified by glycosylation, and a carboxyl-terminal glycosyl phosphatidylinositol (GPI) membrane-anchoring motif [Fig. 3(a)]. Expression of BST2 in permissive cells caused lower yields of Vpu-deficient HIV-1, and Vpu was able to efficiently counteract this effect.[94,95] Interferon treatment also induced BST2 expression and rendered certain cells such as 293T less permissive for Vpu-deficient HIV-1 replication. Accordingly, BST2 knockdown by siRNA rendered HeLa cells more permissive.[95] Overall, these studies indicated that BST2 is an HIV-1 restriction factor that is counteracted by Vpu.

The most striking phenotype of BST2-mediated restriction is the physical tethering of the viral particles to the surface of the cell. The emerging mechanism is that a Vpu-deficient particle is able to bud, but it remains tethered to the surface of the cell by a dimer of BST2, with the cell and the virus each harboring a transmembrane domain from one molecule and a GPI domain from the other [Fig. 3(b)]. The tether is sensitive to digestion by serine endopeptidase subtilisin, and the resulting viral particles are remarkably still infectious.[94] The *in vivo* advantage of such a tethering mechanism is not completely clear, since tethered particles may still be transmitted by direct cell-to-cell contact. However, it is possible that the tethering mechanism is coupled to a clearance pathway such as the endocytic or the lysosomal degradation pathways (see e.g., Ref. 94, and below). Additionally, BST2 may cause an accumulation of viral epitopes on the cell surface, which in turn may trigger direct immune clearance by cytotoxic T lymphocytes (CTL) or natural killer (NK) cells.

Vpu Antagonizes BST2 Function

Most groups agree that Vpu over-expression causes lower levels of membrane-associated BST2, but the details of the underlying mechanism are not clear.[95,97] Three non-exclusive mechanisms are leading candidates to explain this phenomenon: degradation, internalization and sequestration.

A degradation mechanism is supported by observations indicating that Vpu causes a decrease in both surface and total cellular BST2.[95,98,99] For instance, treating cells with proteasome inhibitors ALLN or lactocystin

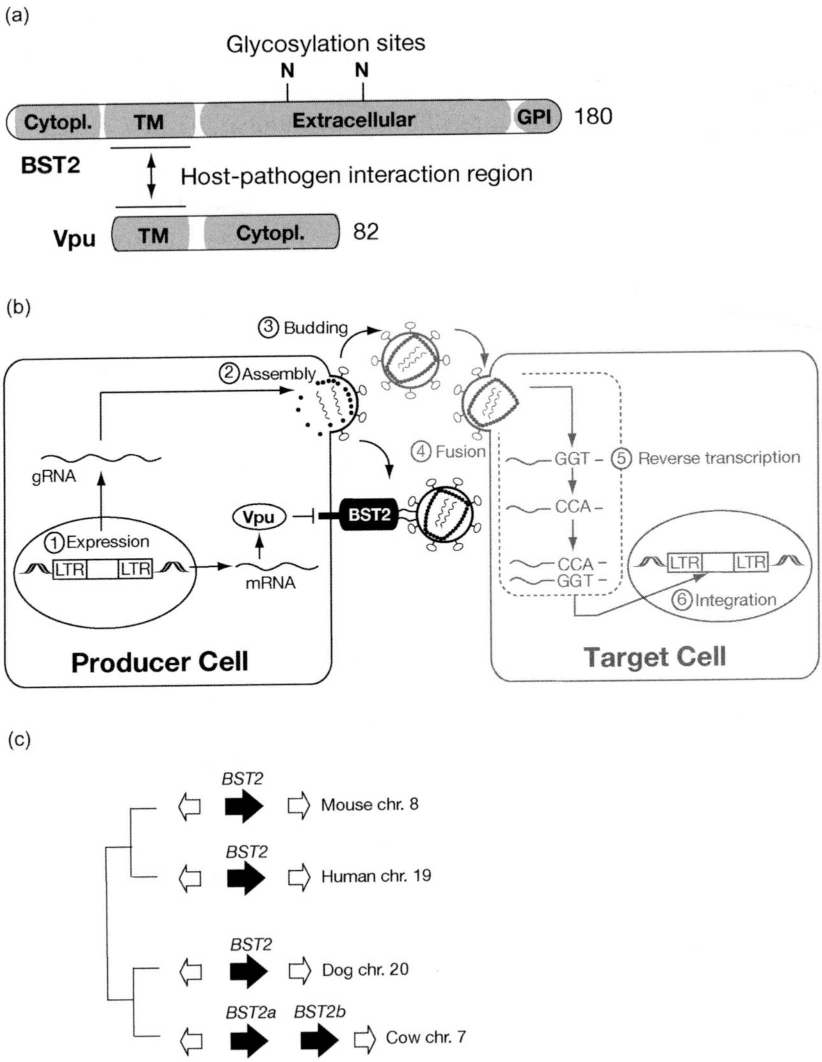

Fig. 3. BST2 protein domains, function and phylogenetic distribution. (a) Schematic of human BST2 (NP_004326) and HIV-1 Vpu (NP_057855.1) indicating some relevant motifs (Cytopl.: cytoplasmic domain; TM: transmembrane domain; N: glycosylation site). (b) A current model for HIV-1 restriction by human BST2 depicted within the context of a simplified six-step viral life cycle. BST2 uses a tethering mechanism to prevent nascent particles from diffusing away from a virus-producing cell. Vpu antagonizes BST2 by causing it to be depleted from the cell surface (see the text for details). The target cell is depicted to facilitate comparisons with Figs. 1 and 2. (c) A schematic of the BST2 genomic region in mice, humans, dogs and cattle (assembled with the NCBI and Ensembl genome browsers). FAM125A and PLVAP flank the BST2 locus in the indicated mammals.

increased the half-life of pulse-labeled BST2.[97] Moreover, another proteasome inhibitor, MG132, partially prevented BST2 degradation by Vpu and it enabled some virus release.[99] Although another group observed no effect with MG132 on the level of surface BST2,[95] these observations can be reconciled if BST2 internalization/endocytosis does not require the proteasome. Moreover, recent studies have indicated that Vpu may recruit the β-TrCP E3 ubiquitin ligation complex to trigger the degradation of BST2.[100] Interestingly, Vpu also uses β-TrCP to degrade CD4, suggesting significant overlap between these two pathways. This neutralization mechanism closely parallels that of Vif and APOBEC3G, suggesting that degradation may be a common defensive strategy employed by viruses.

Other studies have indicated that Vpu may trigger BST2 internalization and/or sequestration and that these events may lead to lysosomal degradation. For instance, localization studies suggested that Vpu's presence in the trans-Golgi network is required for it to antagonize BST2 function.[101] Mis-localizing Vpu mutants were no longer able to prevent BST2 from inhibiting HIV-1 release. Complementary work indicated that Vpu and BST2 might be degraded by lysosomes because degradation was suppressed partially by compounds that inhibit lysosome acidification (chloroquine or bafilomycin[100,101]). An internalization mechanism is further supported by requirements for endocytosis factors RAB5A and AP-2.[94,100] Additional studies will be necessary to determine the relative contributions of the proteasomal and lysosomal degradation pathways.

Expression Profile of BST2

Publicly available microarray data indicate that BST2 is expressed broadly. It is also strongly induced by type-1 interferon, consistent with an innate immune function.[94] BST2 localizes to the cell surface and to internal compartments such as the trans-Golgi network and endosomes.[94,95,101] These localization patterns place it in membranes and in a position to interact directly with Vpu, as described above.

Evolutionary Perspective of BST2

Most mammals have one copy of BST2 positioned between PLVAP and FAM125A (MVB12A) [Fig. 3(c)]. Interestingly, cattle have two copies, indicating that copy number variation is possible. Like the APOBEC3 proteins and select TRIM family members, BST2 shows considerable evidence for positive selection.[99,102] Variations tend to cluster to the transmembrane region, consistent with this part of the protein directly contacting the viral counter-defense factor Vpu. However, other variations occur outside of this region, suggesting that other (past and present) viruses will have distinct BST2 neutralization mechanisms.

Indeed, in the short time since its discovery, BST2 has already been shown to restrict a broad number of viruses, including Lassa, Marburg, Ebola, KSHV, a variety of retroviruses (alpha, beta, delta and spuma) and, of course, HIV-1.[94,103–105] Although the mechanism of virus restriction by BST2 appears broad, there is clear species specificity for the viral counter-defense by Vpu. For instance, HIV-1 Vpu efficiently neutralizes human BST2, but not the homologous protein from rodents.[97] Moreover, there is some variety in the counter-defense mechanism, with SIV Nef, HIV-2 Env and KSHV K5 proteins each able to antagonize BST2.[106,107] Clearly, like APOBEC3G and TRIM5α and some of their family members, these observations are all consistent with BST2 having an integral role in the innate immune response.

Therapeutic Prospects for BST2

The normal cellular function of BST2 is unknown, although the future characterization of knock-out mice will undoubtedly shed light on this important point. If BST2 proves essential, then any future therapeutic will have to be careful not to perturb this endogenous function. If BST2 is non-essential or it has some type of minor immunomodulatory function, then one can envisage the development of compounds that shield BST2 from viral antagonists. Moreover, once the molecular mechanism is better elucidated, it may be possible to develop ways to specifically up-regulate BST2 or develop a method to maintain its surface expression.

CONCLUDING REMARKS

APOBEC3G, TRIM5α and BST2 have become prototypic retrovirus restriction factors. Each fits the definition from the introduction of being (i) dominant-acting, (ii) fast-evolving, (iii) interferon-inducible, and (iv) counter-acted readily by retroviruses. They also represent the diversity of responses that cells have evolved to disrupt retroviral replication and the equally (if not more) diverse viral responses to counteract these blockades. The discovery and characterization of these retrovirus restriction factors has opened the door to many new therapeutic strategies. It is highly likely that some of these will come to fruition within the next decade.

ACKNOWLEDGMENTS

We apologize to our colleagues whose work could not be cited due to space limitations. We thank J. Hultquist and C. Rold for critically reading the manuscript. This work was supported by grants from the National Institutes of Health: AI064046 and GM090437 to R. S. Harris, and AI047770 and GM082545 to T. J. Hope.

REFERENCES

1. Jarmuz A, Chester A, Bayliss J, *et al.* (2002) An anthropoid-specific locus of orphan C to U RNA-editing enzymes on chromosome 22. *Genomics* **79**: 285–296.
2. Harris RS, Petersen-Mahrt SK, Neuberger MS. (2002) RNA editing enzyme APOBEC1 and some of its homologs can act as DNA mutators. *Mol Cell* **10**: 1247–1253.
3. Sheehy AM, Gaddis NC, Choi JD, *et al.* (2002) Isolation of a human gene that inhibits HIV-1 infection and is suppressed by the viral Vif protein. *Nature* **418**: 646–650.
4. Gabuzda DH, Lawrence K, Langhoff E, *et al.* (1992) Role of Vif in replication of human immunodeficiency virus type 1 in CD4+ T lymphocytes. *J Virol* **66**: 6489–6495.

5. Simon JH, Gaddis NC, Fouchier RA, *et al.* (1998) Evidence for a newly discovered cellular anti-HIV-1 phenotype. *Nat Med* **4**: 1397–1400.
6. Harris RS, Bishop KN, Sheehy AM, *et al.* (2003) DNA deamination mediates innate immunity to retroviral infection. *Cell* **113**: 803–809.
7. Lecossier D, Bouchonnet F, Clavel F, *et al.* (2003) Hypermutation of HIV-1 DNA in the absence of the Vif protein. *Science* **300**: 1112.
8. Mangeat B, Turelli P, Caron G, *et al.* (2003) Broad antiretroviral defence by human APOBEC3G through lethal editing of nascent reverse transcripts. *Nature* **424**: 99–103.
9. Zhang H, Yang B, Pomerantz RJ, *et al.* (2003) The cytidine deaminase CEM15 induces hypermutation in newly synthesized HIV-1 DNA. *Nature* **424**: 94–98.
10. Petersen-Mahrt SK, Harris RS, Neuberger MS. (2002) AID mutates *E. coli* suggesting a DNA deamination mechanism for antibody diversification. *Nature* **418**: 99–103.
11. Janini M, Rogers M, Birx DR, *et al.* (2001) Human immunodeficiency virus type 1 DNA sequences genetically damaged by hypermutation are often abundant in patient peripheral blood mononuclear cells and may be generated during near-simultaneous infection and activation of CD4(+) T cells. *J Virol* **75**: 7973–7986.
12. Strebel K, Khan MA. (2008) APOBEC3G encapsidation into HIV-1 virions: Which RNA is it? *Retrovirology* **5**: 55.
13. Svarovskaia ES, Xu H, Mbisa JL, *et al.* (2004) Human apolipoprotein B mRNA-editing enzyme-catalytic polypeptide-like 3G (APOBEC3G) is incorporated into HIV-1 virions through interactions with viral and non-viral RNAs. *J Biol Chem* **279**: 35822–35828.
14. Zennou V, Perez-Caballero D, Gottlinger H, *et al.* (2004) APOBEC3G incorporation into human immunodeficiency virus type 1 particles. *J Virol* **78**: 12058–12061.
15. Schafer A, Bogerd HP, Cullen BR. (2004) Specific packaging of APOBEC3G into HIV-1 virions is mediated by the nucleocapsid domain of the gag polyprotein precursor. *Virology* **328**: 163–168.
16. Chiu YL, Witkowska HE, Hall SC, *et al.* (2006) High-molecular-mass APOBEC3G complexes restrict Alu retrotransposition. *Proc Natl Acad Sci USA* **103**: 15588–15593.

17. Wang T, Tian C, Zhang W, *et al.* (2007) 7SL RNA mediates virion packaging of the antiviral cytidine deaminase APOBEC3G. *J Virol* **81**: 13112–13124.
18. Haché G, Liddament MT, Harris RS. (2005) The retroviral hypermutation specificity of APOBEC3F and APOBEC3G is governed by the C-terminal DNA cytosine deaminase domain. *J Biol Chem* **280**: 10920–10924.
19. Goila-Gaur R, Khan MA, Miyagi E, *et al.* (2007) Targeting APOBEC3A to the viral nucleoprotein complex confers antiviral activity. *Retrovirology* **4**: 61.
20. Aguiar RS, Lovsin N, Tanuri A, *et al.* (2008) Vpr.A3A chimera inhibits HIV replication. *J Biol Chem* **283**: 2518–2525.
21. Yu Q, Konig R, Pillai S, *et al.* (2004) Single-strand specificity of APOBEC3G accounts for minus-strand deamination of the HIV genome. *Nat Struct Mol Biol* **11**: 435–442.
22. Holmes RK, Koning FA, Bishop KN, *et al.* (2007) APOBEC3F can inhibit the accumulation of HIV-1 reverse transcription products in the absence of hypermutation. Comparisons with APOBEC3G. *J Biol Chem* **282**: 2587–2595.
23. Miyagi E, Opi S, Takeuchi H, *et al.* (2007) Enzymatically active APOBEC3G is required for efficient inhibition of human immunodeficiency virus type 1. *J Virol* **81**: 13346–13353.
24. Schumacher AJ, Haché G, MacDuff DA, *et al.* (2008) The DNA deaminase activity of human APOBEC3G is required for Ty1, MusD, and human immunodeficiency virus type 1 restriction. *J Virol* **82**: 2652–2660.
25. Haché G, Shindo K, Albin JS, *et al.* (2008) Evolution of HIV-1 isolates that use a novel Vif-independent mechanism to resist restriction by human APOBEC3G. *Curr Biol* **18**: 819–824.
26. Mbisa JL, Barr R, Thomas JA, *et al.* (2007) HIV-1 cDNAs produced in the presence of APOBEC3G exhibit defects in plus-strand DNA transfer and integration. *J Virol* **81**: 7099–7110.
27. Kaiser SM, Emerman M. (2006) Uracil DNA glycosylase is dispensable for human immunodeficiency virus type 1 replication and does not contribute to the antiviral effects of the cytidine deaminase APOBEC3G. *J Virol* **80**: 875–882.
28. Chiu YL, Greene WC. (2008) The APOBEC3 cytidine deaminases: An innate defensive network opposing exogenous retroviruses and endogenous retroelements. *Annu Rev Immunol* **26**: 317–353.

29. Goila-Gaur R, Strebel K. (2008) HIV-1 Vif, APOBEC, and intrinsic immunity. *Retrovirology* **5**: 51.
30. Malim MH, Emerman M. (2008) HIV-1 accessory proteins — Ensuring viral survival in a hostile environment. *Cell Host Microbe* **3**: 388–398.
31. Chiu YL, Soros VB, Kreisberg JF, *et al.* (2005) Cellular APOBEC3G restricts HIV-1 infection in resting CD4+ T cells. *Nature* **435**: 108–114.
32. Kamata M, Nagaoka Y, Chen IS. (2009) Reassessing the role of APOBEC3G in human immunodeficiency virus type 1 infection of quiescent CD4+ T-cells. *PLoS Pathog* **5**: e1000342.
33. Yu X, Yu Y, Liu B, *et al.* (2003) Induction of APOBEC3G ubiquitination and degradation by an HIV-1 Vif-Cul5-SCF complex. *Science* **302**: 1056–1060.
34. Stopak K, de Noronha C, Yonemoto W, *et al.* (2003) HIV-1 Vif blocks the antiviral activity of APOBEC3G by impairing both its translation and intracellular stability. *Mol Cell* **12**: 591–601.
35. Zhang L, Saadatmand J, Li X, *et al.* (2008) Function analysis of sequences in human APOBEC3G involved in Vif-mediated degradation. *Virology* **370**: 113–121.
36. Kao S, Goila-Gaur R, Miyagi E, *et al.* (2007) Production of infectious virus and degradation of APOBEC3G are separable functional properties of human immunodeficiency virus type 1 Vif. *Virology* **369**: 329–339.
37. Santa-Marta M, Aires da Silva F, Fonseca AM, *et al.* (2007) HIV-1 Vif protein blocks the cytidine deaminase activity of B-cell specific AID in *E. coli* by a similar mechanism of action. *Mol Immunol* **44**: 583–590.
38. Stenglein MD, Matsuo H, Harris RS. (2008) Two regions within the amino-terminal half of APOBEC3G cooperate to determine cytoplasmic localization. *J Virol* **82**: 9591–9599.
39. Tanaka Y, Marusawa H, Seno H, *et al.* (2006) Anti-viral protein APOBEC3G is induced by interferon-alpha stimulation in human hepatocytes. *Biochem Biophys Res Commun* **341**: 314–319.
40. Bonvin M, Achermann F, Greeve I, *et al.* (2006) Interferon-inducible expression of APOBEC3 editing enzymes in human hepatocytes and inhibition of hepatitis B virus replication. *Hepatology* **43**: 1364–1374.
41. Peng G, Lei KJ, Jin W, *et al.* (2006) Induction of APOBEC3 family proteins, a defensive maneuver underlying interferon-induced anti-HIV-1 activity. *J Exp Med* **203**: 41–46.

42. Harris RS, Liddament MT. (2004) Retroviral restriction by APOBEC proteins. *Nat Rev Immunol* **4**: 868–877.
43. LaRue RS, Andrésdóttir V, Blanchard Y, *et al.* (2009) Guidelines for naming nonprimate APOBEC3 genes and proteins. *J Virol* **83**: 494–497.
44. LaRue RS, Jónsson SR, Silverstein KAT, *et al.* (2008) The artiodactyl APOBEC3 innate immune repertoire shows evidence for a multi-functional domain organization that existed in the ancestor of placental mammals. *BMC Mol Biol* **9**: 104.
45. Münk C, Beck T, Zielonka J, *et al.* (2008) Functions, structure, and read-through alternative splicing of feline APOBEC3 genes. *Genome Biol* **9**: R48.
46. Sawyer SL, Emerman M, Malik HS. (2004) Ancient adaptive evolution of the primate antiviral DNA-editing enzyme APOBEC3G. *PLoS Biol* **2**: e275.
47. Peng G, Greenwell-Wild T, Nares S, *et al.* (2007) Myeloid differentiation and susceptibility to HIV-1 are linked to APOBEC3 expression. *Blood* **110**: 393–400.
48. Nathans R, Cao H, Sharova N, *et al.* (2008) Small-molecule inhibition of HIV-1 Vif. *Nat Biotechnol* **26**: 1187–1192.
49. Shibata R, Kawamura M, Sakai H, *et al.* (1991) Generation of a chimeric human and simian immunodeficiency virus infectious to monkey peripheral blood mononuclear cells. *J Virol* **65**: 3514–3520.
50. Shibata R, Sakai H, Kawamura M, *et al.* (1995) Early replication block of human immunodeficiency virus type 1 in monkey cells. *J Gen Virol* **76** (Part 2): 2723–2730.
51. Himathongkham S, Luciw PA. (1996) Restriction of HIV-1 (subtype B) replication at the entry step in rhesus macaque cells. *Virology* **219**: 485–488.
52. Hatziioannou T, Cowan S, Goff SP, *et al.* (2003) Restriction of multiple divergent retroviruses by Lv1 and Ref1. *EMBO J* **22**: 385–394.
53. Owens CM, Yang PC, Gottlinger H, *et al.* (2003) Human and simian immunodeficiency virus capsid proteins are major viral determinants of early, postentry replication blocks in simian cells. *J Virol* **77**: 726–731.
54. Boone LR, Innes CL, Heitman CK. (1990) Abrogation of Fv-1 restriction by genome-deficient virions produced by a retrovirus packaging cell line. *J Virol* **64**: 3376–3381.
55. Lilly F. (1970) Fv-2: Identification and location of a second gene governing the spleen focus response to Friend leukemia virus in mice. *J Natl Cancer Inst* **45**: 163–169.

56. Pincus T, Hartley JW, Rowe WP. (1971) A major genetic locus affecting resistance to infection with murine leukemia viruses. I. Tissue culture studies of naturally occurring viruses. *J Exp Med* **133**: 1219–1233.
57. Pincus T, Rowe WP, Lilly F. (1971) A major genetic locus affecting resistance to infection with murine leukemia viruses. II. Apparent identity to a major locus described for resistance to friend murine leukemia virus. *J Exp Med* **133**: 1234–1241.
58. Jolicoeur P, Baltimore D. (1976) Effect of Fv-1 gene product on proviral DNA formation and integration in cells infected with murine leukemia viruses. *Proc Natl Acad Sci USA* **73**: 2236–2240.
59. Best S, Le Tissier P, Towers G, *et al.* (1996) Positional cloning of the mouse retrovirus restriction gene Fv1. *Nature* **382**: 826–829.
60. Stremlau M, Owens CM, Perron MJ, *et al.* (2004) The cytoplasmic body component TRIM5alpha restricts HIV-1 infection in Old World monkeys. *Nature* **427**: 848–853.
61. Keckesova Z, Ylinen LM, Towers GJ. (2004) The human and African green monkey TRIM5alpha genes encode Ref1 and Lv1 retroviral restriction factor activities. *Proc Natl Acad Sci USA* **101**: 10780–10785.
62. Hatziioannou T, Perez-Caballero D, Yang A, *et al.* (2004) Retrovirus resistance factors Ref1 and Lv1 are species-specific variants of TRIM5alpha. *Proc Natl Acad Sci USA* **101**: 10774–10779.
63. Perron MJ, Stremlau M, Song B, *et al.* (2004) TRIM5alpha mediates the postentry block to N-tropic murine leukemia viruses in human cells. *Proc Natl Acad Sci USA* **101**: 11827–11832.
64. Yap MW, Nisole S, Lynch C, *et al.* (2004) Trim5alpha protein restricts both HIV-1 and murine leukemia virus. *Proc Natl Acad Sci USA* **101**: 10786–10791.
65. Sayah DM, Sokolskaja E, Berthoux L, *et al.* (2004) Cyclophilin A retrotransposition into TRIM5 explains owl monkey resistance to HIV-1. *Nature* **430**: 569–573.
66. Towers GJ, Hatziioannou T, Cowan S, *et al.* (2003) Cyclophilin A modulates the sensitivity of HIV-1 to host restriction factors. *Nat Med* **9**: 1138–1143.
67. Thali M, Bukovsky A, Kondo E, *et al.* (1994) Functional association of cyclophilin A with HIV-1 virions. *Nature* **372**: 363–365.
68. Besnier C, Takeuchi Y, Towers G. (2002) Restriction of lentivirus in monkeys. *Proc Natl Acad Sci USA* **99**: 11920–11925.

69. Cowan S, Hatziioannou T, Cunningham T, *et al.* (2002) Cellular inhibitors with Fv1-like activity restrict human and simian immunodeficiency virus tropism. *Proc Natl Acad Sci USA* **99**: 11914–11919.
70. Yap MW, Nisole S, Stoye JP. (2005) A single amino acid change in the SPRY domain of human Trim5alpha leads to HIV-1 restriction. *Curr Biol* **15**: 73–78.
71. Perez-Caballero D, Hatziioannou T, Zhang F, *et al.* (2005) Restriction of human immunodeficiency virus type 1 by TRIM-CypA occurs with rapid kinetics and independently of cytoplasmic bodies, ubiquitin, and proteasome activity. *J Virol* **79**: 15567–15572.
72. Stremlau M, Perron M, Lee M, *et al.* (2006) Specific recognition and accelerated uncoating of retroviral capsids by the TRIM5alpha restriction factor. *Proc Natl Acad Sci USA* **103**: 5514–5519.
73. Wu X, Anderson JL, Campbell EM, *et al.* (2006) Proteasome inhibitors uncouple rhesus TRIM5alpha restriction of HIV-1 reverse transcription and infection. *Proc Natl Acad Sci USA* **103**: 7465–7470.
74. Anderson JL, Campbell EM, Wu X, *et al.* (2006) Proteasome inhibition reveals that a functional preintegration complex intermediate can be generated during restriction by diverse TRIM5 proteins. *J Virol* **80**: 9754–9760.
75. Campbell EM, Perez O, Anderson JL, *et al.* (2008) Visualization of a proteasome-independent intermediate during restriction of HIV-1 by rhesus TRIM5alpha. *J Cell Biol* **180**: 549–561.
76. Rold CJ, Aiken C. (2008) Proteasomal degradation of TRIM5alpha during retrovirus restriction. *PLoS Pathog* **4**: e1000074.
77. Diaz-Griffero F, Perron M, McGee-Estrada K, *et al.* (2008) A human TRIM5alpha B30.2/SPRY domain mutant gains the ability to restrict and prematurely uncoat B-tropic murine leukemia virus. *Virology* **378**: 233–242.
78. Sakuma R, Mael AA, Ikeda Y. (2007) Alpha interferon enhances TRIM5alpha-mediated antiviral activities in human and rhesus monkey cells. *J Virol* **81**: 10201–10206.
79. Reymond A, Meroni G, Fantozzi A, *et al.* (2001) The tripartite motif family identifies cell compartments. *EMBO J* **20**: 2140–2151.
80. Song B, Diaz-Griffero F, Park DH, *et al.* (2005) TRIM5alpha association with cytoplasmic bodies is not required for antiretroviral activity. *Virology* **343**: 201–211.

81. Campbell EM, Dodding MP, Yap MW, *et al.* (2007) TRIM5 alpha cytoplasmic bodies are highly dynamic structures. *Mol Biol Cell* **18**: 2102–2111.
82. Li X, Sodroski J. (2008) The TRIM5alpha B-box 2 domain promotes cooperative binding to the retroviral capsid by mediating higher-order self-association. *J Virol* **82**: 11495–11502.
83. Sawyer SL, Wu LI, Emerman M, *et al.* (2005) Positive selection of primate TRIM5alpha identifies a critical species-specific retroviral restriction domain. *Proc Natl Acad Sci USA* **102**: 2832–2837.
84. Stremlau M, Perron M, Welikala S, *et al.* (2005) Species-specific variation in the B30.2(SPRY) domain of TRIM5alpha determines the potency of human immunodeficiency virus restriction. *J Virol* **79**: 3139–3145.
85. Liao CH, Kuang YQ, Liu HL, *et al.* (2007) A novel fusion gene, TRIM5-Cyclophilin A in the pig-tailed macaque determines its susceptibility to HIV-1 infection. *Aids* **21**(Suppl 8): S19–26.
86. Virgen CA, Kratovac Z, Bieniasz PD, *et al.* (2008) Independent genesis of chimeric TRIM5-cyclophilin proteins in two primate species. *Proc Natl Acad Sci USA* **105**: 3563–3568.
87. Carthagena L, Parise MC, Ringeard M, *et al.* (2008) Implication of TRIM alpha and TRIMCyp in interferon-induced anti-retroviral restriction activities. *Retrovirology* **5**: 59.
88. Newman RM, Hall L, Kirmaier A, *et al.* (2008) Evolution of a TRIM5-CypA splice isoform in Old World monkeys. *PLoS Pathog* **4**: e1000003.
89. Uchil PD, Quinlan BD, Chan WT, *et al.* (2008) TRIM E3 ligases interfere with early and late stages of the retroviral life cycle. *PLoS Pathog* **4**: e16.
90. Barr SD, Smiley JR, Bushman FD. (2008) The interferon response inhibits HIV particle production by induction of TRIM22. *PLoS Pathog* **4**: e1000007.
91. Kono K, Song H, Shingai Y, *et al.* (2008) Comparison of anti-viral activity of rhesus monkey and cynomolgus monkey TRIM5alphas against human immunodeficiency virus type 2 infection. *Virology* **373**: 447–456.
92. Strebel K, Klimkait T, Martin MA. (1988) A novel gene of HIV-1, Vpu, and its 16-kilodalton product. *Science* **241**: 1221–1223.
93. Strebel K. (2007) HIV accessory genes Vif and Vpu. *Adv Pharmacol* **55**: 199–232.
94. Neil SJ, Zang T, Bieniasz PD. (2008) Tetherin inhibits retrovirus release and is antagonized by HIV-1 Vpu. *Nature* **451**: 425–430.

95. Van Damme N, Goff D, Katsura C, *et al.* (2008) The interferon-induced protein BST-2 restricts HIV-1 release and is downregulated from the cell surface by the viral Vpu protein. *Cell Host Microbe* **3**: 245–252.
96. Varthakavi V, Smith RM, Bour SP, *et al.* (2003) Viral protein U counteracts a human host cell restriction that inhibits HIV-1 particle production. *Proc Natl Acad Sci USA* **100**: 15154–15159.
97. Goffinet C, Allespach I, Homann S, *et al.* (2009) HIV-1 antagonism of CD317 is species specific and involves Vpu-mediated proteasomal degradation of the restriction factor. *Cell Host Microbe* **5**: 285–297.
98. Miyagi E, Andrew AJ, Kao S, *et al.* (2009) Vpu enhances HIV-1 virus release in the absence of Bst-2 cell surface down-modulation and intracellular depletion. *Proc Natl Acad Sci USA* **106**: 2868–2873.
99. Gupta RK, Hue S, Schaller T, *et al.* (2009) Mutation of a single residue renders human tetherin resistant to HIV-1 Vpu-mediated depletion. *PLoS Pathog* **5**: e1000443.
100. Mitchell RS, Katsura C, Skasko MA, *et al.* (2009) Vpu Antagonizes BST-2-Mediated Restriction of HIV-1 Release via beta-TrCP and Endo-Lysosomal Trafficking. *PLoS Pathog* **5**: e1000450.
101. Dube M, Roy BB, Guiot-Guillain P, *et al.* (2009) Suppression of Tetherin-restricting activity upon human immunodeficiency virus type 1 particle release correlates with localization of Vpu in the trans-Golgi network. *J Virol* **83**: 4574–4590.
102. McNatt MW, Zang T, Hatziioannou T, *et al.* (2009) Species-specific activity of HIV-1 Vpu and positive selection of tetherin transmembrane domain variants. *PLoS Pathog* **5**: e1000300.
103. Kaletsky RL, Francica JR, Agrawal-Gamse C, *et al.* (2009) Tetherin-mediated restriction of filovirus budding is antagonized by the Ebola glycoprotein. *Proc Natl Acad Sci USA* **106**: 2886–2891.
104. Jouvenet N, Neil SJ, Zhadina M, *et al.* (2009) Broad-spectrum inhibition of retroviral and filoviral particle release by tetherin. *J Virol* **83**: 1837–1844.
105. Sakuma T, Noda T, Urata S, *et al.* (2009) Inhibition of Lassa and Marburg virus production by tetherin. *J Virol* **83**: 2382–2385.
106. Bartee E, McCormack A, Fruh K. (2006) Quantitative membrane proteomics reveals new cellular targets of viral immune modulators. *PLoS Pathog* **2**: e107.

107. Jia B, Serra-Moreno R, Neidermyer W, *et al.* (2009) Species-specific activity of SIV Nef and HIV-1 Vpu in overcoming restriction by tetherin/BST2. *PLoS Pathog* **5**: e1000429.
108. Tareen SU, Sawyer SL, Malik HS, *et al.* (2009) An expanded clade of rodent Trim5 genes. *Virology* **385**: 473–483.
109. Sawyer SL, Emerman M, Malik HS. (2007) Discordant evolution of the adjacent antiretroviral genes TRIM22 and TRIM5 in mammals. *PLoS Pathog* **3**: e197.

CHAPTER 14

Rev Revisited: Additional Functions of the HIV-1 Rev Protein

Bastian Grewe* and Klaus Überla*,†

SUMMARY

The major function attributed to the lentiviral Rev protein has been the export of viral unspliced and singly spliced mRNAs from the nucleus to the cytoplasm. There is increasing evidence that Rev also enhances translation and encapsidation of the RNA. Unspliced and singly spliced lentiviral RNA reaching the cytoplasm in the absence of Rev is inefficiently translated and encapsidated, but can be rescued by heterologous transport elements, post-transcriptional enhancer elements, or over-expression of certain cellular proteins. Prevention of formation of an inhibitory ribonucleoprotein complex with Rev-dependent viral mRNAs in the nucleus would explain why different effector molecules can enhance nuclear export, translation, and/or encapsidation.

* Department of Molecular and Medical Virology, Ruhr-University Bochum, Universitätsstr. 150, D44780 Bochum, Germany.

†Corresponding author. E-mail: klaus.ueberla@rub.de.

INTRODUCTION

In addition to the *gag*, *pol*, and *env* genes conserved among all retroviruses, lentiviruses contain accessory and regulatory genes that play important roles in virus replication and pathogenesis. One of the regulatory genes essential for virus replication is *rev*. The 116-amino acid Rev protein of HIV-1 is encoded from a multiply spliced transcript (see Fig. 1). By virtue of its nuclear localization (NLS) and export signals (NES), Rev is a nuclear-cytoplasmic shuttle protein. Nuclear import occurs via direct binding to importin beta or other nuclear import factors[1] while the Crm1 export pathway is targeted by the nuclear export signal of Rev.[2] The NLS is an arginine-rich region of Rev which also mediates specific binding to an RNA element of viral transcripts, designated the Rev responsive element (RRE). The RRE, located in the *env* open reading frame, is present on the 9 kb unspliced genomic transcript and on singly spliced approximately 4 kb large transcripts encoding Env and the accessory proteins Vif, Vpr, and Vpu (see Fig. 1). Splicing of HIV is a complex process that relies in general on strong splice donor (SD) and weak splice acceptor (SA) sites which are regulated by *cis*-acting sequences leading to a suboptimal level of splicing and the generation of over 40 RNA species. However, these RNAs can be subdivided in the mentioned 4 kb and fully spliced 2 kb transcripts. These small transcripts encode additional viral regulatory proteins including Rev (see Fig. 1).

Once sufficient Rev has accumulated, Rev triggers nuclear export of the *env* mRNA and the genomic RNA encoding Gag and GagPol. Thus, Rev enhances the ratio of genomic RNA and *env* mRNA to the smaller transcripts present in the cytoplasm. Therefore, the main function attributed to Rev during the viral replication cycle is export of viral RNAs encoding viral structural, enzymatic, and envelope proteins, allowing a switch from the early to the late phase of the viral replication cycle (see Fig. 2). Restricting expression of the late proteins during the early phase of the viral replication cycle might avoid recognition and elimination of the infected cells by the immune system prior to release of progeny viruses. According to a recent model of Rev-mediated export,[3] the viral genomic RNA with its suboptimal splice sites is transcribed and modified with a

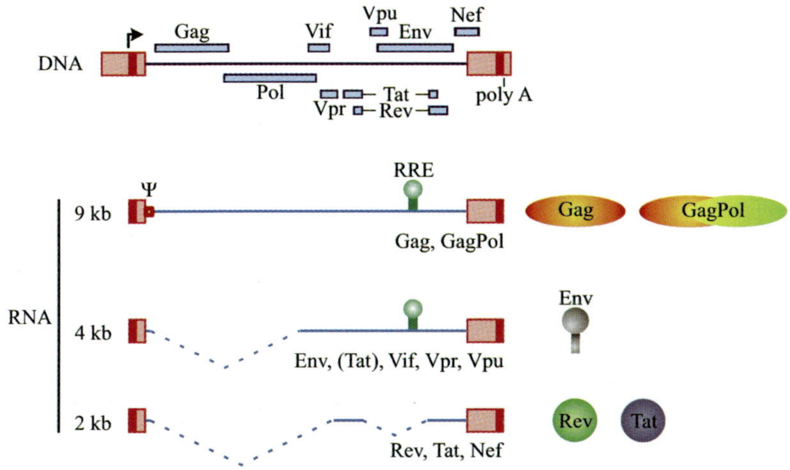

Fig. 1. Map of proviral DNA and mRNA species of HIV-1. To express all nine HIV-1 genes, a complex alternative splicing regimen evolved. The 9 kb primary unspliced transcript which encodes Gag and GagPol is processed into over 40 RNAs. However, all spliced RNAs can be subdivided in 4 kb singly spliced and 2 kb multiply spliced transcripts. Rev, Tat, and Nef are expressed from multiply spliced RNA species that lack all introns. Singly and unspliced transcripts retain intron-sequences including the Rev responsive element (RRE) which is spliced out in smaller RNAs. The 4 kb spliced mRNAs allow expression of Env, Vif, Vpr, or Vpu. The 9 kb RNA leads to Gag and by a ribosomal frameshift event also to GagPol expression. The unspliced RNA serves not only as mRNA but also as viral genome which is incorporated in viral particles. The encapsidation signal (Psi, ψ) overlaps the first splice donor. Therefore, the complete RNA structure for encapsidation is only present in full length RNA. (Figure modified from Ref. 2).

5′ CAP and a 3′ polyA tail. After 5′ and 3′ end modification, but before splicing, Rev binds to the RRE and shuttles the RNA to the cytoplasm.[4–6] Since a direct coupling between polyadenylation and splicing exists, Rev must act in a short time frame between RNA modification and processing. This is further supported by the observation that the Rev export function depends on continued transcription and does not correlate with the amount of RNA present in the nucleus.[7] This implies that once Rev missed the short time frame, the RNA is spliced, permanently sequestered in the nucleus, and/or degraded.

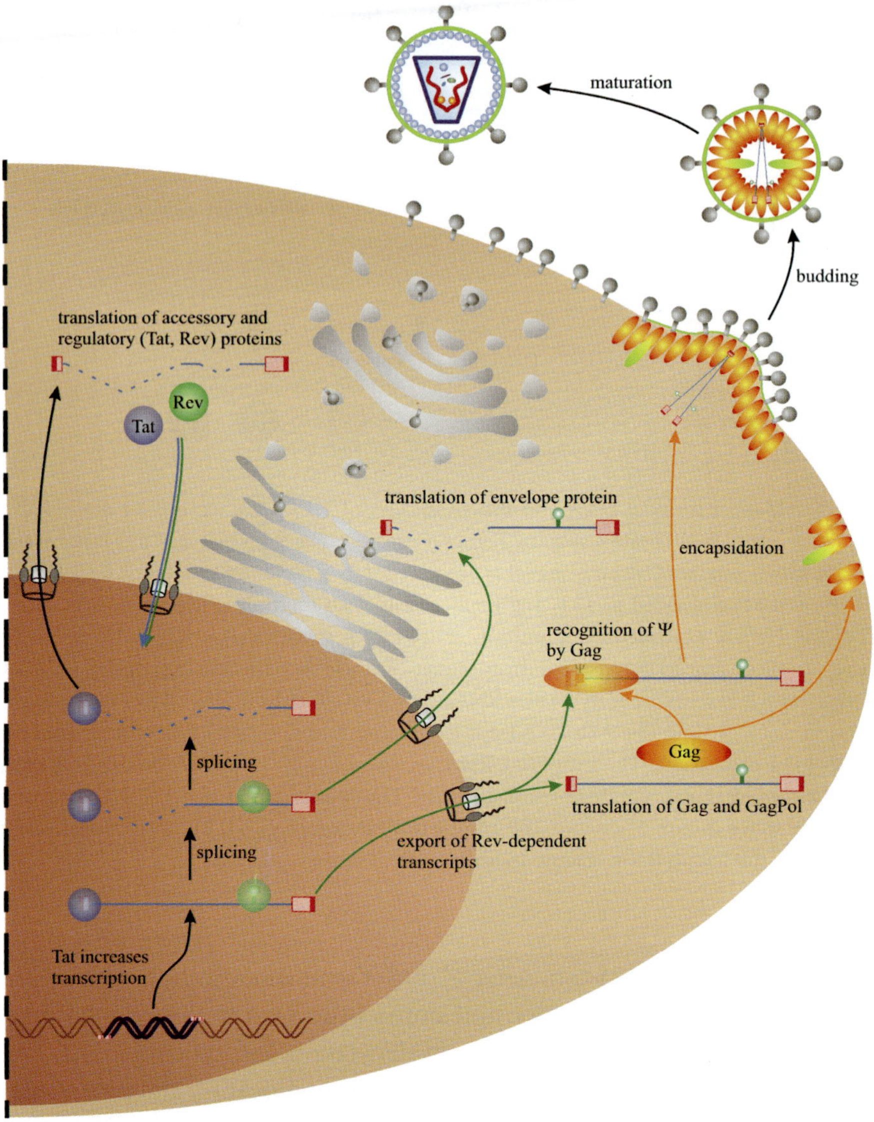

Fig. 2. Export function of Rev. After integration of the proviral DNA into the host cell genome, transcription takes place and leads, after complex splicing events, to fully spliced transcripts which are constitutively exported to the cytoplasm. Rev is expressed from some of the multiply spliced RNAs. After traveling to the nucleus, Rev binds to the RNA element RRE and mediates export of intron-containing transcripts. After translation of Env from singly spliced RNA and Gag as well as GagPol from unspliced RNA, new viral particles can assemble at the plasma membrane.

INFLUENCE OF REV ON SPLICING

Despite extensive evidence for a direct export function of Rev,[8] it remains difficult to formally exclude an additional inhibitory effect of Rev on splicing. Splice inhibition should also enhance the ratio of the cytoplasmic 9 and 4 kb RNAs to the small 2 kb transcripts. It was suggested that Rev actively disassembles the spliceosome, thereby rescuing the export of RNAs entrapped within the splicing machinery.[9] Consistent with this hypothesis, cellular splicing factors bind to Rev[10,11] and Rev-dependently to the RRE,[12] and Rev was reported to inhibit an *in vitro* splice reaction.[13] In addition, the strength of splice sites was shown to influence Rev's ability to enhance cytoplasmic levels of intron-containing RNAs. The strongest enhancement of cytoplasmic viral RNA levels by Rev could be observed with splice sites of intermediate strength.[9] Cytoplasmic levels of intron-containing RNAs with weak splice sites were exported independently of Rev. However, the influence of the splice sites on the magnitude of Rev´s enhancement of cytoplasmic RNA levels is consistent with both an inhibitory effect of Rev on splicing and a Rev export function.[14] If only a minor fraction of the primary transcript is spliced, inhibition of splicing does not enhance the levels of unspliced RNAs substantially. Similarly, primary transcripts with inefficient splice sites might resemble intron-less RNAs, leading to constitutive and efficient export to the cytoplasm independent of Rev. In this scenario, Rev-mediated export would have only modest stimulatory capacity. If splice sites are too strong, the RRE containing intron could be removed prior to Rev binding.[9,15] Again both, a potential splice-inhibitory effect and the export function of Rev, would be blocked by removal of the RRE. Thus, the modulation of the ratio of unspliced and differentially spliced transcripts in the cytoplasm cannot be taken as evidence for Rev-mediated inhibition of splicing. Since binding of Rev to RNA and nuclear shuttling of Rev are essential for Rev-mediated gene expression, it seems that the effect of Rev on the ratio of the differentially spliced transcripts in the cytoplasm is primarily due to Rev's export function and not due to inhibition of splicing.

DETERMINANTS OF REV-DEPENDENCE

The mechanistic details of nuclear retention of incompletely spliced HIV RNAs are still not fully understood. Nuclear sequestration of late HIV transcripts is thought to depend on partial spliceosome assembly and instability sequences called CRS (*cis*-acting repressive sequences) and INS (inhibitory/instability sequences).

As stated above, suboptimal splicing of HIV transcripts is essential for Rev-mediated export. This intermediate splicing is, in general, based on strong splice donor (first and fourth SD) and weak splice acceptor sites.[15–18] It is believed that partial binding of splicing factors is important for Rev function.[9,19] However, Rev-dependent expression does not depend on the splicing event itself. Binding of U1 snRNA via complementary base pairing to a strong 5′ splice donor results in an increased nuclear stability. Remarkably, this stabilization is not dependent on splicing but introduces Rev dependence.[15,20,21]

Furthermore, several sequence motifs throughout the viral genome have been identified as CRS or INS.[22–30] These elements fused to reporter constructs lead to Rev dependence. The exact influence of these sequence elements is still not unraveled, but it is possible that they serve as binding sites for cellular proteins that destabilize the transcript or block further utilization of the RNA in export and translation. Cellular binding partners of the INS-1 element in the matrix part of the *gag* mRNA were identified as polyadenylate-binding protein PABP1 and hnRNP A1.[31,32] Association of hnRNP C with a CRS in *pol* and *env* was also shown[30,33] and, more recently, the heterodimeric factor PSF/p54nrb was found to modulate expression of INS-containing *gag* and *env* mRNA[34] (see also below). Besides specific cellular binding partners, it is interesting to note that many CRS and INS elements are AU-rich.

By introducing silent point mutations in the INSs of *gag*, which decrease the abundance of A and U slightly, the expression of *gag* from subgenomic constructs could be rendered partially Rev-independent.[27,35] In addition to a potential contribution of cellular binding factors, the high AU content of the viral primary transcript also modulates Rev dependence. In the absence of Rev, AU-rich *gag* or *env* RNAs containing the wild-type codons are retained in the nucleus and only negligible amounts of the respective proteins are synthesized. Reducing the AU content of the mRNA by changing the wobble position of the codons without altering the encoded amino acid sequence leads to Rev-independent expression of Gag and Env. This codon optimization strategy was used to synthesize artificial expression constructs which apply the codon usage of mammalian cells but express unmutated HIV proteins totally independently of Rev.[36–39] Mechanistically, codon optimization increased nuclear stability of the RNA and enabled constitutive nuclear export of the synthetic transcript.[20,36] To test whether a high AU content is sufficient for rendering an mRNA Rev-dependent, Graf *et al.* generated an expression plasmid for an AU-rich GFP mRNA which contained the codon usage of HIV. In contrast to the parental GFP mRNA, which was adapted to the human codon usage, the AU-rich transcript was retained in the nucleus with little GFP protein being detectable. Including a 5′ splice donor, the RRE, a 3′ splice acceptor, and Rev rescued cytoplasmic RNA levels of the AU-rich GFP mRNA and GFP expression.[20]

All in all, this indicates that Rev dependence relies on a 5′ splice donor to stabilize the RNA in the nucleus, a high AU content, an RRE, and the presence of weak 3′ splice acceptor sites to allow suboptimal splicing.

TRANSLATIONAL EFFECTS OF REV

Expression of the structural HIV proteins from proviral constructs is dependent on Rev. Using sensitive ELISA-based assays for the quantification of HIV Gag proteins, Rev was found to enhance Gag levels from 100- to nearly 1,000-fold.[40–46] Since cytoplasmic RNA levels of the Gag-encoding RNA were enhanced by Rev only by a factor of 2 to 30,[4,40,43,47–50] it has been concluded that the strong stimulation of protein expression levels cannot be entirely due to Rev's effect on the cytoplasmic RNA levels. However, reports on the magnitude of the enhancement of cytoplasmic levels of Rev-dependent RNAs by Rev range from a 2- to 30-fold increase[4,40,43,47–50] to an absolute requirement for Rev.[51–54] While some of the differences could be attributed to variations of the expression plasmids used and other experimental conditions, such as over-expression by transient transfection and the cell type utilized, the methodologies for quantification of cytoplasmic RNA levels have also been questioned.

Limitations of Experimental Approaches for the Quantification of Rev-dependent Cytoplasmic RNA Levels

The cytoplasmic HIV RNA levels in the presence or absence of Rev have been determined either by (i) subcellular fractionation approaches or (ii) *in situ* hybridization. After subcellular fractionation, cytoplasmic RNA is purified and quantified either by northern blot analyses, RNase-protection assays, or quantitative PCR. A potential bias to all subcellular fractionation approaches is the contamination of the cytoplasmic fraction with the nuclear fraction. Since Rev has a much smaller effect on genomic and incompletely spliced HIV RNA levels in the nucleus, the larger the nuclear contamination, the lower the determined effect of Rev on cytoplasmic RNA levels. It is notable that studies using northern blot analyses or RNase-protection assays frequently report that there are no Rev-dependent RNA species at all in the cytoplasm in the absence of Rev.[51–54] However, in most of these studies, information about the sensitivity and the linear range of the assays and the precise factor by which Rev enhanced the cytoplasmic RNA levels were not provided. More recent publications using quantitative evaluation of northern blots reported a 10- to 40-fold enhancement of export, demonstrating that at least in these experiments a certain amount of cytoplasmic RNA was detectable without Rev.[42,44,55] Nevertheless, despite other drawbacks, the sensitivity and the dynamic range of carefully controlled PCR-based assays seem to be superior. For PCR-based assays, the Rev-effect on Rev-dependent HIV-transcripts is mostly reported to be in the range of 2- to 30-fold.[40,43,47,48,50] As an indicator of proper cell fractionation, the presence or absence of tRNA, ribosomal RNA, and snRNA in the respective nuclear or cytoplasmic fraction was evaluated. More important, the localization of unspliced cellular pre-mRNAs which should be retained in the nucleus was often quantitatively analyzed via RT-PCR. These experiments showed either no contamination[43] or detected minor contamination of pre-mRNA in the cytoplasmic fraction after lysis.[40,47,48] It is not clear if the nuclear contamination detected is sufficient to explain the observed export of unspliced HIV RNA without Rev. It is also not clear if cellular pre-mRNA is a suitable fractionation control in these assays. Cellular transcripts are normally very fast and efficiently spliced and do not accumulate to high amounts in the nucleus.[56–58] HIV transcripts, in contrast, contain suboptimal splice sites, and unspliced or partially spliced forms accumulate, at least to a certain degree, in the nucleus. Thus, the same cytoplasmic preparation might be more heavily contaminated with nuclear HIV RNA than pre-mRNAs of the host cell. However, there is also good evidence that the subcellular fractionation is working correctly and does not bias the results of cytoplasmic levels of Rev-dependent RNAs. In the absence of Rev, the half-life of a Rev-reporter RNA in the nucleus was substantially lower than the half-life of the same RNA in the cytoplasm.[4] Nuclear contamination would result in the same stability of "cytoplasmic" and nuclear RNA. Hence, at least in this study, the RNA detected in the cytoplasm in the absence of Rev cannot be due to contamination with nuclear RNA during fractionation.

In situ hybridization approaches overcome the problem of subcellular fractionation, namely potential nuclear leakage. Using this method, Rev seems to be absolutely required for cytoplasmic localization of Rev-dependent RNA species in many studies.[42,55,59,60] However, information about the sensitivity and the linear range of *in situ* hybridizations and a precise determination of the factor by which Rev enhances the cytoplasmic RNA levels are mostly lacking. In studies that used subcellular fractionation analyses and *in situ* hybridization side by side, a 10- to 40-fold reduced cytoplasmic RNA level was detected in northern blots. However, the hybridization images indicated an absolute requirement for Rev in RNA export.[42,55] Cmarko *et al.* performed a quantitative ultrastructural *in situ* hybridization study. The *env* mRNA levels in the cytoplasm, as determined by *in situ* hybridization and immune electron microscopy, increased by co-expression of Rev from 0.88 grains/μm^2 to 13.11 grains/μm.[2] Since the 0.88 grains/μm^2 value did not differ significantly from the background staining of 0.45 grains/μm,[2] Rev enhanced cytoplasmic RNA levels at least 15-fold.[61] Other studies investigating the importance of Rev in translation also performed *in situ* hybridization experiments and could detect a subpopulation (up to 20 percent) of cells in which cytoplasmic staining (fluorescent or by immunogold labeling) was clearly detectable in the absence of Rev.[43,48] Furthermore, one potential bias that could explain differences in the outcome of *in situ* hybridization and subcellular fractionation approaches has, to our knowledge, not yet been addressed. If Rev enhances the accessibility of the viral RNA by alteration of its ribonucleoprotein (RNP) complex, viral RNA that reached the cytoplasm in the absence of Rev might hybridize to the probes less efficiently, leading to an underestimation of cytoplasmic RNA levels by *in situ* hybridization.

It was also proposed that differences in expression levels between transfected and infected cells might influence the magnitude of the Rev effect, since over-expression could potentially saturate the mechanisms responsible for sequestration of RNA in the nucleus.[52] Therefore, Malim and Cullen infected a lymphoblastoid T-cell line with HIV-1 or an HIV-1 Rev-minus mutant.[52] Since both viruses expressed a selectable marker in place of the *nef* gene, selection of infected cell lines was possible, allowing the RNase-protection assay analysis of viral RNAs in the cytoplasm and the nucleus of the stably-infected cells. In the cytoplasmic fraction of the cell line infected with the Rev-expressing HIV-1, 42 percent of all viral transcripts were unspliced, while less than 1 percent of the transcripts were unspliced in the case of two cell lines infected with the Rev-minus HIV-1 mutant.[52] One caveat of this study is the use of a selectable marker expressed only from multiply spliced transcripts (see Fig. 1). Given the toxicity of viral proteins and selection for a drug resistance gene expressed from multiply spliced transcripts, cells with increased levels of multiply spliced RNAs might have a survival advantage leading to an abnormal cytoplasmic ratio of viral RNAs.

Further Evidence for a Translational Effect of Rev

Despite the limitations of the different assays discussed above, it seems that Rev-dependent viral RNAs can reach the cytoplasm in the absence of Rev. However, usage of these RNAs is inhibited and therefore differs from the same RNAs that reach the cytoplasm in the presence of Rev (see Fig. 3). In one study,[48] the cytoplasmic fraction was subdivided in a first hypertonic and a second detergent lysis fraction. In the presence of Rev, the distribution of unspliced *gagpol* RNAs in both fractions resembles that of cellular spliced RNA (GAPDH mRNA). In the absence of Rev, this profile

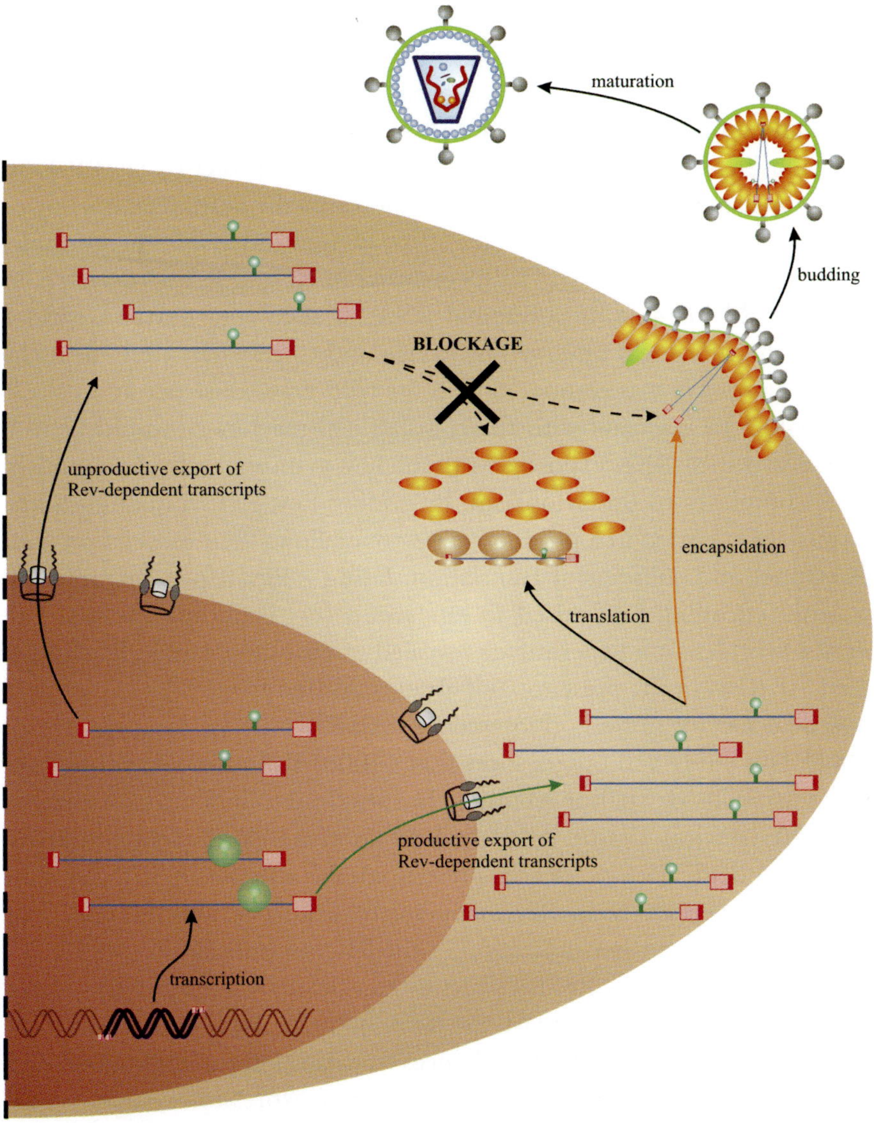

Fig. 3. Unproductive versus productive export of Rev-dependent RNA. There is convincing evidence that RNA export of Rev-dependent transcripts also takes place without Rev. Nonetheless, translation and encapsidation of these cytoplasmic RNAs are still strictly Rev-dependent. Rev-mediated export seems to be "productive" in regard to utilization of the transported RNA. RNAs exported without Rev are not accessible for translation or encapsidation. This export pathway is therefore depicted as "unproductive."

significantly changed, indicating that the RNA was localized in another cytoplasmic compartment. Using biochemical assays, it became further evident that RNAs exported in the absence of Rev were not associated with the translational machinery. These RNA molecules were not present in polyribosomes and did not bind to polyA binding protein 1 (PABP-1) in the cytoplasm.[40,43,62,63] In contrast, Rev increased the polysomal loading of unspliced transcripts dramatically.[40,43,62] D'Agostino *et al.*, for example, used a subgenomic HIV *gag*-RRE construct in HeLa cells and detected an over 16-fold increase in polysomal RNA by supplying Rev.[43] A similar remarkable effect was observed by Arrigo and Chen when a *rev*- or RRE-negative provirus was compared with the wild-type construct in B-cells.[40] The failure to associate with the translational machinery would fit to a model in which RNA exported without Rev is unproductively bound in RNP complexes in the cytoplasm. By adding Rev, RPE-containing transcripts are exported in a productive way that allows their association with ribosomes and subsequent translation. Using a Rev-dependent reporter plasmid encoding CAT fused to *env* sequences (including the RRE) in *in situ* hybridization experiments revealed intense cytoplasmic labeling of the CAT RNA in the presence and absence of Rev, while the CAT protein was detected by immunofluorescence staining only in the presence of Rev.[64] This study and others show that CRS or INS cannot only mediate RNA instability or nuclear retention, but also lead to inhibition of translation after nuclear export.[23,28,64] At least in some experimental systems, Rev-dependent RNA is clearly present in the cytoplasm without Rev. However, the underlying Rev-independent export is unproductive and does not lead to further utilization of the RNA in translation (see Fig. 3). Since RNA exported by Rev is efficiently translated, Rev itself or the export mediated by Rev seems to mark the RNA as "to be used."

To determine whether Rev enhances translation from RNAs directly in the cytoplasm, Perales *et al.* used a cytoplasmic expression system.[50] Cells were infected with a vaccinia virus expressing the T7 RNA polymerase in combination with transfection of an expression plasmid encoding Env under the control of a T7 promoter. Because of its size of 99 kDa, T7 RNA polymerase is localized in the cytoplasm. Transcription and translation of T7 promoter-containing plasmids should therefore take place in the cytoplasm; nuclear export of RNA is not necessary. High amounts of Env were detected without Rev in HeLa and COS cells. Adding Rev further increased the amount of Env in HeLa cells, but not in COS cells.[50] Although these results indicate that Rev can affect expression in the cytoplasm under some experimental conditions, it should be noted that vaccinia virus reprograms the translation machinery, leading to a dramatic host protein shut-off and allowing expression of proteins from RNAs that would not serve as efficient templates for translation in the absence of vaccinia virus infection.[65]

Hence, the emerging picture in nuclear expression systems is that Rev interacts with the RRE in unspliced and singly spliced transcripts in the nucleus and transports these RNA molecules via the cellular Crm1 export pathway to the cytoplasm. However, Rev's function does not end here but seems to deliver the exported RNA to polyribosomes and thus guarantees the proper translation of these RNAs (see Fig. 3). Without Rev, a certain degree of export takes place, but it is unproductive and does not lead to efficient utilization of the RNA in the cytoplasm.

Substitution of Rev by Heterologous Transport and Post-transcriptional Control Elements

The observation that Rev enhances the efficacy of translation in the cytoplasm could lead to the misleading conclusion that the translational effect of Rev described in nuclear expression systems is independent of its export function. Cytoplasmic NLS-deficient Rev mutants which are excluded from the nucleus are not able to promote translation.[66,67] This indicates that the translational effect of Rev is already predetermined in the nucleus.

Furthermore, substitution of Rev by heterologous export elements suggests that export and translational efficacy are coupled. Simple retroviruses like Mason-Pfizer Monkey Virus (MPMV) or Simian Retrovirus Type 1 (SRV-1) also encode Gag, GagPol, and Env but do not contain genes for regulatory proteins like Rev. Nuclear export of the unspliced transcripts was found to be based on *cis*-acting export sequences which were first identified in MPMV. This and all related RNA elements found afterwards were named constitutive transport elements (CTEs) to account for their function.[68] CTEs direct RNA to the cellular Tap export pathway. Tap and Crm1 represent different routes to the cytoplasm that are generally used by the bulk of mRNA and NES-containing shuttle proteins, respectively. Tap directly binds to the structured CTE elements in intron-containing RNAs and mediates their export to the cytoplasm.[69–72] Already in the first study in which the CTE of MPMV was identified, it was shown that the CTE is able to substitute for the Rev/RRE system in late gene expression and even HIV replication. Bray *et al.* used subgenomic expression plasmids for *env* and *gagpolvif* which contained either an RRE or a CTE. The CTE construct led to gene expression independent of the presence of Rev, whereas a clear Rev-dependence was attributed to the RRE constructs.[68] Furthermore, a Rev-deficient provirus was rendered replication competent by inserting a CTE in the 3′ LTR. However, the replication capacity was significantly hampered. Adding multiple CTE elements improved the expression efficiency and viral replication in a context-dependent manner. Nevertheless, the Rev/RRE export remained slightly more effective.[45,73] In all studies, Rev-independent expression could be attributed to constitutive export of CTE-containing RNA. Further, CTE elements in other simple retroviruses were discovered which also confered Rev-independent expression and replication.[72–75] Replacing the Rev/RRE system of SIV by a CTE even rescued the replication defect of the *rev*-negative provirus *in vivo.*[76] Thus, the Rev/RRE system can be functionally replaced with respect to its export function and its translational effects by heterologous RNA transport elements. The Rev protein itself therefore does not appear to mediate association of Rev-dependent RNA with the translational machinery. The translational effects of Rev seem to be secondary to its export function. Potential differences of the productive export pathways targeted by Rev/RRE or CTEs and the unproductive Rev-independent export seem to be associated with the nuclear history of the RNA molecules and will be discussed below.

In addition to heterologous nuclear export elements, a post-transcriptional control element of retroviruses also conferred Rev-independent expression of Gag and GagPol. This post-transcriptional control element (PCE) harbors a structured RNA motif which was shown to associate specifically with RNA helicase A (RHA).[77] Adding the PCE 5′ to *gag* or *gagpol* enhanced Gag and GagPol protein levels up to 1,000-fold without increasing the cytoplasmic levels of the unspliced mRNAs.[62,78,79] Deletion or loss-of-function mutants of the PCE reduced the amount of translated Gag to background levels. Also, downregulation of RHA had a deleterious effect on Gag expression. However, reduction of Gag did not correlate with reduced RNA stability, aberrant splicing, or cytoplasmic diminution of *gag* RNA. Prevention of RHA–RNA interaction instead inhibited polysomal loading of the unspliced *gagpol* RNA.[62,77–82] This indicates that Rev-dependent RNAs indeed reach the cytoplasm in the absence of Rev but that they are poorly translated, a block that can be overcome by a functional PCE.

Cellular Proteins Modulating the Efficacy of Translation of Rev-Dependent RNAs

Expression of the nuclear localized Sam68 protein and Sam68-like proteins SLM-1 and SLM-2 can partially substitute as well as synergize with Rev/RRE-mediated gene expression.[60] Since in some studies Sam68 alone did not alter the cytoplasmic accumulation of intron-containing RNA, nuclear acquisition of Sam68 specifically increased the translatability of RPE-containing unspliced RNA.[5,60,83] It has to be stressed that Sam68 induces expression of cytoplasmic unspliced HIV RNA exported in the absence of Rev, thereby providing evidence that cytoplasmic RNA is indeed present without Rev. This parallels the effect of RHA on PCE-containing *gagpol* expression vectors discussed above. Deletion of the NLS at the C-terminus of Sam68 resulted in a transdominant inhibitor of wild-type Sam68, designated Sam68ΔC, which inhibited Rev function.[60,83] Substitution of the NLS by a heterologous NLS restored the stimulatory capacity of Sam68.[60] Similar to Sam68, Sam68ΔC did not change overall RNA levels, splicing patterns, or cytoplasmic RNA accumulation. The inhibitory effect correlated with perinuclear accumulation of the exported intron-containing RNA.[60] However, in a recent study, Marsh *et al.* could show that perinuclear localization is the consequence rather than the cause of the translational inhibition. Mechanistically, Sam68ΔC inhibited efficient recruitment of *env* RNA to polysomes by blocking cytoplasmic binding of the polyA binding protein.[84] Since an NLS is required for its stimulatory properties, Sam68 seems to modulate the composition of the HIV RNP complex in the nucleus, thereby marking the HIV RNA for efficient translation.

Sam68 and RHA are believed to participate with the Rev-dependent RNA in the nucleus in RNP complexes whose formation allows efficient translation after being exported to the cytoplasm.[60,77,81] Therefore, the nuclear history of the RNA seems to promote or inhibit subsequent utilization of the cytoplasmic RNA. Further proteins implicated in the regulation of HIV RNA translatability that do not change the cytoplasmic unspliced RNA levels are the perinuclear localized hRIP[85,86] and the nuclear localized hnRNP E1 (PCBP-1).[87] In summary, a model emerges in which certain critical cellular factors (proteinous and non-proteinous) are important to form a nuclear RNP complex with the Rev-dependent RNA. Failure to build an appropriate complex in the absence of Rev does not strictly prevent export but inhibits the subsequent cytoplasmic utilization of the RNA. This inhibition could also be due to an inappropriate subcytoplasmic localization of the RNA, a failure of RNA processing, and perhaps inhibitory conformations of the RNA. All of these explanations are not mutually exclusive (see Fig. 4). Although substitution of the Rev/RRE system by heterologous CTEs redirects the RNA to a distinct nuclear export pathway, in this case a productive RNP complex also seems to be generated. A large number of HIV-unrelated examples underline that the nuclear history of an RNA molecule can indeed be important in regard to its cytoplasmic functions.[88,89]

Other nuclear interaction partners of the viral RNAs that bind CRS and INS have been identified, but mechanistic details are mostly lacking. They are believed to diminish RNA stability and export, but potentially also translation. These proteins are hnRNP A1,[90] polypyrimidine tract binding protein (PTB),[90] polypyrimidine tract binding protein-associated splicing factor (PSF) in complex with p54nrb,[34] polyA-binding protein (PABP),[31] and hnRNP C1.[33] PSF and p54nrb seem to act in the nucleus and in the cytoplasm. Over-expression of PSF induced a marked down-regulation of all HIV RNAs with a certain preference for unspliced and singly spliced transcripts, thereby reducing viral protein levels.[34] Binding of PABP-1 to INS-1 seems to contradict the finding of Campbell *et al.* that Rev-independently exported RNA does not associate with PABP-1.[31,63] Afonina *et al.* did not determine in which cellular compartment PABP-1 binds to its target sequence. PolyA binding proteins have nuclear

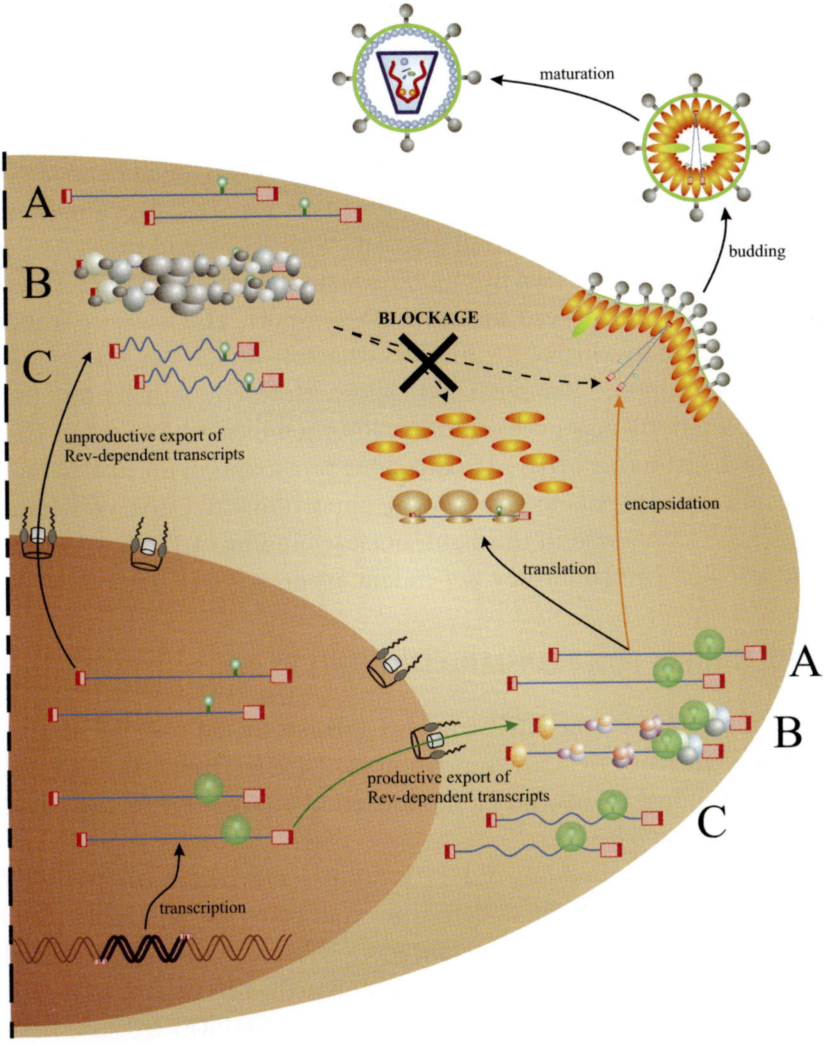

Fig. 4. Potential mechanisms leading to inefficient use of RNA after unproductive export. It is presently unclear what differentiates an unproductive export without Rev from a productive export pathway. Potential, not mutually exclusive, explanations are shown. A: Appropriate localization of the exported transcripts in subcytoplasmic compartments could be essential. B: Composition of ribonucleoprotein (RNP) complexes is known to be crucial for post-transcriptional events. In the case of HIV, lack of factors like hRIP or an excess of bound hnRNP E1 or dysregulation of the amount of hnRNP A2 and Staufen could block the usage of RNA in translation and packaging. C: Productive export could be defined by a special conformation or the proper processing of the RNA which allows translation and encapsidation.

and cytoplasmic functions[91] and in particular PABP-1 is a nuclear-cytoplasmic shuttle protein.[92] Therefore, nuclear binding of PABP-1 to INS-1 could have deleterious effects that are counteracted by Rev. In contrast, cytoplasmic binding of PABP-1 to the polyA tail of unspliced and partially spliced RNAs is necessary for efficient translation and could be markedly reduced after unproductive Rev-independent export. Surprisingly, the protein hnRNP A1 was found to associate with INS-1 and promote Rev-dependent export.[32] At least this binding factor therefore does not seem to counteract Rev but instead increases RNA export. This leads to a more complex model in which CRS and INS cofactors do not necessarily hamper but also support Rev-mediated export. However, without Rev these proteins could still negatively influence RNA stability, export, or translatability. Such factors are believed to direct the transcripts in the nucleus away from splicing hot spots to give Rev the opportunity to associate with the RRE.[25,31,32,34,92] Viewed in this light, nuclear binding of PABP-1 to INS-1 could also be essential for Rev-dependent export.

MODULATION OF POST-TRANSLATIONAL EVENTS BY REV

The composition of the RNP complex formed in the nucleus seems to affect the efficiency of RNA translation in the cytoplasm. However, even post-translational events can be modulated by nuclear events such as the Rev–RRE interaction.

In murine cells, budding of HIV is rather inefficient in comparison to human cell lines due to a defect in trafficking of Rev-dependently expressed Gag to the plasma membrane.[93,94] Transfection of mouse cells with *gagpol* or *gag* expression plasmids equipped with CTEs or a Hepatitis B virus post-transcriptional regulatory element (HPRE) led to efficient Gag expression and release of viral particles into the cell supernatant. Similarly, a codon optimized, Rev-independent *gagpol* vector allowed Gag expression and budding. Expression levels of intracellular Gag were adjusted to be comparable in the Rev-dependent and Rev-independent settings. Budding efficiencies observed with the Rev-independent vectors were restored to levels observed in human cell lines.[93,94] Dissecting this experimental outcome revealed that Rev-independently expressed Gag showed five-fold enhanced intracellular dimerization and a different intracellular localization.[93,94] Expression of *gagpol*-RRE in the presence of Rev resulted in a distribution throughout the cytoplasm whereas the *gagpol*-CTE construct led to accumulation of Gag at the plasma membrane.[94]

It has to be stressed that the amino acid sequences of all these Gag proteins are identical. The only difference among the expression constructs is the RNA export system (Rev/RRE or CTE or HPRE) or the RNA sequence (codon optimization). This clearly demonstrates an influence of the nuclear export route taken by the RNA molecule on the fate of the subsequently translated protein. Variation of the mRNA export pathway leads to variation of protein features. This far-reaching consequence once more underlines the importance of the nuclear history and the exact nature of the chosen nuclear export pathway in regard to subsequent post-transcriptional events.

The finding described above is not restricted to murine cells. Similar experiments in human cells showed that one particular codon optimized *gagpol* expression construct is also impaired in budding.[38,95] Codon optimization normally confers Rev-independent Gag expression and improved protein levels. Furthermore, these constructs are used to produce highly infectious lentiviral particles.[39] Hence, codon optimized expression vectors in general lead to a high amount of functional protein. Nonetheless, the vector mentioned constitutes an exception. After transfection, high amounts of intracellular Gag protein were detected but the supernatant did not contain a corresponding level of Gag particles and vector titers were reduced 130-fold in comparison to Rev-dependent *gagpol* expression plasmids.[96] Direct comparison with a Rev-dependent construct confirmed that Gag expressed from the codon optimized construct was impaired in budding and did not associate intracellularly with the plasma membrane.[95] In addition, Jin *et al.* reported that a *gag*-HPRE construct showed a similar defect.[93] In comparison with a Rev/RPE-containing construct, *gag*-HPRE led to similar expression levels but dimerization was 5-fold diminished and took place in intracellular compartments resulting in an approximately 10-fold budding defect.[93]

In these studies, the RNA was exported from the nucleus and was always efficiently translated. However, the translated proteins showed, dependent on the export of the RNA, some defects. The nuclear history of the transcript is therefore important not only for post-transcriptional but also for post-translational events. In the current model, the failure of Gag to assemble properly is caused by location of the mRNA within subcytoplasmic micro-compartments where oligomerization and trafficking of the translated Gag molecules cannot take place.[95]

ENHANCEMENT OF RNA PACKAGING BY REV

Once exported to the cytoplasm, the unspliced HIV RNA serves not only as mRNA for translation of Gag and GagPol but also represents the viral genome that has to be packaged into new viral particles. These particles are generated by Gag and GagPol through oligomerization at the plasma membrane. Gag furthermore recognizes specifically the encapsidation signal present in the 5′ UTR of the unspliced RNA in order to introduce the genome into the particles (see Fig. 2). The encapsidation signal is a highly structured RNA element that comprises four stem-loops. Deletions or critical point mutations in the sequence of the encapsidation signal reduce RNA encapsidation 2- to 20-fold (see, for example, Refs. 97–99). However, also variations of sequences up- and downstream of the core packaging signal diminish the encapsidation efficiency. The whole encapsidation signal therefore seems to comprise the 5′ UTR and up to 300 nt of *gag* (reviewed in Ref. 100).

Since Rev enhances the efficacy of translation most likely by modulating the RNP complex and/or subcytoplasmic localization of the RNA, Rev might also affect the encapsidation process of the genomic transcript. This was initially suggested by experiments demonstrating that insertion of a 1.1 kb *env* gene fragment containing the RRE into a minimal lentiviral vector enhanced vector titers and encapsidation as measured by the ratio of vector RNA levels in the viral particles and total cellular RNA.[101] However, since cytoplasmic vector RNA levels had not been determined, the enhancing effect could be entirely due to Rev's nuclear export function. Since Rev increases cytoplasmic levels of viral genomic RNA and GagPol expression, it is difficult to analyze the effect of Rev on steps of the viral replication cycle that depend on cytoplasmic RNA and GagPol protein levels, particularly in virus-infected cells or in cells transfected with proviral expression plasmids. However, the use of lentiviral vector systems with codon optimized *gagpol* expression plasmids allowed establishing of experimental conditions in which GagPol expression is completely independent of the

Rev/RRE system.[36,38,39] The titers of RPE-containing lentiviral vectors mobilized by transfection of previously vector-transduced cells with a Rev-independent *gagpol* expression plasmid and a heterologous envelope expression plasmid were increased 10,000-fold by Rev, although Rev only had a minor effect on genomic vector RNA levels in the cytoplasm.[102] Efficient Rev-independent nuclear export of lentiviral vector RNA has been repeatedly observed[38,103–105] and could, in principle, be due to large deletions of the vector compared with the provirus and the corresponding removal of AU-rich inhibitory sequences. In a recent study, it was shown that Rev specifically enhanced RNA encapsidation measured as the ratio of vector RNA levels in the viral particles and the cytoplasm.[106] The packaging effect of Rev correlated well with the enhancement of vector titers by Rev but not with the minor Rev-mediated cytoplasmic accumulation of unspliced vector RNA. It has to be stressed that under these experimental conditions, cytoplasmic vector RNA and high amounts of GagPol were present in the absence of Rev, but this did not lead to efficient encapsidation. Only adding Rev resulted in a highly efficient packaging process and high viral titers.[106] In order to exclude that the slight effect of Rev on cytoplasmic RNA levels and the strong enhancement of encapsidation are due to the artificial nature of the lentiviral vectors, proviral HIV-1 constructs with inactivating point mutations in the *rev* gene were also analyzed for the effect of Rev on cytoplasmic RNA levels and encapsidation of the HIV-1 genomic RNA. Rev supplied in *trans* enhanced cytoplasmic genomic RNA levels 4- to 59-fold depending on the cell line, while packaging of the genomic RNA was enhanced more than 500-fold.[66] This was once more not due to Rev-mediated stimulation of particle production, since high levels of GagPol were expressed by cotransfection of a Rev-independent *gagpol* expression plasmid. Therefore, Rev enhances genomic HIV RNA encapsidation to a much larger extent than cytoplasmic HIV RNA levels. The magnitude of the effect of Rev on packaging is comparable to, or even exceeds, the effect of deletion of parts of the multipartite HIV-1 packaging signal. Thus, in addition to interaction of the nucleocapsid domain of Gag with the packaging signal at the 5′ end of the genome, the Rev/RRE system provides lentiviruses with a second mechanism contributing to preferential encapsidation of genomic RNA.[106] Whether a recently discovered binding site for Rev in the first stem-loop of the encapsidation signal contributes to the packaging function of Rev remains to be investigated further.[107]

The data on the packaging function of Rev are consistent with a model in which export of unspliced HIV RNA without Rev is unproductive. This does not seem to be limited to translation but also extends to the subsequent encapsidation step. It is worth noting that translation is not a prerequisite for packaging.[108] In fact, inhibition of translation after Rev-dependent export slightly stimulated encapsidation. Rev-independently exported RNA is nevertheless not efficiently packaged. The RNP complex associated with the RNA, the localization of the transcript, or its conformation and processing are potential reasons for this defect (see Fig. 4). Since infectious lentiviral vector particles can also be generated by replacing the Rev/RRE system by heterologous export elements,[105,109–113] the packaging function of Rev might also be predetermined in the nucleus, as discussed above for Rev's translational function.

CONCLUDING REMARKS

A high AU content, strong stabilizing 5′ splice donor sites, and weak 3′ splice acceptor sites lead to retention of unspliced and singly spliced HIV transcripts in the nucleus. Although a minor fraction of such mRNAs reaches the cytoplasm in the absence of Rev, these cytoplasmic mRNAs are inefficiently translated and encapsidated. Possible reasons are the formation of an inappropriate ribonucleoprotein complex, failure to localize to specific subcytoplasmic compartments, and features of the RNA molecule itself like 5′ and 3′ end processing and its conformation (see Fig. 4). Rev/RRE, CTEs, PTE, or over-expression of cellular proteins (e.g., Sam68) can overcome this functional impairment of the transcripts in the cytoplasm. Prevention of the generation of an inhibitory RNP complex by a nuclear event would explain why such different effector molecules, which are linked to different export pathways, can enhance nuclear export, translation, and/or encapsidation. Thus, the use of cytoplasmic mRNA seems to be regulated by currently mostly unknown mechanisms. The influence of the nuclear export pathway on Gag protein function even suggests that post-translational events can be modulated at the RNA level in the nucleus. A better understanding of the post-transcriptional regulation of HIV RNA usage will not only shed light onto an important step in the lentiviral replication cycle, but will also provide new insights in fundamental cellular processes.

NOTE ADDED IN PROOF

After submission of the manuscript, three papers were published demonstrating a new and unexpected role of Rev in the early phase of the viral replication cycle of HIV-1. Rev interacts directly with the viral integrase, thereby inhibiting integration of the proviral DNA into the host cell chromosomes after reverse transcription. The authors propose that this mechanism prevents multiple integration events and super-infection.

Levin A, Rosenbluh J, Hayouka Z, *et al.* (2010) Integration of HIV-1 DNA is regulated by interplay between viral Rev and cellular LEDGF/p75 proteins. *Mol Med* **16**: 34–44.

Levin A, Hayouka Z, Brack-Werner R, *et al.* (2009) Novel regulation of HIV-1 replication and pathogenicity: Rev inhibition of integration. *Protein Eng Des Sel* **22**: 753–763.

Levin A, Hayouka Z, Heifer M, *et al.* (2009) Peptides derived from HIV-1 integrase that bind Rev stimulate viral genome integration. *PLoS One* **4**: e4155.

ACKNOWLEDGMENTS

We would like to thank Maik Blißenbach for sharing unpublished data. We also wish to thank Sabine Brandt as well as Bettina Tippler and Klaus Sure for their experimental contributions and excellent technical support, respectively. This work is funded by a grant from the DFG (Ue45/11-1). Bastian Grewe is supported by a fellowship from the DFG graduate school 1045/2.

REFERENCES

1. Arnold M, Nath A, Hauber J, *et al.* (2006) Multiple importins function as nuclear transport receptors for the Rev protein of human immunodeficiency virus type 1. *J Biol Chem* **281**: 20883–20890.
2. Cullen BR. (2003) Nuclear mRNA export: Insights from virology. *Trends Biochem Sci* **28**: 419–424.
3. Cochrane AW, McNally MT, Mouland AJ. (2006) The retrovirus RNA trafficking granule: From birth to maturity. *Retrovirology* **3**: 18.

4. Huang Y, Carmichael GG. (1996) Role of polyadenylation in nucleocytoplasmic transport of mRNA. *Mol Cell Biol* **16**: 1534–1542.
5. McLaren M, Asai K, Cochrane A. (2004) A novel function for Sam68: Enhancement of HIV-1 RNA 3′ end processing. *RNA* **10**: 1119–1129.
6. McLaren M, Marsh K, Cochrane A. (2008) Modulating HIV-1 RNA processing and utilization. *Front Biosci* **13**: 5693–5707.
7. Iacampo S, Cochrane A. (1996) Human immunodeficiency virus type 1 Rev function requires continued synthesis of its target mRNA. *J Virol* **70**: 8332–8339.
8. Pollard VW, Malim MH. (1998) The HIV-1 Rev protein. *Annu Rev Microbiol* **52**: 491–532.
9. Chang DD, Sharp PA. (1989) Regulation by HIV Rev depends upon recognition of splice sites. *Cell* **59**: 789–795.
10. Tange TO, Jensen TH, Kjems J. (1996) *In vitro* interaction between human immunodeficiency virus type 1 Rev protein and splicing factor ASF/SF2-associated protein, p32. *J Biol Chem* **271**: 10066–10072.
11. Zheng YH, Yu HF, Peterlin BM. (2003) Human p32 protein relieves a post-transcriptional block to HIV replication in murine cells. *Nat Cell Biol* **5**: 611–618.
12. Powell DM, Amaral MC, Wu JY, *et al.* (1997) HIV Rev-dependent binding of SF2/ASF to the Rev response element: Possible role in Rev-mediated inhibition of HIV RNA splicing. *Proc Natl Acad Sci USA* **94**: 973–978.
13. Kjems J, Frankel AD, Sharp PA. (1991) Specific regulation of mRNA splicing *in vitro* by a peptide from HIV-1 Rev. *Cell* **67**: 169–178.
14. Hope TJ. (1999) The ins and outs of HIV Rev. *Arch Biochem Biophys* **365**: 186–191.
15. Kammler S, Otte M, Hauber I, *et al.* (2006) The strength of the HIV-1 3′ splice sites affects Rev function. *Retrovirology* **3**: 89.
16. O'Reilly MM, McNally MT, Beemon KL. (1995) Two strong 5′ splice sites and competing, suboptimal 3′ splice sites involved in alternative splicing of human immunodeficiency virus type 1 RNA. *Virology* **213**: 373–385.
17. Si Z, Amendt BA, Stoltzfus CM. (1997) Splicing efficiency of human immunodeficiency virus type 1 *tat* RNA is determined by both a suboptimal 3′ splice site and a 10 nucleotide exon splicing silencer element located within *tat* exon 2. *Nucl Acid Res* **25**: 861–867.

18. Staffa A, Cochrane A. (1994) The *tat/rev* intron of human immunodeficiency virus type 1 is inefficiently spliced because of suboptimal signals in the 3′ splice site. *J Virol* **68**: 3071–3079.
19. Hammarskjold ML, Li H, Rekosh D, *et al.* (1994) Human immunodeficiency virus *env* expression becomes Rev-independent if the *env* region is not defined as an intron. *J Virol* **68**: 951–958.
20. Graf M, Ludwig C, Kehlenbeck S, *et al.* (2006) A quasi-lentiviral green fluorescent protein reporter exhibits nuclear export features of late human immunodeficiency virus type 1 transcripts. *Virology* **352**: 295–305.
21. Lu XB, Heimer J, Rekosh D, *et al.* (1990) U1 small nuclear RNA plays a direct role in the formation of a *rev*-regulated human immunodeficiency virus *env* mRNA that remains unspliced. *Proc Natl Acad Sci USA* **87**: 7598–7602.
22. Brighty DW, Rosenberg M. (1994) A *cis*-acting repressive sequence that overlaps the Rev-responsive element of human immunodeficiency virus type 1 regulates nuclear retention of *env* mRNAs independently of known splice signals. *Proc Natl Acad Sci USA* **91**: 8314–8318.
23. Cochrane AW, Jones KS, Beidas S, *et al.* (1991) Identification and characterization of intragenic sequences which repress human immunodeficiency virus structural gene expression. *J Virol* **65**: 5305–5313.
24. Maldarelli F, Martin MA, Strebel K. (1991) Identification of posttranscriptionally active inhibitory sequences in human immunodeficiency virus type 1 RNA: Novel level of gene regulation. *J Virol* **65**: 5732–5743.
25. Mikaelian I, Krieg M, Gait MJ, *et al.* (1996) Interactions of INS (CRS) elements and the splicing machinery regulate the production of Rev-responsive mRNAs. *J Mol Biol* **257**: 246–264.
26. Nasioulas G, Zolotukhin AS, Tabernero C, *et al.* (1994) Elements distinct from human immunodeficiency virus type 1 splice sites are responsible for the Rev dependence of *env* mRNA. *J Virol* **68**: 2986–2993.
27. Schwartz S, Campbell M, Nasioulas G, *et al.* (1992) Mutational inactivation of an inhibitory sequence in human immunodeficiency virus type 1 results in Rev-independent *gag* expression. *J Virol* **66**: 7176–7182.
28. Schwartz S, Felber BK, Pavlakis GN. (1992) Distinct RNA sequences in the *gag* region of human immunodeficiency virus type 1 decrease RNA stability and inhibit expression in the absence of Rev protein. *J Virol* **66**: 150–159.
29. Seguin B, Staffa A, Cochrane A. (1998) Control of human immunodeficiency virus type 1 RNA metabolism: Role of splice sites and intron sequences in unspliced viral RNA subcellular distribution. *J Virol* **72**: 9503–9513.

30. Suh D, Seguin B, Atkinson S, *et al.* (2003) Mapping of determinants required for the function of the HIV-1 *env* nuclear retention sequence. *Virology* **310**: 85–99.
31. Afonina E, Neumann M, Pavlakis GN. (1997) Preferential binding of poly(A)-binding protein 1 to an inhibitory RNA element in the human immunodeficiency virus type 1 *gag* mRNA. *J Biol Chem* **272**: 2307–2311.
32. Najera I, Krieg M, Karn J. (1999) Synergistic stimulation of HIV-1 *rev*-dependent export of unspliced mRNA to the cytoplasm by hnRNP A1. *J Mol Biol* **285**: 1951–1964.
33. Olsen HS, Cochrane AW, Rosen C. (1992) Interaction of cellular factors with intragenic *cis*-acting repressive sequences within the HIV genome. *Virology* **191**: 709–715.
34. Zolotukhin AS, Michalowski D, Bear J, *et al.* (2003) PSF acts through the human immunodeficiency virus type 1 mRNA instability elements to regulate virus expression. *Mol Cell Biol* **23**: 6618–6630.
35. Schneider R, Campbell M, Nasioulas G, *et al.* (1997) Inactivation of the human immunodeficiency virus type 1 inhibitory elements allows Rev-independent expression of Gag and Gag/protease and particle formation. *J Virol* **71**: 4892–4903.
36. Graf M, Bojak A, Deml L, *et al.* (2000) Concerted action of multiple *cis*acting sequences is required for Rev dependence of late human immunodeficiency virus type 1 gene expression. *J Virol* **74**: 10822–10826.
37. Haas J, Park EC, Seed B. (1996) Codon usage limitation in the expression of HIV-1 envelope glycoprotein. *Curr Biol* **6**: 315–324.
38. Kotsopoulou E, Kim VN, Kingsman AJ, *et al.* (2000) A Rev-independent human immunodeficiency virus type 1 (HIV-1)-based vector that exploits a codon-optimized HIV-1 *gag-pol* gene. *J Virol* **74**: 4839–4852.
39. Wagner R, Graf M, Bieler K, *et al.* (2000) Rev-independent expression of synthetic *gag-pol* genes of human immunodeficiency virus type 1 and simian immunodeficiency virus: Implications for the safety of lentiviral vectors. *Hum Gene Ther* **11**: 2403–2413.
40. Arrigo SJ, Chen IS. (1991) Rev is necessary for translation but not cytoplasmic accumulation of HIV-1 *vif*, *vpr*, and *env/vpu* 2 RNAs. *Genes Dev* **5**: 808–819.
41. Arrigo SJ, Weitsman S, Rosenblatt JD, *et al.* (1989) Analysis of *rev* gene function on human immunodeficiency virus type 1 replication in lymphoid cells by using a quantitative polymerase chain reaction method. *J Virol* **63**: 4875–4881.

42. Coyle JH, Guzik BW, Bor YC, *et al.* (2003) Sam68 enhances the cytoplasmic utilization of intron-containing RNA and is functionally regulated by the nuclear kinase Sik/BRK. *Mol Cell Biol* **23**: 92–103.
43. D'Agostino DM, Felber BK, Harrison JE, *et al.* (1992) The Rev protein of human immunodeficiency virus type 1 promotes polysomal association and translation of *gag/pol* and *vpu/env* mRNAs. *Mol Cell Biol* **12**: 1375–1386.
44. Guzik BW, Levesque L, Prasad S, *et al.* (2001) NXT1 (p15) is a crucial cellular cofactor in TAP-dependent export of intron-containing RNA in mammalian cells. *Mol Cell Biol* **21**: 2545–2554.
45. Wodrich H, Schambach A, Krausslich HG. (2000) Multiple copies of the Mason-Pfizer monkey virus constitutive RNA transport element lead to enhanced HIV-1 Gag expression in a context-dependent manner. *Nucl Acid Res* **28**: 901–910.
46. Yi R, Bogerd HP, Cullen BR. (2002) Recruitment of the Crm1 nuclear export factor is sufficient to induce cytoplasmic expression of incompletely spliced human immunodeficiency virus mRNAs. *J Virol* **76**: 2036–2042.
47. Favaro JP, Arrigo SJ. (1997) Characterization of Rev function using subgenomic and genomic constructs in T and COS cells. *Virology* **228**: 29–38.
48. Favaro JP, Maldarelli F, Arrigo SJ, *et al.* (1999) Effect of *rev* on the cytoplasmic localization of intron-containing human immunodeficiency virus type 1 RNA. *Virology* **255**: 237–249.
49. Kimura T, Hashimoto I, Nishikawa M, *et al.* (1996) A role for Rev in the association of HIV-1 *gag* mRNA with cytoskeletal beta-actin and viral protein expression. *Biochimie* **78**: 1075–1080.
50. Perales C, Carrasco L, Gonzalez ME. (2005) Regulation of HIV-1 *env* mRNA translation by Rev protein. *Biochim Biophys Acta* **1743**: 169–175.
51. Felber BK, Hadzopoulou-Cladaras M, Cladaras C, *et al.* (1989) Rev protein of human immunodeficiency virus type 1 affects the stability and transport of the viral mRNA. *Proc Natl Acad Sci USA* **86**: 1495–1499.
52. Malim MH, Cullen BR. (1993) Rev and the fate of pre-mRNA in the nucleus: Implications for the regulation of RNA processing in eukaryotes. *Mol Cell Biol* **13**: 6180–6189.
53. Malim MH, Hauber J, Fenrick R, *et al.* (1988) Immunodeficiency virus *rev* trans-activator modulates the expression of the viral regulatory genes. *Nature* **335**: 181–183.

54. Malim MH, Hauber J, Le SY, *et al.* (1989) The HIV-1 *rev* trans-activator acts through a structured target sequence to activate nuclear export of unspliced viral mRNA. *Nature* **338**: 254–257.
55. Jin L, Guzik BW, Bor YC, *et al.* (2003) Tap and NXT promote translation of unspliced mRNA. *Genes Dev* **17**: 3075–3086.
56. Izaurralde E, Mattaj IW. (1995) RNA export. *Cell* **81**: 153–159.
57. Kohler A, Hurt E. (2007) Exporting RNA from the nucleus to the cytoplasm. *Nat Rev Mol Cell Biol* **8**: 761–773.
58. Nakielny S, Fischer U, Michael WM, *et al.* (1997) RNA transport. *Annu Rev Neurosci* **20**: 269–301.
59. Emerman M, Vazeux R, Peden K. (1989) The *rev* gene product of the human immunodeficiency virus affects envelope-specific RNA localization. *Cell* **57**: 1155–1165.
60. Soros VB, Carvajal HV, Richard S, *et al.* (2001) Inhibition of human immunodeficiency virus type 1 Rev function by a dominant-negative mutant of Sam68 through sequestration of unspliced RNA at perinuclear bundles. *J Virol* **75**: 8203–8215.
61. Cmarko D, Boe SO, Scassellati C, *et al.* (2002) Rev inhibition strongly affects intracellular distribution of human immunodeficiency virus type 1 RNAs. *J Virol* **76**: 10473–10484.
62. Butsch M, Hull S, Wang Y, *et al.* (1999) The 5′ RNA terminus of spleen necrosis virus contains a novel posttranscriptional control element that facilitates human immunodeficiency virus Rev/RRE-independent Gag production. *J Virol* **73**: 4847–4855.
63. Campbell LH, Borg KT, Haines JK, *et al.* (1994) Human immunodeficiency virus type 1 Rev is required *in vivo* for binding of poly(A)-binding protein to Rev-dependent RNAs. *J Virol* **68**: 5433–5438.
64. Lawrence JB, Cochrane AW, Johnson CV, *et al.* (1991) The HIV-1 Rev protein: A model system for coupled RNA transport and translation. *New Biol* **3**: 1220–1232.
65. Elroy-Stein O, Moss B. (1990) Cytoplasmic expression system based on constitutive synthesis of bacteriophage T7 RNA polymerase in mammalian cells. *Proc Natl Acad Sci USA* **87**: 6743–6747.
66. Blissenbach M, Grewe B, Hoffmann B, *et al.* (2010) Nuclear RNA export and packaging functions of HIV-1 Rev revisited. *J Virol* (in press).
67. Nam YS, Petrovic A, Jeong KS, *et al.* (2001) Exchange of the basic domain of human immunodeficiency virus type 1 Rev for a polyarginine stretch

expands the RNA binding specificity, and a minimal arginine cluster is required for optimal RRE RNA binding affinity, nuclear accumulation, and trans-activation. *J Virol* 75: 2957–2971.

68. Bray M, Prasad S, Dubay JW, *et al.* (1994) A small element from the Mason-Pfizer monkey virus genome makes human immunodeficiency virus type 1 expression and replication Rev-independent. *Proc Natl Acad Sci USA* **91**: 1256–1260.
69. Ernst RK, Bray M, Rekosh D, *et al.* (1997) A structured retroviral RNA element that mediates nucleocytoplasmic export of intron-containing RNA. *Mol Cell Biol* **17**: 135–144.
70. Gruter P, Tabernero C, von Kobbe C, *et al.* (1998) TAP, the human homolog of Mex67p, mediates CTE-dependent RNA export from the nucleus. *Mol Cell* **1**: 649–659.
71. Pasquinelli AE, Ernst RK, Lund E, *et al.* (1997) The constitutive transport element (CTE) of Mason-Pfizer monkey virus (MPMV) accesses a cellular mRNA export pathway. *EMBO J* **16**: 7500–7510.
72. Zolotukhin AS, Valentin A, Pavlakis GN, *et al.* (1994) Continuous propagation of RRE(-) and Rev(-)RRE(-) human immunodeficiency virus type 1 molecular clones containing a *cis*-acting element of simian retrovirus type 1 in human peripheral blood lymphocytes. *J Virol* **68**: 7944–7952.
73. Smulevitch S, Bear J, Alicea C, *et al.* (2006) RTE and CTE mRNA export elements synergistically increase expression of unstable, Rev-dependent HIV and SIV mRNAs. *Retrovirology* **3**: 6.
74. Nappi F, Schneider R, Zolotukhin A, *et al.* (2001) Identification of a novel posttranscriptional regulatory element by using a *rev*- and RRE-mutated human immunodeficiency virus type 1 DNA proviral clone as a molecular trap. *J Virol* **75**: 4558–4569.
75. Wodrich H, Bohne J, Gumz E, *et al.* (2001) A new RNA element located in the coding region of a murine endogenous retrovirus can functionally replace the Rev/Rev-responsive element system in human immunodeficiency virus type 1 Gag expression. *J Virol* **75**: 10670–10682.
76. von Gegerfelt AS, Liska V, Ray NB, *et al.* (1999) Persistent infection of rhesus macaques by the *rev*-independent Nef(-) simian immunodeficiency virus SIVmac239: Replication kinetics and genomic stability. *J Virol* **73**: 6159–6165.
77. Hartman TR, Qian S, Bolinger C, *et al.* (2006) RNA helicase A is necessary for translation of selected messenger RNAs. *Nat Struct Mol Biol* **13**: 509–516.

78. Hull S, Boris-Lawrie K. (2002) RU5 of Mason-Pfizer monkey virus 5′ long terminal repeat enhances cytoplasmic expression of human immunodeficiency virus type 1 *gag-pol* and nonviral reporter RNA. *J Virol* **76**: 10211–10218.
79. Roberts TM, Boris-Lawrie K. (2003) Primary sequence and secondary structure motifs in spleen necrosis virus RU5 confer translational utilization of unspliced human immunodeficiency virus type 1 reporter RNA. *J Virol* **77**: 11973–11984.
80. Bolinger C, Yilmaz A, Hartman TR, *et al.* (2007) RNA helicase A interacts with divergent lymphotropic retroviruses and promotes translation of human T-cell leukemia virus type 1. *Nucl Acid Res* **35**: 2629–2642.
81. Dangel AW, Hull S, Roberts TM, *et al.* (2002) Nuclear interactions are necessary for translational enhancement by spleen necrosis virus RU5. *J Virol* **76**: 3292–3300.
82. Roberts TM, Boris-Lawrie K. (2000) The 5′ RNA terminus of spleen necrosis virus stimulates translation of nonviral mRNA. *J Virol* **74**: 8111–8118.
83. Reddy TR, Xu W, Mau JK, *et al.* (1999) Inhibition of HIV replication by dominant negative mutants of Sam68, a functional homolog of HIV-1 Rev. *Nat Med* **5**: 635–642.
84. Marsh K, Soros V, Cochrane A. (2008) Selective translational repression of HIV-1 RNA by Sam68DeltaC occurs by altering PABP1 binding to unspliced viral RNA. *Retrovirology* **5**: 97.
85. Sanchez-Velar N, Udofia EB, Yu Z, *et al.* (2004) hRIP, a cellular cofactor for Rev function, promotes release of HIV RNAs from the perinuclear region. *Genes Dev* **18**: 23–34.
86. Yu Z, Sanchez-Velar N, Catrina IE, *et al.* (2005) The cellular HIV-1 Rev cofactor hRIP is required for viral replication. *Proc Natl Acad Sci USA* **102**: 4027–4032.
87. Woolaway K, Asai K, Emili A, *et al.* (2007) hnRNP E1 and E2 have distinct roles in modulating HIV-1 gene expression. *Retrovirology* **4**: 28.
88. Giorgi C, Moore MJ. (2007) The nuclear nurture and cytoplasmic nature of localized mRNPs. *Semin Cell Dev Biol* **18**: 186–193.
89. Palacios IM. (2007) How does an mRNA find its way? Intracellular localisation of transcripts. *Semin Cell Dev Biol* **18**: 163–170.
90. Black AC, Luo J, Chun S, *et al.* (1996) Specific binding of polypyrimidine tract binding protein and hnRNP A1 to HIV-1 CRS elements. *Virus Genes* **12**: 275–285.

91. Mangus DA, Evans MC, Jacobson A. (2003) Poly(A)-binding proteins: Multifunctional scaffolds for the post-transcriptional control of gene expression. *Genome Biol* **4**: 223.
92. Afonina E, Stauber R, Pavlakis GN. (1998) The human poly(A)-binding protein 1 shuttles between the nucleus and the cytoplasm. *J Biol Chem* **273**: 13015–13021.
93. Jin J, Sturgeon T, Chen C, *et al.* (2007) Distinct intracellular trafficking of equine infectious anemia virus and human immunodeficiency virus type 1 Gag during viral assembly and budding revealed by bimolecular fluorescence complementation assays. *J Virol* **81**: 11226–11235.
94. Swanson CM, Puffer BA, Ahmad KM, *et al.* (2004) Retroviral mRNA nuclear export elements regulate protein function and virion assembly. *EMBO J* **23**: 2632–2640.
95. Swanson CM, Malim MH. (2006) Retrovirus RNA trafficking: From chromatin to invasive genomes. *Traffic* **7**: 1440–1450.
96. Bahner I, Sumiyoshi T, Kagoda M, *et al.* (2007) Lentiviral vector transduction of a dominant-negative Rev gene into human CD34+ hematopoietic progenitor cells potently inhibits human immunodeficiency virus-1 replication. *Mol Ther* **15**: 76–85.
97. Clever JL, Parslow TG. (1997) Mutant human immunodeficiency virus type 1 genomes with defects in RNA dimerization or encapsidation. *J Virol* **71**: 3407–3414.
98. Harrison GP, Miele G, Hunter E, *et al.* (1998) Functional analysis of the core human immunodeficiency virus type 1 packaging signal in a permissive cell line. *J Virol* **72**: 5886–5896.
99. Houzet L, Paillart JC, Smagulova F, *et al.* (2007) HIV controls the selective packaging of genomic, spliced viral and cellular RNAs into virions through different mechanisms. *Nucl Acid Res* **35**: 2695–2704.
100. D'Souza V, Summers MF. (2005) How retroviruses select their genomes. *Nat Rev Microbiol* **3**: 643–655.
101. Richardson JH, Child LA, Lever AM. (1993) Packaging of human immunodeficiency virus type 1 RNA requires *cis*-acting sequences outside the 5′ leader region. *J Virol* **67**: 3997–4005.
102. Lucke S, Grunwald T, Uberla K. (2005) Reduced mobilization of Rev-responsive element-deficient lentiviral vectors. *J Virol* **79**: 9359–9362.

103. Anson DS, Fuller M. (2003) Rational development of a HIV-1 gene therapy vector. *J Gene Med* 5: 829–838.
104. Cui Y, Iwakuma T, Chang LJ. (1999) Contributions of viral splice sites and *cis*-regulatory elements to lentivirus vector function. *J Virol* 73: 6171–6176.
105. Mautino MR, Ramsey WJ, Reiser J, *et al.* (2000) Modified human immunodeficiency virus-based lentiviral vectors display decreased sensitivity to trans-dominant Rev. *Hum Gene Ther* 11: 895–908.
106. Brandt S, Blissenbach M, Grewe B, *et al.* (2007) Rev proteins of human and simian immunodeficiency virus enhance RNA encapsidation. *PLoS Pathog* 3: e54.
107. Greatorex JS, Palmer EA, Pomerantz RJ, *et al.* (2006) Mutation of the Rev-binding loop in the human immunodeficiency virus 1 leader causes a replication defect characterized by altered RNA trafficking and packaging. *J Gen Virol* 87: 3039–3044.
108. Butsch M, Boris-Lawrie K. (2000) Translation is not required to generate virion precursor RNA in human immunodeficiency virus type 1-infected T cells. *J Virol* 74: 11531–11537.
109. Mautino MR, Keiser N, Morgan RA. (2000) Improved titers of HIV-based lentiviral vectors using the SRV-1 constitutive transport element. *Gene Ther* 7: 1421–1424.
110. Mautino MR, Keiser N, Morgan RA. (2001) Inhibition of human immunodeficiency virus type 1 (HIV-1) replication by HIV-1-based lentivirus vectors expressing transdominant Rev. *J Virol* 75: 3590–3599.
111. Oh T, Bajwa A, Jia G, *et al.* (2007) Lentiviral vector design using alternative RNA export elements. *Retrovirology* 4: 38.
112. Srinivasakumar N, Chazal N, Helga-Maria C, *et al.* (1997) The effect of viral regulatory protein expression on gene delivery by human immunodeficiency virus type 1 vectors produced in stable packaging cell lines. *J Virol* 71: 5841–5848.
113. Srinivasakumar N, Schuening FG. (1999) A lentivirus packaging system based on alternative RNA transport mechanisms to express helper and gene transfer vector RNAs and its use to study the requirement of accessory proteins for particle formation and gene delivery. *J Virol* 73: 9589–9598.

CHAPTER 15

HIV-1 Interactions with Small RNA Induced Silencing Mechanisms

Joost Haasnoot* and Ben Berkhout

SUMMARY

Small non-coding RNA-induced silencing mechanisms such as the RNA interference (RNAi) and microRNA (miRNA) pathways play important roles in the regulation of eukaryotic cellular gene expression. Recent studies show that these mechanisms also affect replication of the human immunodeficiency virus type 1 (HIV-1). For example, cellular miRNAs target and repress the expression of HIV-1 mRNAs. This miRNA-mediated inhibition of HIV-1 is important for establishing HIV-1 latency and determining viral tropism. HIV-1 also encodes an RNA silencing suppressor (RSS) that can counter inhibitory RNAi effects. In addition, the virus encodes its own miRNA that represses the expression of specific cellular genes involved in apoptosis. These and other interactions between HIV-1 and small RNA-induced silencing processes form a complex network of

*Corresponding author: Laboratory of Experimental Virology, Department of Medical Microbiology, Center for Infection and Immunity Amsterdam (CINIMA), Academic Medical Center, University of Amsterdam, Meibergdreef 15, 1105 AZ Amsterdam, The Netherlands, e-mail: p.c.haasnoot@amc.uva.nl.

positive and negative forces affecting virus replication. In this chapter, we summarize the available data on these virus-cell interactions. Furthermore, we discuss the various strategies to use RNAi-based therapeutics for inhibition of HIV-1.

INTRODUCTION

Since HIV-1 was discovered in 1983 as the causative agent of acquired immuno deficiency syndrome (AIDS), researchers have studied its interaction with host cellular processes. It is currently well established that HIV-1 infection affects the expression of many cellular genes, altering physiological aspects of both the infected cell and the infected host as a whole.[1,2] Small RNA-induced gene silencing, involving the RNA interference (RNAi) and the microRNA (miRNA) pathways, has only recently been added to the list of important cellular processes that are influenced by HIV-1 infection, and that in turn affect virus replication.

RNA-induced gene silencing is triggered by small RNA molecules of 20–30 nucleotides that associate with Argonaute or Piwi proteins and serve as guides in the sequence-specific inhibition of gene expression.[3] These mechanisms play important roles in heterochromatin formation, mRNA destabilization and translational control. Thus, RNA silencing is involved in diverse biological processes such as development, cell differentiation and proliferation, cell death, metabolism, transposon silencing and antiviral defences.

Currently, three classes of small RNAs involved in silencing have been identified in mammals: miRNAs, small interfering RNAs (siRNAs) and Piwi-associated RNAs (piRNAs). In mammals, the pathway for miRNA-mediated silencing is best characterized (Fig. 1). This regulatory pathway is triggered by endogenously expressed miRNAs of 21–22 nucleotides.[4] miRNAs are incorporated in the multi-protein RNA-induced silencing complex (RISC), which contains one of four possible (AGO1–4) Argonaute proteins. Once loaded in RISC, miRNAs trigger translational inhibition via imperfect base pairing to sequences in the 3′ untranslated region (3′ UTR) of target mRNAs.[5] Currently, over 600 human miRNAs

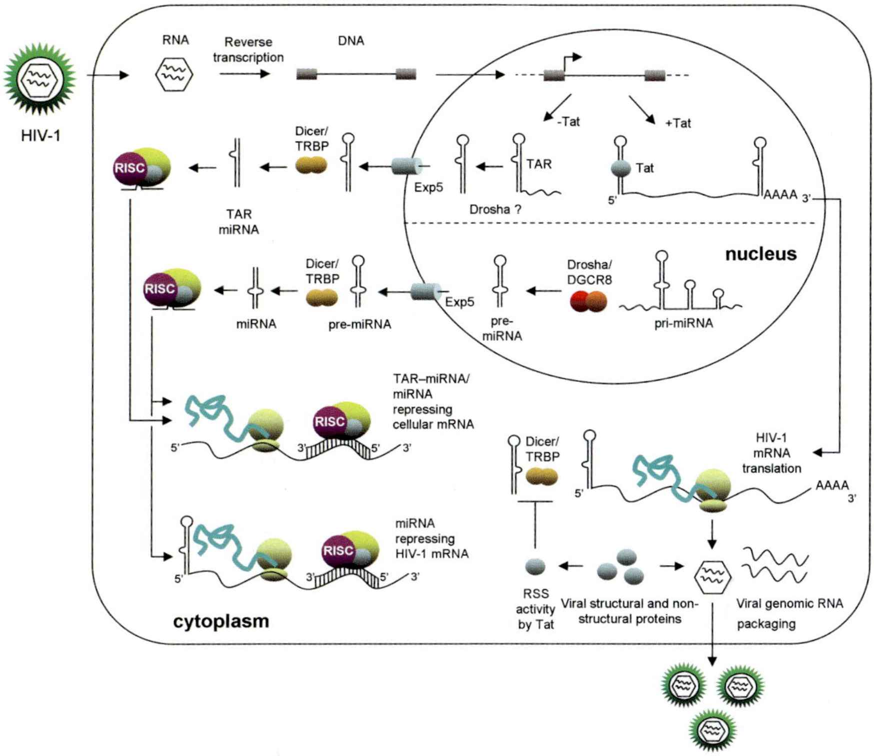

Fig. 1. Overview of the interactions between HIV-1 and cellular RNAi mechanisms. After virus entry and reverse transcription, the proviral DNA integrates into the host genome. From the integrated viral DNA, the HIV-1 transcripts are expressed. In the absence of the viral Tat protein, short abortive transcripts are produced that form the TAR hairpin, which is a substrate for Dicer and is processed into functional TAR-miRNAs. TAR miRNAs can target cellular mRNAs. In the case that Tat interacts with the 5′ TAR element, full-length HIV-1 transcripts are produced that are translated in the viral structural and non-structural proteins that participate in the production of new viral particles. Besides transactivating transcription, the viral Tat protein exhibits RSS activity and can block miRNA processing by interacting with Dicer, or sequestering pre-miRNAs. Alterations in cellular miRNA expression can change the expression of cellular cofactors required for HIV-1 replication. Cellular miRNAs are processed from endogenously expressed pri-miRNA into pre-miRNAs and finally mature miRNAs that can target and inhibit both cellular mRNAs and HIV-1 mRNAs.

have been cloned and these are estimated to regulate the expression of at least 30 percent of human genes.[6]

In plants, worms, insects and fungi, RNA silencing also acts as a potent antiviral mechanism.[7] During infection, virus-specific small interfering RNA (siRNA) duplexes of 21 nucleotides are produced. These siRNAs are fully complementary to the target RNA and guide RISC to catalytic cleavage of the homologous viral transcript. To counter antiviral RNAi, the viruses encode specific RSS factors. In mammals, virus-specific siRNAs could not be easily detected in infected cells.[8] Therefore, the question of whether or not mammalian RNAi serves as an antiviral mechanism has remained controversial, although evidence supporting such a role is accumulating. One possibility is that cellular miRNAs have a role in mammalian antiviral defense responses. Indeed, cellular miRNAs are able to target and inhibit gene expression of several viruses. For example, primate foamy virus type 1 (PFV-1), vesicular stomatitis virus (VSV), hepatitis C virus (HCV) and HIV-1 can all be targeted and repressed by specific host cell miRNAs.[9–13] In the case of HIV-1, miRNA-mediated inhibition is involved in viral latency and has been proposed to be an important determinant of viral tropism.[12,13] In addition, several mammalian viruses including HIV-1 encode RSS factors that inhibit RNAi, which is indicative of a role for RNAi in antiviral defence responses.[14]

The way in which HIV-1 interacts with cellular RNA silencing mechanisms has received a lot of attention. Besides being targeted and repressed by cellular miRNAs, HIV-1 infection also induces significant changes in the cellular miRNA expression profile. This will have important consequences not only for cellular gene expression but also for viral pathogenicity.[15,16] For instance, virus-induced changes in miRNA expression can alter the expression of host factors required for HIV-1 replication.[17] The interaction between HIV-1 and the miRNA machinery becomes even more complex considering that a viral cofactor is also involved in the miRNA pathway.[18] Finally, there are indications that HIV-1 encodes its own miRNA that targets and represses cellular genes involved in apoptosis.[19] Taken together, these interactions suggest a high level of regulation both from the cellular and from the virus perspective. In this chapter, we will first briefly summarize the mechanisms of mammalian RNA silencing, and then give an overview of its interactions with HIV-1. We will also discuss the various strategies in which RNAi can be used to develop an alternative HIV-1 therapeutic approach.

SMALL RNA-MEDIATED SILENCING IN MAMMALS

Compared to the other classes of small regulatory RNAs, the biogenesis of mammalian miRNAs is reasonably well understood.[3,4] Mature miRNAs are single-stranded RNAs of approximately 21–23 nucleotides that are processed from endogenously expressed primary transcripts (pri-miRNAs). These pri-miRNAs are thousands of nucleotides long, generated by polymerase II, and contain imperfectly base paired hairpin structures. In the nucleus, the RNA hairpin structure is excised by RNAse III-like enzyme Drosha and its cofactor DGCR8 (together these proteins form the microprocessor complex), resulting in an imperfect hairpin structure with a two-nucleotide 3′ overhang. This structure, the precursor miRNA (pre-miRNA) of approximately 60 nucleotides, is then exported to the cytoplasm by the nuclear export factor Exportin 5 (Exp5). In the cytoplasm, the pre-miRNA is processed further by the RNAse III-like enzyme Dicer in association with its cofactors TAR binding protein (TRBP) and PACT.[20] The Dicer–TRBP–PACT complex cleaves off the terminal loop resulting in the miRNA duplex with a two-nucleotide 3′ overhang on either side.[21] Depending on the structure, miRNAs associate with one of the four Argonaute proteins to form a miRNA-loaded RISC. miRNAs containing mismatches within the central part of the duplex are preferentially incorporated in AGO1, whereas more perfect duplexes associate with AGO2.[4] Once loaded into RISC, the miRNA duplex is unwound into the mature miRNA strand and the complementary passenger strand, which is degraded. RISC containing the mature miRNA, sometimes referred to as miRISC, can now target mRNAs bearing complementary sequences within the 3′ UTR, resulting in translational repression and in some cases RNA cleavage.[22] Pairing of the 5′ 7–8 nucleotides of the miRNA (seed region) to the 3′ UTR of a target mRNA is of critical importance for target recognition and translational inhibition.[23]

Unlike miRNAs, siRNAs originate from exogenous dsRNA precursors that are processed by Dicer. These dsRNA precursors can be virus-derived replication intermediates or artificially introduced molecules. Dicer cleaves the dsRNA precursors into perfectly complementary duplexes of 21 nucleotides that are preferentially loaded into AGO2-containing RISC. AGO2 is the only member of the Argonaute proteins with slicing activity. Thus, in the case of sufficient complementarity to the target, siRNAs guide RISC towards cleavage of the target RNA rather than translational repression. In plants and insects, siRNAs arise from long virus-derived dsRNAs that are cleaved by Dicer into the virus-specific siRNAs.[24,25] In mammals, virus-specifc siRNAs could not be detected in infected cells. However, endogenous-siRNAs (endo-siRNAs) have been identified that repress the expression of transposable elements and certain protein coding genes in mammalian cells. These endo-siRNAs are processed from naturally occurring dsRNAs formed by hybridization of perfectly complementary transcripts that are made by bidirectional transcription of the genome segments. Loss of Dicer or AGO2 resulted in decreased levels of endo-siRNAs and increased levels of the corresponding retrotransposon and protein-coding transcripts.[26,27] Similarly, siRNAs also play an important role in blocking transposon activity in insects, worms and plants.[27–31]

Besides miRNAs and siRNAs, piRNAs represent a third group of small RNAs involved in RNA silencing. piRNAs are 25–27 nucleotides long and are primarily involved in transposon silencing. These small RNAs are incorporated into Piwi-containing effector complexes, which target transposon transcripts for post-transcriptional silencing.[31] In contrast to siRNAs, piRNAs are generated and active in a Dicer and AGO2-independent manner. Because retroviruses such as HIV-1 are related to retrotransposons, it seems possible that HIV-1-specific siRNAs can arise in HIV-1 infected cells in a similar way to endo-siRNAs and subsequently inhibit virus replication. However, the generation of HIV-1-specific siRNAs by means of bidirectional transcription across the proviral genome needs to be verified experimentally.

HIV-1 GENOME ORGANIZATION AND REPLICATION

HIV-1 belongs to the lentivirus genus within the family of retroviruses. Its genome is encoded by a single positive-stranded RNA molecule of 9.8 kb, two copies of which are present within the enveloped viral particle.

The RNA genome encodes nine open reading frames that give rise to 14 proteins, including matrix, capsid, nucleocapsid, protease, reverse transcriptase, integrase, Tat, Rev, the envelope glycoproteins gp120 and gp41, and four accessory proteins Vif, Vpr, Vpu and Nef. The HIV-1 coding region is flanked by the non-coding long terminal repeats (LTRs), which harbor the promoter and several sequence motifs that are required for viral replication.[32] HIV-1 replicates in CD4-positive T cells, resulting in a direct attack on the immune system. Viral particles enter the host cell via interactions between its envelope protein and the cellular CD4 receptor and the CXCR4 or CCR5 co-receptor. After entry, the RNA is reverse transcribed by the viral reverse transcriptase enzyme into double-stranded DNA that integrates in the host genome. The integrated proviral DNA acts as the template from which the viral genes are expressed. Important for transcription is the transcriptional enhancer protein Tat. The Tat protein stimulates transcription from the viral LTR promoter by interacting with the transactivation response (TAR) element, a stable RNA stem-loop structure present at the 5′ terminus of each viral transcript.[33]

HIV-1 INDUCED CHANGES IN CELLULAR miRNA EXPRESSION

Virus infection generally triggers cells to mount an antiviral response that involves the expression of many different antiviral and stress related genes. To replicate efficiently, the virus encodes specific factors to overcome these inhibitory responses. For HIV-1, the changes also involve changes in the cellular miRNA expression profile. In an early study, Yeung *et al.* determined the miRNA expression profile in HeLa cells transfected with HIV-1 DNA.[16] The authors reported significant down-regulation of miR-93, miR-148b, miR-221 and miR-16. The miRNA profile showed more dramatic changes in peripheral blood mononuclear cells (PBMCs) from HIV-1 infected patients compared to cells from uninfected controls.[15] Depending on the disease stage of the patient, the T-cell specific miR-223, miR-150, miR-146, miR-16 and miR-191 were down-regulated three to nine-fold. These changes are most likely induced by signaling and bystander effects because of the low percentage of infected cells *in vivo*.

The HIV-1-induced changes in the cellular miRNA expression profile can also affect the expression of cofactors that are required for viral replication. Triloubet and co-workers observed increased HIV-1 replication in Drosha and Dicer knock-down cells, suggesting a role for cellular miRNAs in HIV-1 replication.[17] Subsequent miRNA profiling in HIV-1 infected Jurkat cells revealed that 11 miRNAs are up-regulated, whereas expression of the polycistronic miRNA cluster miR17/92 was strongly decreased. The miR17/92 cluster comprises miR-17-(5p/3p), miR-18, miR-19a, miR-20a, miR-19b-1 and miR-92-1 and has been implicated in malignant lymphoma and lung cancer. This miRNA cluster has been used for the design of an anti-HIV-1 RNAi therapeutic.[34] Computer-assisted target prediction showed that the mRNA encoding histone acetylase PCAF has four target sequences in its 3′ UTR for miR-17-5p and miR-20a. PCAF is an important cofactor in HIV-1 transcription via association with the viral Tat protein. Thus, HIV-1-induced down-regulation of miR-17/92 in turn increased PCAF expression, resulting in further enhancement of HIV-1 replication. The mechanism that causes down-regulation of miR-17/92 is currently unclear.

CELLULAR miRNAs TARGETING HIV-1

Binding of the miRNA seed region to a target sequence in the 3′ UTR of a transcript can be sufficient to trigger translational repression.[23] This means that viral RNAs are at risk of being targeted and repressed by cellular miRNAs with a complementary seed sequence. Several mammalian viruses are targeted and repressed by host cell miRNAs. The positive-strand RNA virus HCV is repressed by a specific set of miRNAs that are induced in response to antiviral interferon (IFN) signaling. The negative-strand RNA virus VSV is repressed by miR-24 and miR-93, and the retrovirus PFV-1 is targeted by miR-32.[9–13] Interestingly, PFV-1 requires RSS activity of the viral Tas protein to overcome this inhibition.

Recently, cellular miRNAs were also shown to target and potently repress HIV-1 gene expression.[9] The set of miR-28, miR-125b, miR-150, miR-223 and miR-382 can target sequences in the 3′ part of HIV-1 transcripts and suppress expression of HIV-1 mRNAs in resting CD4-positive T cells. Inhibition of these miRNAs with antisense inhibitors induced virus replication in latently infected cells. In activated CD4-positive T cells, the expression of these miRNAs is reduced, thus allowing virus replication. These results indicated a role for cellular miRNAs in establishing HIV-1 latency. HIV-1 latency occurs in resting CD4 positive T cells when the HIV-1 provirus is integrated into the host genome without producing new viral particles. This latency phase is essential for HIV-1 to propagate *in vivo*. Surprisingly, miRNA-mediated inhibition of HIV-1 replication is therefore apparently beneficial for the virus. This idea is supported by the fact that HIV-1 retains the miRNA target sequences, which otherwise could easily be inactivated by the acquisition of point mutations. HIV-1 latency is a problem for treatment of HIV-1 infected individuals with antiviral therapy. Even though replicating HIV-1 can be cleared, the virus can re-emerge from the latent viral reservoir once therapy is stopped. Therefore, inhibition of the latency-inducing miRNAs has been suggested as a new approach to purge the HIV-1 reservoir.

In addition to the latency-inducing miRNAs, several other miRNAs were predicted to target the HIV-1 RNA genome.[35,36] One of these, miR-29a, was recently shown to target sequences within the HIV-1 Nef gene, thus restricting Nef expression and virus replication.[35] A recent study has shown that HIV-1 mRNAs associate and co-localize with RISC components.[37] Several miRNA effectors, proteins required for miRNA-mediated silencing, negatively regulate HIV-1 gene expression. Knock-down of RCK/p54 or DGCR8 reactivated virus in latently infected PBMCs from patients.

miRNAs AND HIV-1 TROPISM

The cellular miRNAs involved in HIV-1 latency have also been suggested to play a role in the restriction of viral tropism for certain cell types. Wang and co-workers showed that miRNAs determine the poor susceptibility of peripheral blood monocytes for HIV-1 infection.[13] These cells possess the receptors for HIV-1 entry but are infrequently infected. The data suggest that miRNA-mediated suppression protects the cells against a productive HIV-1 infection. miRNAs can also indirectly affect the viral tropism by regulating the expression of host factors that are required for virus replication.[17,38] In monocytes, miR-198 inhibits the expression of Cyclin T1, which is required for HIV-1 transcription elongation.[38] Once monocytes are activated to differentiate into macrophages, miR-198 expression is reduced, resulting in increased Cyclin T1 expression that allows HIV-1 replication.

TAT-MEDIATED RNAi SUPPRESSION

To overcome antiviral RNA silencing, plant-, insect- and fungi-infecting viruses encode specific RSS factors.[24,39] These RSS factors block the anti-viral silencing responses at diverse levels in the pathway. Although virus-specific siRNAs could not thus far be detected in infected mammalian cells, several mammalian viruses do encode factors that inhibit RNAi responses. For example, proteins of influenza virus (NS1), vaccinia virus (E3L) and Ebola virus (VP35) were recently shown to have RSS activity.[40,41] E3L and NS1 were able to replace the RSS activity of the B2 protein of the insect-infecting Flock house virus and to suppress RNAi in luciferase reporter assays. Intriguingly, these proteins were previously identified as potent inhibitors of interferon-mediated antiviral responses, which are also triggered by dsRNA.

So far, two viruses from the family of retroviruses, HIV-1 and PFV-1, have been reported to encode RSS activity.[10,42–44] For both viruses the transcriptional enhancer proteins Tat (HIV-1) and Tas (PFV-1) were shown to exhibit RSS activity. During HIV-1 and PFV-1 replication, these proteins stimulate transcription from the viral LTR promoter. Interestingly, both viruses are controlled by the RNAi mechanism via cellular miRNAs.[9,10] The Tas protein was shown to counter inhibitory effects of the cellular miR-32 and also displayed RSS activity in plants.

Jeang and co-workers were the first to report HIV-1 Tat-mediated RSS activity.[42,45] They showed accumulation of HIV-1 specific siRNAs during virus infection. This siRNA originated from an RNA duplex that is formed by long-distance base pairing of sequences in the *pol* and *env* genes. Tat-mediated RSS activity was required to overcome the inhibitory effects of this siRNA. Tat RSS activity involves direct interaction with Dicer and requires dsRNA binding. The RSS activity appears to be independent of the transcriptional transactivation activity of the HIV-1 Tat protein. We confirmed these data using a designed HIV-1 variant that is not dependent on Tat for the transactivation of transcription.[44] Nevertheless, this variant showed a severe virus production defect when a frameshift mutation was introduced in the Tat gene. Virus production could be restored by heterologous viral RSS factors (influenza virus A NS1 protein, vaccinia virus E3L and Ebola virus VP35). We recently showed that the NS3 protein of rice hoja blanca virus (RHBV), a previously identified RSS factor, is able to complement the RSS function of HIV-1 Tat.[46] NS3 exclusively binds small dsRNA molecules, suggesting that HIV-1 replication is inhibited either by siRNAs or miRNAs and that Tat RSS activity is required to counter this inhibition.[47,48] The Tat RSS function could also be replaced by the P19 protein of tomato bushy stunt virus (TBSV).[43] This study showed that HIV-1 Tat is able to suppress miRNA-mediated silencing. These combined results suggest that Tat may counter the inhibitory effects of cellular miRNAs that target HIV-1 mRNAs.

HIV-1 ENCODED miRNAs

Several viruses, in particular from the family of herpesviruses, encode miRNAs that regulate the expression of host and/or viral genes.[49,50] No HIV-1 encoded miRNAs or siRNAs were identified in initial screens, but computer analyses predicted the presence of potential miRNAs.[8,51] Subsequent studies showed that several of these HIV-1 specific miRNAs/siRNAs are indeed produced. A Nef-derived miRNA (miR-N367) was reported to suppress HIV-1 both transcriptionally and post-transcriptionally.[52] Bennasser and co-workers revealed the accumulation of an HIV-1 derived siRNA corresponding to a dsRNA duplex that could be formed via base pairing of sequences in *pol* and *env*.[42] The best characterized HIV-1 encoded miRNA originates from the TAR element that is located at the extreme 5′ and 3′ termini of all HIV-1 transcripts. During HIV-1 transcription, the cooperative binding of HIV-1 Tat, TRBP and cellular Cyclin T1 to the 5′ TAR motif activates the processivity of the engaged RNA polymerase II. Basal transcription from the HIV-1 LTR in the absence of Tat generates short transcripts containing the 5′ TAR element. The TAR hairpin element structurally resembles pre-miRNAs. *In vitro* experiments revealed that TAR is indeed a substrate for Dicer, giving rise to a viral miRNA of 21 nucleotides.[19] Moreover, miRNAs from both the 5′ and 3′ sides of the TAR stem could be detected in HIV-1 infected CD4-positive T lymphocytes, miR-TAR-3′ being somewhat more abundant than miR-TAR-5′.[53] These TAR-derived miRNAs were subsequently shown to be expressed at all stages of the HIV-1 life cycle and in diverse cell types, including latently infected cells. Recent studies indicate that the TAR-derived miRNAs target and repress the cellular ERCC1 and IER3 genes that are involved in apoptosis.[19] These genes contain targets for both miR-TAR-5′ and miR-TAR-3′. By affecting the process of apoptosis, the TAR miRNAs may extend the life of the infected cell, thus resulting in prolonged and increased virus production.

HIV-1 AND TRBP

Binding of TRBP to TAR is required for transactivation of HIV-1 transcription.[18] However, TRBP is also involved in the RNAi pathway, assisting in Dicer-mediated miRNA processing and the subsequent loading of miRNA duplexes into RISC. Bennasser *et al.* showed that over-expression of TAR can sequester TRBP and thus inhibit RNAi action.[54] The authors suggested that this may provide the virus with an additional strategy to block antiviral RNAi responses. However, because TRBP is required for efficient transcription, knock-down of TRBP decreases HIV-1 replication.[55] Apparently, the requirement for TRBP in HIV-1 transcription is more important than the putative countering of antiviral RNAi processes. The fact that the TAR hairpin is a substrate for Dicer is in agreement with its ability to interact with TRBP. This TAR-Dicer link also fits with the Tat-mediated RSS activity via inhibition of Dicer. TRBP stimulates HIV-1 gene expression and replication by inhibition of dsRNA-activated protein kinase (PKR). In contrast, the cellular protein PACT, a TRBP cofactor in siRNA production, acts as a cellular activator of PKR.[56] Thus, both positive and negative TRBP effects are operational during HIV-1 replication. The TAR hairpin appears to be at the center of all these interactions. It is tempting to speculate that Tat protects the TAR structure from processing by the RNAi machinery.

RNAi-BASED HIV-1 THERAPEUTICS

RNAi as a tool to specifically silence genes has provided new possibilities in antiviral drug design.[41] The strategies used for therapeutic RNAi in mammalian cells tap in on the natural RNAi pathway. For instance, transfected synthetic siRNAs corresponding to the targeted virus are directly loaded into RISC and can trigger a potent and specific inhibitory RNAi response. Another popular strategy is the endogenous expression of short hairpin RNAs (shRNAs) that correspond to viral sequences. These shRNAs are modeled after pre-miRNA stem-loop structures and need to be processed by Dicer into siRNAs in order to become active. Recently, artificial miRNAs have been designed to target viruses for therapeutic purposes. These artifical miRNAs more closely resemble natural pre- or pri-miRNAs, contain internal loops and bulges, and go through Drosha and Dicer processing to yield a mature miRNA duplex.[34,41] Such RNAi-based therapeutics are currently in development for many different viruses. A number of approaches have been tested against HIV-1, some more successful than others.[41] Although HIV-1 can be potently inhibited by a single siRNA or shRNA, the virus easily escapes via the selection of mutations in the target sequence.[57,58] HIV-1 can also escape via mutations which induce an alternative local RNA structure that occludes the target site.[59,60] The combined expression of multiple shRNAs in a combinatorial RNAi approach can significantly delay the emergence of RNAi-resistant variants.[61–64] This combinatorial approach should target essential viral sequences that are conserved among different virus strains. Another option is to target host cell encoded cofactors that are essential for virus replication, for example the CCR5 co-receptor. The inclusion of cellular cofactors as RNAi target is likely to make viral escape more difficult. However, it cannot be ruled out that the virus can eventually escape. In addition, it should be carefully monitored whether the knock-down of such host factors does not result in adverse effects for the cell. Chronic infections such as HIV-1 should preferentially be targeted with a durable RNAi treatment using gene therapy approaches that provide a constant supply of intracellularly expressed antiviral shRNAs. In the case of HIV-1, one could make T cells non-susceptible for the virus by *ex vivo* transduction of blood stem cells to express the RNAi trigger, and give these cells back to the patient.[65]

CONCLUSIONS

Recent studies show that mammalian viruses are targeted by cellular miRNAs, but viruses can, in turn, alter cellular miRNA expression profiles, encode RSS factors and their own miRNAs that regulate both host and viral genes. So far, HIV-1 is one of the few viruses for which all these phenomena have been reported, which may in part be due to the more intense research interest it receives. Although we still know very little about how these multiple events are integrated mechanistically, it is likely that accurate timing will be an essential factor. Because miRNA expression profiles vary for different cell types, it will be important to study these processes in the most biologically relevant test system. Despite the availability of powerful novel sequencing technology, virus-specific siRNAs have so far not been reported for mammalian virus–host combinations. This may suggest that antiviral RNAi activity in mammals originates primarily from cellular miRNAs, although it is too early to rule out a role for virus-specific siRNA. Future studies will need to address these issues in further detail.

ACKNOWLEDGMENTS

RNAi research in the Berkhout lab is sponsored by ZonMw (Vici grant and Translational Gene Therapy program), NWO-CW (Top grant), the European Union (LSHP-CT-2006-037301) and the Technology Foundation STW (grant AGT.7708).

REFERENCES

1. Ringrose J, Jeeninga RE, Berkhout B, Speijer D. (2008) Proteomic studies reveal coordinated changes in T cell expression patterns upon HIV-1 infection. *J Virol* **82:** 4320–4330.
2. Goff SP. (2008) Knockdown screens to knockout HIV-1. *Cell* **135:** 417–420.

3. Rana TM. (2007) Illuminating the silence: Understanding the structure and function of small RNAs. *Nat Rev Mol Cell Biol* **8**: 23–36.
4. Kim VN, Han J, Siomi MC. (2009) Biogenesis of small RNAs in animals. *Nat Rev Mol Cell Biol* **10**: 126–139.
5. Bartel DP. (2004) MicroRNAs: Genomics, biogenesis, mechanism, and function. *Cell* **116**: 281–297.
6. Krek A, Grun D, Poy MN, *et al.* (2005) Combinatorial microRNA target predictions. *Nat Genet* **37**: 495–500.
7. Aliyari R, Ding SW. (2009) RNA-based viral immunity initiated by the Dicer family of host immune receptors. *Immunol Rev* **227**: 176–188.
8. Pfeffer S, Zavolan M, Grasser FA, *et al.* (2004) Identification of virus-encoded microRNAs. *Science* **304**: 734–736.
9. Huang J, Wang F, Argyris E, *et al.* (2007) Cellular microRNAs contribute to HIV-1 latency in resting primary CD4(+) T lymphocytes. *Nat Med* **13**: 1241–1247.
10. Lecellier CH, Dunoyer P, Arar K, *et al.* (2005) A cellular microRNA mediates antiviral defense in human cells. *Science* **308**: 557–560.
11. Otsuka M, George P, *et al.* (2007) Hypersusceptibility to vesicular stomatitis virus infection in Dicer1-deficient mice is due to impaired miR24 and miR93 expression. *Immunity* **27**: 123–134.
12. Pedersen IM, Cheng G, Wieland S, *et al.* (2007) Interferon modulation of cellular microRNAs as an antiviral mechanism. *Nature* **449**: 919–922.
13. Wang X, Ye L, Hou W, *et al.* (2009) Cellular microRNA expression correlates with susceptibility of monocytes/macrophages to HIV-1 infection. *Blood* **113**: 671–674.
14. de Vries W, Haasnoot J, Fouchier R, *et al.* (2009) Differential RNA silencing suppression activity of NS1 proteins from different influenza A strains. *J Gen Virol* **90**: 1916–1922.
15. Houzet L, Yeung ML, de Lame V, *et al.* (2008) MicroRNA profile changes in human immunodeficiency virus type 1 (HIV-1) seropositive individuals. *Retrovirology* **5**: 118.
16. Yeung ML, Bennasser Y, Myers TG, *et al.* (2005) Changes in microRNA expression profiles in HIV-1-transfected human cells. *Retrovirology* **2**: 81.
17. Triboulet R, Mari B, Lin YL, *et al.* (2007) Suppression of microRNA-silencing pathway by HIV-1 during virus replication. *Science* **315**: 1579–1582.

18. Gatignol A, Buckler-White A, Berkhout B, Jeang KT. (1991) Characterization of a human TAR RNA-binding protein that activates the HIV-1 LTR. *Science* **251**: 1597–1600.
19. Klase Z, Winograd R, Davis J, *et al.* (2009) HIV-1 TAR miRNA protects against apoptosis by altering cellular gene expression. *Retrovirology* **6**: 18.
20. Grishok A, Pasquinelli AE, Conte D, *et al.* (2001) Genes and mechanisms related to RNA interference regulate expression of the small temporal RNAs that control *C. elegans* developmental timing. *Cell* **106**: 23–34.
21. Kok KH, Ng MH, Ching YP, Jin DY. (2007) Human TRBP and PACT directly interact with each other and associate with Dicer to facilitate the production of small interfering RNA. *J Biol Chem* **282**: 17649–17657.
22. Hutvagner G, Zamore PD. (2002) A microRNA in a multiple-turnover RNAi enzyme complex. *Science* **297**: 2056–2060.
23. Bartel DP. (2009) MicroRNAs: Target recognition and regulatory functions. *Cell* **136**: 215–233.
24. Ding SW, Voinnet O. (2007) Antiviral immunity directed by small RNAs. *Cell* **130**: 413–426.
25. Molnar A, Csorba T, Lakatos L, *et al.* (2005) Plant virus-derived small interfering RNAs originate predominantly from highly structured single-stranded viral RNAs. *J Virol* **79**: 7812–7818.
26. Tam OH, Aravin AA, Stein P, *et al.* (2008) Pseudogene-derived small interfering RNAs regulate gene expression in mouse oocytes. *Nature* **453**: 534–538.
27. Watanabe T, Totoki Y, Toyoda A, *et al.* (2008) Endogenous siRNAs from naturally formed dsRNAs regulate transcripts in mouse oocytes. *Nature* **453**: 539–543.
28. Czech B, Malone CD, Zhou R, *et al.* (2008) An endogenous small interfering RNA pathway in Drosophila. *Nature* **453**: 729–731.
29. Yang N, Kazazian HH, Jr. (2006) L1 retrotransposition is suppressed by endogenously encoded small interfering RNAs in human cultured cells. *Nat Struct Mol Biol* **13**: 763–771.
30. Sijen T, Plasterk RH. (2003) Transposon silencing in the *Caenorhabditis elegans* germ line by natural RNAi. *Nature* **426**: 310–314.
31. Girard A, Hannon GJ. (2008) Conserved themes in small-RNA-mediated transposon control. *Trends Cell Biol* **18**: 136–148.
32. Berkhout B. (1996) Structure and function of the human immunodeficiency virus leader RNA. *Prog Nucl Acid Res Mol Biol* **54**: 1–34.

33. Berkhout B, Silverman RH, Jeang KT. (1989) Tat trans-activates the human immunodeficiency virus through a nascent RNA target. *Cell* **59**: 273–282.
34. Liu YP, Haasnoot J, Ter Brake O, *et al.* (2008) Inhibition of HIV-1 by multiple siRNAs expressed from a single microRNA polycistron. *Nucl Acid Res* **36**: 2811–2824.
35. Ahluwalia JK, Khan SZ, Soni K, *et al.* (2008) Human cellular microRNA hsa-miR-29a interferes with viral Nef protein expression and HIV-1 replication. *Retrovirology* **5**: 117.
36. Hariharan M, Scaria V, Pillai B. Brahmachari SK. (2005) Targets for human encoded microRNAs in HIV genes. *Biochem Biophys Res Commun* **337**: 1214–1218.
37. Chable-Bessia C, Meziane O, Latreilla D, *et al.* (2009) Suppression of HIV-1 replication by microRNA effectors. *Retrovirology* **6**: 26.
38. Sung TL, Rice AP. (2009) miR-198 inhibits HIV-1 gene expression and replication in monocytes and its mechanism of action appears to involve repression of Cyclin T1. *PLoS Pathog* **5**: e1000263.
39. Li WX, Ding SW. (2001) Viral suppressors of RNA silencing. *Curr Opin Biotechnol* **12**: 150–154.
40. Li WX, Li H, Lu R, *et al.* (2004) Interferon antagonist proteins of influenza and vaccinia viruses are suppressors of RNA silencing. *Proc Natl Acad Sci USA* **101**: 1350–1355.
41. Haasnoot J, Westerhout EM, Berkhout B. (2007) RNA interference against viruses: Strike and counterstrike. *Nat Biotechnol* **25**: 1435–1443.
42. Bennasser Y, Le SY, Benkirane M, Jeang KT. (2005) Evidence that HIV-1 encodes an siRNA and a suppressor of RNA silencing. *Immunity* **22**: 607–619.
43. Qian S, Zhong X, Yu L, *et al.* (2009) HIV-1 Tat RNA silencing suppressor activity is conserved across kingdoms and counteracts translational repression of HIV-1. *Proc Natl Acad Sci USA* **106**: 605–610.
44. Haasnoot J, de Vries W, Geutjes EJ, *et al.* (2007) The Ebola virus VP35 protein is a suppressor of RNA silencing. *PLoS Pathog* **3**: e86.
45. Bennasser Y, Jeang KT. (2006) HIV-1 Tat interaction with Dicer: Requirement for RNA. *Retrovirology* **3**: 95.
46. Schnettler E, de VW, Hemmes H, *et al.* (2009) The NS3 protein of rice hoja blanca virus complements the RNAi suppressor function of HIV-1 Tat. *EMBO Rep.* **10**: 258–263.

47. Hemmes H, Lakatos L, Goldbach R, *et al.* (2007) The NS3 protein of rice hoja blanca tenuivirus suppresses RNA silencing in plant and insect hosts by efficiently binding both siRNAs and miRNAs. *RNA* **13**: 1079–1089.
48. Schnettler E, Hemmes H, Goldbach R, Prins M. (2008) The NS3 protein of rice hoja blanca virus suppresses RNA silencing in mammalian cells. *J Gen Virol* **89**: 336–340.
49. Gottwein E, Cullen BR. (2008) Viral and cellular microRNAs as determinants of viral pathogenesis and immunity. *Cell Host Microbe* **3**: 375–387.
50. Cullen BR. (2009) Viral and cellular messenger RNA targets of viral microRNAs. *Nature* **457**: 421–425.
51. Bennasser Y, Le SY, Yeung ML, Jeang KT. (2004) HIV-1 encoded candidate micro-RNAs and their cellular targets. *Retrovirology* **1**: 43.
52. Omoto S, Fuji YR. (2005) Regulation of human immunodeficiency virus 1 transcription by *nef* microRNA. *J Gen Virol* **86**: 751–755.
53. Ouellet DL, Plante I, Landry P, *et al.* (2008) Identification of functional microRNAs released through asymmetrical processing of HIV-1 TAR element. *Nucl Acid Res* **36**: 2353–2365.
54. Bennasser Y, Yeung ML, Jeang KT. (2006) HIV-1 TAR RNA subverts RNA interference in transfected cells through sequestration of TAR RNA-binding protein, TRBP. *J Biol Chem* **281**: 27674–27678.
55. Christensen HS, Daher A, Soye KJ, *et al.* (2007) Small interfering RNAs against the TAR RNA binding protein, TRBP, a Dicer cofactor, inhibit human immunodeficiency virus type 1 long terminal repeat expression and viral production. *J Virol* **81**: 5121–5131.
56. Daher A, Laraki G, Singh M, *et al.* (2009) TRBP control of PACT-induced phosphorylation of protein kinase R is reversed by stress. *Mol Cell Biol* **29**: 254–265.
57. Boden D, Pusch O, Lee F, *et al.* (2003) Human immunodeficiency virus type 1 escape from RNA interference. *J Virol* **77**: 11531–11535.
58. Das AT, Brummelkamp TR, Westerhout EM, *et al.* (2004) Human immunodeficiency virus type 1 escapes from RNA interference mediated inhibition. *J Virol* **78**: 2601–2605.
59. Westerhout EM, Ooms M, Vink M, *et al.* (2005) HIV-1 can escape from RNA interference by evolving an alternative structure in its RNA genome. *Nucl Acid Res* **33**: 796–804.

60. Westerhout EM, Berkhout B. (2007) A systematic analysis of the effect of target RNA structure on RNA interference. *Nucl Acid Res* **35**: 4322–4330.
61. Ter Brake O, Konstantinova P, Ceylan M, Berkhout B. (2006) Silencing of HIV-1 with RNA interference: A multiple shRNA approach. *Mol Ther* **14**: 883–892.
62. Grimm D, Kay MA. (2007) Combinatorial RNAi: A winning strategy for the race against evolving targets? *Mol Ther* **15**: 878–888.
63. Liu YP, Berkhout B. (2008) Combinatorial RNAi strategies against HIV-1 and other escape-prone viruses. *Int J Biosci Technol* **1**: 1–10.
64. Ter Brake O, 't Hooft K, Liu YP, *et al.* (2008) Lentiviral vector design for multiple shRNA expression and durable HIV-1 Inhibition. *Mol Ther* **16**: 557–564.
65. Rossi JJ. (2006) RNAi as a treatment for HIV-1 infection. *Biotechniques* **40**: S25–S29.

Index